Anthology of Statistics in Sports

ASA-SIAM Series on Statistics and Applied Probability

The ASA-SIAM Series on Statistics and Applied Probability is published jointly by the American Statistical Association and the Society for Industrial and Applied Mathematics. The series consists of a broad spectrum of books on topics in statistics and applied probability. The purpose of the series is to provide inexpensive, quality publications of interest to the intersecting membership of the two societies.

Editorial Board

Albert, J., Bennett, J., and Cochran, J. J., eds., *Anthology of Statistics in Sports*

Smith, W. F., *Experimental Design for Formulation*

Baglivo, J. A., *Mathematica Laboratories for Mathematical Statistics: Emphasizing Simulation and Computer Intensive Methods*

Lee, H. K. H., *Bayesian Nonparametrics via Neural Networks*

O'Gorman, T. W., *Applied Adaptive Statistical Methods: Tests of Significance and Confidence Intervals*

Ross, T. J., Booker, J. M., and Parkinson, W. J., eds., *Fuzzy Logic and Probability Applications: Bridging the Gap*

Nelson, W. B., *Recurrent Events Data Analysis for Product Repairs, Disease Recurrences, and Other Applications*

Mason, R. L. and Young, J. C., *Multivariate Statistical Process Control with Industrial Applications*

Smith, P. L., *A Primer for Sampling Solids, Liquids, and Gases: Based on the Seven Sampling Errors of Pierre Gy*

Meyer, M. A. and Booker, J. M., *Eliciting and Analyzing Expert Judgment: A Practical Guide*

Latouche, G. and Ramaswami, V., *Introduction to Matrix Analytic Methods in Stochastic Modeling*

Peck, R., Haugh, L., and Goodman, A., *Statistical Case Studies: A Collaboration Between Academe and Industry, Student Edition*

Peck, R., Haugh, L., and Goodman, A., *Statistical Case Studies: A Collaboration Between Academe and Industry*

Barlow, R., *Engineering Reliability*

Czitrom, V. and Spagon, P. D., *Statistical Case Studies for Industrial Process Improvement*

Anthology of Statistics in Sports

Edited by

Jim Albert
Bowling Green State University
Bowling Green, Ohio

Jay Bennett
Telcordia Technologies
Piscataway, New Jersey

James J. Cochran
Louisiana Tech University
Ruston, Louisiana

Society for Industrial and Applied Mathematics
Philadelphia, Pennsylvania

American Statistical Association
Alexandria, Virginia

The correct bibliographic citation for this book is as follows: Albert, Jim, Jay Bennett, and James J. Cochran, eds., *Anthology of Statistics in Sports*, ASA-SIAM Series on Statistics and Applied Probability, SIAM, Philadelphia, ASA, Alexandria, VA, 2005.

10 9 8 7 6 5 4 3 2 1

Library of Congress Cataloging-in-Publication Data

Anthology of statistics in sports / [compiled by] Jim Albert, Jay Bennett, James J. Cochran.
p. cm. – (ASA-SIAM series on statistics and applied probability)
On cover: The ASA Section on Statistics in Sports.
Includes bibliographical references.
ISBN 0-89871-587-3 (pbk.)
1. Sports–United States–Statistics. 2. Sports–United States–Statistical methods. I. Albert, Jim, 1953- II. Bennett, Jay. III. Cochran, James J. IV. American Statistical Association. Section on Statistics in Sports. V. Series.

GV741.A694 2005
796'.021–dc22 2005042540

Contents

Acknowledgments

Original Sources of Contributed Articles

Chapter 4 originally appeared in *Chance*, vol. 12, no. 3, 1999, pp. 51–56.
Chapter 5 originally appeared in *Journal of the American Statistical Association*, vol. 93, no. 441, 1998, pp. 25–35.
Chapter 6 originally appeared in *Journal of the American Statistical Association*, vol. 75, no. 371, 1980, pp. 516–524.
Chapter 7 originally appeared in *Chance*, vol. 6, no. 3, 1993, pp. 30–37.
Chapter 8 originally appeared in *The American Statistician*, vol. 45, no. 3, 1991, pp. 179–183.
Chapter 10 originally appeared in *Journal of the American Statistical Association*, vol. 89, no. 427, 1994, pp. 1066–1074.
Chapter 11 originally appeared in *The American Statistician*, vol. 47, no. 4, 1993, pp. 241–250.
Chapter 12 originally appeared in *American Statistical Association Proceedings of the Section on Statistics in Sports*, 1992, pp. 64–66.
Chapter 13 originally appeared in *Journal of the American Statistical Association*, vol. 89, no. 427, 1994, pp. 1080–1090.
Chapter 14 originally appeared in *Chance*, vol. 7, no. 3, 1994, pp. 24–30.
Chapter 15 originally appeared in *Chance*, vol. 6, no. 2, 1993, pp. 17–22, 30.
Chapter 16 originally appeared in *American Statistical Association Journal*, September 1961, pp. 703–728.
Chapter 18 originally appeared in *The American Statistician*, vol. 50, no. 1, 1996, pp. 39–43.
Chapter 19 originally appeared in *Chance*, vol. 2, no. 4, 1989, pp. 22–30.
Chapter 20 originally appeared in *The American Statistician*, vol. 50, no. 1, 1996, pp. 34–38.
Chapter 21 originally appeared in *Chance*, vol. 2, no. 1, 1989, pp. 16–21.
Chapter 22 originally appeared in *The American Statistician*, vol. 49, no. 1, 1995, pp. 24–28.
Chapter 24 originally appeared in *American Statistical Association Proceedings of the Section on Statistics in Sports*, 1993, pp. 4–9.
Chapter 25 originally appeared in *Chance*, vol. 8, no. 1, 1995, pp. 19–22.
Chapter 26 originally appeared in *Chance*, vol. 11, no. 1, 1998, pp. 3–7.
Chapter 28 originally appeared in *Journal of the American Statistical Association*, vol. 94, no. 447, 1999, pp. 661–676.
Chapter 29 originally appeared in *Journal of the American Statistical Association*, vol. 70, no. 350, 1975, pp. 311–319.
Chapter 30 originally appeared in *Journal of the American Statistical Association*, vol. 68, no. 342, 1973, pp. 312–316.
Chapter 31 originally appeared in *Chance*, vol. 2, no. 4, 1989, pp. 35–37.
Chapter 32 originally appeared in *The American Statistician*, vol. 51, no. 4, 1997, pp. 305–310.
Chapter 33 originally appeared in *Chance*, vol. 6, no. 3, 1993, pp. 25–29, 69.
Chapter 34 originally appeared in *Journal of the American Statistical Association*, vol. 89, no. 427, 1994, pp. 1128–1134.
Chapter 36 originally appeared in *Chance*, vol. 10, no. 3, 1997, pp. 16–19.
Chapter 37 originally appeared in *Chance*, vol. 12, no. 4, 1999, pp. 50–55.
Chapter 38 originally appeared in *American Statistical Association Proceedings of the Section on Statistics in Sports*, 1995, pp. 1–5.

Chapter 39 originally appeared in Journal of the American Statistical Association, vol. 89, no. 427, 1994, pp. 1075–1079.
Chapter 40 originally appeared in Chance, vol. 10, no. 1, 1997, pp. 15–19.
Chapter 41 originally appeared in Journal of the American Statistical Association, vol. 89, no. 427, 1994, pp. 1124–1127.
Chapter 42 originally appeared in Chance, vol. 10, no. 2, 1997, pp. 27–34.
Chapter 43 originally appeared in The American Statistician, vol. 51, no. 2, 1997, pp. 106–111.
Chapter 44 originally appeared in Chance, vol. 7, no. 1, 1994, pp. 20–25.

Figure Permissions

The artwork on page 111 in Chapter 15 is used with permission of the artist, John Gampert.
The photograph on page 170 in Chapter 21 is used with permission of the photographer, Marcy Dubroff.
The photographs on pages 197 and 198 in Chapter 26 are used with permission of Getty Images.
The photograph on page 310 in Chapter 42 is used with permission of Time, Inc.

Chapter 1

Introduction

Jim Albert, Jay Bennett, and James J. Cochran

1.1 The ASA Section on Statistics in Sports (SIS)

The 1992 Joint Statistical Meetings (JSM) saw the creation of a new section of the American Statistical Association (ASA). Joining the host of traditional areas of statistics such as Biometrics, Survey Research Methods, and Physical and Engineering Sciences was the Section on Statistics in Sports (SIS). As stated in its charter, the section is dedicated to promoting high professional standards in the application of statistics to sports and fostering statistical education in sports both within and outside the ASA.

Statisticians worked on sports statistics long before the founding of SIS. Not surprisingly, some of the earliest sports statistics pieces in the *Journal of the American Statistical Association* (JASA) were about baseball and appeared in the 1950s. One of the first papers was Frederick Mosteller's 1952 analysis of the World Series (JASA, 47 (1952), pp. 355–380). Through the years, Mosteller has continued his statistical research in baseball as well as other sports. Fittingly, his 1997 paper "Lessons from Sports Statistics" (*The American Statistician*, 51-4 (1997), pp. 305–310) is included in this volume (Chapter 32) and provides a spirited example of the curiosity and imagination behind all of the works in this volume.

Just as the nation's sporting interests broadened beyond baseball in the 1960s, 1970s, and 1980s, so did the topics of research in sports statistics. Football, basketball, golf, tennis, ice hockey, and track and field were now being addressed (sometimes very passionately). Perhaps no question has been so fiercely debated as the existence of the "hot hand" in basketball. The basketball section (Part III) of this volume provides some examples of this debate, which remains entertainingly unresolved.

Until the creation of SIS, research on sports statistics had to be presented and published in areas to which it was only tangentially related—one 1984 paper published in this volume (Chapter 12) was presented at the JSM under the auspices of the Social Statistics Section since it had no other home. The continued fervent interest in sports statistics finally led to the creation of SIS in the 1990s. Since its creation, the section has provided a forum for the presentation of research at the JSM. This in turn has created an explosion of sports statistics papers in ASA publications as well as an annual volume of proceedings from the JSM (published by the section). The September 1994 issue of JASA devoted a section exclusively to sports statistics. *The American Statistician* typically has a sports statistics paper in each issue. *Chance* often has more than one article plus the regular column "A Statistician Reads the Sports Pages."

What lies in the future? Research to date has been heavily weighted in the areas of competition (rating players/ teams and evaluating strategies for victory). This differs greatly from the research being performed in other parts of the world. Papers in publications of the International Sports Statistics Committee of the International Statistical Institute (ISI) have emphasized analysis of participation and popularity of sports. This is certainly one frontier of sports statistics that North American statisticians should explore in future work. Given the ongoing debate about the social and economic values of professional sports franchises, research in this area may become more common as well.

1.2 The Sports Anthology Project

Given this rich history of research, one goal of SIS since its inception has been to produce a collection of papers that best represents the research of statisticians in sports. In 2000, SIS formed a committee to review sports statistics papers from ASA publications. The committee was charged with producing a volume of reasonable length with a broad representation of sports, statistical concepts, and authorship. While some emphasis was placed on including recent research, the committee also wished to include older seminal works that have laid the foundation for current research.

This volume's basic organization is by sport. Each major spectator sport (football, baseball, basketball, and ice hockey) has its own section. Sports with less representation in the ASA literature (such as golf, soccer, and track and field) have been collected into a separate section. Another, separate, section presents research that has greater breadth and generality, with each paper addressing several sports. Each section has been organized (with an introduction) by a notable contributor to that area.

1.3 Organization of the Papers

Sport provides a natural organization of the papers, and many readers will start reading papers on sports that they are particularly interested in as a participant or a fan. However, there are several alternative ways of grouping the papers that may be useful particularly for the statistics instructor.

1.4 Organization by Technical Level

All of these articles have been published in the following journals of the American Statistical Association (ASA): *Chance*, *Journal of the American Statistical Association* (JASA), *The American Statistician* (TAS), and the *Proceedings of the Statistics in Sports Section of the American Statistical Association*. A graph of the distribution of the papers in the four journals is displayed in Figure 1.1.

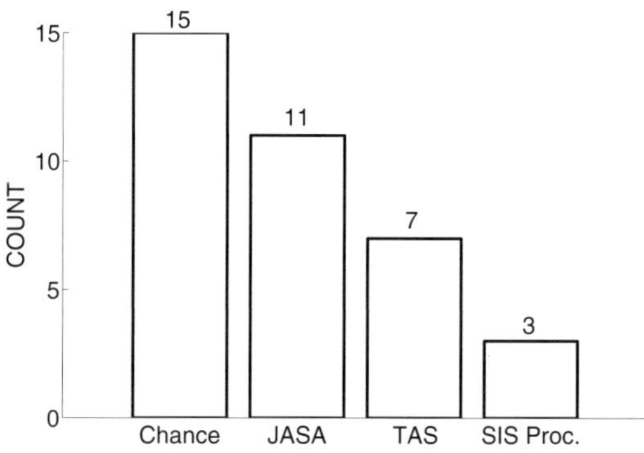

Figure 1.1. *Distribution of papers in four journals.*

The *Chance* and TAS articles in this volume are written at a relatively modest technical level and are accessible for both beginner and advanced students of statistics. *Chance* is a magazine, jointly published by ASA and Springer-Verlag, about statistics and the use of statistics in society. *Chance* features articles that showcase the use of statistical methods and ideas in the social, biological, physical, and medical sciences. One special feature of *Chance* is the regular column "A Statistician Reads the Sports Pages" and some of the articles in this volume are taken from this column. TAS is the "general interest" journal of the ASA and publishes articles on statistical practice, the teaching of statistics, the history of statistics, and articles of a general nature. JASA is the main research journal of the ASA. Although the technical level of JASA articles is typically higher than that of *Chance* and TAS articles, JASA contains an "Application and Case Studies" section that features interesting applications of statistical methodology. As mentioned earlier, JASA featured in 1994 a special collection of papers on the application of statistics to sports. The SIS *Proceedings* contains papers that were presented at the annual JSM under the sponsorship of SIS. The three proceedings papers appearing here in Chapters 12, 24, and 38 are also very accessible for a broad range of statisticians with different levels of training. As a general rule of thumb for the purposes of this collection, the spectrum of technical sophistication from high to low for these journals is JASA, TAS, *Chance*, and SIS *Proceedings*.

1.5 Organization by Inferential Method

An alternative organization scheme is by the inferential method. Some of the papers deal with particular inferential methods that are relevant to answering statistical questions related to sports. Indeed, one could actually create an undergraduate or graduate course in statistics devoted to particular inferential topics that are important in the context of sports. Below we describe the methods and list the corresponding papers that are included in this anthology.

1.5.1 Prediction of Sports Events

(Papers by Glickman and Stern (Chapter 5), Harville (Chapter 6), Stern (Chapter 8), Carlin (Chapter 18), and

Schwertman et al. (Chapter 20).)

A general statistical problem is to predict the winner of a particular sports game. This is obviously an important problem for the gambling industry, as millions of dollars are bet annually on the outcomes of sports events. An important problem is the development of good prediction methods. The articles in this volume focus on predicting the results of professional American football and college basketball games.

1.5.2 Hot Hand Phenomena

(Papers by Larkey et al. (Chapter 19), Tversky and Gilovich (Chapter 21), Wardrop (Chapter 22), Hooke (Chapter 31), Morrison and Schmittlein (Chapter 26), and Jackson and Mosurski (Chapter 42).)

Many people believe that the abilities of athletes and teams can go through short periods of highs and lows. They propose that these hot hand/cold hand phenomena account for streaky performance in sports. Psychologists generally believe that people misinterpret the streaky patterns in coin-tossing and sports data and think that players and teams are streaky when, really, they aren't. A general statistical problem is to detect true streakiness from sports data. The articles in this volume discuss the "hot hand" in basketball, baseball, hockey, and tennis.

1.5.3 Probability of Victory

(Papers by Bennett and Flueck (Chapter 12), Bennett (Chapter 11), Lindsey (Chapter 16), and Stern (Chapter 34).)

These papers address an interesting problem for sports fans. Suppose you are watching a game in progress and your team is leading by a particular margin. What is the probability that your team will win? This general question is addressed by these authors for different sports. Obviously the answer to this question has an impact on the fan's enjoyment of the game and also on the team's strategy in winning the game.

1.5.4 Learning from Selected Data

(Papers by Albert (Chapter 10) and Casella and Berger (Chapter 13).)

Often sports announcers will report "selected data." Specifically, they might talk about how a player or team performs in a given situation or report on an unusually high or low player or team performance in a short time period. There is an inherent bias in this information, since it has been reported because of its "interesting" nature. What has the fan actually learned about the ability of the player or team from this information? These two articles discuss this problem within the context of baseball, although it applies to many other sports.

1.5.5 Rating Players or Teams

(Papers by Danehy and Lock (Chapter 24), Bassett and Persky (Chapter 39), Morrison and Kalwani (Chapter 7), and Berry (Chapters 4 and 37).)

An interesting statistical problem is how to rate or rank players or teams based on their performances. In some sports, such as American college football, it is not possible to have a complete round-robin competition, thus presenting a challenge in ranking teams based on these incomplete data. This is not a purely academic exercise, as various ranking systems are used to select contenders for national championships (as in American college football) or to seed tournament positions. There are also interesting statistical issues in evaluating sports performers. How does one effectively compare two sports players when there are many measurements of a player's ability? How does one compare players from different eras? These papers illustrate statistical issues in ranking teams (college hockey teams) and individuals (professional football kickers, skaters, golfers, and hockey and baseball players).

1.5.6 Unusual Outcome

(Paper by Frohlich (Chapter 14).)

The sports media pays much attention to "unusual" outcomes. For example, in baseball, we are surprised by a no-hitter or if a batter gets a hit in a large number of consecutive games. From a statistical viewpoint, should we be surprised by these events, or are these events to be expected using standard probability models?

1.5.7 The Rules of Sports

(Papers by Hurley (Chapter 25), Scheid and Calvin (Chapter 38), and Wainer and De Veaux (Chapter 44).)

Statistical thinking can be useful in deciding on rules in particular sports. For example, statistics can help determine a reasonable method of breaking a tie game in soccer or a good way of setting a golf player's handicap.

1.5.8 Game Strategy

(Paper by Harville (Chapter 30).)

Statistics can be useful in determining proper strategies in sports. In baseball, strategy is often based on

"The Book," that is, a collection of beliefs based on anecdotal evidence absorbed by players and managers throughout baseball history. Through a careful analysis of baseball data, one can investigate the wisdom of particular strategies such as the sacrifice bunt, the intentional walk, and stealing of bases. It is not surprising that managers' intuition about the correct strategy is often inconsistent with the results of a careful statistical analysis.

1.5.9 Illustrating Statistical Methods

(Papers by Efron and Morris (Chapter 29) and Roberts (Chapter 33).)

Sports data can provide an accessible and interesting way of introducing new statistical methodology. The paper by Efron and Morris illustrates the advantages of shrinkage estimators using baseball batting averages as an example. The paper by Roberts illustrates the use of techniques in Total Quality Management in improving a student's ability in golf (specifically, putting) and in the game of pool.

1.5.10 Modeling Team Competition

(Papers by Lee (Chapter 40) and James, Albert, and Stern (Chapter 15).)

After a season of team sports competition has been played, fans assume that the winner of the championship game or series is truly the best team. But is this a reasonable conclusion? What have we really learned about the abilities of the teams based on these observed competition results? These two articles explore these issues for the sports of soccer and baseball.

1.5.11 The Use of These Articles in Teaching Statistics

Sports examples provide an attractive way of introducing statistical ideas, at both elementary and advanced levels. Many people are involved in sports, either as a participant or a fan, and so are naturally interested in statistical problems that are framed within the context of sports. Chapter 2 in this anthology gives an overview of the use of sports examples in teaching statistics. It describes statistics courses that have been offered with a sports theme and discusses a number of other articles that illustrate statistical analyses on interesting sports datasets. From this teaching perspective, the reader will see that many of the research papers in this volume are very appropriate for use in the classroom.

1.6 The Common Theme

Ignoring the differences in sport topics and statistical techniques applied, we can see that the papers presented in this volume have one noteworthy quality in common: All research was performed for the sheer joy of it. These statisticians did not perform this research to fulfill a contract, advance professionally, or promote their employers. They initiated this research out of an inner dissatisfaction with standard presentations of sports statistics in various media and a personal need to answer questions that were either being addressed inadequately or not at all. We hope that you enjoy reading their work as much as they enjoyed producing it.

Chapter 2

The Use of Sports in Teaching Statistics

Jim Albert and James J. Cochran

2.1 Motivation to Use Sports Examples in Teaching Statistics

Teaching introductory statistics is a challenging endeavor because the students have little prior knowledge about the discipline of statistics and many of them are anxious about mathematics and computation. Many students come to the introductory statistics course with a general misunderstanding of the science of statistics. These students regard "statistics" as a collection of numbers and they believe that the class will consist of a series of computations on these numbers. Indeed, although statistics texts will discuss the wide application of statistical methodology, many students will believe that they will succeed in the class if they are able to recall and implement a series of statistical recipes.

Statistical concepts and examples are usually presented in a particular context. However, one obstacle in teaching this introductory class is that we often describe the statistical concepts in a context (such as medicine, law, or agriculture) that is completely foreign to the student. The student is much more likely to understand concepts in probability and statistics if they are described in a familiar context. Many students are familiar with sports either as a participant or a spectator. They know of popular athletes, such as Tiger Woods and Barry Bonds, and they are generally knowledgeable about the rules of major sports, such as baseball, football, and basketball. To many students, sports is a familiar context in which an instructor can describe statistical thinking. Students often know and have some intuition about the issues in a particular sport. Since they are knowledgeable in sports, they can see the value of statistics in gaining additional insight on particular issues.

Statistical methodology is useful in addressing problems in many disciplines. The use of statistical thinking in the context of baseball is called sabermetrics. James (1982) defines sabermetrics as follows (note that his comments apply equally if the word "sabermetrics" is replaced with "statistics"):

> "Sabermetrics does not begin with the numbers. It begins with issues. The numbers, the statistics are not the subject of the discussion . . . The subject is baseball. The numbers bear a relationship to that subject and to us which is much like the relationship of tools to a machine and to the mechanic who uses them. The mechanic does not begin with a monkey wrench; basically, he is not even interested in the damn monkey wrench. All that he wants from the monkey wrench is that it do its job and not give him any trouble. He begins with the machine, with the things which he sees and hears there, and from those he forms an idea—a thesis—about what must be happening in that machine. The tools are a way of taking the thing apart so he can see if he was right or if he needs to try something else."

Sports provides a familiar setting where instructors can discuss questions or issues related to sports and show how statistical thinking is useful in answering these questions. Using James' analogy, we can use sports examples to emphasize that the main subject of the discussion is sports and the statistical methodology is a tool that contributes to our understanding of sports.

2.2 Mosteller's Work in Statistics and Sports

Frederick Mosteller was one of the first statisticians to extensively work on statistical problems relating to sports. Mosteller was also considered one of the earlier innovators in the teaching of statistics; his "one-minute survey" is currently used by many instructors to get quick feedback from the students on the "muddiest point in the lecture." The article by Mosteller (1997), reprinted in this volume as Chapter 32, summarizes his work on problems in baseball, football, basketball, and hockey. He describes several lessons he learned from working on sports, many of which are relevant to the teaching of statistics. First, if many reviewers (you can substitute the word "students" for "reviewers") are knowledgeable about the materials and interested in the findings, they will drive the author crazy with the volume, perceptiveness, and relevance of their suggestions. Mosteller's first lesson is relevant to classroom teaching: teaching in the sports context will encourage class discussion. A second lesson is that any inferential method in statistics can likely be applied to a sports example. Mosteller believes that statisticians can learn from newswriters, who are able to communicate with the general public. Perhaps statisticians (or perhaps statistics instructors or writers of statistics texts) could learn from newswriters on how to communicate technical content to the general public. Given the large number of sports enthusiasts in the United States, Mosteller believes that sports provides an opportunity for statisticians to talk to young people about statistics. In a paper on broadening the scope of statistical education, Mosteller (1988) talks about the need for books aimed at the general public that communicate statistical thinking in an accessible way. Albert and Bennett (2003) and Ross (2004) are illustrations of books written with a sports theme that are intended to teach ideas of probability and statistics to the general public.

2.3 The Use of Interesting Sports Datasets in Teaching

The use of sports data in teaching has several advantages. First, sports data are easily obtained from almanacs and the Internet. For example, there are a number of websites, such as the official Major League Baseball site (www.mlb.com) and the Baseball Reference site (www.baseball-reference.com) that give detailed data on baseball players and teams over the last 100 seasons of Major League Baseball competition. There are similar sites that provide data for other sports. Extensive statistics for professional football, basketball, hockey, golf, and soccer can be found at www.nfl.com, www.nba.com, www.nhl.com, www.pga.com, and www.mlsnet.com, respectively.

The second advantage of using sports in teaching statistics is that there are a number of interesting questions regarding sports players and teams that can be used to motivate statistical analyses, such as: When does an athlete peak in his/her career? What is the role of chance in determining the outcomes of sports events? Who is the best player in a given sport? How can one best rank college football teams who play a limited schedule? Are sports participants truly streaky?

The references to this chapter contain a number of articles in the *Journal of Statistics Education* (JSE), *The American Statistician* (TAS), the *Proceedings of the Section on Statistics in Sports*, and the *Proceedings of the Section on Statistical Education* that illustrate statistical analyses on interesting sports datasets. In addition, the regular column "A Statistician Reads the Sports Page" (written by Hal Stern and Scott Berry) in *Chance* and the column "The Statistical Sports Fan" (written by Robin Lock) in *STATS* (the magazine for students of statistics) contain accessible statistical analyses on questions dealing with sports. (Some of these columns are reprinted in this volume.) The JSE articles are specifically intended to equip the instructors of statistics with interesting sports examples and associated questions that can be used for presentation or homework. Following are five representative "sports stories" from JSE that can be used to motivate statistical analyses. We conclude this section with a reference to an extensive dataset available at the Chance Course website (www.dartmouth.edu/~chance) to detect streakiness in sports performance.

2.3.1 Exploring Attendance at Major League Baseball Games (Cochran (2003))

Economic issues in major league sports can be addressed using the data provided here. This dataset contains measurements on team performance (wins and losses, runs scored and allowed, games behind division winner, rank finish within division), league and division affiliation, and total home game attendance for every Major League Baseball team for the 1969–2000 seasons. The author explains how he uses these data to teach regression and econometrics by assigning students the task of modeling the relationship between team performance and annual home attendance.

2.3.2 How Do You Measure the Strength of a Professional Football Team? (Watnik and Levine (2001))

This dataset gives a large number of performance variables for National Football League (NFL) teams for the 2000 regular season. The authors use principal components methodology to identify components of a football team's strength. The first principal component is identified with a team's offensive capability and the second component with a team's defensive capability. These components were then related to the number of wins and losses for the teams. A number of questions can be answered with these data, including: Which conference (NFC or AFC) tends to score more touchdowns? What variables are most helpful in predicting a team's success? How important are the special teams, such as the kicking and punting teams, in a team's success?

2.3.3 Who Will Be Elected to the Major League Baseball Hall of Fame? (Cochran (2000))

Election to the Hall of Fame is Major League Baseball's greatest honor, and its elections are often controversial and contentious. The author explains how he uses the data provided to teach descriptive statistics, classification, and discrimination by assigning students the task of using the various career statistics to model whether the players have been elected to the Hall of Fame. Career totals for standard baseball statistics (number of seasons played, games played, official at-bats, runs scored, hits, doubles, triples, home runs, runs batted in, walks, strikeouts, batting average, on base percentage, slugging percentage, stolen bases and times caught stealing, fielding average, and primary position played) are provided for each position player eligible for the Major League Baseball Hall of Fame as of 2000. In addition, various sabermetric composite measures (adjusted production, batting runs, adjusted batting runs, runs created, stolen base runs, fielding runs, and total player rating) are provided for these players. Finally, an indication is made of whether or not each player has been admitted into the Major League Baseball Hall of Fame and, if so, under what set of rules he was admitted.

2.3.4 Hitting Home Runs in the Summer of 1998 (Simonoff (1998))

The 1998 baseball season was one of the most exciting in history due to the competition between Mark McGwire and Sammy Sosa to break the season home run record established by Roger Maris in 1961. This article contains the game-to-game home run performance for both McGwire and Sosa in this season and gives an overview of different types of exploratory analyses that can be done with these data. In particular, the article explores the pattern of home run hitting of both players during the season; sees if McGwire's home runs helped his team win games during this season; and sees if the home run hitting varied between games played at home and games played away from home. In a closing section, the author outlines many other questions that can be addressed with these data.

2.3.5 Betting on Football Games (Lock (1997))

This dataset contains information for all regular season and playoff games placed by professional football games for a span of five years. This dataset provides an opportunity for the instructor to introduce casino bets on sports outcomes presented by means of point spreads and over/under values. Also, the dataset can be used to address a number of questions of interest to sports-minded students, such as: What are typical professional game scores and margins of victory? Is there a home-field advantage in football and how can we measure it? Is there a correlation between the scores of the home and away teams? Is there a relationship between the point spread and the actual game results? Can one develop useful rules for predicting the outcomes of football games?

2.3.6 Data on Streaks (from *Chance* Datasets, http://www.dartmouth.edu/~chance/teaching_aids/data.html)

Albright (1993) performed an extensive analysis to detect the existence of streakiness in the sequences of bats for all players in Major League Baseball during a season. The dataset available on the *Chance* datasets site provides 26 bits of information on the situation and outcome for each time at bat for a large number of players in both the American and National Leagues during the time period 1987–1990. Albright (1993) and Albert (1993) give illustrations of the types of exploratory and confirmatory analyses that can be performed to detect streakiness in player performances.

2.4 Special Probability and Statistics Classes Focused on Sports

A number of authors have created special undergraduate classes with a focus on the application of statistics to sports. One of these classes is designed to provide a complete introduction to the basic topics in data analysis and probability within the context of sports. Other courses are examples of seminar-type classes that use statistics from sports articles and books to motivate critical discussions of particular issues in sports.

2.4.1 An Introductory Statistics Course Using Baseball

Albert (2002, 2003) discusses a special section of an introductory statistics course where all of the course material is taught from a baseball perspective. This course was taken by students fulfilling a mathematics elective requirement. The topics of this class were typical of many introductory classes: data analysis for one and two variables, elementary probability, and an introduction to inference. This class was distinctive in that all of the material was taught within the context of baseball. Data analysis was discussed using baseball data on players and teams, both current and historical. Probability was introduced by the description of several tabletop baseball games and an all-star game was simulated using spinners constructed from players' hitting statistics. The distinction between a player's performance and his ability was used to introduce statistical inference, and the significance of situational hitting statistics and streaky data was discussed.

2.4.2 Using Baseball Board Games to Teach Probability Concepts

Cochran (2001) uses baseball simulation board games to introduce a number of concepts in probability. Comparing playing cards for two players from the Strat-O-Matic baseball game, he explains different approaches to assigning probabilities to events, various discrete probability distributions, transformations and sample spaces, the laws of multiplication and addition, conditional probability, randomization and independence, and Bayes' Theorem. Cochran finds that this twenty-minute demonstration dramatically improves the understanding of basic probability achieved by most students (even those who don't like or know anything about baseball).

2.4.3 A Sabermetrics Course

Costa and Huber (2003) describe a one-credit-hour course, developed and taught by Costa at Seton Hall University in 1998, that focused on sabermetrics (the science of studying baseball records). This course has evolved into a full three-credit-hour, baseball-oriented introductory statistics course similar to the course offered by Albert (2002, 2003) that is taught at both Seton Hall University and the U.S. Air Force Academy. This course also incorporates field trips to various baseball sites (such as the National Baseball Hall of Fame and Museum in Cooperstown, NY) and is team-taught.

2.4.4 A Course on Statistics and Sports

Reiter (2001) taught a special course on statistics and sports at Williams College. This course was offered during a three-week winter term period, where students were encouraged to enroll in a class outside of their major. The students had strong interests in sports but had varying backgrounds in statistics. Each student made two oral presentations in this class. In the first presentation, the student gave some background concerning the statistics used in a sports article of his/her choosing. For the second presentation, the student presented research from a second article or described his or her own data analysis. The seminar helped the students learn the concept of random variability and the dependence of the statistical methodology on assumptions such as independence and normality.

2.4.5 A Freshmen Seminar: Mathematics and Sports

Gallian (2001) taught a liberal arts freshmen seminar at the University of Minnesota at Duluth that utilized statistical concepts to analyze achievements and strategies in sports. The intent of the freshmen seminar is to provide an atypical course by a senior professor where all of the students actively participate. The students, working primarily in groups of two or three, give written and oral reports from a list of articles that apply statistical methodology to sports. Students are encouraged to ask questions of the speakers and participate in class discussion. Biographical and historical information about athletes and their special accomplishments (such as Joe DiMaggio's 56-game hitting streak) is provided via videotapes.

2.4.6 A Freshmen Seminar: Statistics and Mathematics of Baseball

Ken Ross taught a similar freshmen seminar at the University of Oregon on the statistics and mathematics of baseball. The goal of this seminar is to provide an interesting course with much student interaction and critical thinking. Most of the students were very knowledgeable about baseball. The books by Hoban (2000) and Schell (1999) were used for this seminar. Hoban (2000) seemed appropriate for this class since it is relatively easy to read and its conclusions are controversial, which provoked student discussion. Schell (1999) was harder to read for these students since it is more sophisticated statistically, and thus the instructor supplemented this book with additional discussion. A range of statistical topics was discussed in this class, including Simpson's paradox, correlation, and chi-square tests. Ross recently completed a book (Ross, 2004) that was an outgrowth of material from this freshmen seminar.

2.5 Summary

Sports examples have been used successfully by many statistics instructors and now comprise the entire basis of many introductory statistics courses. Furthermore, entire introductory statistics textbooks and "popular-style" books that exclusively use sports examples have recently been published. (Examples of these popular-style books are Albert and Bennett (2003), Haigh (2000), Ross (2004), Skiena (2001), and Watts and Bahill (2000).) Statistics instructors who use sports examples extensively have found a corresponding increase in comprehension and retention of concepts that students usually find to be abstract and obtuse. Instructors have found that most students, even those who are not knowledgeable about sports, enjoy and appreciate sports examples in statistics courses. Due to the popularity of sports in the general public and the easy availability of useful sports datasets, we anticipate that sports will continue to be a popular medium for communicating concepts of probability and statistics.

References

Albert, J. (1993), Discussion of "A statistical analysis of hitting streaks in baseball" by S. C. Albright, Journal of the American Statistical Association, 88, 1184–1188.

Albert, J. (2002), "A baseball statistics course," Journal of Statistics Education, 10, 2.

Albert, J. (2003), *Teaching Statistics Using Baseball*, Washington, DC: Mathematical Association of America.

Albert, J. and Bennett, J. (2003), *Curve Ball*, New York: Copernicus Books.

Albright, S. C. (1993), "A statistical analysis of hitting streaks in baseball," Journal of the American Statistical Association, 88, 1175–1183.

Bassett, G. W. and Hurley, W. J. (1998), "The effects of alternative HOME–AWAY sequences in a best-of-seven playoff series," The American Statistician, 52, 51–53.

Cochran, J. (2000), "Career records for all modern position players eligible for the Major League Baseball Hall of Fame," Journal of Statistics Education, 8, 2.

Cochran, J. (2001), "Using Strat-O-Matic baseball to teach basic probability for statistics," ASA Proceedings of the Section on Statistics in Sports.

Cochran, J. (2003), "Data management, exploratory data analysis, and regression analysis with 1969–2000 Major League Baseball attendance," Journal of Statistics Education, 10, 2.

Costa, G. and Huber, M. (2003), *Whaddya Mean? You Get Credit for Studying Baseball?* Technical report.

Gallian, J. (2001), "Statistics and sports: A freshman seminar," ASA Proceedings of the Section on Statistics in Sports.

Gould, S. J. (2003), *Triumph and Tragedy in Mudville: A Lifelong Passion for Baseball*, New York: W. W. Norton.

Haigh, J. (2000), *Taking Chances*, Oxford, UK: Oxford University Press.

Hoban, M. (2000), *Baseball's Complete Players*, Jefferson, NC: McFarland.

James, B. (1982), *The Bill James Baseball Abstract*, New York: Ballantine Books.

Lackritz, James R. (1981), "The use of sports data in the teaching of statistics," ASA Proceedings of the Section on Statistical Education, 5–7.

Lock, R. (1997), "NFL scores and pointspreads," Journal of Statistics Education, 5, 3.

McKenzie, John D., Jr. (1996), "Teaching applied statistics courses with a sports theme," ASA Proceedings of the Section on Statistics in Sports, 9–15.

Morris, Pamela (1984), "A course work project for examination," Teaching Statistics, 6, 42–47.

Mosteller, F. (1988), "Broadening the scope of statistics and statistical education," The American Statistician, 42, 93–99.

Mosteller, F. (1997), "Lessons from sports statistics," The American Statistician, 51, 305–310.

Nettleton, D. (1998), "Investigating home court advantage," Journal of Statistics Education, 6, 2.

Quinn, Robert J. (1997), "Investigating probability with the NBA draft lottery," Teaching Statistics, 19, 40–42.

Quinn, Robert J. (1997), "Anomalous sports performances," Teaching Statistics, 19, 81–83.

Reiter, J. (2001), "Motivating students' interest in statistics through sports," ASA Proceedings of the Section on Statistics in Sports.

Ross, K. (2004), *A Mathematician at the Ballpark*, New York: Pi Press.

Schell, M. (1999), *Baseball All-Time Best Hitters*, Princeton, NJ: Princeton University Press.

Simonoff, J. (1998), "Move over, Roger Maris: Breaking baseball's most famous record," Journal of Statistics Education, 6, 3.

Skiena, S. (2001), *Calculated Bets*, Cambridge, UK: Cambridge University Press.

Starr, Norton (1997), "Nonrandom risk: The 1970 draft lottery," Journal of Statistics Education, 5.

Watnik, M. (1998), "Pay for play: Are baseball salaries based on performance?" Journal of Statistics Education, 6, 2.

Watnik, M. and Levine, R. (2001), "NFL Y2K PCA," Journal of Statistics Education, 9, 3.

Watts, R. G. and Bahill, A. T. (2000), *Keep Your Eye on the Ball*, New York: W. H. Freeman.

Wiseman, Frederick and Chatterjee, Sangit (1997), "Major League Baseball player salaries: Bringing realism into introductory statistics courses," The American Statistician, 51, 350–352.

Part I
Statistics in Football

Chapter 3

Introduction to the Football Articles

Hal Stern

This brief introduction, a sort of "pregame show," provides some background information on the application of statistical methods in football and identifies particular research areas. The main goal is to describe the history of research in football, emphasizing the place of the five selected articles.

3.1 Background

Football (American style) is currently one of the most popular sports in the U.S. in terms of fan interest. As seems to be common in American sports, large amounts of quantitative information are recorded for each football game. These include summaries of both team and individual performances. For professional football, such data can be found for games played as early as the 1930s and, for college football, even earlier years. Somewhat surprisingly, despite the large amount of data collected, there has been little of the detailed statistical analysis that is common for baseball.

Two explanations for the relatively small number of statistics publications related to football are obvious. First, despite the enormous amount of publicity given to professional football, it is actually difficult to obtain detailed (play-by-play) information in computer-usable form. This is not to say that the data do not exist—they do exist and are used by the participating teams. The data have not been easily accessible to those outside the sport. Now play-by-play listings can be found on the World Wide Web at the National Football League's own site (www.nfl.com). These data are not in convenient form for research use, but one can work with them.

A second contributing factor to the shortage of research results is the nature of the game itself. Here are four examples of the kinds of things that can complicate statistical analyses: (1) scores occur in steps of size 2, 3, 6, 7, and 8 rather than just a single scoring increment; (2) the game is time-limited with each play taking a variable amount of time; (3) actions (plays) move the ball over a continuous surface; and (4) several players contribute to the success or failure of each play. Combined, these properties of football make the number of possible score-time-location situations that can occur extremely large and this considerably complicates analysis.

The same two factors that conspire to limit the amount of statistical research related to football also impact the types of work that are feasible. Most of the published research has focused on aspects of the game that can be easily separated from the ordinary progress of the game. For example, it can be difficult to rate the contribution of players because several are involved in each play. As a result, more work has been done evaluating the performance of kickers (including two articles published here) than of running backs. In the same way, the question of whether to try a one-point or two-point conversion after touchdown has received more attention than any other strategy-related question. In the remainder of this introduction, the history of statistical research related to football is surveyed and the five football articles in our collection are placed in context.

Since selecting articles for a collection like this invariably omits some valuable work, it is important to mention sources containing additional information. Stern (1998) provides a more comprehensive review of the work that has been carried out in football. The book by Carroll, Palmer, and Thorn (1988) is a sophisticated analysis of the game by three serious researchers with access to play-by-play data. Written for a general audience, the book does not provide many statistical details but is quite thought provoking. The collections of articles edited by Ladany and Machol (1977) and Machol, Ladany, and Morrison (1976) are also good academic sources.

3.2 Information Systems

Statistical contributions to the study of football have appeared in four primary areas: information systems, player performance evaluation, football strategy, and team performance evaluation. The earliest significant contribution of statistical reasoning to football was the development of computerized systems for studying opponents' performances. This remains a key area of research activity. Professional and college football teams prepare reports detailing the types of plays and formations favored by opponents in a variety of situations. The level of detail in the reports can be quite remarkable, e.g., they might indicate that Team A runs to the left side of the field 70% of the time on second down with five or fewer yards required for a new first down. These reports clearly influence team preparation and game-time decision making. These data could also be used to address strategic issues (e.g., whether a team should try to maintain possession when facing fourth down or kick the ball over to its opponent) but that would require more formal analysis than is typically done. It is interesting that much of the early work applying statistical methods to football involved people affiliated with professional or collegiate football (i.e., players and coaches) rather than statisticians. The author of one early computerized play-tracking system was 1960s professional quarterback Frank Ryan (Ryan, Francia, and Strawser, 1973). Data from such a system would be invaluable for statistically minded researchers but the data have not been made available.

3.3 Player Performance Evaluation

Evaluation of football players has always been important for selecting teams and rewarding players. Formally evaluating players, however, is a difficult task because several players contribute to each play. A quarterback may throw the ball 5 yards down the field and the receiver, after catching the ball, may elude several defensive players and run 90 additional yards for a touchdown. Should the quarterback get credit for the 95-yard touchdown pass or just the 5 yards the ball traveled in the air? What credit should the receiver get? The difficulty in apportioning credit to the several players that contribute to each play has meant that a large amount of research has focused on aspects of the game that are easiest to isolate, such as kicking. Kickers contribute to their team's scoring by kicking field goals (worth 3 points) and points-after-touchdowns (worth 1 point). On fourth down a coach often has the choice of (1) attempting an offensive play to gain the yards needed for a new first down, (2) punting the ball to the opposition, or (3) attempting a field goal. A number of papers have concerned the performance of field goal kickers. The article of Berry (Chapter 4) is the latest in a series of attempts to model the probability that a field goal attempted from a given distance will be successful. His model builds on earlier geometric models for field goal kick data. Our collection also includes a second article about kickers, though focused on a very different issue. Morrison and Kalwani (Chapter 7) examine the performances of all professional kickers over a number of seasons and ask whether there are measurable differences in ability. Their somewhat surprising result is that the data are consistent with the hypothesis that all kickers have equivalent abilities. Though strictly speaking this is not likely to be true, the data are indicative of the large amount of variability in field goal kicking. As more play-by-play data are made available to researchers, one may expect further advances in the evaluation of performance by nonkickers.

3.4 Football Strategy

An especially intriguing goal for statisticians is to discover optimal strategies for football. Examples of the kinds of questions that one might address include optimal decision-making on fourth down (to try to maintain possession, try for a field goal, or punt the ball to the opposing team), optimal point-after-touchdown conversion strategy (one-point or two-point conversion attempt), and perhaps even optimal offensive and defensive play calling. As is the case with evaluation of players, the complexities of football have meant that the majority of the research work has been carried out on the most isolated components of the game, especially point-after-touchdown conversion decisions. Attempts to address broader strategy considerations require being able to place a value on game situations, for example, to have the probability of winning the game for a given score, time remaining, and field position. Notable attempts in this regard include the work of Carter and Machol (1971)—that is, former NFL quarterback Virgil Carter—and Carroll, Palmer, and Thorn (1988).

3.5 Team Performance Evaluation

The assessment of team performance has received considerably more attention than the assessment of individual players. This may be partly a result of the enormous betting market for professional and college football in the U.S.

A key point is that teams usually play only a single game each week. This limits the number of games per team per season to between 10 and 20 (depending on whether we are thinking of college or professional football) and thus many teams do not play one another in a season. Because teams play unbalanced schedules, an unequivocal determination of the best team is not possible. This means there is an opportunity for developing statistical methods to rate teams and identify their relative strengths.

For a long time there has been interest in rating college football teams with unbalanced schedules. The 1920s and 1930s saw the development of a number of systems. Early methods relied only on the record of which teams had defeated which other teams (with no use made of the game scores). It has become more popular to make use of the scores accumulated by each team during its games, as shown in two of the articles included here. Harville (Chapter 6), in the second of his two papers on this topic, describes a linear model approach to predicting National Football League games. The linear model approach incorporates parameters for team strength and home field advantage. In addition, team strengths are allowed to vary over time. Building on Harville's work, Glickman and Stern (Chapter 5) apply a model using Bayesian methods to analyze the data. The methods of Harville and Glickman and Stern rate teams on a scale such that the difference in estimated ratings for two teams is a prediction of the outcome for a game between them. The methods suggest that a correct prediction percentage of about 67% is possible; the game itself is too variable for better prediction.

This last observation is the key to the final chapter of Part I of this book. In Chapter 8, Stern examines the results of football games over several years and finds that the point differentials are approximately normal with mean equal to the Las Vegas betting point spread and standard deviation between 13 and 14 points. Stern's result means that it is possible to relate the point spread and the probability of winning. This relationship, and similar work for other sports, have made formal probability calculations possible in a number of settings for which it had previously been quite difficult.

3.6 Summary

Enough of the pregame show. Now on to the action! The following five football articles represent how statistical thinking can influence the way that sports are watched and played. In the case of football, we hope these articles foreshadow future developments that will touch on more difficult work involving player evaluation and strategy questions.

References

Berry, S. (1999), "A geometry model for NFL field goal kickers," Chance, 12 (3), 51–56.

Carroll, B., Palmer, P., and Thorn, J. (1988), *The Hidden Game of Football*, New York: Warner Books.

Carter, V. and Machol, R. E. (1971), "Operations research on football," Operations Research, 19, 541–545.

Glickman, M. E. and Stern, H. S. (1998), "A state-space model for National Football League scores," Journal of the American Statistical Association, 93, 25–35.

Harville, D. (1980), "Predictions for National Football League games via linear-model methodology," Journal of the American Statistical Association, 75, 516–524.

Ladany, S. P. and Machol, R. E. (editors) (1977), *Optimal Strategy in Sports*, Amsterdam: North–Holland.

Machol, R. E., Ladany, S. P., and Morrison, D. G. (editors) (1976), *Management Science in Sports*, Amsterdam: North–Holland.

Morrison, D. G. and Kalwani, M. U. (1993), "The best NFL field goal kickers: Are they lucky or good?" Chance, 6 (3), 30–37.

Ryan, F., Francia, A. J., and Strawser, R. H. (1973), "Professional football and information systems," Management Accounting, 54, 43–47.

Stern, H. S. (1991), "On the probability of winning a football game," The American Statistician, 45, 179–183.

Stern, H. S. (1998), "American football," in *Statistics in Sport*, edited by J. Bennett, London: Arnold, 3–23.

Chapter 4

A STATISTICIAN READS THE SPORTS PAGES

Scott M. Berry, Column Editor

A Geometry Model for NFL Field Goal Kickers

My dad always made it a point whenever we had a problem to discuss it openly with us. This freedom to discuss ideas had its limits — there was one thing that could not be discussed. This event was too painful for him to think about — a weak little ground ball hit by Mookie Wilson that rolled through the legs of Bill Buckner. This error by Buckner enabled the New York Mets to beat the Boston Red Sox in the 1986 World Series. My dad is a life-long Red Sox fan and was crushed by the loss. This loss was devastating because it appeared as though the Red Sox were going to win and end the "Curse of the Babe" by winning their first championship since the trade of Babe Ruth after the 1919 season.

The Minnesota Vikings have suffered similar championship futility. They have lost four Super Bowls without winning one. The 1998 season appeared to be different. They finished the season 15–1, set the record for the most points scored in a season, and were playing the Atlanta Falcons for a chance to go to the 1999 Super Bowl. The Vikings had the ball on the Falcon's 21-yard line with a 7-point lead and just over two minutes remaining. Their place kicker, Gary Anderson, had not missed a field-goal attempt all season. A successful field goal and the Vikings would earn the right to play the Denver Broncos in the Super Bowl. Well, in Buckner-like fashion, he missed the field goal — wide left ... for a life-long Viking fan, it was painful. The Falcons won the game, and the Vikings were eliminated. I took the loss much the same as my dad had taken the Mets beating the Red Sox — not very well.

When the pain subsided I got to thinking about modeling field goals. In an interesting geometry model, Berry and Berry (1985) decomposed each kick attempt with a distance and directional component. In this column I use similar methods, but the data are more detailed and the current computing power enables a more detailed model. I model the direction and distance of each kick with separate distributions. They are combined to find estimates for the probability of making a field goal for each player, from each distance. By modeling distance and accuracy the intuitive result is that different kickers are better from different distances. There has also been some talk in the NFL that field goals are too easy. I address the effect of reducing the distance between the uprights. Although it may be painful, I also address the probability that Gary Anderson would miss a 38-yard field goal!

Column Editor: Scott M. Berry, Department of Statistics, Texas AM University, 410B Blocker Building, College Station, TX 77843–3143, USA; E-mail *berry@stat.tamu.edu*.

Background Information

A field goal attempt in the NFL consists of a center hiking the ball to a holder, seven yards from the line of scrimmage. The holder places the ball on the ground, where the place kicker attempts to kick it through the uprights. The uprights are 18 feet, 6 inches apart and are centered 10 yards behind the goal line. Therefore, if the line of scrimmage is the 21-yard line, the ball is "snapped" to the holder at the 28-yard line and the uprights are 38 yards from the position of the attempt, which is considered a 38-yard attempt. For the kick attempt to be successful, the place kicker must kick the ball through the uprights and above a crossbar, which is 10 feet off the ground. The information recorded by the NFL for each kick is whether it was successful, short, or wide, and if it is wide, which side it misses on. I could not find any official designation of how a kick is categorized if it is both short and wide. I assume that if a kick is short that is the first categorization used for a missed kick. Therefore, a kick that is categorized as wide must have had enough distance to clear the crossbar. After a touchdown is scored, the team has an option to kick a point after touchdown (PAT). This is a field-goal try originating from the 2-yard line — thus a 19-yard field-goal try.

I collected the result of every place kick during the 1998 regular season. Each kick is categorized by the kicker, distance, result, and the reason for any unsuccessful kicks. Currently NFL kickers are ranked by their field-goal percentage. The trouble with these rankings is that the distance of the kick is ignored. A good field-goal kicker is generally asked to attempt longer and more difficult kicks. When a kicker has a 38-yard field goal attempt, frequently the TV commentator will provide the success proportion for the kicker from that decile

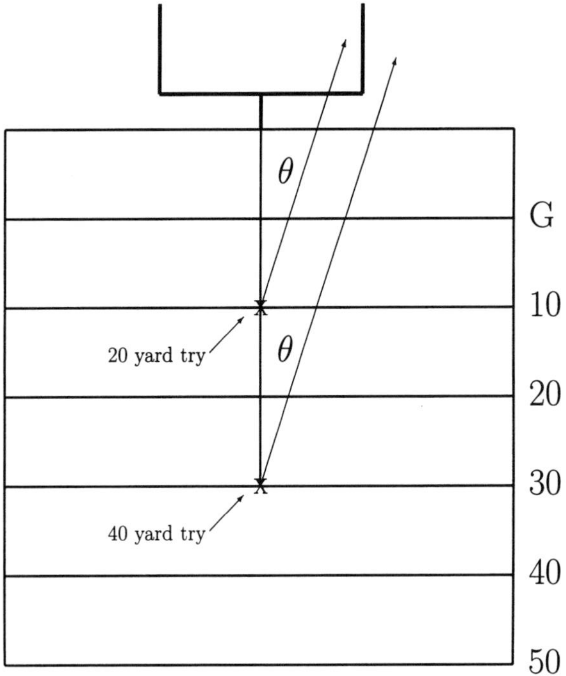

Figure 1. The accuracy necessary for a kicker is proportional to the distance of the kick. For a field-goal try of 20 yards, you can have an angle twice as big as that for a try from 40 yards.

yardage — that is, 30–39 yards. These percentages are generally based on small sample sizes. Information from longer kicks can also be informative for a 38-yard attempt. A success from a longer distance would be successful from shorter distances. Moreover, misses from shorter distances would also be misses from the longer distances. I model the accuracy and length for each of the place kickers and update from all the place kicks.

Mathematical Model

To measure the accuracy of a place kicker, I model the angle of a kick from the center of the uprights. Refer to Fig. 1 for a description of the accuracy aspect of the model. For a shorter kick, there is a larger margin of error for the kicker. Let the angle from the center line for a kick be θ. Rather than model the angle explicitly, I model the distance from the center line, for an angle θ, if the kick were a 40-yard attempt. Let the random variable W be the number of feet from the center line if the kick were from 40 yards. I assume that each kicker is unbiased — that is, on average they hit the middle of the uprights. It may be that some kickers are biased to one side or the other, but the data are insufficient to investigate "biasedness" of a kicker. A kicked ball must be within 9.25 feet of the center line when it crosses the plane of the uprights for the kick to be successful (assuming it is far enough). Therefore, if the attempt is from 40 yards, $|W| < 9.25$ implies that the kick will be successful if it is long enough. For a 20-yard field goal the angle can be twice as big as for the 40-yard attempt to be successful. Thus, a 20-yard field goal will be successful if $|W| < 18.5$. In general, a field goal from X yards will be successful if $|W| < (9.25)(40/X)$.

I assume that the resulting distance from the center line, 40 yards from the holder, is normally distributed with a mean of 0 and a standard deviation of σ_i, for player i. I refer to σ_i as the "accuracy" parameter for a kicker. For kicker i, if a kick from X yards is long enough, the probability that it is successful is

$$P\left(-\frac{(9.25)(40)}{X} < W < \frac{(9.25)(40)}{X}\right) = 2\Phi\left[\left(\frac{40}{X}\right)\left(\frac{9.25}{\sigma_i}\right)\right] - 1$$

where Φ is the cumulative distribution function of a standard normal random variable.

For a kick to be successful, it also has to travel far enough to clear the crossbar 10 feet off the ground. For each attempt the distance for which it would be successful if it were straight enough is labeled d. I model the distance traveled for a kick attempt with a normal distribution. The mean distance for each kicker is μ_i and the standard deviation is τ, which is considered the same for all kickers. The parameter μ_i is referred to as the "distance" parameter for kicker i.

The distance traveled, d, for each kick attempt is assumed to be independent of the accuracy of each kick, W. Again this assumption is hard to check. It may be that on long attempts a kicker will try to kick it harder, thus affecting both distance and accuracy. Another concern may be that kickers try to "hook" or "slice" their kicks. All I am concerned with is the angle at which the ball passes the plane of the uprights. If the ball is kicked at a small angle with the center line and it hooks directly through the middle of the uprights, then for my purposes that is considered to have had an angle of 0 with the center line. Another possible complication is the spot of the ball on the line of scrimmage. The ball is always placed between

Where are the Data?

There are many sources on the Web for NFL football data. The official NFL Web site, *nfl.com*, has "gamebooks" for every game. It also maintains a play-by-play account of every game. These are not in a great format to download, but they can be handled. This site also provides the usual statistics for the NFL. It has individual statistics for every current NFL player.

ESPN (*espn.go.com*), CNNSI (*cnnsi.com*), *The Sporting News* (*sportingnews.com*) and *USA Today* (*www.usatoday.com*) also have statistics available. These sites also have live updates of games while they are in process. Each have different interesting features.

I have made the data I used for this article available on the Web at *stat.tamu.edu/berry*.

the hash marks on the field. When the ball is placed on the hash marks this can change slightly the range of angles that will be successful. The data for ball placement are not available. I assume that there is no difference in distance or accuracy depending on the placement of the ball on the line of scrimmage.

The effect of wind is ignored. Clearly kicking with the wind or against the wind would affect the distance traveled. A side wind would also affect the accuracy of a kick. Data are not available on the wind speed and direction for each kick. Adverse weather such as heavy rain and/or snow could also have adverse impacts on both distance and accuracy. This is relatively rare, and these effects are ignored. Bilder and Loughlin (1998) investigated factors that affect field-goal success. Interestingly they found that whether a kick will cause a lead change is a significant factor.

I use a hierarchical model for the distribution of the accuracy parameters and the distance parameters for the k kickers (see sidebar). A normal distribution is used for each. For the distance parameters, $\mu_1,\ldots,\mu_k$, the distribution is $N(\delta_\mu, \tau_\mu^2)$. The accuracy parameters, $\sigma_1,\ldots,\sigma_k$ are $N(\delta_\sigma, \tau_\sigma^2)$. The normal distribution for the σs is for convenience — technically the σs must be nonnegative, but the resulting distribution has virtually no probability less than 0. Morrison and Kalwani (1993) analyzed three years worth of data for place kickers and found no evidence for differences in ability of those kickers.

Table 1 — The Posterior Mean for σ and μ for Each Kicker

Rank	Kicker	σ_i	St. Dev.	μ_i	St. Dev.	*P*(30)	*P*(50)
1	G. Anderson	5.51	.88	56.82	2.54	.97	.78
2	Brien	5.95	.95	56.39	2.01	.96	.73
3	Vanderjagt	6.11	.87	56.04	1.89	.96	.71
4	Johnson	6.27	.85	51.30	1.70	.95	.47
5	Hanson	6.36	.79	56.37	2.13	.95	.70
6	Delgreco	6.30	.76	56.22	2.53	.95	.70
7	Elam	6.54	.76	58.50	2.31	.94	.73
8	Cunningham	6.71	.74	52.79	1.80	.93	.54
9	Peterson	6.78	.80	54.71	1.79	.93	.63
10	Stoyanovich	6.91	.72	54.12	2.05	.93	.60
11	Kasay	6.99	.76	57.39	2.04	.92	.68
12	Hall	7.03	.73	50.51	1.64	.92	.39
13	M. Andersen	7.20	.70	56.68	2.55	.91	.66
14	Vinatieri	7.25	.71	51.88	1.68	.91	.46
15	Carney	7.30	.73	57.21	2.41	.91	.66
16	Daluiso	7.34	.74	56.23	2.64	.91	.64
17	Stover	7.36	.77	55.99	2.63	.91	.63
18	Mare	7.36	.76	56.90	2.48	.91	.46
19	Nedney	7.36	.81	57.19	2.44	.91	.65
20	Hollis	7.39	.72	55.90	2.61	.90	.63
21	Pelfrey	7.42	.78	54.33	1.76	.90	.58
22	Blanton	7.43	.96	52.18	2.57	.90	.47
23	Boniol	7.47	.82	51.38	1.95	.90	.42
24	Akers	7.61	1.00	55.79	2.83	.89	.61
25	Christie	7.62	.66	57.21	2.46	.89	.64
26	Jacke	7.62	.89	56.43	2.61	.89	.62
27	Richey	7.68	.70	53.48	2.17	.89	.53
28	Longwell	7.74	.68	55.93	2.65	.89	.61
29	Wilkins	7.80	.76	58.99	2.19	.89	.65
30	Jaeger	7.90	.75	52.62	2.18	.88	.48
31	Blanchard	7.92	.76	55.05	2.18	.88	.57
32	Davis	8.12	.75	54.90	2.21	.87	.56
33	Husted	8.30	.74	56.20	2.48	.86	.58

Note: The P(30) and P(50) columns present the estimated probability that the kicker makes a 30-yard and 50-yard field goal.

Data and Results

During the 1998 regular season, there were $k = 33$ place kickers combining for 1,906 attempts (including PATs). For each kicker i we have the distance X_{ij} for his jth kick. I ignore kicks that were blocked. Although, in some cases this is an indication that the kicker has performed poorly, it is more likely a fault of the offensive line. This is a challenging statistical problem because the distance d_{ij} and accuracy W_{ij} for each kick are not observed. Instead, we observe censored observations. This means that we don't observe W or d, but we do learn a range for them. If the kick is successful we know that $d_{ij} > X_{ij}$ and $|W_{ij}| < 9.25(40/X_{ij})$. If the kick is short we know that $d_{ij} < X_{ij}$ and we learn nothing about W_{ij}. If the kick is wide we learn that $|W_{ij}| > 9.25(40/X_{ij})$ and that $d_{ij} > X_{ij}$. A Markov-chain Monte Carlo algorithm is used to find the joint posterior distribution of all of the parameters.

Intuitively we learn about σ_i by the frequency of made kicks. If the kicker has a high probability of making kicks then they must have a small value of σ_i. The same relationship happens for the distance parameter μ_i. If a kicker kicks it far enough when the distance is X_{ij}, that gives information about the mean distance kicked. Likewise, if the kick is short, a good deal of information is learned about μ_i.

Table 1 presents the order of kickers in terms of their estimated accuracy parameter. The mean and standard deviation are reported for each σ_i and each μ_i. The probability of making a 30-yard and a 50-yard field goal are also presented.

Gary Anderson is rated as the best kicker from an accuracy standpoint. His mean σ_i is 5.51, which means that from 40 yards away he has a mean of 0 feet from the center line with a standard deviation of 5.51 feet. He is estimated to have a 97% chance of making a 30-yard field goal with a 78% chance of making a 50-yard field goal. The impact of the hierarchical model shows in the estimates for Anderson. Using just data from Anderson, who was successful on every attempt, it would be natural to claim he has perfect accuracy and a distance guaranteed to be greater than his longest kick. The hierarchical model regresses his performance toward the mean of the other kickers. This results in him being estimated to be very good, but not perfect.

Although Anderson is the most accurate, Jason Elam is estimated to be the longest kicker. The best kicker from 52 yards and closer is Gary Anderson, but Jason Elam is the best from 53 and farther. There can be large differences between kickers from different distances. Hall is estimated to be a very precise kicker with a 91% chance of making a 30-yard field goal. Brien has a 90% chance of making a 30-yard field goal. But from 50 yards Brien has a 51% chance, but Hall has only a 28% chance. Hall is an interesting case. Football experts would claim he has a strong leg, but the model estimates he has a relatively weak leg. Of his 33 attempts he was short from 60, 55, 46, and 32 yards. The 32 has a big influence on the results because of the normal assumption for distance. I don't know the circumstances behind this miss. There may have been a huge wind or he may have just "shanked" it. A more appropriate model might have a small probability of a shanked kick, which would be short from every distance. The fact that Hall was asked to attempt kicks from 60 and 55 yards is probably a strong indication that he does have a strong leg.

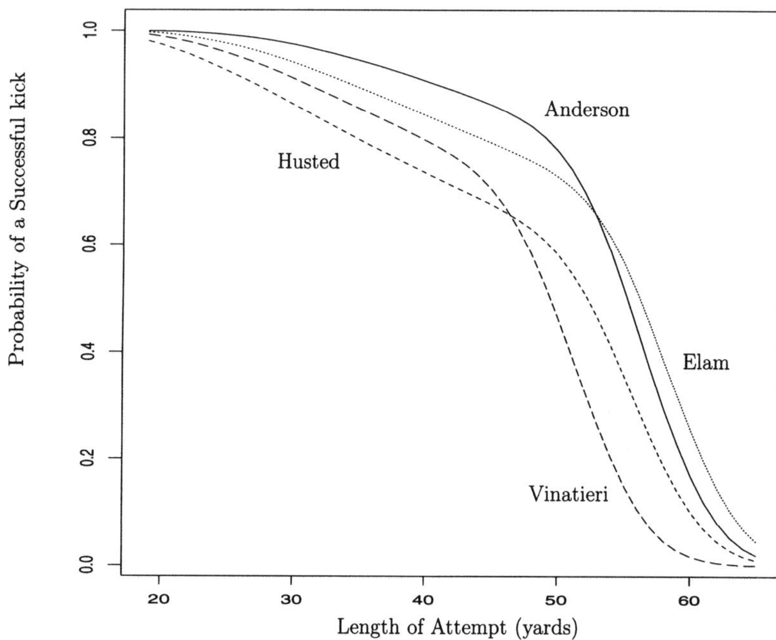

Figure 2. The probability of a successful field goal for Gary Anderson, Jason Elam, Mike Husted and Adam Vinatieri.

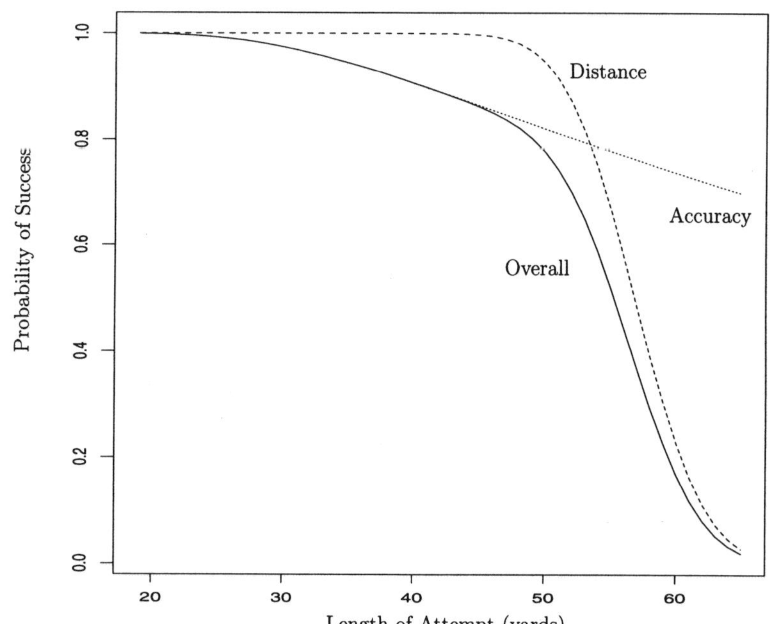

Figure 3. The probability of success for Gary Anderson decomposed into distance and accuracy. The curve labeled "distance" is the probability that the kick travels far enough to be successful, and the curve labeled "accuracy" is the probability that the kick would go through the uprights if it is long enough.

Figure 2 shows the probability of making a field goal for four kickers with interesting profiles. Elam and Anderson are terrific kickers from short and long distances. Adam Vinatieri is a precise kicker with poor distance. Husted is not very precise but has a good leg. Figure 3 shows the decomposition of Anderson's probability of making a field goal into the accuracy component and the distance component. The probability of making a field goal is the product of the probabilities of the kick being far enough and accurate enough.

Table 2 — The Predicted Number of Successes From the Model for Each Kicker for Their Field Goal Attempts of 1998

Rank	Kicker	Attempts	Made	Predicted
1	G. Anderson	35	35	32
2	Brien	21	20	18
3	Vanderjagt	30	27	25
4	Johnson	31	26	25
5	Hanson	33	29	27
6	Delgreco	39	36	35
7	Elam	26	23	22
8	Cunningham	35	29	29
9	Peterson	24	19	19
10	Stoyanovich	32	27	27
11	Kasay	25	19	19
12	Hall	33	25	25
13	M. Andersen	25	23	21
14	Vinatieri	38	30	31
15	Carney	30	26	25
16	Daluiso	26	21	22
17	Stover	27	21	22
18	Mare	27	22	23
19	Nedney	18	13	14
20	Hollis	25	21	22
21	Pelfrey	27	19	20
22	Blanton	4	2	3
23	Boniol	19	13	15
24	Akers	2	0	1
25	Christie	41	33	34
26	Jacke	14	10	11
27	Richey	26	18	21
28	Longwell	33	29	27
29	Wilkins	26	20	19
30	Jaeger	26	21	22
31	Blanchard	16	11	12
32	Davis	27	17	21
33	Husted	27	21	21

The hierarchical distributions are estimated as follows:

$$\sigma_i \sim N(7.17, 1.01)$$
$$\mu_i \sim N(55.21, 3.00)$$

The common standard deviation in the distance random variable, τ, is estimated to be 4.37.

To investigate goodness of fit for the model, I calculated the probability of success for each of the kicker's attempts. From these I calculated the expected number of successful kicks for each kicker. Table 2 presents these results. The predicted values are very close to the actual outcomes. Only 3 of the 33 kickers had a predicted number of successes that deviated from the actual by more than 2 (Davis, 4; Anderson, 3; Richey, 3). I also characterized each of the field-goal attempts by the model-estimated probability of success. In Table 3 I grouped these into the 10 deciles for the probability of success. The actual proportion of success for these deciles are presented. Again there is a very good fit. The last 5 deciles, which are the only ones with more than 12 observations, have incredibly good fit. It certainly appears as though kickers can be well modeled by their distance and accuracy.

Table 3 — Actual Rates of Success for 1998 Field-Goal Attempts for Categories Defined By the Estimated Probability of Success

Predicted %	Attempts	Made	Actual %
0–10	4	1	25
11–20	3	2	67
21–30	4	1	25
31–40	7	5	71
41–50	12	4	33
51–60	43	24	56
61–70	81	49	60
71–80	194	143	74
81–90	228	193	85
91–100	292	284	97

During the 1998 season, 82% of all field-goal attempts were successful. Even from 50+ yards, there was a 55% success rate. Frequently, teams use less risky strategies because the field goal is such a likely event. Why should an offense take chances when they have an almost sure 3 points? This produces a less exciting, more methodical game. There has been talk (and I agree with it) that the uprights should be moved closer together. From the modeling of the accuracy of the kickers I can address what the resulting probabilities would be from the different distances if the uprights were moved closer.

Using the hierarchical models, I label the "average" kicker as one with $\mu = 55.21$ and $\sigma = 7.17$. Figure 4 shows the probability of an average kicker making a field goal from the different distances. The label for each of the curves represents the number of feet each side of the upright is reduced. With the current uprights, from 40 yards, the average kicker has about an 80% chance of a successful kick. If the uprights are moved in 3 feet on each side, this probability will be reduced to 62%. There is not a huge difference in success probability for very long field goals, but this is because distance becomes more important. Reducing each upright by 5 feet would make field goals too unlikely and would upset the current balance between field goals and touchdowns. Although a slight change in the balance is desirable, a change of 5 would have too large an impact. I would suggest a reduction of between 1 and 3 feet on each upright.

Well, I have been dreading this part of the article. What is the probability that Gary Anderson would make a 38-yard field goal? The model probability is 92.3%. So close to a Super Bowl! Did Gary Anderson "choke"? Clearly that depends on your definition of choke. I don't think he choked. Actually, in all honesty, the Viking defense is more to blame than Anderson. They had a chance to stop the Falcon offense from going 72 yards for the touchdown and they failed. As a Viking fan I am used to it. Maybe the Vikings losing to the Falcons saved me from watching Anderson miss a crucial 25-yarder in the Super Bowl. Misery loves company, and thus there is comfort in knowing that Bill's fans have felt my pain ... wide right!

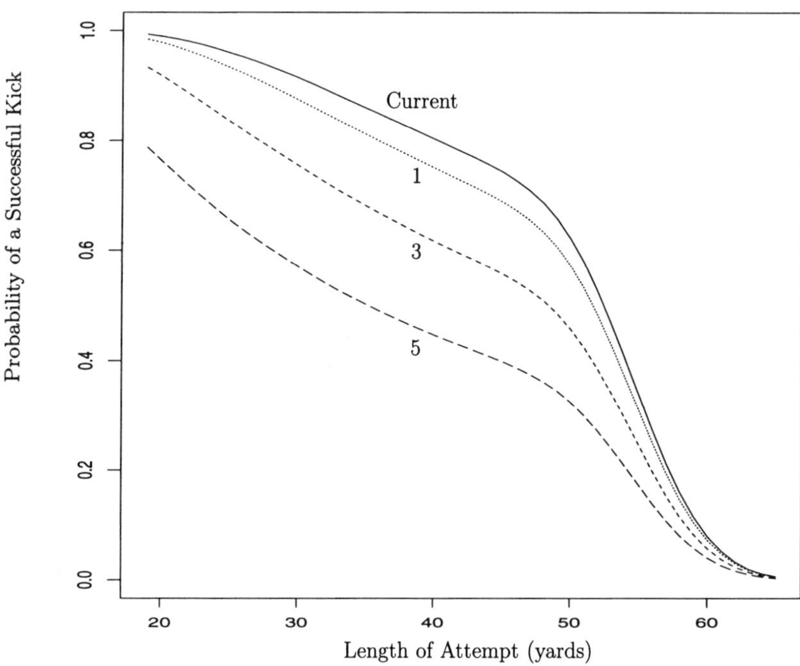

Figure 4. The probability of a successful kick for an average kicker for different size uprights. The curve labeled "current" refers to the current rule, which has each upright 9 feet, 3 inches, from the center of the uprights. The curves labeled "1," "3," and "5" are for each upright being moved in 1, 3, and 5 feet, respectively.

References and Further Reading

Berry, D.A., and Berry, T.D. (1985), "The Probability of a Field Goal: Rating Kickers," *The American Statistician*, 39, 152–155.

Bilder, C.R., and Loughlin, T.M. (1998), "'It's Good!' An Analysis of the Probability of Success for Placekicks," *Chance*, 11(2), 20–24.

Morrison, D.G., and Kalwani, M.U. (1993), "The Best NFL Field Goal Kickers: Are They Lucky or Good?" *Chance*, 6(3), 30–37.

Hierarchical Models

A hierarchical model is commonly used to model the individual parameters of subjects from a common population. The notion is that there is information about one subject from the others. These models are powerful for modeling the performance of athletes in all sports. They are being used increasingly in every branch of statistics but are commonly used in biostatistics, environmental statistics, and educational testing. The general idea is that each of the subjects has its own parameter θ_i. This parameter describes something about that subject — for example, ability to hit a home run, kick a field goal, or hit a golf ball. For each subject a random variable $X_i \sim f(x|\theta_i)$ is observed that is informative about θ_i. The distribution of the subject's θs is explicitly modeled — $\theta_i \sim g(x|\alpha)$. This distribution g, which is indexed by a parameter α, describes the distribution of θs in the population. This distribution is generally unknown to the researcher.

We learn about θ_i from X_i but also from g. The distribution g is unknown, but we learn about it from the estimated θs (based on each X_i). Thus, there is information about θ_i from X_i but also from the other Xs. This is called "shrinkage," "borrowing strength," or "regression to the mean." In this article, the field-goal kickers are all fantastic and are similar. By observing one kicker, information is gathered about all kickers. This creates the effect that, even though Gary Anderson never missed a single kick, I still believe that he is not perfect — he is very good, but not perfect. This information comes from using a hierarchical model for the population of kickers. The mathematical calculation of the posterior distribution from these different sources of information is found using Bayes theorem.

Chapter 5

A State-Space Model for National Football League Scores

Mark E. GLICKMAN and Hal S. STERN

This article develops a predictive model for National Football League (NFL) game scores using data from the period 1988–1993. The parameters of primary interest—measures of team strength—are expected to vary over time. Our model accounts for this source of variability by modeling football outcomes using a state-space model that assumes team strength parameters follow a first-order autoregressive process. Two sources of variation in team strengths are addressed in our model; week-to-week changes in team strength due to injuries and other random factors, and season-to-season changes resulting from changes in personnel and other longer-term factors. Our model also incorporates a home-field advantage while allowing for the possibility that the magnitude of the advantage may vary across teams. The aim of the analysis is to obtain plausible inferences concerning team strengths and other model parameters, and to predict future game outcomes. Iterative simulation is used to obtain samples from the joint posterior distribution of all model parameters. Our model appears to outperform the Las Vegas "betting line" on a small test set consisting of the last 110 games of the 1993 NFL season.

KEY WORDS: Bayesian diagnostics; Dynamic models; Kalman filter; Markov chain Monte Carlo; Predictive inference.

1. INTRODUCTION

Prediction problems in many settings (e.g., finance, political elections, and in this article, football) are complicated by the presence of several sources of variation for which a predictive model must account. For National Football League (NFL) games, team abilities may vary from year to year due to changes in personnel and overall strategy. In addition, team abilities may vary within a season due to injuries, team psychology, and promotion/demotion of players. Team performance may also vary depending on the site of a game. This article describes an approach to modeling NFL scores using a normal linear state-space model that accounts for these important sources of variability.

The state-space framework for modeling a system over time incorporates two different random processes. The distribution of the data at each point in time is specified conditional on a set of time-indexed parameters. A second process describes the evolution of the parameters over time. For many specific state-space models, including the model developed in this article, posterior inferences about parameters cannot be obtained analytically. We thus use Markov chain Monte Carlo (MCMC) methods, namely Gibbs sampling (Gelfand and Smith 1990; Geman and Geman 1984), as a computational tool for studying the posterior distribution of the parameters of our model. Pre-MCMC approaches to the analysis of linear state-space models include those of Harrison and Stevens (1976) and West and Harrison (1990). More recent work on MCMC methods has been done by Carter and Kohn (1994), Fruhwirth-Schnatter (1994), and Glickman (1993), who have developed efficient procedures for fitting normal linear state-space models. Carlin, Polson, and Stoffer (1992), de Jong and Shephard (1995), and Shephard (1994) are only a few of the recent contributors to the growing literature on MCMC approaches to non-linear and non-Gaussian state-space models.

The Las Vegas "point spread" or "betting line" of a game, provided by Las Vegas oddsmakers, can be viewed as the "experts" prior predictive estimate of the difference in game scores. A number of authors have examined the point spread as a predictor of game outcomes, including Amoako-Adu, Marmer, and Yagil (1985), Stern (1991), and Zuber, Gandar, and Bowers (1985). Stern, in particular, showed that modeling the score difference of a game to have a mean equal to the point spread is empirically justifiable. We demonstrate that our model performs at least as well as the Las Vegas line for predicting game outcomes for the latter half of the 1993 season.

Other work on modeling NFL football outcomes (Stefani 1977, 1980; Stern 1992; Thompson 1975) has not incorporated the stochastic nature of team strengths. Our model is closely related to one examined by Harville (1977, 1980) and Sallas and Harville (1988), though the analysis that we perform differs in a number of ways. We create prediction inferences by sampling from the joint posterior distribution of all model parameters rather than fixing some parameters at point estimates prior to prediction. Our model also describes a richer structure in the data, accounting for the possibility of shrinkage towards the mean of team strengths over time. Finally, the analysis presented here incorporates model checking and sensitivity analysis aimed at assessing the propriety of the state-space model.

2. A MODEL FOR FOOTBALL GAME OUTCOMES

Let $y_{ii'}$ denote the outcome of a football game between team i and team i' where teams are indexed by the integers from 1 to p. For our dataset, $p = 28$. We take $y_{ii'}$ to be the difference between the score of team i and the score of team i'. The NFL game outcomes can be modeled as approximately normally distributed with a mean that depends on the relative strength of the teams involved in the game and the site of the game. We assume that at week j of season

Mark E. Glickman is Assistant Professor, Department of Mathematics, Boston University, Boston, MA 02215. Hal S. Stern is Professor, Department of Statistics, Iowa Sate University, Ames, IA 50011. The authors thank the associate editor and the referees for their helpful comments. This work was partially supported by National Science Foundation grant DMS94-04479.

Journal of the American Statistical Association
March 1998, Vol. 93, No. 441, Applications and Case Studies

k, the strength or ability of team i can be summarized by a parameter $\theta_{(k,j)i}$. We let $\boldsymbol{\theta}_{(k,j)}$ denote the vector of p team-ability parameters for week j of season k. An additional set of parameters, $\alpha_i, i = 1, \ldots, p$, measures the magnitude of team i's advantage when playing at its home stadium rather than at a neutral site. These home-field advantage (HFA) parameters are assumed to be independent of time but may vary across teams. We let $\boldsymbol{\alpha}$ denote the vector of p HFA parameters. The mean outcome for a game between team i and team i' played at the site of team i during week j of season k is assumed to be

$$\theta_{(k,j)i} - \theta_{(k,j)i'} + \alpha_i.$$

We can express the distribution for the outcomes of all $n_{(k,j)}$ games played during week j of season k as

$$\mathbf{y}_{(k,j)}|\tilde{\mathbf{X}}_{(k,j)}, \tilde{\theta}_{(k,j)}, \phi \sim \mathrm{N}(\tilde{\mathbf{X}}_{(k,j)}\tilde{\theta}_{(k,j)}, \phi^{-1}\mathbf{I}_{n_{(k,j)}}),$$

where $\mathbf{y}_{(k,j)}$ is the vector of game outcomes, $\tilde{\mathbf{X}}_{(k,j)}$ is the $n_{(k,j)} \times 2p$ design matrix for week j of season k (described in detail later), $\tilde{\theta}_{(k,j)} = (\boldsymbol{\theta}_{(k,j)}, \boldsymbol{\alpha})$ is the vector of p team-ability parameters and p HFA parameters, and ϕ is the regression precision of game outcomes. We let $\tau^2 = \phi^{-1}$ denote the variance of game outcomes conditional on the mean. The row of the matrix $\tilde{\mathbf{X}}_{(k,j)}$ for a game between team i and team i' has the value 1 in the ith column (corresponding to the first team involved in the game), -1 in the i'th column (the second team), and 1 in the $(p+i)$th column (corresponding to the HFA) if the first team played on its home field. If the game were played at the site of the second team (team i'), then home field would be indicated by a -1 in the $(p+i')$th column. Essentially, each row has entries 1 and -1 to indicate the participants and then a single entry in the column corresponding to the home team's HFA parameter (1 if it is the first team at home; -1 if it is the second team). The designation of one team as the first team and the other as the second team is arbitrary and does not affect the interpretation of the model, nor does it affect inferences.

We take K to be the number of seasons of available data. For our particular dataset there are $K = 6$ seasons. We let g_k, for $k = 1, \ldots, K$, denote the total number of weeks of data available in season k. Data for the entire season are available for season $k, k = 1, \ldots, 5$, with g_k varying from 16 to 18. We take $g_6 = 10$, using the data from the remainder of the sixth season to perform predictive inference. Additional details about the structure of the data are provided in Section 4.

Our model incorporates two sources of variation related to the evolution of team ability over time. The evolution of strength parameters between the last week of season k and the first week of season $k+1$ is assumed to be governed by

$$\boldsymbol{\theta}_{(k+1,1)}|\beta_s, \boldsymbol{\theta}_{(k,g_k)}, \phi, \omega_s \sim \mathrm{N}(\beta_s\mathbf{G}\boldsymbol{\theta}_{(k,g_k)}, (\phi\omega_s)^{-1}\mathbf{I}_p),$$

where $\mathbf{G}$ is the matrix that maps the vector $\boldsymbol{\theta}_{(k,g_k)}$ to $\boldsymbol{\theta}_{(k,g_k)} -$ ave$(\boldsymbol{\theta}_{(k,g_k)})$, β_s is the between-season regression parameter that measures the degree of shrinkage ($\beta_s < 1$) or expansion ($\beta_s > 1$) in team abilities between seasons, and the product $\phi\omega_s$ is the between-season evolution precision. This particular parameterization for the evolution precision simplifies the distributional calculus involved in model fitting. We let $\sigma_s^2 = (\phi\omega_s)^{-1}$ denote the between-season evolution variance. Then ω_s is the ratio of variances, τ^2/σ_s^2.

The matrix $\beta_s\mathbf{G}$ maps the vector $\boldsymbol{\theta}_{(k,g_k)}$ to another vector centered at 0, and then shrunk or expanded around 0. We use this mapping because the distribution of the game outcomes $\mathbf{y}_{(k,j)}$ is a function only of differences in the team ability parameters; the distribution is unchanged if a constant is added to or subtracted from each team's ability parameter. The mapping $\mathbf{G}$ translates the distribution of team strengths to be centered at 0, though it is understood that shrinkage or expansion is actually occurring around the mean team strength (which may be drifting over time). The season-to-season variation is due mainly to personnel changes (new players or coaches). One would expect $\beta_s < 1$, because the player assignment process is designed to assign the best young players to the teams with the weakest performance in the previous season.

We model short-term changes in team performance by incorporating evolution of ability parameters between weeks,

$$\boldsymbol{\theta}_{(k,j+1)}|\beta_w, \boldsymbol{\theta}_{(k,j)}, \phi, \omega_w \sim \mathrm{N}(\beta_w\mathbf{G}\boldsymbol{\theta}_{(k,j)}, (\phi\omega_w)^{-1}\mathbf{I}_p),$$

where the matrix $\mathbf{G}$ is as before, β_w is the between-week regression parameter, and $\phi\omega_w$ is the between-week evolution precision. Analogous to the between-season component of the model, we let $\sigma_w^2 = (\phi\omega_w)^{-1}$ denote the variance of the between-week evolution, so that $\omega_w = \tau^2/\sigma_w^2$. Week-to-week changes represent short-term sources of variation; for example, injuries and team confidence level. It is likely that $\beta_w \approx 1$, because there is no reason to expect that such short-term changes will tend to equalize the team parameters ($\beta_w < 1$) or accentuate differences ($\beta_w > 1$).

Several simplifying assumptions built into this model are worthy of comment. We model differences in football scores, which can take on integer values only, as approximately normally distributed conditional on team strengths. The rules of football suggest that some outcomes (e.g., 3 or 7) are much more likely than others. Rosner (1976) modeled game outcomes as a discrete distribution that incorporates the rules for football scoring. However, previous work (e.g., Harville 1980, Sallas and Harville 1988, Stern 1991) has shown that the normality assumption is not an unreasonable approximation, especially when one is not interested in computing probabilities for exact outcomes but rather for ranges of outcomes (e.g., whether the score difference is greater than 0). Several parameters, no ly the regression variance τ^2 and the evolution variances σ_w^2 and σ_s^2, are assumed to be the same for all teams and for all seasons. This rules out the possibility of teams with especially erratic performance. We explore the adequacy of these modeling assumptions using posterior predictive model checks (Gelman, Meng, and Stern 1996; Rubin 1984) in Section 5.

Prior distributions of model parameters are centered at values that seem reasonable based on our knowledge of football. In each case, the chosen distribution is widely dispersed, so that before long the data will play a dominant role. We assume the following prior distributions:

$$\phi \sim \text{gamma}(.5, .5(100)),$$

$$\omega_{\text{w}} \sim \text{gamma}(.5, .5/60),$$

$$\omega_{\text{s}} \sim \text{gamma}(.5, .5/16),$$

$$\beta_{\text{s}} \sim \text{N}(.98, 1),$$

and

$$\beta_{\text{w}} \sim \text{N}(.995, 1).$$

Our prior distribution on ϕ corresponds to a harmonic mean of 100, which is roughly equivalent to a 10-point standard deviation, τ, for game outcomes conditional on knowing the teams' abilities. This is close to, but a bit lower than, Stern's (1991) estimate of $\hat{\tau} = 13.86$ derived from a simpler model. In combination with this prior belief about ϕ, the prior distributions on ω_{w} and ω_{s} assume harmonic means of σ_{w}^2 and σ_{s}^2 equal to 100/60 and 100/16, indicating our belief that the changes in team strength between seasons are likely to be larger than short-term changes in team strength. Little information is currently available about σ_{w}^2 and σ_{s}^2, which is represented by the .5 df. The prior distributions on the regression parameters assume shrinkage toward the mean team strength, with a greater degree of shrinkage for the evolution of team strengths between seasons. In the context of our state-space model, it is not necessary to restrict the modulus of the regression parameters (which are assumed to be equal for every week and season) to be less than 1, as long as our primary concern is for parameter summaries and local prediction rather than long-range forecasts.

The only remaining prior distributions are those for the initial team strengths in 1988, $\boldsymbol{\theta}_{(1,1)}$, and the HFA parameters, $\boldsymbol{\alpha}$. For team strengths at the onset of the 1988 season, we could try to quantify our knowledge perhaps by examining 1987 final records and statistics. We have chosen instead to use an exchangeable prior distribution as a starting point, ignoring any pre-1988 information:

$$\boldsymbol{\theta}_{(1,1)} | \phi, \omega_{\text{o}} \sim \text{N}(\mathbf{0}, (\omega_{\text{o}}\phi)^{-1}\mathbf{I}_p), \tag{1}$$

where we assume that

$$\omega_{\text{o}} \sim \text{gamma}(.5, .5/6). \tag{2}$$

Let $\sigma_{\text{o}}^2 = (\omega_{\text{o}}\phi)^{-1}$ denote the prior variance of initial team strengths. Our prior distribution for ω_{o} in combination with the prior distribution on ϕ implies that σ_{o}^2 has prior harmonic mean of 100/6 based on .5 df. Thus the a priori difference between the best and worst teams would be about $4\sigma_{\text{o}} = 16$ points.

We assume that the α_i have independent prior distributions

$$\boldsymbol{\alpha} | \phi, \omega_{\text{h}} \sim \text{N}(3, (\omega_{\text{h}}\phi)^{-1}\mathbf{I}_p) \tag{3}$$

with

$$\omega_{\text{h}} \sim \text{gamma}(.5, .5/6). \tag{4}$$

We assume a prior mean of 3 for the α_i, believing that competing on one's home field conveys a small but persistent advantage. If we let $\sigma_{\text{h}}^2 = (\omega_{\text{h}}\phi)^{-1}$ denote the prior variance of the HFA parameters, then our prior distributions for ω_{h} and ϕ imply that σ_{h}^2 has prior harmonic mean of 100/6 based on .5 df.

3. MODEL FITTING AND PREDICTION

We fit and summarize our model using MCMC techniques, namely the Gibbs sampler (Gelfand and Smith 1990; Geman and Geman 1984). Let $\mathbf{Y}_{(K,g_K)}$ represent all observed data through week (K, g_K). The Gibbs sampler is implemented by drawing alternately in sequence from the following three conditional posterior distributions:

$$f(\boldsymbol{\theta}_{(1,1)}, \ldots, \boldsymbol{\theta}_{(K,g_K)}, \boldsymbol{\alpha}, \phi | \omega_{\text{o}}, \omega_{\text{h}}, \omega_{\text{w}}, \omega_{\text{s}}, \beta_{\text{s}}, \beta_{\text{w}}, \mathbf{Y}_{(K,g_K)}),$$

$$f(\omega_{\text{o}}, \omega_{\text{h}}, \omega_{\text{w}}, \omega_{\text{s}} | \boldsymbol{\theta}_{(1,1)}, \ldots, \boldsymbol{\theta}_{(K,g_K)}, \boldsymbol{\alpha}, \phi, \beta_{\text{s}}, \beta_{\text{w}}, \mathbf{Y}_{(K,g_K)}),$$

and

$$f(\beta_{\text{s}}, \beta_{\text{w}} | \boldsymbol{\theta}_{(1,1)}, \ldots, \boldsymbol{\theta}_{(K,g_K)}, \boldsymbol{\alpha}, \phi, \omega_{\text{o}}, \omega_{\text{h}}, \omega_{\text{w}}, \omega_{\text{s}}, \mathbf{Y}_{(K,g_K)}).$$

A detailed description of the conditional distributions appears in the Appendix. Once the Gibbs sampler has converged, inferential summaries are obtained by using the empirical distribution of the simulations as an estimate of the posterior distribution.

An important use of the fitted model is in the prediction of game outcomes. Assume that the model has been fit via the Gibbs sampler to data through week g_K of season K, thereby obtaining m posterior draws of the final team-ability parameters $\boldsymbol{\theta}_{(K,g_K)}$, the HFA parameters $\boldsymbol{\alpha}$, and the precision and regression parameters. Denote the entire collection of the parameters by $\boldsymbol{\eta}_{(K,g_K)} = (\omega_{\text{w}}, \omega_{\text{s}}, \omega_{\text{o}}, \omega_{\text{h}}, \beta_{\text{w}}, \beta_{\text{s}}, \phi, \boldsymbol{\theta}_{(K,g_K)}, \boldsymbol{\alpha})$. Given the design matrix for the next week's games, $\tilde{\mathbf{X}}_{(K,g_K+1)}$, the posterior predictive distribution of next week's game outcomes, $\mathbf{y}_{(K,g_K+1)}$, is given by

$$\begin{aligned} &\mathbf{y}_{(K,g_K+1)} | \mathbf{Y}_{(K,g_K)}, \boldsymbol{\eta}_{(K,g_K)} \\ &\quad \sim \text{N}\left(\tilde{\mathbf{X}}_{(K,g_K+1)} \begin{pmatrix} \beta_{\text{w}}\mathbf{G}\theta_{(K,g_K)} \\ \boldsymbol{\alpha} \end{pmatrix}, \quad \tau^2\mathbf{I}_{n_{(K,g_K+1)}} \right. \\ &\qquad \left. + \sigma_{\text{w}}^2 \tilde{\mathbf{X}}_{(K,g_K+1)} \tilde{\mathbf{X}}'_{(K,g_K+1)} \right). \end{aligned} \tag{5}$$

A sample from this distribution may be simulated by randomly selecting values of $\boldsymbol{\eta}_{(K,g_K)}$ from among the Gibbs sampler draws and then drawing $\mathbf{y}_{(K,g_K+1)}$ from the distribution in (5) for each draw of $\boldsymbol{\eta}_{(K,g_K)}$. This process may be repeated to construct a sample of desired size. To obtain point predictions, we could calculate the sample average of these posterior predictive draws. It is more efficient, however, to calculate the sample average of the means in (5) across draws of $\boldsymbol{\eta}_{(K,g_K)}$.

4. POSTERIOR INFERENCES

We use the model described in the preceding section to analyze regular season results of NFL football games for the years 1988–1992 and the first 10 weeks of 1993 games.

The NFL comprised a total of 28 teams during these seasons. During the regular season, each team plays a total of 16 games. The 1988–1989 seasons lasted a total of 16 weeks, the 1990–1992 seasons lasted 17 weeks (each team had one off week), and the 1993 season lasted 18 weeks (each team had two off weeks). We use the last 8 weeks of 1993 games to assess the accuracy of predictions from our model. For each game we recorded the final score for each team and the site of the game. Although use of covariate information, such as game statistics like rushing yards gained and allowed, might improve the precision of the model fit, no additional information was recorded.

4.1 Gibbs Sampler Implementation

A single "pilot" Gibbs sampler with starting values at the prior means was run to determine regions of the parameter space with high posterior mass. Seven parallel Gibbs samplers were then run with overdispersed starting values relative to the draws from the pilot sampler. Table 1 displays the starting values chosen for the parameters in the seven parallel runs. Each Gibbs sampler was run for 18,000 iterations, and convergence was diagnosed from plots and by examining the potential scale reduction (PSR), as described by Gelman and Rubin (1992), of the parameters $\omega_w, \omega_s, \omega_o, \omega_h, \beta_w$, and β_s; the HFA parameters; and the most recent team strength parameters. The PSR is an estimate of the factor by which the variance of the current distribution of draws in the Gibbs sampler will decrease with continued iterations. Values near 1 are indicative of convergence. In diagnosing convergence, parameters that were restricted to be positive in the model were transformed by taking logs. Except for the parameter σ_w, all of the PSRs were less than 1.2. The slightly larger PSR for σ_w could be explained from the plot of successive draws versus iteration number; the strong autocorrelation in simulations of ω_w slowed the mixing of the different series. We concluded that by iteration, 17,000 the separate series had essentially converged to the stationary distribution. For each parameter, a sample was obtained by selecting the last 1,000 values of the 18,000 in each series. This produced the final sample of 7,000 draws from the posterior distribution for our analyses.

4.2 Parameter Summaries

Tables 2 and 3 show posterior summaries of some model parameters. The means and 95% central posterior intervals for team parameters describe team strengths after the 10th week of the 1993 regular season. The teams are ranked according to their estimated posterior means. The posterior means range from 9.06 (Dallas Cowboys) to −7.73 (New England Patriots), which suggests that on a neutral field, the best team has close to a 17-point advantage over the worst team. The 95% intervals clearly indicate that a considerable amount of variability is associated with the team-strength parameters, which may be due to the stochastic nature of team strengths. The distribution of HFAs varies from roughly 1.6 points (Dallas Cowboys, Cleveland Browns) to over 7 points (Houston Oilers). The 7-point HFA conveyed to the Oilers is substantiated by the numerous "blowouts" they have had on their home field. The HFA parameters are centered around 3.2. This value is consistent with the results of previous modeling (Glickman 1993; Harville 1980; Sallas and Harville 1988).

The distributions of the standard deviation parameters $\tau, \sigma_o, \sigma_h, \sigma_w$, and σ_s are shown in Figures 1 and 2. The plots show that each of the standard deviations is approximately

Table 1. Starting Values for Parallel Gibbs Samplers

Parameter	Gibbs sampler series 1	2	3	4	5	6	7
ω_w	10.0	100.0	200.0	500.0	1,000.0	1,000.0	10.0
ω_s	1.0	20.0	80.0	200.0	800.0	1.0	800.0
ω_o	.5	5.0	15.0	100.0	150.0	150.0	.5
ω_h	100.0	20.0	6.0	1.0	.6	.3	100.0
β_w	.6	.8	.99	1.2	1.8	.6	1.8
β_s	.5	.8	.98	1.2	1.8	1.8	.6

Table 2. Summaries of the Posterior Distributions of Team Strength and HFA Parameters After the First 10 Weeks of the 1993 Regular Season

Parameter	Mean strength	Mean HFA
Dallas Cowboys	9.06 (2.26, 16.42)	1.62 (−1.94, 4.86)
San Francisco 49ers	7.43 (.29, 14.40)	2.77 (−.76, 6.19)
Buffalo Bills	4.22 (−2.73, 10.90)	4.25 (.91, 7.73)
New Orleans Saints	3.89 (−3.04, 10.86)	3.44 (−.01, 6.87)
Pittsburgh Steelers	3.17 (−3.66, 9.96)	3.30 (.00, 6.68)
Miami Dolphins	2.03 (−4.79, 8.83)	2.69 (−.81, 6.14)
Green Bay Packers	1.83 (−4.87, 8.66)	2.19 (−1.17, 5.45)
San Diego Chargers	1.75 (−5.02, 8.62)	1.81 (−1.70, 5.12)
New York Giants	1.43 (−5.38, 8.21)	4.03 (.75, 7.53)
Denver Broncos	1.18 (−5.75, 8.02)	5.27 (1.90, 8.95)
Philadelphia Eagles	1.06 (−5.98, 7.80)	2.70 (−.75, 6.06)
New York Jets	.98 (−5.95, 8.00)	1.86 (−1.51, 5.15)
Kansas City Chiefs	.89 (−5.82, 7.77)	4.13 (.75, 7.55)
Detroit Lions	.80 (−5.67, 7.49)	3.12 (−.31, 6.48)
Houston Oilers	.72 (−6.18, 7.51)	7.28 (3.79, 11.30)
Minnesota Vikings	.25 (−6.57, 6.99)	3.34 (−.01, 6.80)
Los Angeles Raiders	.25 (−6.43, 7.10)	3.21 (−.05, 6.55)
Phoenix Cardinals	−.15 (−6.64, 6.56)	2.67 (−.69, 5.98)
Cleveland Browns	−.55 (−7.47, 6.25)	1.53 (−2.04, 4.81)
Chicago Bears	−1.37 (−8.18, 5.37)	3.82 (.38, 7.27)
Washington Redskins	−1.46 (−8.36, 5.19)	3.73 (.24, 7.22)
Atlanta Falcons	−2.94 (−9.89, 3.85)	2.85 (−.55, 6.23)
Seattle Seahawks	−3.17 (−9.61, 3.43)	2.21 (−1.25, 5.52)
Los Angeles Rams	−3.33 (−10.18, 3.37)	1.85 (−1.61, 5.23)
Indianapolis Colts	−5.29 (−12.11, 1.63)	2.45 (−.97, 5.81)
Tampa Bay Buccaneers	−7.43 (−14.38, −.68)	1.77 (−1.69, 5.13)
Cincinnati Bengals	−7.51 (−14.74, −.68)	4.82 (1.53, 8.33)
New England Patriots	−7.73 (−14.54, −.87)	3.94 (.55, 7.34)

NOTE: Values within parentheses represent central 95% posterior intervals.

Table 3. Summaries of the Posterior Distributions of Standard Deviations and Regression Parameters After the First 10 Weeks of the 1993 Regular Season

Parameter	Mean
τ	12.78 (12.23, 13.35)
σ_o	3.26 (1.87, 5.22)
σ_w	.88 (.52, 1.36)
σ_s	2.35 (1.14, 3.87)
σ_h	2.28 (1.48, 3.35)
β_w	.99 (.96, 1.02)
β_s	.82 (.52, 1.28)

NOTE: Values within parentheses represent central 95% posterior intervals.

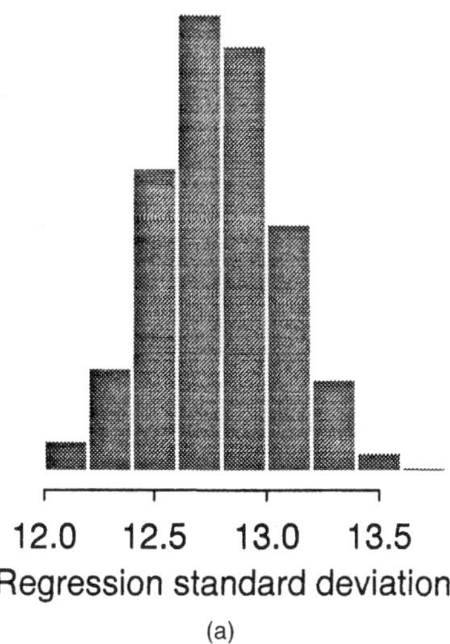

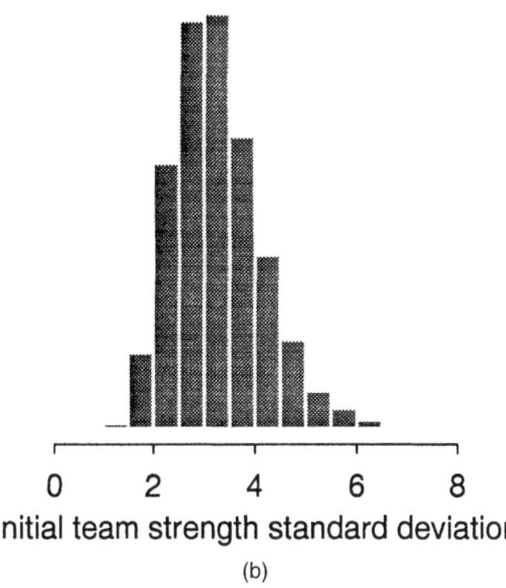

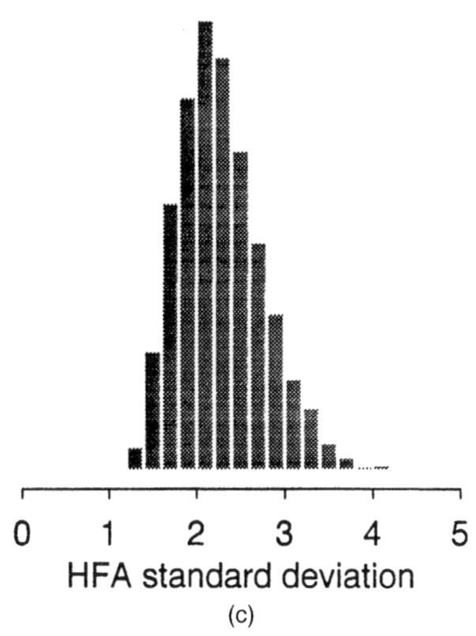

Figure 1. Estimated Posterior Distributions. (a) Regression Standard Deviation (τ); (b) the Initial Team Strength Standard Deviation (σ_o); (c) HFA Standard Deviation (σ_h).

symmetrically distributed around its mean. The posterior distribution of τ is centered just under 13 points, indicating that the score difference for a single game conditional on team strengths can be expected to vary by about $4\tau \approx 50$ points. The posterior distribution of σ_{o} shown in Figure 1 suggests that thc normal distribution of teams' abilities prior to 1988 have a standard deviation somewhere between 2 and 5, so that the a priori difference between the best and worst teams is near $4\sigma_{\text{o}} \approx 15$. This range of team strength appears to persist in 1993, as can be calculated from Table 2. The distribution of σ_{h} is centered near 2.3, suggesting that teams' HFAs varied moderately around a mean of 3 points.

As shown in the empirical contour plot in Figure 2, the posterior distribution of the between-week standard deviation, σ_{w}, is concentrated on smaller values and is less dispersed than that of the between-season evolution standard deviation, σ_{s}. This difference in magnitude indicates that the types of changes that occur between weeks are likely to have less impact on a team's ability than are the changes that occur between seasons. The distribution for the between-week standard deviation is less dispersed than that for the between-season standard deviation, because the data provide much more information about weekly innovations than about changes between seasons. Furthermore, the contour plot shows a slight negative posterior correlation between the standard deviations. This is not terribly surprising if we consider that the total variability due to the passage of time over an entire season is the composition of between-week variability and between-season variability. If between-week variability is small, then between-season variability must be large to compensate. An interesting fea-

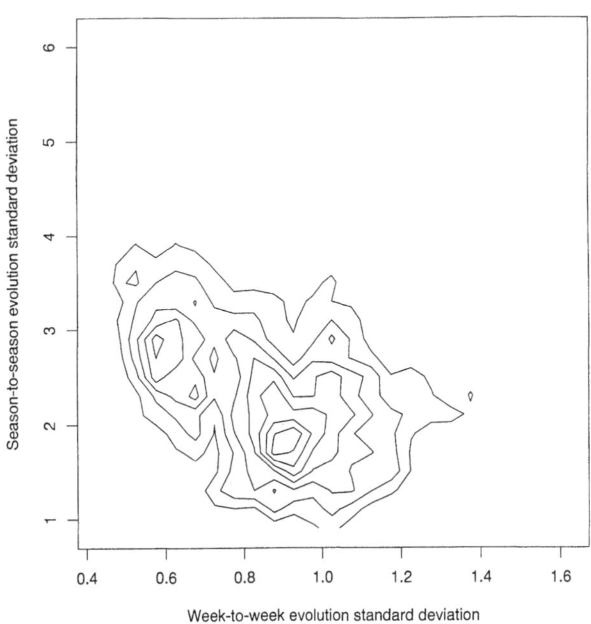

Figure 2. Estimated Joint Posterior Distribution of the Week-to-Week Evolution Standard Deviation (σ_w) and the Season-to-Season Evolution Standard Deviation (σ_s).

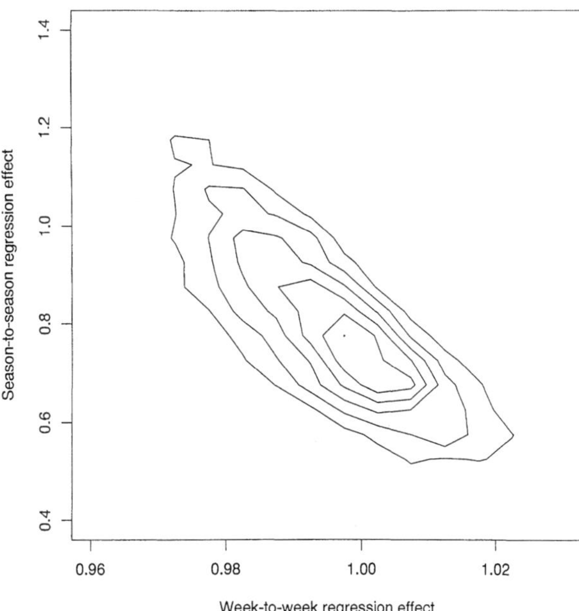

Figure 3. Estimated Joint Posterior Distribution of the Week-to-Week Regression Effect (β_w) and the Season-to-Season Regression Effect (β_s).

ture revealed by the contour plot is the apparent bimodality of the joint distribution. This feature was not apparent from examining the marginal distributions. Two modes of (σ_w, σ_s) appear at (.6, 3) and (.9, 2).

Figure 3 shows contours of the bivariate posterior distribution of the parameters β_w and β_s. The contours of the plot display a concentration of mass near $\beta_s \approx .8$ and $\beta_w \approx 1.0$, as is also indicated in Table 3. The plot shows a more marked negative posterior correlation between these two parameters than between the standard deviations. The negative correlation can be explained in an analogous manner to the negative correlation between standard deviations, viewing the total shrinkage over the season as being the composition of the between-week shrinkages and the between-season shrinkage. As with the standard deviations, the data provide more precision about the between-week regression parameter than about the between-season regression parameter.

4.3 Prediction for Week 11

Predictive summaries for the week 11 games of the 1993 NFL season are shown in Table 4. The point predictions were computed as the average of the mean outcomes across all 7,000 posterior draws. Intervals were constructed empirically by simulating single-game outcomes from the predictive distribution for each of the 7,000 Gibbs samples. Of the 13 games, six of the actual score differences were contained in the 50% prediction intervals. All of the widths of the intervals were close to 18–19 points. Our point predictions were generally close to the Las Vegas line. Games where predictions differ substantially (e.g., Oilers at Bengals) may reflect information from the previous week that our model does not incorporate, such as injuries of important players.

4.4 Predictions for Weeks 12 Through 18

Once game results for a new week were available, a single-series Gibbs sampler was run using the entire dataset to obtain a new set of parameter draws. The starting values for the series were the posterior mean estimates of $\omega_w, \omega_s, \omega_o, \omega_h, \beta_w$, and β_s, from the end of week 10. Because the posterior variability of these parameters is small, the addition of a new week's collection of game outcomes is not likely to have substantial impact on posterior inferences. Thus our procedure takes advantage of knowing regions of the parameter space a priori that will have high posterior mass. Having obtained data from the results of week 11, we ran a single-series Gibbs sampler for 5,000 iterations, saving the last 1,000 for predictive inferences. We repeated this procedure for weeks 12–17 in an analogous manner. Point predictions were computed as described earlier. In practice, the model could be refit periodically using a multiple-chain procedure as an alternative to using this one-chain updating algorithm. This might be advantageous in reassessing the propriety of the model or determining whether significant shifts in parameter values have occurred.

4.5 Comparison with Las Vegas Betting Line

We compared the accuracy of our predictions with those of the Las Vegas point spread on the 110 games beyond the 10th week of the 1993 season. The mean squared error (MSE) for predictions from our model for these 110 games was 165.0. This is slightly better than the MSE of 170.5 for the point spread. Similarly, the mean absolute error (MAE) from our model is 10.50, whereas the analogous result for the point spread is 10.84. Our model correctly predicted the winners of 64 of the 110 games (58.2%), whereas the Las Vegas line predicted 63. Out of the 110 predictions from our model, 65 produced mean score differences that "beat the point spread"; that is, resulted in predictions that were greater than the point spread when the actual score difference was larger than the point spread, or resulted in predictions that were lower than the point spread when the actual score difference was lower. For this small sample, the model fit outperforms the point spread, though the difference is not large enough to generalize. However, the results

Table 4. Forecasts for NFL Games During Week 11 of the 1993 Regular Season

Week 11 games	*Predicted score difference*	*Las Vegas line*	*Actual score difference*
Packers at Saints	−5.49	−6.0	2 (−14.75, 3.92)
Oilers at Bengals	3.35	8.5	35 (−6.05, 12.59)
Cardinals at Cowboys	−10.77	−12.5	−5 (−20.25, −1.50)
49ers at Buccaneers	13.01	16.0	24 (3.95, 22.58)
Dolphins at Eagles	−1.74	4.0	5 (−10.93, 7.55)
Redskins at Giants	−6.90	−7.5	−14 (−16.05, 2.35)
Chiefs at Raiders	−2.57	−3.5	11 (−11.63, 6.86)
Falcons at Rams	−1.46	−3.5	13 (−10.83, 7.93)
Browns at Seahawks	.40	−3.5	−17 (−8.98, 9.72)
Vikings at Broncos	−6.18	−7.0	3 (−15.42, 3.25)
Jets at Colts	3.78	3.5	14 (−5.42, 13.19)
Bears at Chargers	−4.92	−8.5	3 (−14.31, 4.28)
Bills at Steelers	−2.26	−3.0	−23 (−11.53, 7.13)

NOTE: Values within parentheses represent central 50% prediction intervals.

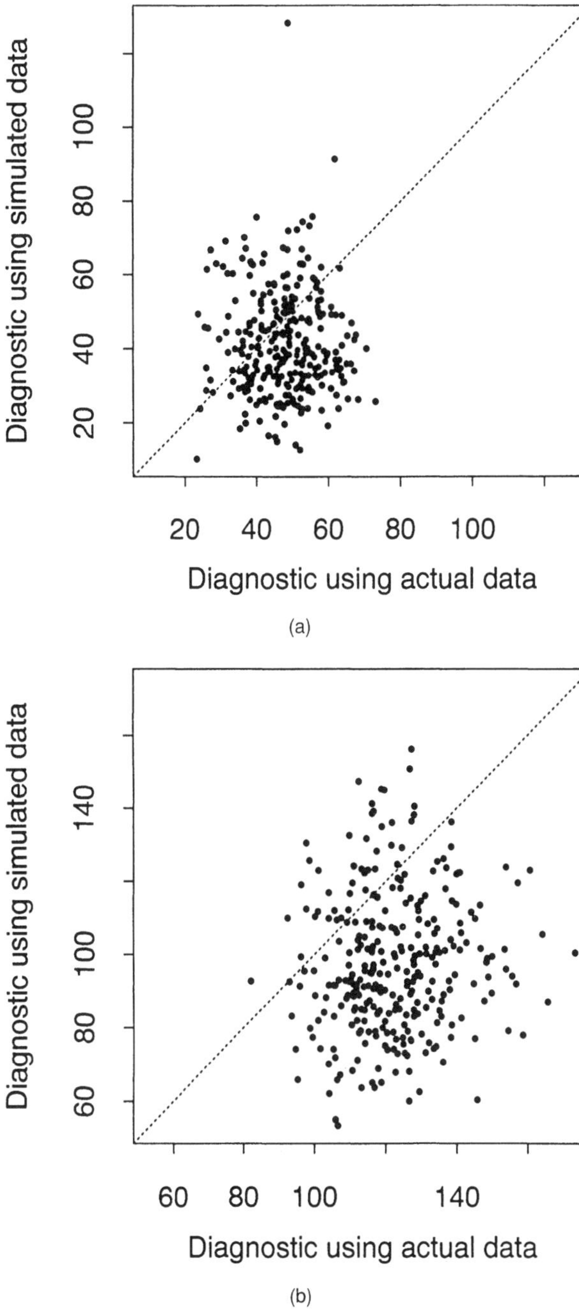

Figure 4. Estimated Bivariate Distributions for Regression Variance Diagnostics. (a) Scatterplot of the joint posterior distribution of $D_1(\mathbf{y}; \theta^*)$ and $D_1(\mathbf{y}^*; \theta^*)$ where $D_1(\cdot\ ;\ \cdot)$ is a discrepancy measuring the range of regression variance estimates among the six seasons and $(\theta^*, \mathbf{y}^*)$ are simulations from their posterior and posterior predictive distributions; (b) scatterplot of the joint posterior distribution of $D_2(\mathbf{y}; \theta^*)$ and $D_2(\mathbf{y}^*; \theta^*)$ for $D_2(\cdot\ ;\ \cdot)$ a discrepancy measuring the range of regression variance estimates among the 28 teams.

here suggest that the state-space model yields predictions that are comparable to those implied by the betting line.

5. DIAGNOSTICS

Model validation and diagnosis is an important part of the model-fitting process. In complex models, however, diagnosing invalid assumptions or lack of fit often cannot be carried out using conventional methods. In this section we examine several model assumptions through the use of posterior predictive diagnostics. We include a brief description of the idea behind posterior predictive diagnostics. We also describe how model diagnostics were able to suggest an improvement to an earlier version of the model.

The approach to model checking using posterior predictive diagnostics has been discussed in detail by Gelman et al. (1996), and the foundations of this approach have been described by Rubin (1984). The strategy is to construct discrepancy measures that address particular aspects of the data that one suspects may not be captured by the model. Discrepancies may be ordinary test statistics, or they may depend on both data values and parameters. The discrepancies are computed using the actual data, and the resulting values are compared to the reference distribution obtained using simulated data from the posterior predictive distribution. If the actual data are "typical" of the draws from the posterior predictive distribution under the model, then the posterior distribution of the discrepancy measure evaluated at the actual data will be similar to the posterior distribution of the discrepancy evaluated at the simulated datasets. Otherwise, the discrepancy measure provides some indication that the model may be misspecified.

To be concrete, we may construct a "generalized" test statistic, or discrepancy, $D(\mathbf{y}; \boldsymbol{\theta})$, which may be a function not only of the observed data, generically denoted by $\mathbf{y}$, but also of model parameters, generically denoted by $\boldsymbol{\theta}$. We compare the posterior distribution of $D(\mathbf{y}; \boldsymbol{\theta})$ to the posterior predictive distribution of $D(\mathbf{y}^*; \boldsymbol{\theta})$, where we use $\mathbf{y}^*$ to denote hypothetical replicate data generated under the model with the same (unknown) parameter values. One possible summary of the evaluation is the tail probability, or p value, computed as

$$\Pr(D(\mathbf{y}^*; \boldsymbol{\theta}) \geq D(\mathbf{y}; \boldsymbol{\theta}) | \mathbf{y})$$

or

$$\Pr(D(\mathbf{y}^*; \boldsymbol{\theta}) \leq D(\mathbf{y}; \boldsymbol{\theta}) | \mathbf{y}),$$

depending on the definition of the discrepancy. In practice, the relevant distributions or the tail probability can be approximated through Monte Carlo integration by drawing samples from the posterior distribution of $\boldsymbol{\theta}$ and then the posterior predictive distribution of $\mathbf{y}^*$ given $\boldsymbol{\theta}$.

The choice of suitable discrepancy measures, D, depends on the problem. We try to define measures that evaluate the fit of the model to features of the data that are not explicitly accounted for in the model specification. Here we consider diagnostics that assess the homogeneity of variance assumptions in the model and diagnostics that assess assumptions concerning the HFA. The HFA diagnostics were useful in detecting a failure of an earlier version of the model. Our summary measures D are functions of Bayesian residuals as defined by Chaloner and Brant (1988) and Zellner (1975). As an alternative to focusing on summary measures D, the individual Bayesian residuals can be used to search for outliers or to construct a "distribution" of residual plots; we do not pursue this approach here.

5.1 Regression Variance

For a particular game played between teams i and i' at team i's home field, let

$$e_{ii'}^2 = (y_{ii'} - (\theta_i - \theta_{i'} + \alpha_i))^2$$

be the squared difference between the observed outcome and the expected outcome under the model, which might be called a squared residual. Averages of $e_{ii'}^2$ across games can be interpreted as estimates of τ^2 (the variance of $y_{ii'}$ given its mean). The model assumes that τ^2 is constant across seasons and for all competing teams. We consider two discrepancy measures that are sensitive to failures of these assumptions. Let $D_1(\mathbf{y}; \boldsymbol{\theta})$ be the difference between the largest of the six annual average squared residuals and the smallest of the six annual average squared residuals. Then $D_1(\mathbf{y}^*; \boldsymbol{\theta}^*)$ is the value of this diagnostic evaluated at simulated parameters $\boldsymbol{\theta}^*$ and simulated data $\mathbf{y}^*$, and $D_1(\mathbf{y}; \boldsymbol{\theta}^*)$ is the value evaluated at the same simulated parameters but using the actual data. Based on 300 samples of parameters from the posterior distribution and simulated data from the posterior predictive distribution, the approximate bivariate posterior distribution of $(D_1(\mathbf{y}; \boldsymbol{\theta}^*), D_1(\mathbf{y}^*; \boldsymbol{\theta}^*))$ is shown on Figure 4a. The plot shows that large portions of the distribution of the discrepancies lie both above and below the line $D_1(\mathbf{y}; \boldsymbol{\theta}^*) = D_1(\mathbf{y}^*; \boldsymbol{\theta}^*)$, with the relevant tail probability equal to .35. This suggests that the year-to-year variation in the regression variance of the actual data is quite consistent with that expected under the model (as evidenced by the simulated datasets).

As a second discrepancy measure, we can compute the average $e_{ii'}^2$ for each team and then calculate the difference between the maximum of the 28 team-specific estimates and the minimum of the 28 team-specific estimates. Let $D_2(\mathbf{y}^*; \boldsymbol{\theta}^*)$ be the value of this diagnostic measure for the simulated data $\mathbf{y}^*$ and simulated parameters $\boldsymbol{\theta}^*$, and let $D_2(\mathbf{y}; \boldsymbol{\theta}^*)$ be the value for the actual data and simulated parameters. The approximate posterior distribution of $(D_2(\mathbf{y}; \boldsymbol{\theta}^*), D_2(\mathbf{y}^*; \boldsymbol{\theta}^*))$ based on the same 300 samples of parameters and simulated data are shown on Figure 4b.

The value of D_2 based on the actual data tends to be larger than the value based on the posterior predictive simulations. The relevant tail probability $P(D_2(\mathbf{y}^*; \boldsymbol{\theta}) > D_2(\mathbf{y}; \boldsymbol{\theta})|\mathbf{y})$ is not terribly small (.14), so we conclude that there is no evidence of heterogeneous regression variances for different teams. Thus we likely would not be interested in extending our model in the direction of a nonconstant regression variance.

5.2 Site Effect: A Model Diagnostics Success Story

We can use a slightly modified version of the game residuals,

$$r_{ii'} = y_{ii'} - (\theta_i - \theta_{i'}),$$

to search for a failure of the model in accounting for HFA. The $r_{ii'}$ are termed site-effect residuals, because they take the observed outcome and subtract out estimated team strengths but do not subtract out the HFA. As we did with the regression variance, we can examine differences in the magnitude of the HFA over time by calculating the average value of the $r_{ii'}$ for each season, and then examining the range of these averages. Specifically, for a posterior predictive dataset $\mathbf{y}^*$ and for a draw $\boldsymbol{\theta}^*$ from the posterior distribution of all parameters, let $D_3(\mathbf{y}^*; \boldsymbol{\theta}^*)$ be the differ-

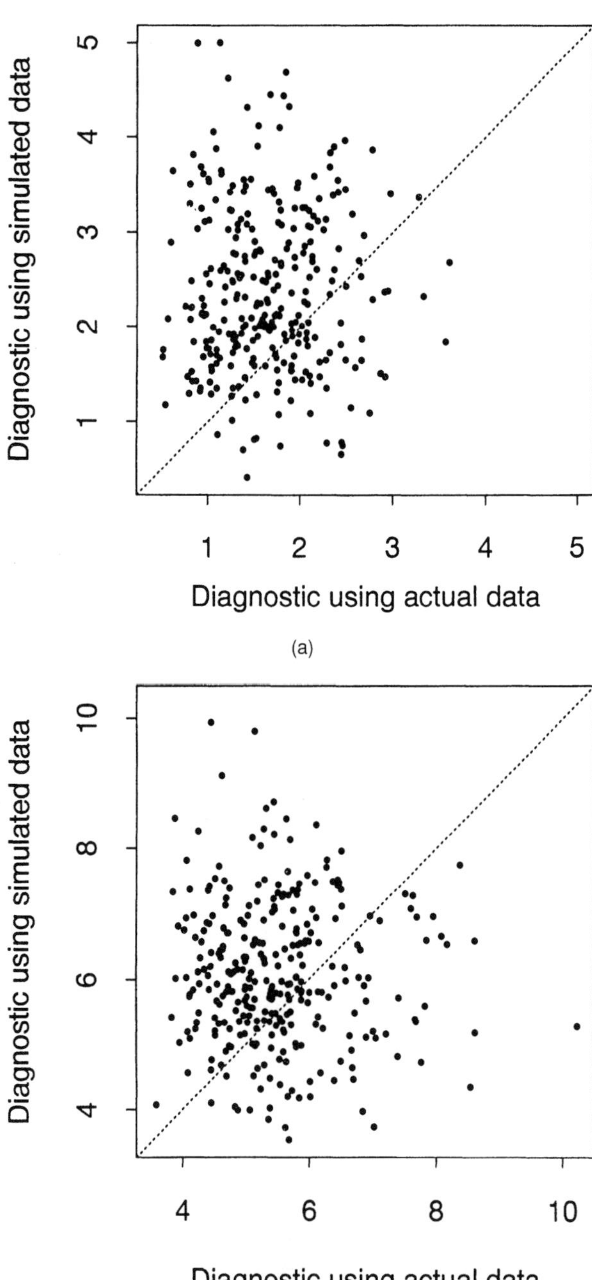

Figure 5. Estimated Bivariate Distributions for Site-Effect Diagnostics. (a) Scatterplot of the joint posterior distribution of $D_3(\mathbf{y}; \boldsymbol{\theta}^*)$ and $D_3(\mathbf{y}^*; \boldsymbol{\theta}^*)$ where $D_3(\cdot\,;\cdot)$ is a discrepancy measuring the range of HFA estimates among the six seasons and $(\boldsymbol{\theta}^*, \mathbf{y}^*)$ are simulations from their posterior and posterior predictive distributions; (b) scatterplot of the joint posterior distribution of $D_4(\mathbf{y}; \boldsymbol{\theta}^*)$ and $D_4(\mathbf{y}^*; \boldsymbol{\theta}^*)$ for $D_4(\cdot\,;\cdot)$ a discrepancy measuring the range of HFA estimates among the 28 teams.

ence between the maximum and the minimum of the average site-effect residuals by season. Using the same 300 values of $\boldsymbol{\theta}^*$ and $\mathbf{y}^*$ as before, we obtain the estimated bivariate distribution of $(D_3(\mathbf{y}; \boldsymbol{\theta}^*), D_3(\mathbf{y}^*; \boldsymbol{\theta}^*))$ shown on Figure 5a.

The plot reveals no particular pattern, although there is a tendency for $D_3(\mathbf{y}; \boldsymbol{\theta}^*)$ to be less than the discrepancy evaluated at the simulated datasets. This seems to be a chance occurrence (the tail probability equals .21).

We also include one other discrepancy measure, although it will be evident that our model fits this particular aspect of the data. We examined the average site-effect residuals across teams to assess whether the site effect depends on team. We calculated the average value of $r_{ii'}$ for each team. Let $D_4(\mathbf{y}^*; \boldsymbol{\theta}^*)$ be the difference between the maximum and minimum of these 28 averages for simulated data $\mathbf{y}^*$. It should be evident that the model will fit this aspect of the data, because we have used a separate parameter for each team's advantage. The approximate bivariate distribution of $(D_4(\mathbf{y}; \boldsymbol{\theta}^*), D_4(\mathbf{y}^*; \boldsymbol{\theta}^*))$ is shown on Figure 5b. There is no evidence of lack of fit (the tail probability equals .32).

This last discrepancy measure is included here, despite the fact that it measures a feature of the data that we have explicitly addressed in the model, because the current model was not the first model that we constructed. Earlier, we fit a model with a single HFA parameter for all teams. Figure 6 shows that for the single HFA parameter model, the observed values of $D_4(\mathbf{y}; \boldsymbol{\theta}^*)$ were generally greater than the values of $D_4(\mathbf{y}^*; \boldsymbol{\theta}^*)$, indicating that the average site-effect residuals varied significantly more from team to team than was expected under the model (tail probability equal to .05).

This suggested the model presented here in which each team has a separate HFA parameter.

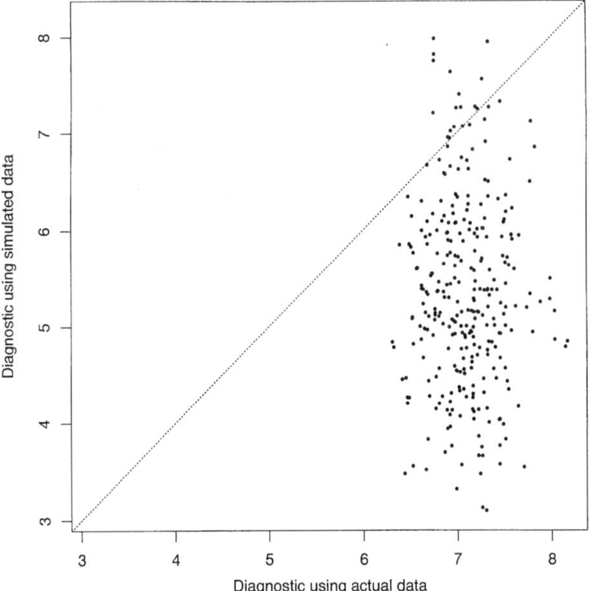

Figure 6. Estimated Bivariate Distribution for Site-Effect Diagnostic from a Poor-Fitting Model. The scatterplot shows the joint posterior distribution of $D_4(\mathbf{y}; \boldsymbol{\theta}^*)$ and $D_4(\mathbf{y}^*; \boldsymbol{\theta}^*)$ for a model that includes only a single parameter for the site-effect rather than 28 separate parameters, one for each team. The values of $D_4(\mathbf{y}; \boldsymbol{\theta}^*)$ are generally larger than the values of $D_4(\mathbf{y}^*; \boldsymbol{\theta}^*)$, suggesting that the fitted model may not be capturing a source of variability in the observed data.

5.3 Sensitivity to Heavy Tails

Our model assumes that outcomes are normally distributed conditional on the parameters, an assumption supported by Stern (1991). Rerunning the model with t distributions in place of normal distributions is straightforward, because t distributions can be expressed as scale mixtures of normal distributions (see, e.g., Gelman, Carlin, Stern, and Rubin 1995; and Smith 1983). Rather than redo the entire analysis, we checked the sensitivity of our inferences to the normal assumption by reweighting the posterior draws from the Gibbs sampler by ratios of importance weights (relating the normal model to a variety of t models). The reweighting is easily done and provides information about how inferences would be likely to change under alternative models. Our conclusion is that using a robust alternative can slightly alter estimates of team strength but does not have a significant effect on the predictive performance. It should be emphasized that the ratios of importance weights can be unstable, so that a more definitive discussion of inference under a particular t model (e.g., 4 df) would require a complete reanalysis of the data.

6. CONCLUSIONS

Our model for football game outcomes assumes that team strengths can change over time in a manner described by a normal state-space model. In previous state-space modeling of football scores (Harville 1977, 1980; Sallas and Harville 1988), some model parameters were estimated and then treated as fixed in making inferences on the remaining parameters. Such an approach ignores the variability associated with these parameters. The approach taken here, in contrast, is fully Bayesian in that we account for the uncertainty in all model parameters when making posterior or predictive inferences.

Our data analysis suggests that the model can be improved in several different dimensions. One could argue that teams' abilities should not shrink or expand around the mean from week to week, and because the posterior distribution of the between-week regression parameter β_w is not substantially different from 1, the model may be simplified by setting it to 1. Also, further exploration may be necessary to assess the assumption of a heavy-tailed distribution for game outcomes. Finally, as the game of football continues to change over time, it may be necessary to allow the evolution regression and variance parameters or the regression variance parameter to vary over time.

Despite the room for improvement, we feel that our model captures the main components of variability in football game outcomes. Recent advances in Bayesian computational methods allow us to fit a realistic complex model and diagnose model assumptions that would otherwise be difficult to carry out. Predictions from our model seem to perform as well, on average, as the Las Vegas point spread, so our model appears to track team strengths in a manner similar to that of the best expert opinion.

APPENDIX: CONDITIONAL DISTRIBUTIONS FOR MCMC SAMPLING

A.1 Conditional Posterior Distribution of $(\boldsymbol{\theta}_{(1,1)}, \ldots, \boldsymbol{\theta}_{(K, g_K)}, \boldsymbol{\alpha}, \phi)$

The conditional posterior distribution of the team strength parameters, HFA parameters, and observation precision ϕ, is normal-gamma—the conditional posterior distribution of ϕ given the evolution precision and regression parameters $(\omega_{\rm o}, \omega_{\rm h}, \omega_{\rm w}, \omega_{\rm s}, \beta_{\rm s}, \beta_{\rm w})$ is gamma, and the conditional posterior distribution of the team strengths and home-field parameters given all other parameters is an $(M+1)p$-variate normal distribution, where p is the number of teams and $M = \sum_{k=1}^{K} g_k$ is the total number of weeks for which data are available. It is advantageous to sample using results from the Kalman filter (Carter and Kohn 1994; Fruhwirth-Schnatter 1994; Glickman 1993) rather than consider this $(M+1)p$-variate conditional normal distribution as a single distribution. This idea is summarized here.

The Kalman filter (Kalman 1961; Kalman and Bucy 1961) is used to compute the normal-gamma posterior distribution of the final week's parameters,

$$f(\boldsymbol{\theta}_{(K,g_K)}, \boldsymbol{\alpha}, \phi | \omega_{\rm o}, \omega_{\rm h}, \omega_{\rm w}, \omega_{\rm s}, \beta_{\rm s}, \beta_{\rm w}, \mathbf{Y}_{(K,g_K)}),$$

marginalizing over the previous weeks' vectors of team strength parameters. This distribution is obtained by a sequence of recursive computations that alternately update the distribution of parameters when new data are observed and then update the distribution reflecting the passage of time. A sample from this posterior distribution is drawn. Samples of team strengths for previous weeks are drawn by using a back-filtering algorithm. This is accomplished by drawing recursively from the normal distributions for the parameters from earlier weeks,

$$f(\boldsymbol{\theta}_{(k,j)} | \boldsymbol{\theta}_{(k,j+1)}, \ldots, \boldsymbol{\theta}_{(K,g_K)}, \boldsymbol{\alpha}, \phi, \omega_{\rm o}, \omega_{\rm h}, \omega_{\rm w}, \omega_{\rm s}, \beta_{\rm s}, \beta_{\rm w}, \mathbf{Y}_{(K,g_K)}).$$

The result of this procedure is a sample of values from the desired conditional posterior distribution.

A.2 Conditional Posterior Distribution of $(\omega_{\rm o}, \omega_{\rm h}, \omega_{\rm w}, \omega_{\rm s})$

Conditional on the remaining parameters and the data, the parameters $\omega_{\rm o}, \omega_{\rm h}, \omega_{\rm w}$, and $\omega_{\rm s}$ are independent gamma random variables with

$$\omega_{\rm o} | \boldsymbol{\theta}_{(1,1)}, \ldots, \boldsymbol{\theta}_{(K,g_K)}, \boldsymbol{\alpha}, \phi, \beta_{\rm s}, \beta_{\rm w}, \mathbf{Y}_{(K,g_K)} \sim \text{gamma}\left(1 + \frac{(p+1)}{2}, \frac{1}{2}\left(\frac{1}{6} + \phi \boldsymbol{\theta}'_{(1,1)} \boldsymbol{\theta}_{(1,1)}\right)\right),$$

$$\omega_{\rm h} | \boldsymbol{\theta}_{(1,1)}, \ldots, \boldsymbol{\theta}_{(K,g_K)}, \boldsymbol{\alpha}, \phi, \beta_{\rm s}, \beta_{\rm w}, \mathbf{Y}_{(K,g_K)} \sim \text{gamma}\left(1 + \frac{(p+1)}{2}, \frac{1}{2}\left(\frac{1}{6} + \phi(\boldsymbol{\alpha} - 3)'(\boldsymbol{\alpha} - 3)\right)\right),$$

$$\omega_{\rm w} | \boldsymbol{\theta}_{(1,1)}, \ldots, \boldsymbol{\theta}_{(K,g_K)}, \boldsymbol{\alpha}, \phi, \beta_{\rm s}, \beta_{\rm w}, \mathbf{Y}_{(K,g_K)} \sim \text{gamma}\left(2 + \frac{p\sum_{k=1}^{K}(g_k - 1)}{2}, \frac{1}{2}\left(\frac{1}{60} + \phi \sum_{k=1}^{K} \sum_{j=1}^{g_k - 1} (\boldsymbol{\theta}_{(k,j+1)} - \beta_{\rm w} \mathbf{G} \boldsymbol{\theta}_{(k,j)})' \times (\boldsymbol{\theta}_{(k,j+1)} - \beta_{\rm w} \mathbf{G} \boldsymbol{\theta}_{(k,j)})\right)\right),$$

and

$$\omega_{\rm s} | \boldsymbol{\theta}_{(1,1)}, \ldots, \boldsymbol{\theta}_{(K,g_K)}, \boldsymbol{\alpha}, \phi, \beta_{\rm s}, \beta_{\rm w}, \mathbf{Y}_{(K,g_K)} \sim \text{gamma}\left(1 + \frac{(1 + p(K-1))}{2}, \frac{1}{2}\left(\frac{1}{16} + \phi \sum_{k=1}^{K-1} (\boldsymbol{\theta}_{(k+1,1)} - \beta_{\rm s} \mathbf{G} \boldsymbol{\theta}_{(k,g_k)})' \times (\boldsymbol{\theta}_{(k+1,1)} - \beta_{\rm s} \mathbf{G} \boldsymbol{\theta}_{(k,g_k)})\right)\right).$$

A.3 Conditional Posterior Distribution of $(\beta_{\rm s}, \beta_{\rm w})$

Conditional on the remaining parameters and the data, the distributions of $\beta_{\rm w}$ and $\beta_{\rm s}$ are independent random variables with normal distributions. The distribution of $\beta_{\rm w}$ conditional on all other parameters is normal, with

$$\beta_{\rm w} | \omega_{\rm o}, \omega_{\rm h}, \omega_{\rm w}, \omega_{\rm s}, \boldsymbol{\theta}_{(1,1)}, \ldots, \boldsymbol{\theta}_{(K,g_K)}, \boldsymbol{\alpha}, \phi, \mathbf{Y}_{(K,g_K)} \sim \text{N}(M_{\rm w}, V_{\rm w}),$$

where

$$V_{\rm w} = (1 + \phi \omega_{\rm w} A)^{-1},$$

$$M_{\rm w} = V_{\rm w}(.995 + \phi \omega_{\rm w} B),$$

$$A = \sum_{k=1}^{K} \sum_{j=1}^{g_k - 1} \boldsymbol{\theta}'_{(k,j)} \mathbf{G} \boldsymbol{\theta}_{(k,j)},$$

and

$$B = \sum_{k=1}^{K} \sum_{j=1}^{g_k - 1} \boldsymbol{\theta}'_{(k,j+1)} \mathbf{G} \boldsymbol{\theta}_{(k,j)}.$$

The distribution of $\beta_{\rm s}$ conditional on all other parameters is also normal, with

$$\beta_{\rm s} | \omega_{\rm o}, \omega_{\rm h}, \omega_{\rm w}, \omega_{\rm s}, \boldsymbol{\theta}_{(1,1)}, \ldots, \boldsymbol{\theta}_{(K,g_K)}, \boldsymbol{\alpha}, \phi, \mathbf{Y}_{(K,g_K)} \sim \text{N}(M_{\rm s}, V_{\rm s}),$$

where

$$V_{\rm s} = (1 + \phi \omega_{\rm w} C)^{-1},$$

$$M_{\rm s} = V_{\rm s}(.98 + \phi \omega_{\rm w} D),$$

$$C = \sum_{k=1}^{K-1} \boldsymbol{\theta}'_{(k,g_k)} \mathbf{G} \boldsymbol{\theta}_{(k,g_k)},$$

and

$$D = \sum_{k=1}^{K-1} \boldsymbol{\theta}'_{(k+1,1)} \mathbf{G} \boldsymbol{\theta}_{(k,g_k)}.$$

[Received December 1996. Revised August 1997.]

REFERENCES

Amoako-Adu, B., Marmer, H., and Yagil, J. (1985), "The Efficiency of Certain Speculative Markets and Gambler Behavior," *Journal of Economics and Business*, 37, 365–378.

Carlin, B. P., Polson, N. G., and Stoffer, D. S. (1992), "A Monte Carlo Approach to Nonnormal and Nonlinear State-Space Modeling," *Journal of the American Statistical Association*, 87, 493–500.

Carter, C. K., and Kohn, R. (1994), "On Gibbs Sampling for State-Space Models," *Biometrika*, 81, 541–553.

Chaloner, K., and Brant, R. (1988), "A Bayesian Approach to Outlier Detection and Residual Analysis," *Biometrika*, 75, 651–659.

de Jong, P., and Shephard, N. (1995), "The Simulation Smoother for Time Series Models," *Biometrika*, 82, 339–350.

Fruhwirth-Schnatter, S. (1994), "Data Augmentation and Dynamic Linear Models," *Journal of Time Series Analysis*, 15, 183–202.

Gelfand, A. E., and Smith, A. F. M. (1990), "Sampling-Based Approaches to Calculating Marginal Densities," *Journal of the American Statistical Association*, 85, 972–985.

Gelman, A., Carlin, J. B., Stern, H. S., and Rubin, D. B. (1995), *Bayesian Data Analysis*, London: Chapman and Hall.

Gelman, A., Meng, X., and Stern, H. S. (1996), "Posterior Predictive Assessment of Model Fitness via Realized Discrepancies" (with discussion), *Statistica Sinica*, 6, 733–807.

Gelman, A., and Rubin, D. B. (1992), "Inference From Iterative Simulation Using Multiple Sequences," *Statistical Science*, 7, 457–511.

Geman, S., and Geman, D. (1984), "Stochastic Relaxation, Gibbs Distributions, and the Bayesian Restoration of Images," *IEEE Transactions on Pattern Analysis and Machine Intelligence*, 6, 721–741.

Glickman, M. E. (1993), "Paired Comparison Models With Time-Varying Parameters," unpublished Ph.D. dissertation, Harvard University, Dept. of Statistics.

Harrison, P. J., and Stevens, C. F. (1976), "Bayesian Forecasting," *Journal of the Royal Statistical Society*, Ser. B, 38, 240–247.

Harville, D. (1977), "The Use of Linear Model Methodology to Rate High School or College Football Teams," *Journal of the American Statistical Association*, 72, 278–289.

——— (1980), "Predictions for National Football League Games via Linear-Model Methodology," *Journal of the American Statistical Association*, 75, 516–524.

Kalman, R. E. (1960), "A New Approach to Linear Filtering and Prediction Problems," *Journal of Basic Engineering*, 82, 34–45.

Kalman, R. E., and Bucy, R. S. (1961), "New Results in Linear Filtering and Prediction Theory," *Journal of Basic Engineering*, 83, 95–108.

Rosner, B. (1976), "An Analysis of Professional Football Scores," in *Management Science in Sports*, eds. R. E. Machol, S. P. Ladany, and D. G. Morrison, New York: North-Holland, pp. 67–78.

Rubin, D. B. (1984), "Bayesianly Justifiable and Relevant Frequency Calculations for the Applied Statistician," *The Annals of Statistics*, 12, 1151–1172.

Sallas, W. M., and Harville, D. A. (1988), "Noninformative Priors and Restricted Maximum Likelihood Estimation in the Kalman Filter," in *Bayesian Analysis of Time Series and Dynamic Models*, ed. J. C. Spall, New York: Marcel Dekker, pp. 477–508.

Shephard, N. (1994), "Partial Non-Gaussian State Space," *Biometrika*, 81, 115–131.

Smith, A. F. M. (1983), "Bayesian Approaches to Outliers and Robustness," in *Specifying Statistical Models From Parametric to Nonparametric, Using Bayesian or Non-Bayesian Approaches*, eds. J. P. Florens, M. Mouchart, J. P. Raoult, L. Simer, and A. F. M. Smith, New York: Springer-Verlag, pp. 13–35.

Stern, H. (1991), "On the Probability of Winning a Football Game," *The American Statistician*, 45, 179–183.

——— (1992), "Who's Number One? Rating Football Teams," in *Proceedings of the Section on Statistics in Sports, American Statistical Association*, pp. 1–6.

Stefani, R. T. (1977), "Football and Basketball Predictions Using Least Squares," *IEEE Transactions on Systems, Man, and Cybernetics*, 7, 117–120.

——— (1980), "Improved Least Squares Football, Basketball, and Soccer Predictions," *IEEE Transactions on Systems, Man, and Cybernetics*, 10, 116–123.

Thompson, M. (1975), "On Any Given Sunday: Fair Competitor Orderings With Maximum Likelihood Methods," *Journal of the American Statistical Association*, 70, 536–541.

West, M., and Harrison, P. J. (1990), *Bayesian Forecasting and Dynamic Models*, New York: Springer-Verlag.

Zellner, A. (1975), "Bayesian Analysis of Regression Error Terms," *Journal of the American Statistical Association*, 70, 138–144.

Zuber, R. A., Gandar, J. M., and Bowers, B. D. (1985), "Beating the Spread: Testing the Efficiency of the Gambling Market for National Football League Games," *Journal of Political Economy*, 93, 800–806.

Chapter 6

Predictions for National Football League Games Via Linear-Model Methodology

DAVID HARVILLE*

Results on mixed linear models were used to develop a procedure for predicting the outcomes of National Football League games. The predictions are based on the differences in score from past games. The underlying model for each difference in score takes into account the home-field advantage and the difference in the yearly characteristic performance levels of the two teams. Each team's yearly characteristic performance levels are assumed to follow a first-order autoregressive process. The predictions for 1,320 games played between 1971 and 1977 had an average absolute error of 10.68, compared with 10.49 for bookmaker predictions.

KEY WORDS: Football predictions; Mixed linear models; Variance components; Maximum likelihood; Football ratings.

1. INTRODUCTION

Suppose that we wish to predict the future price of a common stock or to address some other complex real-life prediction problem. What is the most useful role for statistics?

One approach is to use the available information rather informally, relying primarily on intuition and on past experience and employing no statistical methods or only relatively simple statistical methods. A second approach is to rely exclusively on some sophisticated statistical algorithm to produce the predictions from the relevant data. In the present article, these two approaches are compared in the context of predicting the outcomes of National Football League (NFL) games.

The statistical algorithm to be used is set forth in Section 2. It is closely related to an algorithm devised by Harville (1977b) for rating high school or college football teams.

The essentially nonstatistical predictions that are to be compared with the statistical predictions are those given by the betting line. The betting line gives the favored team for each game and the point spread, that is, the number of points by which the favorite is expected to win.

If a gambler bets on the favorite (underdog), he wins (loses) his bet when the favorite wins the game by more than the point spread, but he loses (wins) his bet when the favorite either loses or ties the game or wins the game by less than the point spread. On a $10 bet, the gambler pays the bookmaker an additional dollar (for a total of $11) when he loses his bet and receives a net of $10 when he wins. If the favorite wins the game by exactly the point spread, the bet is in effect cancelled (Merchant 1973). To break even, the gambler must win 52.4 percent of those bets that result in either a win or a loss (assuming that the bets are for equal amounts).

Merchant described the way in which the betting line is established. A prominent bookmaker devises an initial line, which is known as the outlaw line, the early line, or the service line. This line is such that, in his informed opinion, the probability of winning a bet on the favorite equals the probability of winning a bet on the underdog.

A select group of knowledgeable professional gamblers are allowed to place bets (in limited amounts) on the basis of the outlaw line. A series of small adjustments is made in the outlaw line until an approximately equal amount of the professionals' money is being attracted on either side. The betting line that results from this process is the official opening line, which becomes available on Tuesday for public betting.

Bets can be placed until the game is played, which is generally on Sunday but can be as early as Thursday or as late as Monday. If at any point during the betting period the bookmaker feels that there is too big a discrepancy between the amount being bet on the favorite and the amount being bet on the underdog, he may make a further adjustment in the line.

The nonstatistical predictions used in the present study are those given by the official opening betting line. These predictions can be viewed as the consensus opinion of knowledgeable professional gamblers.

The statistical algorithm that is set forth in Section 2 can be used to rate the various NFL teams as well as to make predictions. While the prediction and rating problems are closely related, there are also some important differences, which are discussed in Section 5.

2. STATISTICAL PREDICTIONS

Each year's NFL schedule consists of three parts: preseason or exhibition games, regular-season games, and postseason or playoff games. The statistical algorithm presented in Sections 2.2 through 2.4 translates scores from regular-season and playoff games that have already been played into predictions for regular-season and play-

* David Harville is Professor, Department of Statistics, Iowa State University, Ames, IA 50011. This article is based on an invited paper (Harville 1978) presented at the 138th Annual Meeting of the American Statistical Association, San Diego, CA.

© Journal of the American Statistical Association
September 1980, Volume 75, Number 371
Applications Section

off games to be played in the future. The model that serves as the basis for this algorithm is described in Section 2.1.

The scores of exhibition games were not used in making the statistical predictions, and predictions were not attempted for future exhibition games. The rationale was that these games are hard to incorporate into the model and thus into the algorithm and that they have very little predictive value anyhow. Merchant (1973) argues that, in making predictions for regular-season games, it is best to forget about exhibition games.

2.1 Underlying Model

Suppose that the scores to be used in making the predictions date back to the Year F. Ultimately, F is to be chosen so that the interlude between the beginning of Year F and the first date for which predictions are required is long enough that the effect of including earlier scores is negligible. For each year, number the regular-season and playoff games 1, 2, 3, ... in chronological order.

Number the NFL teams 1, 2, 3, ... New teams are formed by the NFL from time to time. If Team i is added after Year F, let $F(i)$ represent the year of addition. Otherwise, put $F(i) = F$. The home team and the visiting team for the kth game in Year j are to be denoted by $h(j, k)$ and $v(j, k)$, respectively. (If the game were played on a neutral field, $h(j, k)$ is taken arbitrarily to be one of the two participating teams, and $v(j, k)$ is taken to be the other.)

Let S_{jk} equal the home team's score minus the visiting team's score for the kth game in Year j. The prediction algorithm presented in Sections 2.2 through 2.4 depends only on the scores and depends on the scores only through the S_{jk}'s.

Our model for the S_{jk}'s involves conceptual quantities H and T_{im} $(i = 1, 2, \ldots; m = F(i), F(i) + 1, \ldots)$. The quantity H is an unknown parameter that represents the home-field advantage (in points) that accrues to a team from playing on its own field rather than a neutral field. The quantity T_{im} is a random effect that can be interpreted as the characteristic performance level (in points) of Team i in Year m relative to that of an "average" team in Year m.

The model equation for S_{jk} is

$$S_{jk} = T_{h(j,k),j} - T_{v(j,k),j} + R_{jk}$$

if the game is played on a neutral field, or

$$S_{jk} = H + T_{h(j,k),j} - T_{v(j,k),j} + R_{jk}$$

if it is not. Here, R_{jk} is a random residual effect. Assume that $E(R_{jk}) = 0$, that $\text{var}(R_{jk}) = \sigma_R^2$ where σ_R^2 is an unknown strictly positive parameter, and that the R_{jk}'s are uncorrelated with each other and with the T_{im}'s.

Suppose that $\text{cov}(T_{im}, T_{i'm'}) = 0$ if $i' \neq i$; that is, that the yearly characteristic performance levels of any given team are uncorrelated with those of any other team. The yearly characteristic performance levels of Team i are assumed to follow a first-order autoregressive process

$$T_{i,m+1} = \rho T_{im} + U_{im} \quad (m = F(i), F(i) + 1, \ldots) ,$$

where $U_{i,F(i)}, U_{i,F(i)+1}, \ldots$ are random variables that have zero means and common unknown variance σ_U^2 and that are uncorrelated with each other and with $T_{i,F(i)}$, and where ρ is an unknown parameter satisfying $0 \leq \rho < 1$.

It remains to specify assumptions, for each i, about $E[T_{i,F(i)}]$ and $\text{var}[T_{i,F(i)}]$, that is, about the mean and variance of the first yearly characteristic performance level for Team i. The sensitivity of the prediction procedure to these specifications depends on the proximity (in time) of the predicted games to the beginning of Year $F(i)$ and on whether the predicted games involve Team i. For i such that $F(i) = F$, that is, for teams that date back to Year F, the prediction procedure will be relatively insensitive to these specifications, provided that F is sufficiently small, that is, provided the formation of the data base was started sufficiently in advance of the first date for which predictions are required.

Put $\sigma_T^2 = \sigma_U^2/(1 - \rho^2)$, and for convenience assume that

$$E[T_{i,F(i)}] = 0 \quad \text{and} \quad \text{var}[T_{i,F(i)}] = \sigma_T^2 \tag{2.1}$$

for i such that $F(i) = F$. Then, for any given year, the yearly characteristic performance levels of those teams that date back to Year F have zero means and common variance σ_T^2, as would be the case if they were regarded as a random sample from an infinite population having mean zero and variance σ_T^2. Moreover, for any such team, the correlation between its characteristic performance levels for any two years m and m' is $\rho^{|m'-m|}$, which is a decreasing function of elapsed time.

For i such that $F(i) > F$, that is, for teams that came into being after Year F, it is assumed that

$$E[T_{i,F(i)}] = \mu_{F(i)} \quad \text{and} \quad \text{var}[T_{i,F(i)}] = \tau_{F(i)}^2 ,$$

where $\mu_{F(i)}$ and $\tau_{F(i)}^2$ are quantities that are to be supplied by the user of the prediction procedure. The quantities $\mu_{F(i)}$ and $\tau_{F(i)}^2$ can be regarded as the mean and variance of a common prior distribution for the initial yearly characteristic performance levels of expansion teams. Information on the performance of expansion teams in their first year that predates Year $F(i)$ can be used in deciding on values for $\mu_{F(i)}$ and $\tau_{F(i)}^2$.

The model for the S_{jk}'s is similar to that applied to high school and college football data by Harville (1977b). One distinguishing feature is the provision for data from more than one year.

2.2 Preliminaries

We consider the problem of predicting S_{JK} from $S_{F1}, S_{F2}, \ldots, S_{LG}$, where either $J = L$ and $K > G$ or $J > L$; that is, the problem of predicting the winner and the margin of victory for a future game based on the information accumulated as of Game G in Year L. This problem is closely related to that of estimating or predicting H and T_{im} ($i = 1, 2, \ldots$; $m = F(i), F(i) + 1, \ldots$).

Take $\lambda = \sigma_T^2/\sigma_R^2$. If λ and ρ were given, H would have a unique minimum-variance linear unbiased (Aitken) estimator, which we denote by $\tilde{H}(\lambda, \rho)$. Define $\tilde{T}_{im}(\lambda, \rho, H)$ by

$$\tilde{T}_{im}(\lambda, \rho, H) = E(T_{im}|S_{F1}, S_{F2}, \ldots, S_{LG}) \ , \quad (2.2)$$

where the conditional expectation is taken under the assumption that T_{im} and $S_{F1}, S_{F2}, \ldots, S_{LG}$ are jointly normal or, equivalently, is taken to be Hartigan's (1969) linear expectation. Put

$$\tilde{T}_{im}(\lambda, \rho) = \tilde{T}_{im}(\lambda, \rho, \tilde{H}(\lambda, \rho)) \ . \quad (2.3)$$

The quantity $\tilde{T}_{im}(\lambda, \rho)$ is the best linear unbiased predictor (BLUP) of T_{im} (in the sense described by Harville 1976) for the case in which λ and ρ are given; and, for that same case, the quantity $\tilde{S}_{JK}(\lambda, \rho)$, defined as follows is the BLUP of S_{jk}:

$$\tilde{S}_{JK}(\lambda, \rho) = \tilde{T}_{h(J,K),J}(\lambda, \rho) - \tilde{T}_{v(J,K),J}(\lambda, \rho) \quad (2.4)$$

if the JKth game is played on a neutral field; or

$$\tilde{S}_{JK}(\lambda, \rho) = \tilde{H}(\lambda, \rho) + \tilde{T}_{h(J,K),J}(\lambda, \rho) - \tilde{T}_{v(J,K),J}(\lambda, \rho) \quad (2.5)$$

if it is not.

Let $M_{JK}(\sigma_R^2, \lambda, \rho)$ denote the mean squared difference between $\tilde{S}_{JK}(\lambda, \rho)$ and S_{JK}, that is,

$$M_{JK}(\sigma_R^2, \lambda, \rho) = E[\tilde{S}_{JK}(\lambda, \rho) - S_{JK}]^2 \ . \quad (2.6)$$

Specific representations for $\tilde{H}(\lambda, \rho)$, $\tilde{T}_{im}(\lambda, \rho)$, and $M_{JK}(\sigma_R^2, \lambda, \rho)$ can be obtained as special cases of representations given, for example, by Harville (1976).

2.3 Estimation of Model Parameters

In practice, λ and ρ (and σ_R^2) are not given and must be estimated. One approach to the estimation of these parameters is to use Patterson and Thompson's (1971) restricted maximum likelihood procedure. (See, e.g., Harville's (1977a) review article for a general description of this procedure.)

Suppose that $S_{F1}, S_{F2}, \ldots, S_{LG}$ constitute the data available for estimating λ, ρ, and σ_R^2. For purposes of estimating these parameters, we assume that the data are jointly normal, and we take $E[T_{i,F(i)}]$ and $\text{var}[T_{i,F(i)}]$ to be of the form (2.1) for *all* i; however, we eliminate from the data set any datum S_{jk} for which $h(j, k)$ or $v(j, k)$ corresponds to an expansion team formed within I years of Year j (where I is to be specified by the user). Extending the assumption (2.1) to all i simplifies the estimation procedure, while the exclusion of games involving expansion teams in their early years desensitizes the procedure to the effects of this assumption.

We write $\tilde{H}$ and $\tilde{T}_{im}$ for the quantities $\tilde{H}(\lambda, \rho)$ and $\tilde{T}_{im}(\lambda, \rho)$ defined in Section 2.2 (with allowances for the deletions in the data set and the change in assumptions). Let X_{jk} equal 0 or 1 depending on whether or not the jkth game is played on a neutral field.

The likelihood equations for the restricted maximum likelihood procedure can be put into the form

$$Q_i - E(Q_i) = 0 \quad (i = 1, 2, 3) \ , \quad (2.7)$$

with

$$Q_1 = (\tfrac{1}{2})[\sigma_R^2\lambda(1 - \rho^2)]^{-1} \sum_i [(1 - \rho^2)\tilde{T}_{i,F(i)} + \sum_{k=F(i)}^{L-1} (\tilde{T}_{i,k+1} - \rho\tilde{T}_{ik})^2] \ , \quad (2.8)$$

$$Q_2 = (\tfrac{1}{2})[\sigma_R^2\lambda(1 - \rho^2)]^{-1} \sum_i \sum_{k=F(i)}^{L-1} (\tilde{T}_{ik} - \rho\tilde{T}_{i,k+1})(\tilde{T}_{i,k+1} - \rho\tilde{T}_{ik}) \ , \quad (2.9)$$

and

$$Q_3 = \sum_{j=F}^{L} \sum_k [S_{jk} - X_{jk}\tilde{H} - \tilde{T}_{h(j,k),j} + \tilde{T}_{v(j,k),j}]^2 \ . \quad (2.10)$$

It can be shown that

$$E[S_{jk} - X_{jk}\tilde{H} - \tilde{T}_{h(j,k),j} + \tilde{T}_{v(j,k),j}]^2 = \sigma_R^2 - (X_{jk}, 1, -1) \cdot \text{var}[\tilde{H} - H, \tilde{T}_{h(j,k),j} - T_{h(j,k),j}, \tilde{T}_{v(j,k),j} - T_{v(j,k),j}] \cdot (X_{jk}, 1, -1)' \quad (2.11)$$

and that

$$E(\tilde{T}_{im}) = E(T_{im}) = 0 \quad (2.12)$$

and

$$\text{cov}(\tilde{T}_{im}, \tilde{T}_{im'}) = \rho^{|m'-m|}\sigma_T^2 - \text{cov}(\tilde{T}_{im} - T_{im}, \tilde{T}_{im'} - T_{im'}) \ . \quad (2.13)$$

Equations (2.7) can be solved numerically by the same iterative numerical algorithm used by Harville (1977b, p. 288). This procedure calls for the repeated evaluation of the Q_i's and their expectations for various trial values of λ and ρ. Making use of (2.11), (2.12), and (2.13) reduces the problem of evaluating the Q_i's and their expectations for particular values of λ and ρ to the problem of evaluating $\tilde{H}$ and the $\tilde{T}_{im}$'s and various elements of their dispersion matrix. Kalman filtering and smoothing algorithms (suitably modified for mixed models, as described by Harville 1979) can be used for maximum efficiency in carrying out the computations associated with the latter problem.

The amount of computation required to evaluate $\tilde{H}$ and the $\tilde{T}_{im}$'s and the relevant elements of their dispersion matrix, for fixed values of λ and ρ, may not be feasible if data from a large number of years are being used. The procedure for estimating λ, ρ, and σ_R^2 can be modified in these instances by, for example, basing the "estimates" $\tilde{T}_{i,k+1}$ and $\tilde{T}_{ik}$ in the term $(\tilde{T}_{i,k+1} - \rho\tilde{T}_{ik})^2$ of (2.8) on only those data accumulated through Year $k + Y$, for some Y, rather than on all the data. Such modifications can significantly reduce the amount of storage and computation required to evaluate the Q_i's and their expectations. (The modifications reduce the amount of smoothing that must be carried out in the Kalman algorithm.)

This modified estimation procedure can be viewed as a particular implementation of the approximate restricted maximum likelihood approach outlined by Harville (1977a, Sec. 7). This approach seems to have produced a reasonable procedure even though it is based on the assumption of a distributional form (multivariate normal) that differs considerably from the actual distributional form of the S_{jk}'s.

2.4 Prediction Algorithm

Let $\hat{\lambda}$, $\hat{\rho}$, and $\hat{\sigma}_R^2$ represent estimates of λ, ρ, and σ_R^2, respectively. In particular, we can take $\hat{\lambda}$, $\hat{\rho}$, and $\hat{\sigma}_R^2$ to be the estimates described in Section 2.3.

Let

$$\hat{H} = \tilde{H}(\hat{\lambda}, \hat{\rho}) \quad \text{and} \quad \hat{T}_{im} = \tilde{T}_{im}(\hat{\lambda}, \hat{\rho}) \quad (i = 1, 2, \ldots; m = F(i), F(i) + 1, \ldots) .$$

The quantity $\hat{H}$ gives an estimate of H, and $\hat{T}_{im}$ gives an estimate or prediction of T_{im}.

It can be shown that

$$\hat{H} = N^{-1}(G - \sum_i \sum_{m=F(i)}^{L} D_{im}\hat{T}_{im}) , \quad (2.14)$$

where N equals the total number of games played minus the number of games played on neutral fields, G equals the grand total of all points scored by home teams minus the grand total for visiting teams, and D_{im} equals the number of games played in Year m by Team i on its home field minus the number played on its opponents' fields. The NFL schedule is such that, if it were not for playoff games and for games not yet played in Year L, all of the D_{im}'s would equal zero, and $\hat{H}$ would coincide with the ordinary average $N^{-1}G$.

It can also be shown that, if $F(i) = L$; that is, if only one season of data or a partial season of data is available on Team i, then

$$\hat{T}_{iL} = E(T_{iL}) + (N_{iL} + \hat{\gamma})^{-1} \cdot [G_{iL} - N_{iL}E(T_{iL}) - D_{iL}\hat{H} + \sum_j \hat{T}_{r(j),L}] , \quad (2.15)$$

where N_{iL} equals the number of games played (in Year L) by Team i, G_{iL} equals the total points scored (in Year L) by Team i minus the total scored against it by its opponents, $r(j)$ equals Team i's opponent in its jth game (of Year L), and

$$\hat{\gamma} \text{ equals } \hat{\lambda}^{-1} \text{ if } F(i) = F \text{ and equals } \hat{\sigma}_R^2/\tau_{F(i)}^2 \text{ if } F(i) > F .$$

Thus, if $F(i) = L$, the estimator $\hat{T}_{iL}$ is seen to be a "shrinker"; that is, instead of the "corrected total" for the ith team being divided by N_{iL}, it is divided by $N_{iL} + \hat{\gamma}$. If $F(i) < L$, that is, if more than one season of data is available on Team i, the form of the estimator is similar.

The prediction for the outcome S_{JK} of a future game is taken to be

$$\hat{S}_{JK} = \tilde{S}_{JK}(\hat{\lambda}, \hat{\rho}) = X_{JK}\hat{H} + \hat{T}_{h(J,K),J} - \hat{T}_{v(J,K),J} .$$

An estimate of the mean squared error of this prediction is given by $\hat{M}_{JK} = M_{JK}(\hat{\sigma}_R^2, \hat{\lambda}, \hat{\rho})$. The estimate $\hat{M}_{JK}$ underestimates the mean squared error to an extent that depends on the precision of the estimates $\hat{\lambda}$ and $\hat{\rho}$.

The quantities $\tilde{S}_{JK}(\lambda, \rho)$ and $M_{JK}(\sigma_R^2, \lambda, \rho)$ can be interpreted as the mean and the variance of a posterior distribution for S_{JK} (Harville 1976, Sec. 4). Depending on the precision of the estimates $\hat{\lambda}$, $\hat{\rho}$, and $\hat{\sigma}_R^2$, it may be reasonable to interpret $\hat{S}_{JK}$ and $\hat{M}_{JK}$ in much the same way.

Take B_{JK} to be a constant that represents the difference in score given by the betting line for Game K of Year J. Relevant posterior probabilities for gambling purposes are $\Pr(S_{JK} < B_{JK})$ and $\Pr(S_{JK} > B_{JK})$. These posterior probabilities can be obtained from the posterior probabilities $\Pr(S_{JK} = s)$ $(s = \ldots, -2, -1, 0, 1, 2, \ldots)$, which we approximate by $\Pr(s - .5 < S_{JK}^* \leq s + .5)$, where S_{JK}^* is a normal random variable with mean $\hat{S}_{JK}$ and variance $\hat{M}_{JK}$.

Due to the "lumpiness" of the actual distribution of the differences in score (as described, for college football, by Mosteller 1970) these approximations to the posterior probabilities may be somewhat crude; however, it is not clear how to improve on them by other than ad hoc procedures. Rosner (1976) took a somewhat different approach to the prediction problem in an attempt to accommodate this feature of the distribution.

Our prediction algorithm can be viewed as consisting of two stages. In the first stage, λ, ρ, and σ_R^2 are estimated. Then, in the second stage, $\hat{H}$ and $\hat{T}_{iL}$ $(i = 1, 2, \ldots)$ and their estimated dispersion matrix are computed. By making use of the Kalman prediction algorithm (as described by Harville 1979), the output of the second stage can easily be converted into a prediction $\hat{S}_{JK}$ and an estimated mean squared prediction error $\hat{M}_{JK}$ for any future game.

The second-stage computations, as well as the first-stage computations, can be facilitated by use of the Kalman filtering algorithm. This is especially true in instances where the second-stage computations were previously carried out based on S_{jk}'s available earlier and where λ and ρ have not been reestimated.

3. EMPIRICAL EVALUATION OF PREDICTIONS

The statistical algorithm described in Section 2 was used to make predictions for actual NFL games. These predictions were compared for accuracy with those given by the betting line. The games for which the comparisons were made were 1,320 regular-season and playoff games played between 1971 and 1977, inclusive.

The betting line for each game was taken to be the opening line. The primary source for the opening line was the *San Francisco Chronicle*, which, in its Wednesday

1. Parameter Estimates Over Each of Seven Time Periods

Last Year in Period	$\hat{\lambda}$	$\hat{\rho}$	$\hat{\sigma}_R^2$	$\hat{H}$
1970	.29	.79	185	—
1971	.26	.83	181	2.19 ± .50
1972	.27	.81	180	2.03 ± .44
1973	.28	.82	182	2.27 ± .40
1974	.26	.78	175	2.31 ± .37
1975	.27	.80	175	2.18 ± .34
1976	.25	.79	171	2.32 ± .32
1977	—	—	—	2.42 ± .30

editions, ordinarily reported the opening line listed in *Harrah's Tahoe Racebook*. There were 31 games played between 1971 and 1977 for which no line could be found and for which no comparisons were made.

The statistical prediction for each of the 1,320 games was based on the outcomes of all NFL regular-season and playoff games played from the beginning of the 1968 season through the week preceding the game to be predicted. The estimates $\hat{\lambda}$, $\hat{\rho}$, and $\hat{\sigma}_R^2$ used in making the predictions were based on the same data but were recomputed yearly rather than weekly. (In estimating λ, ρ, and σ_R^2 at the ends of Years 1970–1975, I was taken to be zero, so that all accumulated games were used. However, in estimating these parameters at the end of 1976, games played during 1968 were excluded as were games involving the expansion teams, Seattle and Tampa Bay, that began play in 1976.)

Games that were tied at the end of regulation play and decided in an overtime period were counted as ties when used in estimating λ, ρ, and σ_R^2 and in making predictions. The values assigned to $\mu_{F(i)}$ and $\tau_{F(i)}^2$ for prediction purposes were -11.8 and 17.0, respectively.

The values obtained for $\hat{\lambda}$, $\hat{\rho}$, and $\hat{\sigma}_R^2$ at the end of each year are listed in Table 1. The values of the estimate $\hat{H}$ of the home-field advantage and the estimated standard error of $\hat{H}$ as of the end of each year (based on the values $\hat{\lambda}$, $\hat{\rho}$, and $\hat{\sigma}_R^2$ obtained at the end of the previous year) are also given.

Some decline in the variability of the outcomes of the games (both "among teams" and "within teams") appears to have taken place beginning in about 1974. The estimates of the among-teams variance σ_T^2 and the within-teams variance σ_R^2 are both much smaller than those obtained by Harville (1977b) for Division-I college football (42 and 171 vs. 104 and 214, respectively). The estimate of the home-field advantage is also smaller than for college football (2.42 vs. 3.35). Merchant (1973) conjectured that, in professional football, the home-field advantage is becoming a thing of the past. The results given in Table 1 seem to indicate otherwise.

Table 2 provides comparisons, broken down on a year-by-year basis, between the accuracy of the statistical predictions and the accuracy of the predictions given by the betting line. Three criteria were used to assess accuracy: the frequency with which the predicted winners actually won, the average of the absolute values of the prediction errors, and the average of the squares of the prediction errors, where prediction error is defined to be the actual (signed) difference in score between the home team and the visiting team minus the predicted difference. With regard to the first criterion, a predicted tie was counted as a success or half a success depending on whether the actual outcome was a tie, and, if a tie occurred but was not predicted, credit for half a success was given.

The statistical predictions are seen to be somewhat less accurate on the average than the predictions given by the betting line. Comparisons with Harville's (1977b) results indicate that both types of predictions tend to be more accurate for professional football than for college football. The average absolute difference between the statistical predictions and the predictions given by the betting line was determined to be 2.48.

Table 3 gives the accuracy, as measured by average absolute error, of the two types of predictions for each of the 14 weeks of the regular season and for the playoff games. Both types of predictions were more accurate over a midseason period, extending approximately from Week 6 to 13, than at the beginning or end of the season. Also, the accuracy of the statistical predictions compared more favorably with that of the betting line during midseason than during the rest of the season. Specifically, the average absolute prediction error for Weeks 6 through 13 was 10.37 for the statistical predictions and 10.35 for the betting line (vs. the overall figures of 10.68 and 10.49, respectively).

2. Accuracy of the Statistical Procedure Versus That of the Betting Line

		Percentage of Winners		Average Absolute Error		Average Squared Error	
Year(s)	Number of Games	Statistical Procedure	Betting Line	Statistical Procedure	Betting Line	Statistical Procedure	Betting Line
1971	164	66.2	68.6	10.28	10.61	172.1	181.0
1972	189	66.4	71.4	11.35	10.94	201.8	192.6
1973	187	74.6	75.7	11.89	11.36	228.2	205.6
1974	187	65.5	68.5	10.07	10.16	159.8	161.6
1975	187	73.8	76.2	10.76	10.38	201.0	183.6
1976	203	72.7	72.9	10.79	10.60	185.9	182.6
1977	203	72.4	70.9	9.67	9.46	166.1	168.0
All	1,320	70.3	72.1	10.68	10.49	187.8	182.0

3. Week-by-Week Breakdown for Prediction Accuracy

Week(s)	Number of Games	Average Absolute Error: Statistical Procedure	Betting Line
1	92	11.55	11.20
2	80	10.46	9.79
3	93	11.31	11.31
4	92	10.97	10.39
5	92	11.02	10.52
6	93	10.41	10.27
7	92	10.99	10.72
8	93	10.48	10.78
9	93	8.95	8.94
10	93	9.82	9.95
11	91	10.34	10.26
12	93	10.02	10.12
13	93	11.96	11.75
14	84	11.73	11.18
Playoffs	46	9.96	9.73
All	1,320	10.68	10.49

It is not surprising that the statistical predictions are more accurate, relative to the betting line, during mid-season than during earlier and later periods. The statistical predictions are based only on differences in score from previous games. A great deal of additional information is undoubtedly used by those whose opinions are reflected in the betting line. The importance of taking this additional information into account depends on the extent to which it is already reflected in available scores. Early and late in the season, the additional information is less redundant than at mid-season. At the beginning of the season, it may be helpful to supplement the information on past scores with information on roster changes, injuries, exhibition-game results, and so on. During the last week or two of the regular season, it may be important to take into account which teams are still in contention for playoff berths.

The statistical predictions were somewhat more similar to the predictions given by the betting line during mid-season than during earlier and later periods. For Weeks 6 through 13, the average absolute difference between the two types of predictions was found to be 2.27 (vs. the overall figure of 2.48).

There remains the question of whether the statistical predictions could serve as the basis for a successful betting scheme. Suppose that we are considering betting on a future game, say, Game K of Year J.

If $\hat{S}_{JK} > B_{JK}$, that is, if the statistical prediction indicates that the chances of the home team are better than those specified by the betting line, then we might wish to place a bet on the home team. The final decision on whether to make the bet could be based on the approximation (discussed in Section 2.4) to the ratio

$$\Pr(S_{JK} > B_{JK}) / [\Pr(S_{JK} > B_{JK}) + \Pr(S_{JK} < B_{JK})] , \quad (3.1)$$

that is, on the approximation to the conditional probability (as defined in Section 2.4) that the home team will win by more (or lose by less) than the betting line would indicate given that the game does not end in a tie relative to the betting line. The bet would be made if the approximate value of this conditional probability were sufficiently greater than .5.

If $\hat{S}_{JK} < B_{JK}$, then depending on whether the approximation to the ratio (3.1) were sufficiently smaller than .5, we would bet on the visiting team. The actual success of such a betting scheme would depend on the frequency with which bets meeting our criteria arise and on the relative frequency of winning bets among those bets that do qualify.

In Table 4a, the predictions for the 1,320 games are divided into six categories depending on the (approximate) conditional probability that the team favored by

4. Theoretical Versus Observed Frequency of Success for Statistical Predictions Relative to the Betting Line

Probability Interval	Number of Games and Number of Ties	Average Probability	Observed Relative Frequency	Number of Games and Number of Ties (Cumulative)	Average Probability (Cumulative)	Observed Relative Frequency (Cumulative)
			a. All weeks			
[.50, .55)	566 (16)	.525	.525	1320 (48)	.570	.528
[.55, .60)	429 (17)	.574	.534	754 (32)	.604	.530
[.60, .65)	221 (12)	.621	.483	325 (15)	.643	.526
[.65, .70)	78 (3)	.671	.627	104 (3)	.688	.614
[.70, .75)	18 (0)	.718	.556	26 (0)	.736	.577
>.75	8 (0)	.778	.625	8 (0)	.778	.625
			b. Weeks 6–13			
[.50, .55)	337 (7)	.525	.503	741 (26)	.564	.541
[.55, .60)	248 (11)	.573	.570	404 (19)	.598	.574
[.60, .65)	112 (6)	.620	.528	156 (8)	.638	.581
[.65, .70)	39 (2)	.672	.730	44 (2)	.682	.714
[.70, .75)	3 (0)	.728	.333	5 (0)	.758	.600
>.75	2 (0)	.805	1.000	2 (0)	.805	1.000

the statistical algorithm relative to the betting line will win by more (or lose by less) than indicated by the betting line (given that the game does not end in a tie relative to the betting line). For each category, the table gives the total number of games or predictions, the number of games that actually ended in a tie relative to the betting line, the average of the conditional probabilities, and the observed frequency with which the teams favored by the statistical algorithm relative to the betting line actually won by more (or lost by less) than predicted by the line (excluding games that actually ended in a tie relative to the line). Cumulative figures, starting with the category corresponding to the highest conditional probabilities, are also given. Table 4b gives the same information for the 741 games of Weeks 6 through 13.

The motivation for the proposed betting scheme is a suspicion that the relative frequency with which the teams favored by the statistical algorithm relative to the line actually beat the line might be a strictly increasing (and possibly approximately linear) function of the "theoretical" frequency (approximate conditional probability), having a value of .50 at a theoretical relative frequency of .50. The results given in Table 4 tend to support this suspicion and to indicate that the rate of increase is greater for the midseason than for the entire season.

Fitting a linear function to overall observed relative frequencies by iterative weighted least squares produced the following equations:

$$\text{relative frequency} = .50 + .285\,(\text{theoretical frequency} - .50)\ .$$

The fitted equation for Weeks 6 through 13 was:

$$\text{relative frequency} = .50 + .655\,(\text{theoretical frequency} - .50)\ .$$

The addition of quadratic and cubic terms to the equations resulted in only negligible improvements in fit.

The proposed betting scheme would generally have shown a profit during the 1971–1977 period. The rate of profit would have depended on whether betting had been restricted to midseason games and on what theoretical frequency had been used as the cutoff point in deciding whether to place a bet. Even if bets (of equal size) had been placed on every one of the 1,320 games, some profit would have been realized (since the overall observed relative frequency was .528 vs. the break-even point of .524).

4. DISCUSSION

4.1 Modification of the Prediction Algorithm

One approach to improving the accuracy of the statistical predictions would be to modify the underlying model. In particular, instead of assuming that the residual effects are uncorrelated, we could, following Harville (1977b), assume that

$$R_{ik} = C_{h(j,k),j,w(j,k)} - C_{v(j,k),j,w(j,k)} + F_{jk}\ , \quad (4.1)$$

where, taking the weeks of each season to be numbered 1, 2, 3, ..., $w(j, k) = m$ if Game k of Year j were played during Week m. The quantities C_{imn} and F_{jk} represent random variables such that $E(C_{imn}) = E(F_{jk}) = 0$,

$$\operatorname{var}(F_{jk}) = \sigma_F^2\ , \quad \operatorname{cov}(F_{jk}, F_{j'k'}) = 0$$

if

$$j' \neq j \quad \text{or} \quad k' \neq k\ , \quad \operatorname{cov}(C_{imn}, F_{jk}) = 0\ ,$$

and

$$\begin{aligned} \operatorname{cov}(C_{imn}, C_{i'm'n'}) &= \alpha^{|n'-n|}\sigma_C^2\ , && \text{if } i'=i \text{ and } m'=m\ , \\ &= 0\ , && \text{if } i'\neq i \text{ or } m'\neq m\ . \end{aligned}$$

Here, σ_F^2, σ_C^2, and α are unknown parameters.

The correlation matrix of $C_{im1}, C_{im2}, \ldots$ is that for a first-order autoregressive process. The quantity C_{imn} can be interpreted as the deviation in the performance level of Team i in Week n of Year m from the level that is characteristic of Team i in Year m. The assumption (4.1) allows the weekly performance levels of any given team to be correlated to an extent that diminishes with elapsed time.

If the assumption (4.1) were adopted and positive values were used for α and σ_C^2, the effect on the statistical prediction algorithm would be an increased emphasis on the most recent of those games played in the year for which the prediction was being made. The games played early in that year would receive less emphasis.

The parameters σ_T^2, ρ, σ_C^2, α, and σ_F^2 associated with the modified model were actually estimated from the 1968–1976 NFL scores by an approximate restricted maximum likelihood procedure similar to that described in Section 2.3 for estimating parameters of the original model. As in Harville's (1977b) study of college football scores, there was no evidence that α differed from zero.

A second way to improve the accuracy of the statistical algorithm would be to supplement the information in the past scores with other quantitative information. A mixed linear model could be written for each additional variate. The random effects or the residual effects associated with each variate could be taken to be correlated with those for the other variates. At least in principle, the new variates could be incorporated into the prediction algorithm by following the same approach used in Section 2 in devising the original algorithm. In practice, depending on the number of variates that are added and the complexity of the assumed linear models, the computations could be prohibitive.

One type of additional variate would be the (signed) difference for each regular-season and playoff game between the values of any given statistic for the home team and the visiting team. For example, the yards gained by the home team minus the yards gained by the visiting team could be used. The linear model for a variate of this type could be taken to be of the same form as that applied to the difference in score.

There are two ways in which the incorporation of an additional variate could serve to improve the accuracy of the statistical prediction for the outcome S_{JK} of a future game. It could contribute additional information about the yearly characteristic performance levels $T_{h(J,K),J}$ and $T_{v(J,K),J}$ of the participating teams, or it could contribute information about the residual effect R_{JK}. Comparison of the estimates of σ_R^2 given in Table 1 with the figures given in Table 2 for average squared prediction error indicate that the first type of contribution is unlikely to be important, except possibly early in the season. Variates that quantify injuries are examples of variates that might contribute in the second way.

4.2 Other Approaches

The statistical prediction algorithm presented in Section 2 is based on procedures for mixed linear models described, for example, by Harville (1976, 1977a). These procedures were derived in a frequentist framework; however, essentially the same algorithm could be arrived at by an empirical Bayes approach like that described by Haff (1976) and Efron and Morris (1975) and used by the latter authors to predict batting averages of baseball players.

Essentially statistical algorithms for predicting the outcomes of NFL games were developed previously by Goode (as described in a nontechnical way by Marsha 1974), Rosner (1976), and Stefani (1977). Mosteller (1973) listed some general features that seem desirable in such an algorithm.

Comparisons of the results given in Section 3 with Stefani's results indicate that the predictions produced by the algorithm outlined in Section 2 tend to be more accurate than those produced by Stefani's algorithm. Moreover, Stefani reported that the predictions given by his algorithm compare favorably with those given by Goode's algorithm and with various other statistical and nonstatistical predictions.

There is some question whether it is possible for a bettor who takes an intuitive, essentially nonstatistical approach to beat the betting line (in the long run) more than 50 percent of the time. DelNagro (1975) reported a football prognosticator's claim that in 1974 he had made predictions for 205 college and NFL games relative to the betting line with 184 (89.8 percent) successes (refer also to Revo 1976); however, his claim must be regarded with some skepticism in light of subsequent well-documented failures (DelNagro 1977).

Winkler (1971) found that the collective rate of success of sportswriters' predictions for 153 college and NFL games was only 0.476, while Pankoff (1968), in a similar study, reported somewhat higher success rates.

Merchant (1973) followed the betting activities of two professional gamblers during the 1972 NFL season. He reported that they bet on 109 and 79 games and had rates of success of .605 and .567, respectively.

5. THE RATING PROBLEM

A problem akin to the football prediction problem is that of rating, ranking, or ordering the teams or a subset of the teams from first possibly to last. The rating may be carried out simply as a matter of interest, or it may be used to honor or reward the top team or teams.

The NFL has used what might be considered a rating system in picking its playoff participants. The NFL consists of two conferences, and each conference is divided into three divisions. Ten teams (eight before 1978) enter the playoffs: the team in each division with the highest winning percentage and the two teams in each conference that, except for the division winners, have the highest winning percentages. A tie for a playoff berth is broken in accordance with a complex formula. The formula is based on various statistics including winning percentages and differences in score for games between teams involved in the tie.

The prediction procedure described in Section 2 can also be viewed as a rating system. The ratings for a given year, say Year P, are obtained by ordering the teams in accordance with the estimates $\hat{T}_{1P}$, $\hat{T}_{2P}$, ... of their Year P characteristic performance levels. However, as a rating system, this procedure lacks certain desirable characteristics.

To insure that a rating system will be fair and will not affect the way in which the games are played, it should depend only on knowing the scores from the given season (Year P), should reward a team for winning per se, and should not reward a team for "running up the score" (Harville 1977b). The procedure in Section 2 can be converted into a satisfactory system by introducing certain modifications.

Define a "truncated" difference in score for Game k of Year P by

$$\begin{aligned} \dot{S}_{Pk}(M) &= M\ , && \text{if } S_{Pk} > M - \tfrac{1}{2}\ , \\ &= S_{Pk}\ , && \text{if } -(M - \tfrac{1}{2}) \le S_{Pk} \le M - \tfrac{1}{2}\ , \\ &= -M\ , && \text{if } S_{Pk} < -(M - \tfrac{1}{2})\ , \end{aligned}$$

where M is some number of points. Let

$$\begin{aligned} &\ddot{S}_{Pk}(M;\, T_{h(P,k),P},\, T_{v(P,k),P}) \\ &\quad = \bar{S}_{Pk}(M;\, T_{h(P,k),P},\, T_{v(P,k),P};\, \hat{\sigma}_R^2, \hat{\lambda}, \hat{H})\ , \quad (5.1) \end{aligned}$$

where $\bar{S}_{Pk}(M;\, T_{h(P,k),P},\, T_{v(P,k),P};\, \sigma_R^2, \lambda, H)$ is the conditional expectation (based on an assumption that S_{P1}, S_{P2}, ..., T_{1P}, T_{2P}, ... are jointly normal) of S_{Pk} given $\dot{S}_{P1}(M)$, $\dot{S}_{P2}(M)$, ..., T_{1P}, T_{2P}, ..., that is, given all of the available truncated differences in score and all of the characteristic performance levels for Year P, or, equivalently, given $\dot{S}_{Pk}(M)$, $T_{h(P,k),P}$, and $T_{v(P,k),P}$. Note that

$$\ddot{S}_{Pk}(M;\, T_{h(P,k),P},\, T_{v(P,k),P}) = S_{Pk} \quad \text{if} \quad |S_{Pk}| < M\ .$$

Let $\bar{T}_{iP}(\lambda, H;\, S_{P1}, S_{P2}, \ldots)$ represent the conditional expectation of T_{iP} given S_{P1}, S_{P2}, ... In the modified rating procedure for Year P, the ratings are obtained by ordering estimates $\hat{T}_{1P}$, $\hat{T}_{2P}$, ..., where these estimates are based only on differences in score from Year P and where, instead of putting $\hat{T}_{iP} = \bar{T}_{iP}(\hat{\lambda}, \hat{H};\, S_{P1}, S_{P2}, \ldots)$

(as we would with the procedure in Section 2), we put

$$\hat{T}_{iP} = \bar{T}_{iP}(\lambda, \hat{H}; \ddot{S}_{P1}(\hat{T}_{h(P,1),P}, \hat{T}_{v(P,1),P}), \ddot{S}_{P2}(\hat{T}_{h(P,2),P}, \hat{T}_{v(P,2),P}), \ldots) \quad (i = 1, 2, \ldots) \ . \qquad (5.2)$$

Here, $\ddot{S}_{Pk}(\hat{T}_{h(P,k),P}, \hat{T}_{v(P,k),P})$ is an "estimated difference in score" given by

$$\beta\ddot{S}(1; \hat{T}_{h(P,k),P}, \hat{T}_{v(P,k),P}) + (1 - \beta)\ddot{S}(U; \hat{T}_{h(P,k),P}, \hat{T}_{v(P,k),P}) \ , \qquad (5.3)$$

where U is some number of points (say $U = 21$) and β $(0 < \beta < 1)$ is some weight (say $\beta = \frac{1}{3}$) specified by the user.

Equations (5.2) can be solved iteratively for $\hat{T}_{1P}$, $\hat{T}_{2P}$, ... by the method of successive approximations. On each iteration, we first compute new estimates of T_{1P}, T_{2P}, ... by using the procedure in Section 2 with the current estimated differences in score in place of the "raw differences." Expressions (5.1) and (5.3) are then used to update the estimated differences in score.

The underlying rationale for the proposed rating system is essentially the same as that given by Harville (1977b) for a similar, but seemingly less satisfactory, scheme.

If the proposed rating system is to be accepted, it should be understandable to the public, at least in general terms. Perhaps (2.15) could be used as the basis for a fairly simple description of the proposed rating system. Our basic procedure is very similar to a statistical procedure developed by Henderson (1973) for use in dairy cattle sire selection. It is encouraging to note that it has been possible to develop an intuitive understanding of this procedure among dairy cattle farmers and to sell them on its merits.

It can be shown that taking $\hat{T}_{1P}$, $\hat{T}_{2P}$, ... to be as defined by (5.2) is equivalent to choosing them to maximize the function

$$\beta L_1(T_{1P}, T_{2P}, \ldots; \dot{S}_{P1}(1), \dot{S}_{P2}(1), \ldots; \hat{\sigma}_R^2, \lambda, \hat{H}) + (1 - \beta)L_U(T_{1P}, T_{2P}, \ldots; \dot{S}_{P1}(U), \dot{S}_{P2}(U), \ldots; \hat{\sigma}_R^2, \lambda, \hat{H})$$

where $L_M(T_{1P}, T_{2P}, \ldots; \dot{S}_{P1}(M), \dot{S}_{P2}(M), \ldots; \sigma_R^2, \lambda, H)$ is the logarithm of the joint probability "density" function of T_{1P}, T_{2P}, ..., $\dot{S}_{P1}(M)$, $\dot{S}_{P2}(M)$, ... that results from taking S_{P1}, S_{P2}, ..., T_{1P}, T_{2P}, ... to be jointly normal. When viewed in this way, the proposed rating system is seen to be similar in spirit to a system devised by Thompson (1975).

[Received July 1978. Revised June 1979.]

REFERENCES

DelNagro, M. (1975), "Tough in the Office Pool," *Sports Illustrated*, 43, 74–76.

——— (1977), "Cashing in a Sure Thing," *Sports Illustrated*, 47, 70–72.

Efron, Bradley, and Morris, Carl (1975), "Data Analysis Using Stein's Estimator and Its Generalizations," *Journal of the American Statistical Association*, 70, 311–319.

Haff, L.R. (1976), "Minimax Estimators of the Multinormal Mean: Autoregressive Priors," *Journal of Multivariate Analysis*, 6, 265–280.

Hartigan, J.A. (1969), "Linear Bayesian Methods," *Journal of the Royal Statistical Society*, Ser. B, 31, 446–454.

Harville, David A. (1976), "Extension of the Gauss-Markov Theorem to Include the Estimation of Random Effects," *Annals of Statistics*, 4, 384–395.

——— (1977a), "Maximum Likelihood Approaches to Variance Component Estimation and to Related Problems," *Journal of the American Statistical Association*, 72, 320–338.

——— (1977b), "The Use of Linear-Model Methodology to Rate High School or College Football Teams," *Journal of the American Statistical Association*, 72, 278–289.

——— (1978), "Football Ratings and Predictions Via Linear Models," with discussion by Carl R. Morris, *Proceedings of the American Statistical Association*, Social Statistics Section, 74–82; 87–88.

——— (1979), "Recursive Estimation Using Mixed Linear Models With Autoregressive Random Effects," in *Proceedings of the Variance Components and Animal Breeding Conference in Honor of Dr. C.R. Henderson*, Ithaca, N.Y.: Cornell University, Biometrics Unit.

Henderson, Charles R. (1973), "Sire Evaluation and Genetic Trends," in *Proceedings of the Animal Breeding and Genetics Symposium in Honor of Dr. Jay L. Lush*, Champaign, Ill.: American Society of Animal Science, 10–41.

Marsha, J. (1974), "Doing It by the Numbers," *Sports Illustrated*, 40, 42–49.

Merchant, Larry (1973), *The National Football Lottery*, New York: Holt, Rinehart & Winston.

Mosteller, Frederick (1970), "Collegiate Football Scores, U.S.A.," *Journal of the American Statistical Association*, 65, 35–48.

——— (1973), "A Resistant Adjusted Analysis of the 1971 and 1972 Regular Professional Football Schedule," Memorandum EX-5, Harvard University, Dept. of Statistics.

Pankoff, Lyn D. (1968), "Market Efficiency and Football Betting," *Journal of Business*, 41, 203–214.

Patterson, H.D., and Thompson, Robin (1971), "Recovery of Inter-Block Information When Block Sizes Are Unequal," *Biometrika*, 58, 545–554.

Revo, Larry T. (1976), "Predicting the Outcome of Football Games or Can You Make a Living Working One Day a Week," in *Proceedings of the American Statistical Association*, Social Statistics Section, Part II, 709–710.

Rosner, Bernard (1976), "An Analysis of Professional Football Scores," in *Management Science in Sports*, eds. R.E. Machol, S.P. Ladany, and D.G. Morrison, Amsterdam: North-Holland Publishing Co., 67–78.

Stefani, R.T. (1977), "Football and Basketball Predictions Using Least Squares," *IEEE Transactions on Systems, Man, and Cybernetics*, SMC-7, 117–121.

Thompson, Mark (1975), "On Any Given Sunday: Fair Competitor Orderings With Maximum Likelihood Methods," *Journal of the American Statistical Association*, 70, 536–541.

Winkler, Robert L. (1971), "Probabilistic Prediction: Some Experimental Results," *Journal of the American Statistical Association*, 66, 675–685.

Chapter 7

Data suggest that decisions to hire and fire kickers are often based on overreaction to random events.

The Best NFL Field Goal Kickers: Are They Lucky or Good?

Donald G. Morrison and Manohar U. Kalwani

The Question

In the September 7, 1992 issue of *Sports Illustrated* (SI) an article titled "The Riddle of the Kickers" laments the inability of teams to sign kickers who consistently make clutch field goals late in the game. Why do some kickers score better than others? In our article, we propose an answer to SI's "riddle." More formally, we ask the question: Do the observed data on the success rates across NFL field goal kickers suggest significant skill differences across kickers? Interestingly, we find that the 1989–1991 NFL field goal data are consistent with the hypothesis of no skill difference across NFL kickers. It appears then that, in searching for clutch field goal kickers, the NFL teams may be seeking a species that does not exist.

The Kicker and His Coach

In its February 18, 1992 issue, the *New York Times* reported that Ken Willis, the place kicker who had been left unsigned by the Dallas Cowboys, had accepted an offer of almost $1 million for two years to kick for Tampa Bay. The Cowboys, Willis's team of the previous season, had earlier agreed that they would pay him $175,000 for one year if Willis would not sign with another team during the Plan B free agency period. They were infuriated that Willis had gone back on his word.

> "That scoundrel," said Dallas Coach Jimmy Johnson, when informed of Willis's decision. "He broke his word."
> "I'm not disappointed as much about losing a player because we can find another kicker," Johnson told The Associated Press, "but I am disappointed that an individual compromised trust for money. When someone gives me their word, that's stronger to me than a contract."
> "I did give the Cowboys my word," said Willis, "but under the circumstances, to not leave would have been ludicrous."

Is Ken Willis really a "scoundrel"? Was Jimmy Johnson more

Photo courtesy of Philadelphia Soul, member of the Arena Football League © 2004.

of a scoundrel than his former kicker? We leave these questions to the reader. Rather, we focus on Johnson's assessment

"... we can find another kicker."

Just how differentiated are NFL kickers? Are some really better than others—or are they interchangeable parts? Inquiring (statistical) minds want to know!

Some Caveats

We begin with an observation: Some NFL kickers have stronger legs than others. Morton Andersen of the New Orleans Saints, for example, kicked a 60-yard field goal in 1991; virtually none of the other kickers even try one that long. His kickoffs also consistently go deep into the end zone for touchbacks; many kickers rarely even reach the goal line. Thus, we concede that, everything considered, some NFL kickers really are better than others.

The specific question we are asking, however, is: Given the typical length of field goal attempts, is there any statistical evidence of skill difference across kickers?

Before giving our statistical analysis, a few more anecdotes are in order.

Tony Z Has a Perfect Season!

In 1989, Tony Zendejas, kicking for Houston, with a success rate of 67.6%, was among the NFL's bottom fourth of 28 kickers. In 1990, he was at the very bottom with 58.3%. In 1991, though, Tony did something no other kicker in NFL history has ever done—he was successful on *all* of his field goal attempts! (Note: Zendejas was then kicking for the inept Los Angeles Rams, and he had only 17 attempts all season.) Early in the 1992 season, however, Tony missed three field goals *in one game*!

One Miss and He's History

In 1991, Brad Daluiso, a rookie out of UCLA, won the kicking job for the Atlanta Falcons. Brad made his first two kicks and missed a game-winning 27 yarder late in the second half. The next day, Daluiso was cut. Did the Falcons have enough "evidence" for firing Brad?

Portrait of a Slump?

Jeff Jaeger of the Los Angeles Raiders had a great 1991 season. He made 85.3% of his kicks—second only to Tony Zendejas's unprecedented perfect season. After 4 games in 1992, Jeff was 5 for 11 and the media were all over him for his "slump." Two of the misses, however, were blocked and not Jaeger's fault. Three of the misses were 48-, 51-, and 52-yard attempts—hardly "sure shots." In fact, of Jeff's 11 attempts, a badly hooked 29-yard miss was the only really poor kick. This is a slump?

Wandering in the NFL Wilderness

Finally, we give our favorite kicker story. Nick Lowery of the Kansas City Chiefs is sure to be in the Hall of Fame. He has made more field goals of 50 yards or more than anyone and holds the all-time high career field goal percentage—just under 80%. Were the Chiefs clever observers who saw Lowery kick in college and signed him before others discovered his talent? No, almost every team in the NFL saw him either as a teammate or an opponent. The road to NFL fame for Lowery was rocky indeed; he was cut 12 times by 9 different teams before landing his present decade-long job with Kansas City. Was this just luck for the Chiefs or did something else play a role?

Lucky or Good?

These anecdotes—and remember the plural of "anecdote" is not "data"—suggest that the observed performance of NFL kickers depends a lot more on luck than skill. (Scott Norwood, the Buffalo Bills, and the whole city of Buffalo would be different if Norwood's 47-yard game-ending 1991 Super Bowl field goal attempt had not been 2 feet wide to the right.) A simple correlation analysis for all of the 1989, 1990, and 1991 NFL data will demonstrate this "luck overwhelming skill" conjecture in a qualitative manner. Later, a more appropriate analysis will quantify these effects.

Year-to-Year Correlations

The Appendix gives the number of attempts and successes for all kickers for three seasons, beginning in 1989. These data come from an annual publication called the *Official National Football League Statistics.* (A call to the NFL Office in New York City is all that is required to receive a copy by mail.) For our first analysis, we calculate the correlation across kickers for each pair of years, where the X variable is percentage made in, say, 1989, and the Y variable is the percentage made in, say, 1990. We used only those kickers who "qualified," namely, who had attempted at least 16 field goals in both the 16-game NFL seasons of a pair. The three resulting correlations are as follows:

Pair of years	Correlation	p-value	Sample size
1989, 1990	−.16	.48	22
1990, 1991	+.38	.07	24
1989, 1991	−.53	.02	20

Two of the three correlations are negative, suggesting that an NFL kicker's year of above-average performance is at least as likely to be followed by a year of below-average performance rather than another year of above-average performance. In other words, NFL kickers' records do not exhibit consistency from one year to another.

Admittedly, this is a very *un*sophisticated analysis; the number of attempts can vary greatly between years. Some kickers may try lots of long field goals one year and mostly short ones the next year. Our next analysis will take these and other factors into account. However, this naive analysis has already let the cat out of the bag. The imperfect—but reasonable—performance measure of percentage made is a very poor predictor of this same measure in the following year or two.

An Illustrative Thought Experiment

Consider the following hypothetical experiment: We have 300 subjects each trying to kick 2 field goals from 30 yards away. Each subject is successful zero, one, or two times. Three possible scenarios for the results for all 300 are given in Table 1.

Scenario C—All Luck

Let's model each subject (kicker) as a Bernoulli process with some unobservable probability of successfully making each kick. Now recall what happens when you flip a fair coin ($p = .5$) twice. You will get zero, one, or two heads with probabilities .25, .50, and .25, respectively. Scenario C, therefore, is what we would expect if there were no skill differences across kickers and if each and every kicker had the same probability of success of $p = .5$. In Scenario C there is no skill difference—rather, the "0 for 2" kickers were "unlucky," whereas the "2 for 2's" were simply "lucky."

Scenario A—All Skill

The reader's intuition will lead to the obvious analysis of Scenario A. These data are consistent with half the kickers being perfect, that is, $p = 1$, and the other half being totally inept, with $p = 0$.

The intermediate Scenario B is consistent with the unobservable p-values being distributed uniformly between 0 and 1 across kickers. The spirit of our analysis on the NFL kicking data is to see which of these scenarios is most consistent with the data.

The Binomial All-Luck Benchmark

Assume, for illustration, every NFL kicker had the same number of attempts each year and all field goals were of the same length (as in our hypothetical example). Our analysis would proceed as follows: We would compute for each kicker, x_i, the number of successful kicks out of n, the common number of attempts. If the success rates, $\hat{p}_i = x_i/n$, for most kickers turn out to be close to $\bar{p}$, which is the average success rate across all kickers, we would have the analogue of Scenario C, that is, indicating very little skill difference. The observed data would be consistent with each kicker having the same (or very close to) common $\bar{p}$-value. As the data display greater than binomial variance across kickers (e.g., more toward Scenarios B and A), they would indicate more and more skill differences across kickers.

Table 1—Three Possible Results of a Hypothetical Test of 300 Field Goal Kickers

	Frequency distribution		
No. of successes	Scenario A	Scenario B	Scenario C
0	150	100	75
1	0	100	150
2	150	100	75
Total	300	300	300
Average percentages of successes	50	50	50

Note that all three scenarios yield an overall success rate of 50%.

A Beta Binomial Analysis

To formalize the argument just illustrated, we construct a probability model for the performance of field goal kickers. The assumptions of our model are:

1. With respect to each field goal attempt, each kicker is a Bernoulli process with an unobservable probability p of making each kick.
2. The unobservable p-value has a probability density function $f(p)$ across kickers.

The first assumption literally says that for a given kicker all field goals are equally difficult and whether or not he makes or misses a kick does not affect the probability of making any future kicks. The second assumption merely allows skill differences across kickers.

All Field Goals—or Segmented by Distance?

Table 2 displays the average proportions of field goals made by the NFL kickers during the 1989, 1990, and 1991 seasons. A quick look at the figures shows that short field goals are made over 90% of the time, whereas long ones are made less than 60% of the time. Thus, given the yardage of the attempts made by a kicker, he is not a Bernoulli process, that is, the probability of success is not constant. But if we only know that a kick is attempted, we would have to weight all of these yardage-dependent p-values by the probability of each particular yardage attempt. This would give a common overall (weighted) p-value for each attempt. Thus, knowing only that a kicker tried, say, 30 field goals, the number made would have a binomial distribution, with n = 30, and this overall p-value. This is the context in which we model each kicker as a Bernoulli process for all kicks, irrespective of the distance from the goal posts at which the field goal attempt is made.

We also do separate analyses by yardage groups, namely, under 29 yards, 30–39 yards, and 40–49 yards. In these analyses, to the extent we have reduced the effect of the varying field goal lengths across kickers, differences in success proportions are more likely to be due to skill differences.

Beta Skill Variability

Because of its flexibility to describe each of the anticipated scenarios, we use the beta distribution to represent the heterogeneity in the p-values (or true probabilities of success) across kickers (see sidebar). For example, the beta distribution can be chosen to represent the situation charac-

Table 2—NFL Field Goal Kickers' Success Proportions

	Season over 3 years			
	1989	1990	1991	Aggregate
All kicks	.731	.749	.737	.740
< 29 yards	.945	.958	.937	.950
30–39 yards	.801	.795	.787	.793
40–49 yards	.563	.635	.578	.587

terized by Scenario C, in which all kickers have similar abilities. Or, it can be chosen to represent Scenario A, in which there are two types of kickers—good and bad. Conveniently, it turns out that a parameter of the beta distribution, namely, the polarization index, φ, can serve as an indicator of the amount and nature of heterogeneity in the *p*-values across kickers.

Lucky or Good Redux

Given our very reasonable assumptions of Bernoulli kickers with the probabilities of success having a beta distribution across kickers, all we have to do is estimate the parameters, μ (the mean success rate), and the polarization index, φ (a measure of variability), of the distribution to completely tell our story:

$$\mu = \frac{\alpha}{\alpha+\beta}, \quad \varphi = \frac{1}{\alpha+\beta+1}$$

For each analysis, μ will give the average skill level (e.g., the mean *p*-value) and φ will say how much of the observed variability in performance is due to skill. It is all skill when $\varphi = 1$, and all luck at the other extreme when $\varphi = 0$.

Results

We report the maximum likelihood estimates of μ and φ for the field goal data from the 1989, 1990, and 1991 NFL seasons. Table 3 contains the results, which are very compelling and tell a very simple story. (Please recall our earlier caveat about the distance dimension of these kickers.) For the field goals that are attempted, the data overwhelmingly support the hypothesis of no skill differences across the elite group of NFL kickers. Half of the analyses have $\varphi = 0$. [In these cells, the data show slightly less variability than would be expected under a homogeneous (no skill difference) Bernoulli population of kickers.] The other half of the cells have positive, but very tiny, φ values; for example, they are all less than .03.

Table 3–Mean and Polarization for NFL Field Goals

Year	Distance	Maximum likelihood estimates μ	φ
1989	All	.732	0
	<29 yards	.943	0
	30–39 yards	.804	0
	40–49 yards	.566	0
1990	All	.748	0
	<29 yards	.958	.026
	30–39 yards	.799	.002
	40–49 yards	.638	0
1991	All	.737	.003
	<29 yards	.936	.027
	30–39 yards	.788	.024
	40–49 yards	.579	.005
Aggregate	All	.740	0
	<29 yards	.950	.003
	30–39 yards	.793	0
	40–49 yards	.587	0

Table 3 also displays the maximum likelihood estimates of μ and φ for field goal data aggregated across each kicker for the 1989, 1990, and 1991 seasons. The total number of field goals made and attempted, and the corresponding success rates are included in the Appendix. For the 38 kickers who kicked 16 or more field goals during at least 1 of the 3 seasons, the numbers of field goals attempted varied from 18 to 123, with an average of about 62. As Table 3 reveals, even in these aggregate data with larger sample size, the estimates of φ are very close to 0 for all kicks or kicks segmented by field goal length. These findings from aggregate data provide further support for our inference of a lack of skill differences among the NFL kickers.

Reliability of Model Results

Our inference of little, if any, skill differences among the elite group of NFL kickers relies on our estimate of the polarization index, φ, being 0 or close to it in almost all the cases analyzed. The question arises: How accurate are our estimates of φ, particularly since our sample sizes are not large? The number of qualified NFL kickers (with an average of at least one kick per game) in any given year is about 28. These kickers on the average attempt about 28 field goals over the course of a 16-game season. Our maximum likelihood estimators of φ and μ have good statistical properties but only when the sample sizes are large. Simulation results indicate that the sample sizes are adequate in the present problem for us to have great confidence in these results.

Should We Be Surprised?

If the readers of this article were lined up to attempt 20 yard field goals, we would find big skill

The Beta Distribution

The beta distribution is frequently used to describe the distribution of probabilities across a population. Areas of application include biometrics, genetics, market research, opinion research, and psychometrics. The functional form of the beta distribution is given by

$$f(p;\alpha,\beta) = (\Gamma(\alpha+\beta)/\Gamma(\alpha)\Gamma(\beta))\,p^{\alpha-1}(1-p)^{\beta-1},\quad 0<p<1,\ \alpha,\beta \geq 0$$

where p denotes an individual kicker's true probability of success, a and b are parameters of the beta distribution, and $\Gamma()$ denotes the gamma function. The mean and variance of this distribution are

$$E[p] = \alpha/\alpha+\beta \text{ and } var[p] = \alpha\beta/(\alpha+\beta+1)(\alpha+\beta)^2$$

The beta distribution is flexible and can take different shapes depending on the values of the parameters α and β:

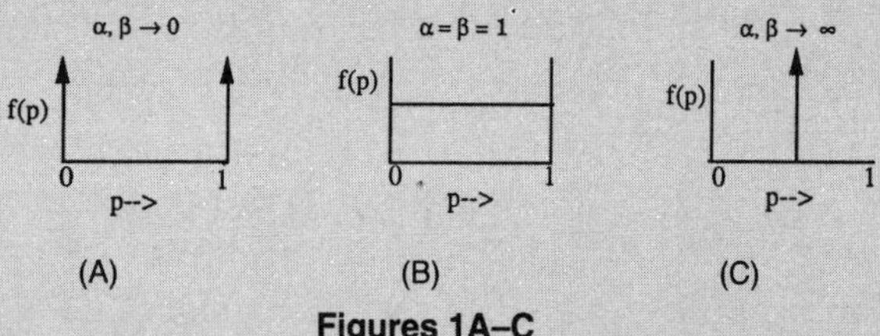

Figures 1A–C

The bimodal form of the beta distribution (an extreme version of which is displayed in Fig. 1A would be appropriate if the NFL kickers could be classified as either very good or very poor. The bell-shaped form of the beta distribution (again, an extreme form of which is shown in Fig. 1C) would imply that most NFL kickers have very similar true probabilities of success. The intermediate case displayed in Fig. 1B would imply that the true probabilities of success of the NFL kickers are distributed uniformly between 0 and 1. The three different shapes depicted in Figs. 1A, 1B, and 1C, of course, correspond to the Scenarios A, B, and C presented in the text.

Often the beta distribution is expressed in terms of the parameters $\mu = \alpha/(\alpha+\beta)$, the mean success rate across the population, and $\varphi = 1/(\alpha+\beta+1)$, a polarization index. It turns out that the variance of the beta distribution can be written as

$$var[p] = \varphi\,\mu\,(1-\mu)$$

Thus, when $\varphi = 0$, there is zero variance in the p-values across kickers, implying no skill differences (see Fig. 1C). When $\varphi = 1$, the variance of p (given a mean, μ) is the maximum possible value of $\mu(1-\mu)$ (see Fig. 1A).

Working with the Beta Binomial Model

The parameters μ and φ can be estimated by either the method of moments or the maximum likelihood approach; we used both approaches in fitting the model to the field goal data. Those interested in more details should read Colombo and Morrison (1988), which contains an analogous application of the beta binomial model to the success or failure of British Ph.D. students across universities. The appendix of Colombo and Morrison gives the formulas for the maximum likelihood estimators of μ and φ. To obtain method-of-moment estimates of μ and φ, we used an iterative approach due to Kleinman (1973). Reassuringly, they turned out to be very close to the maximum likelihood estimates of μ and φ in all the 12 cases considered in the text.

How good are the estimates in small samples? Evidence on the reliability of Kleinman's moment estimators in small samples is available in the simulation results of Lim (1992). The simulations were carried out in settings similar to our field goal data, in that the number of sampling units (kickers, in our case) was set at 30 and the number of trials per sampling unit was allowed to vary between 20 and 40, with a mean of about 30 trials. The value of the μ parameter was set at 0.75 and the polarization index φ was allowed to vary between 0.25 to 0.75. In these simulations, Lim was able to recover the polarization index parameter within 10% of the true value 95% of the time. In sum, we feel comfortable about the reliability of our estimates of φ for the field goal data because the maximum likelihood estimates and Kleinman's moment estimates agree, and Lim's simulation results suggest that the sample sizes are adequate for the moment estimators.

differences. But we analyzed the "best of the best." Each year thousands of high school kickers get filtered into a few hundred college kickers. These few hundred then compete for the 28 NFL kicking jobs. No one can make every kick, but these guys come close. Even if some of these elite kickers are a little better than others, variability in performance is caused by a poor snap from center, an imperfect hold, a gust of wind, and so on. Upon reflection, we would be surprised if the results had shown even moderate skill differences. Undoubtedly, some small skill differences do exist across these NFL kickers. However, over the course of a season or two, there simply are not enough field goal attempts to separate the best of the kickers from the remaining NFL kickers.

Discussion

In the fall of 1989, only 8 of the 28 NFL kickers were kicking for the team that originally signed them. We have already documented Nick Lowery's odyssey through the NFL training camps. Matt Bahr has kicked for two Super Bowl winners, the Steelers and Giants, going from Pittsburgh to New York via the Cleveland Browns. What accounts for this mobility? Over a long career, an 80% kicker is great, a 70% kicker is a little below average, and a 60% kicker is terrible. But in one season each kicker makes about 30 attempts—sometimes many less. A $p = .7$ kicker has an expected value of 21 successes (70%) out of 30 kicks, but 18 successes (60%) and 24 successes (80%) are only slightly more than one standard deviation from the mean. Obviously, a good kicking coach can assess a kicker's skill level by watching *how* he kicks as well as seeing whether or not the kick went through the uprights. Nevertheless, it is our conjecture that kickers are very often hired and fired based solely on binomial luck variance. Our advice to NFL kickers is: Rent—don't buy.

We conclude by returning to the Jimmy Johnson/Ken Willis episode. If the Tampa Bay team had had access to this article, would they have paid Willis $500,000 a year to kick for them? Probably. Most coaches keep searching for that elusive kicker who will never miss in crunch time, and the Bucs also liked the strength of Willis's leg (he made more 50+ yarders than anyone in 1991). Also, getting "one up" on Jimmy Johnson must have pleased Tampa Bay. So, although this article may not have changed Tampa Bay's behavior, from a normative point of view, we think the Bucs made a financial mistake. Jimmy Johnson, after all, was correct when he said "we can find another kicker."

So to the ethics of Coach Johnson's statement, "He broke his word," we can only note that Johnson himself was not acting in the spirit of Plan B free agency. Was Ken Willis smart to take the money and run? Well, when you are competing against kicking colleagues, all of whom are essentially interchangeable parts, and an owner offers to triple your salary . . . Willis made the right call. (We just hope he did not sign more than a two-year lease.)

APPENDIX

NFL FIELD GOALS DATA
(Source: Elias Sports Bureau)

OBS. #	LEAGUE	KICKER	1989 STATISTICS FG	1989 FGA	1989 PROP.	1990 STATISTICS FG	1990 FGA	1990 PROP.
1	A	Zandejas,T.	25	37	0.676	.	.	.
2	A	Staurovsky	14	17	0.824	16	22	0.727
3	A	Treadwell	27	33	0.818	25	34	0.735
4	A	Biasucci	21	27	0.778	17	24	0.708
5	A	Norwood	23	30	0.767	20	29	0.690
6	A	Stoyanovich	19	26	0.731	21	25	0.840
7	A	Lowery	24	33	0.727	34	37	0.919
8	A	Anderson	21	30	0.700	20	25	0.800
9	A	Bahr-C.	17	25	0.680	.	.	.
10	A	Jaeger	23	34	0.676	15	20	0.750
11	A	Bahr-M.	16	24	0.667	17	23	0.739
12	A	Leahy	14	21	0.667	23	26	0.885
13	A	Johnson	15	25	0.600	23	32	0.719
14	N	Murray	20	21	0.952	13	19	0.684
15	N	Cofer	29	36	0.806	24	36	0.667
16	N	Karlis	31	39	0.795	.	.	.
17	N	Butler	15	19	0.789	26	37	0.703
18	N	Igwebuiki	22	28	0.786	14	16	0.875
19	N	Jacke	22	28	0.786	23	30	0.767
20	N	Allegre	20	26	0.769	.	.	.
21	N	Lansford	23	30	0.767	15	24	0.625
22	N	McFadden	15	20	0.750	.	.	.
23	N	Lohmiller	29	40	0.725	30	40	0.750
24	N	DelGreco	18	26	0.692	17	27	0.630
25	N	Andersen	20	29	0.690	21	27	0.778
26	AN	Davis	23	34	0.676	22	33	0.667
27	AN	Ruzek	13	22	0.591	21	29	0.724
28	AN	Zandejas,L.	14	24	0.583	.	.	.
29	A	Carney	.	.	.	19	21	0.905
30	A	Breech	.	.	.	17	21	0.810
31	A	Garcia	.	.	.	14	20	0.700
32	A	Kauric	.	.	.	14	20	0.700
33	N	Christie	.	.	.	23	27	0.852
34	N	Willis	.	.	.	18	25	0.720
35	N	Reveiz	.	.	.	13	19	0.684
36	A	Kasay	.	.	.	.	.	.
37	A	Stover	.	.	.	.	.	.
38	A	Howfield	.	.	.	.	.	.

Epilogue

As the review process for this article was concluding, the NFL office released the 1992 data. The means (μ), polarization indices (φ), and correlations for 1992 are all consistent with the 1989–1991 results. The only slight de-

1991 STATISTICS			AGGREGATE STATISTICS		
FG	FGA	PROP.	FG	FGA	PROP.
17	17	1.000	42	54	0.778
13	19	0.684	43	58	0.741
27	36	0.750	79	103	0.767
15	26	0.577	53	77	0.688
18	29	0.621	61	88	0.693
31	37	0.838	71	88	0.807
25	30	0.833	83	100	0.830
23	33	0.697	64	88	0.727
.	.	.	17	25	0.680
29	34	0.853	67	88	0.761
22	29	0.759	55	76	0.724
26	37	0.703	63	84	0.750
19	23	0.826	57	80	0.713
19	28	0.679	52	68	0.765
14	28	0.500	67	100	0.670
.	.	.	31	39	0.795
19	29	0.655	60	85	0.706
.	.	.	36	44	0.818
18	24	0.750	63	82	0.768
.	.	.	20	26	0.769
.	.	.	38	54	0.704
.	.	.	15	20	0.750
31	43	0.721	90	123	0.732
.	.	.	35	53	0.660
25	32	0.781	66	88	0.750
21	30	0.700	66	97	0.680
28	33	0.848	62	84	0.738
.	.	.	14	24	0.583
19	29	0.655	38	50	0.760
23	29	0.793	40	50	0.800
.	.	.	14	20	0.700
.	.	.	14	20	0.700
15	20	0.750	38	47	0.809
27	39	0.692	45	64	0.703
17	24	0.708	30	43	0.698
25	31	0.806	25	31	0.806
16	22	0.727	16	22	0.727
13	18	0.722	13	18	0.722

viation is $\mu = .910$ and $\varphi = .046$ for the field goals of 29 yards or less. Although still very small, this polarization index is the highest in the whole study. The success rate of 91% for these short kicks is about three points below the other years. It turns out that one kicker, Greg Davis of Phoenix, caused most of these deviations. In the 3 previous years, Greg was a perfect 18 for 18 from 29 yards or less. In 1992, he missed 4 out of 10 of these short kicks. (We expect to see a lot of kickers in the Phoenix training camp this summer.) The spirit of the 1989–1991 results is clearly maintained in 1992.

So what do we, the authors, conclude? The data are consistent with no skill differences across NFL field goal kickers. Have we proved no skill difference? No, but if there is some true skill difference, it is certainly small compared to the within-kicker binomial variance. Do we believe some kickers are better than others? Yes. We would like to have either the veteran Morton Andersen or the rookie Jason Hanson kicking for our team—but mostly because of how far they kick compared to the typical NFL kicker. When it comes to accuracy per se for the typical attempt of less than 50 yards, the addition of the 1992 results only reinforces our belief that the NFL caliber kickers are, indeed, interchangeable parts. There are certainly numerous coaches, fans, and especially kickers who will disagree with us. But with the data so overwhelmingly on our side, the burden of proof would appear to be on those who disagree with us.

Additional Reading

Colombo, R.A., and Morrison, D.G. (1988), "Blacklisting Social Sciences Departments With Poor Ph.D. Submission Rates," *Management Science*, 34, 696–706.

Irving, G. W., and Smith, H. A. (1976), "A Model of Football Field Goal Kicker," in *Management Science in Sports*, TIMS Studies in the Management Sciences, eds. R.E. Machol and S.P. Ladany, Amsterdam: North-Holland, Vol. 4, pp. 47–58.

Kleinman, J. C. (1973), "Proportions With Extraneous Variance: Single and Independent Samples," *Journal of the American Statistical Association*, 68, 46–54.

Lim, B. (1992), "The Application of Stochastic Models to Study Buyer Behavior in Consumer Durable Product Categories," Ph.D. dissertation, Purdue University, Krannert Graduate School of Management.

Chapter 8

On the Probability of Winning a Football Game

HAL STERN*

Based on the results of the 1981, 1983, and 1984 National Football League seasons, the distribution of the margin of victory over the point spread (defined as the number of points scored by the favorite minus the number of points scored by the underdog minus the point spread) is not significantly different from the normal distribution with mean zero and standard deviation slightly less than fourteen points. The probability that a team favored by p points wins the game can be computed from a table of the standard normal distribution. This result is applied to estimate the probability distribution of the number of games won by a team. A simulation is used to estimate the probability that a team qualifies for the championship playoffs.

KEY WORDS: Goodness-of-fit tests; Normal distribution.

1. INTRODUCTION

The perceived difference between two football teams is measured by the point spread. For example, New York may be a three-point favorite to defeat Washington. Bets can be placed at fair odds (there is a small fee to the person handling the bet) on the event that the favorite defeats the underdog by more than the point spread. In our example, if New York wins by more than three points, then those who bet on New York would win their bets. If New York wins by less than three points (or loses the game), then those who bet on New York would lose their bets. If New York wins by exactly three points then no money is won or lost. The point spread is set so that the amount bet on the favorite is approximately the same as the amount bet against the favorite. This limits the risk of the people who handle the bets.

The point spread is of natural interest as a predictor of the outcome of a game. Although it is not necessarily an estimate of the difference in scores, the point spread has often been used in this capacity. Pankoff (1968), Vergin and Scriabin (1978), Tryfos, Casey, Cook, Leger, and Pylypiak (1984), Amoako-Adu, Marmer, and Yagil (1985), and Zuber, Gandar, and Bowers (1985) considered statistical tests of the relationship between the point spread and the outcome of the game. Due to the large variance in football scores, they typically found that significant results (either proving or disproving a strong relationship) are difficult to obtain. Several of these authors then searched for profitable wagering strategies based on the point spread. The large variance makes such strategies difficult to find. Other authors (Thompson 1975; Stefani 1977, 1980; Harville 1980) attempted to predict game outcomes or rank football teams using information other than the point spread.

The results of National Football League (NFL) games seem to indicate that the true outcome of a game can be modeled as a normal random variable with mean equal to the point spread. This approximation is developed in some detail, and two applications of this approach are described.

*Hal Stern is Assistant Professor, Department of Statistics, Harvard University, Cambridge, MA 02138. The author thanks Thomas M. Cover and a referee for helpful comments on the development and presentation of this article.

2. DATA ANALYSIS

The data set consists of the point spread and the score of each NFL game during the 1981, 1983, and 1984 seasons. Many newspapers list the point spread each day under the heading "the betting line." The sources of the point spread for this data set are the *New York Post* (1981) and the *San Francisco Chronicle* (1983, 1984). There is some variability in the published point spreads (from day to day and from newspaper to newspaper), however, that variability is small (typically less than one point) and should not have a large impact on the results described here. An attempt was made to use point spreads from late in the week since these incorporate more information (e.g., injuries) than point spreads from early in the week. For reasons of convenience, the day on which the data were collected varied between Friday and Saturday. The 1982 results are not included because of a player's strike that occurred that year. The total number of games in the data set is 672. More recent data (from 1985 and 1986) are used later to validate the results of this section. For each game the number of points scored by the favorite (F), the number of points scored by the underdog (U), and the point spread (P) are recorded. The margin of victory over the point spread (M) is defined by

$$M = F - U - P$$

for each game. The distribution of M is concentrated on multiples of one-half since F and U are integers, while P is a multiple of one-half.

A histogram of the margin of victory over the point spread appears in Figure 1. Each bin of the histogram has a width of 4.5 points. The chi-squared goodness-of-fit test indicates that the distribution of M is not significantly different from a Gaussian distribution with mean zero and standard deviation 13.86 (computed from the data). The sample mean of M is .07. This has been rounded to zero because it simplifies the interpretation of the formula for the probability of winning in the next section. All observations of M larger than 33.75 in magnitude are grouped together, leading to a chi-squared test on 17 bins. The chi-squared test statistic is 15.05, between the .5 and .75 quantile of the limiting chi-squared

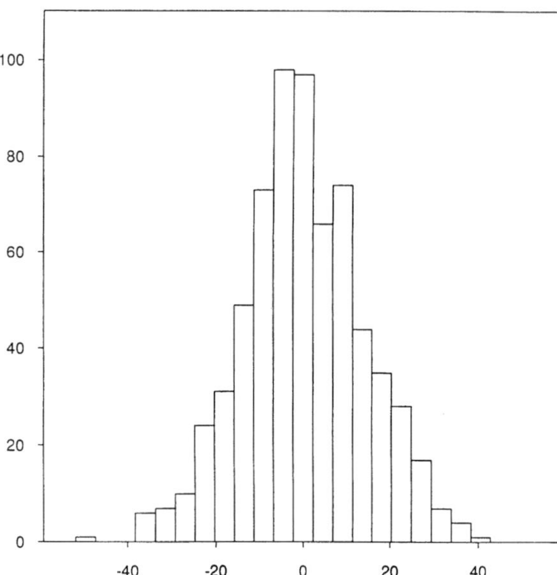

Figure 1. Histogram of the Margin of Victory Over the Point Spread (M). Goodness-of-fit tests indicate that the distribution of M is approximately Gaussian.

distribution (14 degrees of freedom—17 bins with two estimated parameters). The hypothesis of normality is also consistent with histograms having larger and smaller bin widths. Naturally the normal distribution is just an approximation. The variable M is concentrated on multiples of one half, and integer values occur twice as often as noninteger values. This would not be the case if normality provided a more exact fit.

The Kolmogorov–Smirnov test is more powerful than the chi-squared test. The value of this test statistic is .913. Since the parameters of the normal distribution have been estimated from the data, the usual Kolmogorov–Smirnov significance levels do not apply. Using tables computed by Lilliefors (1967), we reject normality at the .05 significance level but not at the .01 significance level. This test is sensitive to the fact that the mode of the data does not match the mode of the normal distribution. We continue with the normal approximation despite this difference.

The results of the 1985 and 1986 seasons, collected after the initial analysis, provide additional evidence in favor of the normal approximation. The chi-squared statistic, using the parameters estimated from the 1981–1984 data, is 16.96. The p value for the chi-squared test is larger than .25. The Kolmogorov–Smirnov test statistic is .810, indicating a better fit than the original data set (the p value is approximately .10). More recent data may be used to verify that the approximation continues to hold.

3. THE PROBABILITY OF WINNING A GAME

What is the probability that a p-point favorite wins a football game? The natural estimate is the proportion of p-point favorites in the sample that have won their game. This procedure leads to estimates with large standard errors because of the small number of games with any particular point spread. The normal approximation of the previous section can be used to avoid this problem.

The probability that a team favored by p points wins the game is

$$\Pr(F > U \mid P = p) = \Pr(F - U - P > -P \mid P = p).$$

The argument in Section 2 shows that $M = F - U - P$ is approximately normal. A more detailed analysis indicates that normality appears to be a valid approximation for $F - U - P$ conditional on each value of P. This is difficult to demonstrate since there are few games with any particular value of P. A series of chi-squared tests were performed for games with similar point spreads. The smallest sample size was 69 games; the largest was 112 games. Larger bins were used in the chi-squared test (a bin width of 10.5 points instead of the 4.5 points used in Fig. 1) because of the size of the samples. Neighboring bins were combined so that each bin had an expected count of at least five. None of the eight tests was significant; the smallest p value was greater than .10. These tests seem to indicate that normality is an adequate approximation for each range of point spreads. If we apply normality for a particular point spread, p, then $F - U$ is approximately normal with mean p and standard deviation 13.861. The probability of winning a game is then computed as

$$\Pr(F > U \mid P = p) = 1 - \Phi\left(-\frac{p}{13.861}\right)$$
$$= \Phi\left(\frac{p}{13.861}\right),$$

where $\Phi(\cdot)$ is the cumulative distribution function of the standard normal random variable.

The normal approximation for the probability of victory is given for some sample point spreads (the odd numbers) in Table 1. The observed proportion of p-point favorites that won their game, $\hat{P}$, and an estimated standard error are also computed. The estimates from the normal formula are consistent with the estimates made directly from the data. In addition, they are monotone increasing in the point spread. This is consistent with the interpretation of the point spread as a measure of the difference between two teams. The empirical estimates do not have this property. A linear approximation to the probability of winning is

$$\Pr(F > U \mid P = p) \approx .50 + .03p.$$

This formula is accurate to within .0175 for $|p| < 6$.

Table 1. The Normal Approximation and the Empirical Probability of Winning

Point spread	Pr(F > U | P)	$\hat{P}$	Standard error
1	.529	.571	.071
3	.586	.582	.055
5	.641	.615	.095
7	.693	.750	.065
9	.742	.650	.107

4. APPLICATIONS

Conditional on the value of the point spread, the outcome of each game (measured by $F - U$) can be thought of as the sum of the point spread and a zero-mean Gaussian random variable. This is a consequence of the normal distribution of M. We assume that the zero-mean Gaussian random variables associated with different games are independent. Although successive football games are almost certainly not independent, it seems plausible that the random components (performance above or below the point spread) may be independent. The probability of a sequence of events is computed as the product of the individual event probabilities.

For example, the New York Giants were favored by two points in their first game and were a five-point underdog in their second game. The probability of winning both games is $\Phi(2/13.861)\Phi(-5/13.861) = .226$. Adding the probabilities for all $\binom{16}{k}$ sequences of game outcomes that have k wins leads to the probability distribution in Table 2. The point spreads used to generate Table 2 are:

$$2, -5, -6, 6, -3, -3.5, -5, 0, \quad -6, -7, 3, -1, 7, 3.5, -4, 9.$$

The Giants actually won nine games. This is slightly higher than the mean of the distribution, which is 7.7. Since this is only one observation, it is difficult to test the fit of the estimated distribution.

We use the results of all 28 teams over three years to assess the fit of the estimated distribution. Let

$p_{ij}(x)$ = probability that team i wins x games during season j for $x = 0, \ldots, 16$

$$F_{ij}(x) = \text{estimated cdf for the number of wins by team } i \text{ during season } j = \sum_{t \le x} p_{ij}(t),$$

Table 2. Distribution of the Number of Wins by the 1984 New York Giants

Number of wins	Probability
0	.0000
1	.0002
2	.0020
3	.0099
4	.0329
5	.0791
6	.1415
7	.1928
8	.2024
9	.1642
10	.1028
11	.0491
12	.0176
13	.0046
14	.0008
15	.0001
16	.0000

and

X_{ij} = observed number of wins for team i during season j

for $i = 1, \ldots, 28$ and $j = 1, 2, 3$. The index i represents the team and j the season (1981, 1983, or 1984). The distribution $p(\cdot)$ and the cdf $F(\cdot)$ represent the distribution of the number of wins when the normal approximation to the distribution of M is applied. Also, let U_{ij} be independent random variables uniformly distributed on (0, 1). According to a discrete version of the probability integral transform, if $X_{ij} \sim F_{ij}$, then $F_{ij}(X_{ij}) - U_{ij}p_{ij}(X_{ij})$ has the uniform distribution on the interval (0, 1). The U_{ij} represent auxiliary randomization needed to attain the uniform distribution. A chi-squared test is used to determine whether the transformed X_{ij} are consistent with the uniform distribution and therefore determine whether the X_{ij} are consistent with the distribution $F_{ij}(\cdot)$. The chi-squared statistic is computed from 84 observations grouped into 10 bins between 0 and 1. Four different sets of uniform variates were used, and in each case the data were found to be consistent with the uniform distribution. The maximum observed chi-squared statistic in the four trials was 13.1, between the .75 and the .90 quantiles of the limiting distribution. The actual records of NFL teams are consistent with predictions made using the normal approximation for the probability of winning a game.

Using the point spreads of the games for an entire season as input, it is possible to determine the probability of a particular outcome of the season. This type of analysis is necessarily retrospective since the point spreads for the entire season are not available until the season has been completed. To find the probability that a particular team qualifies for the postseason playoffs, we could consider all possible outcomes of the season. This would involve extensive computations. Instead, the probability of qualifying for the playoffs is estimated by simulating the NFL season many times. In a simulated season, the outcome of each game is determined by generating a Bernoulli random variable with probability of success determined by the point spread of that game. For each simulated season, the 10 playoff teams are determined. Six playoff teams are determined by selecting the teams that have won each of the six divisions (a division is a collection of four or five teams). The winning team in a division is the team that has won the most games. If two or more teams in a division are tied, then the winner is selected according to the following criteria: results of games between tied teams, results of games within the division, results of games within the conference (a collection of divisions), and finally random selection. It is not possible to use the scores of games, since scores are not simulated. Among the teams in each conference that have not won a division, the two teams with the most wins enter the playoffs as "wildcard" teams (recently increased to three teams). Tie-breaking procedures for wildcard teams are similar to those mentioned above.

The 1984 NFL season has been simulated 10,000 times. For each team, the probability of winning its division

Table 3. Results of 10,000 Simulations of the 1984 NFL Season

Team	Pr(win division)	Pr(qualify for playoffs)	1984 actual result
National Conference—Eastern Division			
Washington	.5602	.8157	division winner
Dallas	.2343	.5669	
St. Louis	.1142	.3576	
New York	.0657	.2291	wildcard playoff team
Philadelphia	.0256	.1209	
National Conference—Central Division			
Chicago	.3562	.4493	division winner
Green Bay	.3236	.4170	
Detroit	.1514	.2159	
Tampa Bay	.1237	.1748	
Minnesota	.0451	.0660	
National Conference—Western Division			
San Francisco	.7551	.8771	division winner
Los Angeles	.1232	.3306	wildcard playoff team
New Orleans	.0767	.2291	
Atlanta	.0450	.1500	
American Conference—Eastern Division			
Miami	.7608	.9122	division winner
New England	.1692	.4835	
New York	.0607	.2290	
Buffalo	.0051	.0248	
Indianapolis	.0042	.0205	
American Conference—Central Division			
Pittsburgh	.4781	.5774	division winner
Cincinnati	.3490	.4574	
Cleveland	.1550	.2339	
Houston	.0179	.0268	
American Conference—Western Division			
Los Angeles	.4555	.7130	wildcard playoff team
Seattle	.2551	.5254	wildcard playoff team
Denver	.1311	.3405	division winner
San Diego	.1072	.2870	
Kansas City	.0511	.1686	

has been computed. The probability of being selected for the playoffs has also been determined. The results appear in Table 3. Each estimated probability has a standard error that is approximately .005. Notice that over many repetitions of the season, the eventual Super Bowl champion San Francisco would not participate in the playoffs approximately 12% of the time.

5. SUMMARY

What is the probability that a team favored to win a football game by p points does win the game? It turns out that the margin of victory for the favorite is approximated by a Gaussian random variable with mean equal to the point spread and standard deviation estimated at 13.86. The normal cumulative distribution function can be used to compute the probability that the favored team wins a football game. This approximation can also be used to estimate the distribution of games won by a team or the probability that a team makes the playoffs. These results are based on a careful analysis of the results of the 1981, 1983, and 1984 National Football League seasons. More recent data (1985 and 1986) indicate that the normal approximation is valid outside of the original data set.

[*Received June 1989. Revised December 1989.*]

REFERENCES

Amoako-Adu, B., Marmer, H., and Yagil, J. (1985), "The Efficiency of Certain Speculative Markets and Gambler Behavior," *Journal of Economics and Business*, 37, 365–378.

Harville, D. (1980), "Predictions for National Football League Games via Linear-Model Methodology," *Journal of the American Statistical Association*, 75, 516–524.

Lilliefors, H. W. (1967), "On the Kolmogorov–Smirnov Test for Normality With Mean and Variance Unknown," *Journal of the American Statistical Association*, 62, 399–402.

Pankoff, L. D. (1968), "Market Efficiency and Football Betting," *Journal of Business*, 41, 203–214.

Stefani, R. T. (1977), "Football and Basketball Predictions Using Least Squares," *IEEE Transactions on Systems, Man, and Cybernetics*, 7, 117–121.

——— (1980), "Improved Least Squares Football, Basketball, and Soccer Predictions," *IEEE Transactions on Systems, Man, and Cybernetics*, 10, 116–123.

Thompson, M. L. (1975), "On Any Given Sunday: Fair Competitor Orderings With Maximum Likelihood Methods," *Journal of the American Statistical Association,* 70, 536–541.

Tryfos, P., Casey, S., Cook, S., Leger, G., and Pylypiak, B. (1984), "The Profitability of Wagering on NFL Games," *Management Science,* 30, 123–132.

Vergin, R. C., and Scriabin, M. (1978), "Winning Strategies for Wagering on Football Games," *Management Science,* 24, 809–818.

Zuber, R. A., Gandar, J. M., and Bowers, B. D. (1985), "Beating the Spread: Testing the Efficiency of the Gambling Market for National Football League Games," *Journal of Political Economy,* 93, 800–806.

Part II
Statistics in Baseball

Chapter 9

Introduction to the Baseball Articles

Jim Albert and James J. Cochran

In this introduction we provide a brief background on the application of statistical methods in baseball and we identify particular research areas. We use the articles selected for this volume to describe the history of statistical research in baseball.

9.1 Background

Baseball, often referred to as the national pastime, is one of the most popular sports in the United States. Baseball began in the eastern United States in the mid 1800s. Professional baseball ensued near the end of the 19th century; the National League was founded in 1876 and the American League in 1900. Currently, in the United States there are 28 professional teams that make up the American and National Leagues, and millions of fans watch games in ballparks and on television.

Baseball is played between two teams, each consisting of nine players. A game of baseball is comprised of nine innings, each of which is divided into two halves. In the top half of the inning, one team plays in the field and the other team comes to bat; the teams reverse their roles in the bottom half of the inning. The team that is batting during a particular half-inning is trying to score runs. The team with the higher number of runs at the end of the nine innings is the winner of the game. If the two teams have the same number of runs at the end of nine innings, additional or "extra" innings are played until one team has an advantage in runs scored at the conclusion of an inning.

During an inning, a player on the team in the field (called the pitcher) throws a baseball toward a player of the team at-bat (who is called the batter). The batter will try to hit the ball using a wooden stick (called a bat) in a location out of the reach of the players in the field. By hitting the ball, the batter has the opportunity to run around four bases that lie in the field. If a player advances around all of the bases, he has scored a run. If a batter hits a ball that can be caught before it hits the ground, hits a ball that can be thrown to first base before he runs to that base, or is tagged with the ball while attempting to advance to any base beyond first base, he is said to be out and cannot score a run. A batter is also out if he fails to hit the baseball three times or if three good pitches (called strikes) have been thrown. The objective of the batting team during an inning is to score as many runs as possible before the defense records three outs.

9.2 Standard Performance Measures and Sabermetrics

One notable aspect of the game of baseball is the wealth of numerical information that is recorded about the game. The effectiveness of batters and pitchers is typically assessed by particular numerical measures. The usual measure of hitting effectiveness for a player is the batting average, computed by dividing the number of hits by the number of at-bats. This statistic gives the proportion of opportunities (at-bats) in which the batter succeeds (gets a hit). The batter with the highest batting average during a baseball season is called the best hitter that year. Batters are also evaluated on their ability to reach one, two, three, or four bases on a single hit; these hits are called, respectively, singles, doubles, triples, and home runs. The slugging average, a measure of this ability, is computed by dividing the total number of bases (in short, total bases) by the number of opportunities. Since it weights hits by the number of bases reached, this measure reflects the ability of a batter to hit a long ball for distance. The most valued hit in baseball, the home run, allows a player to advance four bases on one hit (and allows all other players occupying bases to score as well). The number of home runs is recorded for all players and the batter with the largest number of home runs at the end of the season is given special recognition.

A number of statistics are also used in the evaluation of pitchers. For a particular pitcher, one counts the number of games in which he was declared the winner or loser and the number of runs allowed. Pitchers are usually rated in terms of the average number of "earned" runs (runs scored without direct aid of an error or physical mistake by one of the pitcher's teammates) allowed for every nine innings pitched. Other statistics are useful in understanding pitching ability. A pitcher records a strikeout when the batter fails to hit the ball in the field and records a walk when he throws four inaccurate pitches (balls) to the batter. A pitcher who can throw the ball very fast can record a high number of strikeouts. A pitcher who lacks control over his pitches is said to be "wild" and will record a relatively large number of walks.

Sabermetrics is the mathematical and statistical study of baseball records. One goal of researchers in this field is to find good measures of hitting and pitching performance. Bill James (1982) compares the batting records of two players, Johnny Pesky and Dick Stuart, who played in the 1960s. Pesky was a batter who hit a high batting average but hit few home runs. Stuart, in contrast, had a modest batting average, but hit a high number of home runs. Who was the more valuable hitter? James argues that a hitter should be evaluated by his ability to create runs for his team. From an empirical study of a large collection of team hitting data, he established the following formula for predicting the number of runs scored in a season based on the number of hits, walks, at-bats, and total bases recorded in a season:

$$\text{RUNS} = \frac{(\text{HITS} + \text{WALKS})(\text{TOTAL BASES})}{\text{AT-BATS} + \text{WALKS}}$$

This formula reflects two important aspects in scoring runs in baseball. The number of hits and walks of a team reflects the team's ability to get runners on base, while the number of total bases of a team reflects the team's ability to move runners that are already on base. James' runs created formula can be used at an individual level to compute the number of runs that a player creates for his team. In 1942, Johnny Pesky had 620 at-bats, 205 hits, 42 walks, and 258 total bases; using the formula, he created 96 runs for his team. Dick Stuart in 1960 had 532 at-bats with 160 at-bats, 34 walks, and 309 total bases for 106 runs created. The conclusion is that Stuart in 1960 was a slightly better hitter than Pesky in 1942 since Stuart created a few more runs for his team (and in far fewer plate appearances). An alternative approach to evaluating batting performance is based on a linear weights formula. George Lindsey (1963) was the first person to assign run values to each event that could occur while a team was batting. By the use of recorded data from baseball games and probability theory, he developed the formula

$$\text{RUNS} = (.41)\text{1B} + (.82)\text{2B} + (1.06)\text{3B} + (1.42)\text{HR}$$

where 1B, 2B, 3B, and HR are, respectively, the number of singles, doubles, triples, and home runs hit in a game. One notable aspect of this formula is that it recognizes that a batter creates a run three ways. There is a direct run potential when a batter gets a hit and gets on base. In addition, the batter can advance runners that are already on base. Also, by not getting an out, the hitter allows a new batter a chance of getting a hit, and this produces an indirect run potential. Thorn and Palmer (1993) present a more sophisticated version of the linear weights formula which predicts the number of runs produced by an average baseball team based on all of the offensive events recorded during the game. Like James' runs created formula, the linear weights rule can be used to evaluate a player's batting performance. Although scoring runs is important in baseball, the basic objective is for a team to outscore its opponent. To learn about the relationship between runs scored and the number of wins, James (1982) looked at the number of runs produced, the number of runs allowed, the number of wins, and the number of losses during a season for a number of major league teams. James noted that the ratio of a team's wins to losses was approximately equal to the square of the ratio of runs scored to the runs allowed. Equivalently,

$$\begin{aligned}\text{RUNS} &= \frac{\text{WINS}}{\text{WINS} + \text{LOSSES}} \\ &= \frac{\text{RUNS}^2}{\text{RUNS}^2 + (\text{OPPOSITION RUNS})^2}\end{aligned}$$

This relationship can be used to measure a batter's performance in terms of the number of wins that he creates for his team.

Sabermetrics has also developed better ways of evaluating pitching ability. The standard pitching statistics, the number of wins and the earned runs per game (ERA), are flawed. The number of wins of a pitcher can just reflect the fact that he pitches for a good offensive (run-scoring) team. The ERA does measure the rate of a pitcher's efficiency, but it does not measure the actual benefit of this pitcher over an entire season. Thorn and Palmer (1993) developed the pitching runs formula

$$\begin{aligned}&\text{PITCHING RUNS}\\&=(\text{INNINGS PITCHED})\left(\frac{\text{LEAGUE ERA}}{9}\right)\\&\quad-\text{EARNED RUNS ALLOWED}\end{aligned}$$

The factor (LEAGUE ERA/9) measures the average runs allowed per inning for all teams in the league. This value is multiplied by the number of innings pitched by that pitcher—this product represents the number of runs that pitcher would allow over the season if he was average. Last, one subtracts the actual earned runs the pitcher allowed for that season. If the number of pitching runs is larger than zero, then this pitcher is better than average. This new measure appears to be useful in determining the efficiency and durability of a pitcher.

9.3 Modeling Events

Probability models can be very helpful in understanding the observed patterns in baseball data. One of the earliest papers of this type was that of Lindsey (1961) (Chapter 16 in this volume) who modeled the runs scored in a game. The first step of his approach was to construct an empirical probability function for the number of runs scored in a half-inning. Lindsey used this probability function to model the progression of a game. He modeled the length of the game (in innings), the total runs scored by both teams, and the likelihood that one team would be leading the other by a particular number of runs after a certain number of innings. Suppose that a team is losing to another by two runs after seven innings—does the team have any chance of winning? From Lindsey (1961, Figure 5B), we would estimate this probability to be about 12%. Lindsey is careful to verify his modeling results with data from two baseball seasons. Several important assumptions are made in this paper that greatly simplify the modeling: teams are assumed to be homogeneous with respect to their ability to score, and runs scored in different half-innings are assumed independent. In his summary remarks, Lindsey gives several instances in which these results are helpful to both fans and team managers. For example, in the 1960 World Series, the Pirates defeated the Yankees despite being outscored by 16-3, 10-0, and 12-0 in the three losing games. Lindsey calculates that the probability of three one-sided games like these in seven games is approximately .015, so one can conclude that this series was an unusual occurrence.

Modeling is generally useful for understanding the significance of "rare" events in baseball. One of the most exciting rare events is the no-hitter, a game in which one pitcher pitches a complete game (typically nine innings) and doesn't allow a single base hit. Frohlich (1994) (Chapter 14 in this volume) notes that a total of 202 no-hitters have occurred in Major League Baseball in the period 1900–1993 and uses probability modeling to see if this is an unusually high number. In Frohlich's simple probability (SP) model, he assumes that any batter gets a hit with fixed probability p throughout history—he shows that this model predicts only 135 no-hitters since 1900. Of course, this model is an oversimplication since batters and pitchers have different abilities. Frohlich makes a major improvement to his SP model by assuming that the expected number of hits allowed in a nine-inning game is not constant, but instead varies across pitchers according to a normal distribution with a given spread. This variable pitcher (VP) model is shown to do a much better job in predicting the number of no-hitters.

Using these models, Frohlich comes to some interesting conclusions regarding the pitchers and teams that are involved in no-hitters. He notes that 23% of all no-hitters have been pitched by the best 60 pitchers in history (ranked by the number of hits allowed per nine innings), and this percentage is significantly higher than the percentages predicted using either the SP or VP model. On the other hand, weak-hitting teams are not more likely to receive no hits. Moreover, the assumption of batter variation appears to be much less important than the assumption of pitcher variability in predicating the number of no-hitters. One factor that may be important in a no-hitter is the decision of the official scorer—plays may be scored as "errors" instead of "hits" and these decisions affect the statistics of the pitcher. Using some reasonable probability calculations, Frohlich predicts that this scoring bias may increase the number of no-hitters about 5–10% over the number predicted by the SP model. The rate of no-hitters across time is also explored. There is no general evidence that no-hitters are more common now than in the past. However, Frohlich notes that an unusually large number of no-hitters (16) occurred in 1990 and 1991 and there is no explanation found for this number—we just happened to observe a rare event.

9.4 Comparing Performances of Individual Players

Useful measures of hitting, pitching, and fielding performances of baseball players have been developed. However, these statistics do not directly measure a player's contribution to a win for his team. Bennett and Flueck (1984) (Chapter 12 in this volume) developed the Player

Game Percentage (PGP) method for evaluating a player's game contribution. This work extends work of Lindsey (1963) and Mills and Mills (1970). Using observed game data culled from several seasons, Lindsey (1963) was able to estimate the expected numbers of runs scored in the remainder of a half-inning given the number of outs and the on-base situation (runners present at first, second, and third base). By taking the difference between the expected runs scored before and after a plate appearance, one can judge the benefit of a given batting play. Mills and Mills (1970) took this analysis one step further by estimating the probability that a team would win at a particular point during a baseball game. Bennett and Flueck (1984) extend the methodology of Mills and Mills (1970) in several ways. First, they developed tables of win probabilities for each inning given the run differential. Second, they measured the impact of a play directly by the change in win probabilities, allocating half of this change to the offensive performer and half to the defensive player. One can measure a player's contribution to winning a game by summing the changes in win probabilities for each play in which the player has participated. The PGP statistic is used by Bennett (1993) (Chapter 11 in this volume) to evaluate the batting performance of Joe Jackson. This player was banished from baseball for allegedly throwing the 1919 World Series. A statistical analysis using the PGP showed that Jackson played to his full potential during this series. Bennett looks further to see if Jackson failed to perform well in "clutch" situations. He looks at traditional clutch hitting statistics, investigates whether Jackson's PGP measure was small given his slugging percentage, and does a resampling analysis to see if Jackson's PGP value was unusual for hitters with similar batting statistics. In his conclusions, Bennett highlights a number of players on the team that had weak performances during this World Series.

Baseball fans are often interested in comparing batters or pitchers from different eras. In making these comparisons, it is important to view batting or pitching statistics in the context in which they were achieved. For example, Bill Terry led the National League in 1930 with a batting average of .401, a mark that has been surpassed since by only one hitter. In 1968, Carl Yastrzemski led the American League in hitting with an average of .301. It appears on the surface that Terry clearly was the superior hitter. However, when viewed relative to the hitters that played during the same time, both hitters were approximately 27% better than the average hitter (Thorn and Palmer, 1993). The hitting accomplishments of Terry in 1930 and Yastrzemski in 1968 were actually very similar. Likewise, there are significant differences in hitting in different ballparks, and hitting statistics need to be adjusted for the ballpark played to make accurate comparisons between players.

9.5 Streaks

Another interesting question concerns the existence of streakiness in hitting data. During a season it is observed that some ballplayers will experience periods of "hot" hitting, where they will get a high proportion of hits. Other hitters will go through slumps or periods of hitting with very few hits. However, these periods of hot and cold hitting may be just a reflection of the natural variability observed in coin tossing. Is there statistical evidence for a "hot hand" among baseball hitters where the probability of obtaining a hit is dependent on recent at-bats? Albright (1993) looked at a large collection of baseball hitting data and used a number of statistics, such as the number of runs, to detect streakiness in hitting data. His main conclusion was that there is little statistical evidence generally for a hot hand in baseball hitting.

Currently there is great interest among fans and the media in situational baseball data. The hitting performance of batters is recorded for a number of different situations, such as day versus night games, grass versus artificial turf fields, right-handed versus left-handed pitchers, and home versus away games. There are two basic questions in the statistical analysis of this type of data. First, are there particular situations that can explain a significant amount of variation in the hitting data? Second, are there ballplayers who perform particularly well or poorly in a given situation? Albert (1994) (Chapter 10 in this volume) analyzes a large body of published situational data and used Bayesian hierarchical models to combine data from a large group of players. His basic conclusion is that there do exist some important situational differences. For example, batters hit on average 20 points higher when facing a pitcher of the opposite arm, and hit 8 points higher when they are playing in their home ballpark. Many of these situational differences appear to affect all players in the same way. Coors Field in Colorado is a relatively easy ballpark in which to hit home runs, and all players' home run hitting abilities will be increased in this ballpark by the same amount. It is relatively unusual to see a situation that is a so-called ability effect, where players' hitting abilities are changed by a different amount depending on the situation. One example of this type of ability effect occurs in the pitch count. Good contact hitters, such as Tony Gwynn, hit approximately the same when they are behind or ahead in the count, and other hitters (especially those who strike out frequently) have significantly smaller batting averages when behind in the count. Because of the small sample sizes inherent in situational data, most of the observed variation in situational data is essentially noise, and it is difficult to detect situational abilities based on a single season of data.

9.6 Projecting Player and Team Performances

Watching a baseball game raises questions that motivate interesting statistical analyses. During the broadcast of a game, a baseball announcer will typically report selected hitting data for a player. For example, it may be reported that Barry Bonds has 10 hits in his most recent 20 at-bats. What have you learned about Bonds' batting average on the basis of this information? Clearly, Bonds' batting average can't be as large as $10/20 = .500$ since this data was chosen to maximize the reported percentage. Casella and Berger (1994) (Chapter 13 in this volume) show how one can perform statistical inference based on this type of selected data. Suppose that one observes a sequence of n Bernoulli trials, and one is given the positive integers k^* and n^*, where the ratio k^*/n^* is the largest ratio of hits to at-bats in the entire batting sequence. Casella and Berger construct the likelihood function for a player's true batting average on the basis of this selected information and use modern sampling methodology to find the maximum likelihood (ml) estimate. They are interested in comparing this estimate with the ml estimate of the complete data set. They conclude that this selected data provides relatively little insight into the batting average that is obtained from batting records over the entire season. However, the complete data ml estimate is generally within one standard deviation of the selected data ml estimate. Also the total number of at-bats n is helpful in understanding the significance of the ratio k^*/n^*. As one might expect, Bonds' reported performance of 10 hits in his last 20 at-bats would be less impressive if he had 600 at-bats instead of 400 at-bats in a season.

James, Albert, and Stern (1993) (Chapter 15 in this volume) discuss the general problem of interpreting baseball statistics. There are several measures of batting and pitching performance that define excellence. For example, a pitcher is considered "great" if he wins 20 games or a batter is "great" if he hits 50 home runs or has over 120 runs batted in. But there is usually no consideration of the role of chance variability in these observed season performances. One pitcher who wins 20 games in a season will be considered superior to another pitcher who wins only 15. But it is very plausible that the first pitcher won more games than the second pitcher due solely to chance variability. In other words, if two players with equal ability pitch an equal number of games, it is plausible that, by chance variability, one pitcher will win five games more than the second. To illustrate this point, James, Albert, and Stern focus on the issue of team competition. Suppose that the Yankees win the World Series—are they truly the "best" team in baseball that year? To answer this question, the authors construct a simple model for baseball competition. Teams are assumed to possess abilities that are normally distributed; the spread of this normal curve is set so that the performances of the team match the performances of modern-day teams. Then a Bradley–Terry choice model is used to represent team competition. This model is used to simulate 1000 baseball seasons, and the relationship between the participating teams' abilities and their season performances is explored. The authors reach some interesting conclusions from their simulation. There is a 30% chance that "good" teams (in the top 10% of the ability distribution) will have less than good seasons. Of the World Series winners in the simulations, only 46% corresponded to good teams. The cream will generally rise to the top, but teams of "average" or "above-average" abilities can have pretty good seasons.

Cochran (2000) also uses a Bradley–Terry choice model as the basis for simulating final divisional standing. Expected win/loss percentages for each team in a four-team division are systematically changed in accordance with a complete-block design, and 10,000 seasons are simulated for each unique set of expected win/loss percentages. He then conducts a logistic regression of the probability that the team with the best win/loss percentage in the division actually wins the division, using the differences between the best win/loss percentage in the division and the expected win/loss percentages of the other three teams in the division as independent variables. He found that there is very little marginal benefit (with regard to winning a division title) to improving a team once its expected regular season wins exceed the expected wins for every other team in its division by five games.

9.7 Summary

Currently, Major League Baseball games are recorded in very fine detail. Information about every single ball pitched, fielded, and hit during a game is noted, creating a large database of baseball statistics. This database is used in a number of ways. Public relations departments of teams use the data to publish special statistics about their players. The statistics are used to help determine the salaries of major league ballplayers. Specifically, statistical information is used as evidence in salary arbitration, a legal proceeding which sets salaries. A number of teams have employed full-time professional statistical analysts, and some managers use statistical information in deciding on strategy during a game. Bill James and other baseball statisticians have shown that it is possible to answer a variety of questions about the game of baseball by means of statistical analyses.

The seven baseball articles included in this volume characterize how statistical thinking can influence baseball strategy and performance evaluation. Statistical research in baseball continues to grow rapidly as the level of detail in the available data increases and the use of statistical analyses by professional baseball franchises proliferates.

References

Albert, J. (1994), "Exploring baseball hitting data: What about those breakdown statistics?" Journal of the American Statistical Association, 89, 1066–1074.

Albright, S. C. (1993), "A statistical analysis of hitting streaks in baseball," Journal of the American Statistical Association, 88, 1175–1183.

Bennett, J. M. (1993), "Did Shoeless Joe Jackson throw the 1919 World Series?" The American Statistician, 47, 241–250.

Bennett, J. M. and Flueck, J. A. (1984), "Player game percentage," Proceedings of the Social Science Section, American Statistical Association, 378–380.

Casella, G. and Berger, R. L. (1994), "Estimation with selected binomial information or do you really believe that Dave Winfield is batting .471?" Journal of the American Statistical Association, 89, 1080–1090.

Cochran, J. J. (2000), "A power analysis of the 162 game Major League Baseball schedule," in *Joint Statistical Meetings*, Indianapolis, IN, August 2000.

Frohlich, C. (1994), "Baseball: Pitching no-hitters," Chance, 7, 24–30.

Harville, D. (1980), "Predictions for National Football League games via linear-model methodology," Journal of the American Statistical Association, 75, 516–524.

James, B. (1982), *The Bill James Baseball Abstract*, New York: Ballantine Books.

James, B., Albert, J., and Stern, H. S. (1993), "Answering questions about baseball using statistics," Chance, 6, 17–22, 30.

Lindsey, G. R. (1961), "The progress of the score during a baseball game," American Statistical Association Journal, September, 703–728.

Lindsey, G. R. (1963), "An investigation of strategies in baseball," Operations Research, 11, 447–501.

Mills, E. and Mills, H. (1970), *Player Win Averages*, South Brunswick, NJ: A. S. Barnes.

Thorn, J. and Palmer, P. (1993), *Total Baseball*, New York: HarperCollins.

Chapter 10

Exploring Baseball Hitting Data: What About Those Breakdown Statistics?

Jim ALBERT*

During a broadcast of a baseball game, a fan hears how baseball hitters perform in various situations, such as at home and on the road, on grass and on turf, in clutch situations, and ahead and behind in the count. From this discussion by the media, fans get the misleading impression that much of the variability in players' hitting performance can be explained by one or more of these situational variables. For example, an announcer may state that a particular player struck out because he was behind in the count and was facing a left-handed pitcher. In baseball one can now investigate the effect of various situations, as hitting data is recorded in very fine detail. This article looks at the hitting performance of major league regulars during the 1992 baseball season to see which situational variables are "real" in the sense that they explain a significant amount of the variation in hitting of the group of players. Bayesian hierarchical models are used in measuring the size of a particular situational effect and in identifying players whose hitting performance is very different in a particular situation. Important situational variables are identified together with outstanding players who make the most of a given situation.

KEY WORDS: Hierarchical modeling; Outliers; Situational variables.

1. INTRODUCTION

After the end of every baseball season, books are published that give detailed statistical summaries of the batting and pitching performances of all major league players. In this article we analyze baseball hitting data that was recently published in Cramer and Dewan (1992). This book claims to be the "most detailed statistical account of every major league player ever published," which enables a fan to "determine the strengths and weaknesses of every player."

For hitters, this book breaks down the usual set of batting statistics (e.g., hits, runs, home runs, doubles) by numerous different situations. Here we restrict discussion to the fundamental hitting statistics—hits, official at-bats, and batting average (hits divided by at-bats)—and look at the variation of this data across situations. To understand the data that will be analyzed, consider the breakdowns for the 1992 season of Wade Boggs presented in Table 1. This table shows how Boggs performed against left- and right-handed pitchers and pitchers that induce mainly groundballs and flyballs. In addition, the table gives hitting statistics for day and night games, games played at and away from the batter's home ballpark, and games played on grass and artificial turf. The table also breaks down hits and at-bats by the pitch count ("ahead on count" includes 1-0, 2-0, 3-0, 2-1, and 3-1) and the game situation ("scoring position" is having at least one runner at either second or third, and "none on/out" is when there are no outs and the bases are empty). Finally, the table gives statistics for the batting position of the hitter and different time periods of the season.

What does a fan see from this particular set of statistical breakdowns? First, several situational variables do not seem very important. For example, Boggs appears to hit the same for day and night games and before and after the All-Star game. But other situations do appear to matter. For example, Boggs hit .243 in home games and .274 in away games—a 31-point difference. He appears to be more effective against flyball pitchers compared to groundball pitchers, as the difference in batting averages is 57 points. The most dramatic situation appears to be pitch count. He hit .379 on the first pitch, .290 when he was ahead in the count, but only .197 when he had two strikes on him.

What does a fan conclude from this glance at Boggs's batting statistics? First, it is difficult to gauge the significance of these observed situational differences. It seems that Boggs bats equally well during day or night games. It also appears that there are differences in Boggs's "true" batting behavior during different pitch counts (the 93-point difference between the "ahead in count" and "two strikes" averages described earlier). But consider the situation "home versus away." Because Boggs bats 31 points higher in away games than in home games, does this mean that he is a better hitter away from Fenway Park? Many baseball fans would answer "yes." Generally, people overstate the significance of seasonal breakdown differences. Observed differences in batting averages such as these are often mistakenly interpreted as real differences in true batting behavior. Why do people make these errors? Simply, they do not understand the general variation inherent in coin tossing experiments. There is much more variation in binomial outcomes than many people realize, and so it is easy to confuse this common form of random variation with the variation due to real situational differences in batting behavior.

How can fans gauge the significance of differences of situational batting averages? A simple way is to look at the 5-year hitting performance of a player for the same situations. If a particular observed seasonal situational effect is real for Boggs, then one might expect him to display a similar situational effect during recent years. Cramer and Dewan (1992) also gave the last 5 years' (including 1992) hitting performance of each major league player for all of the same situations of Table 1. Using these data, Table 2 gives situational differences in batting averages for Boggs for 1992 and the previous 4-year period (1988–1991).

Table 2 illustrates the volatility of the situational differences observed in the 1992 data. For example, in 1992 Boggs

* Jim Albert is Professor, Department of Mathematics and Statistics, Bowling Green State University, Bowling Green, OH 43403. The author is grateful to Bill James, the associate editor, and two referees for helpful comments and suggestions.

Journal of the American Statistical Association
September 1994, Vol. 89, No. 427, Statistics in Sports

Table 1. Situational 1992 Batting Record of Wade Boggs

	AVG	AB	H
1992 season	.259	514	133
versus left	.272	158	43
versus right	.253	356	90
groundball	.235	136	32
flyball	.292	144	42
home	.243	251	61
away	.274	263	72
day	.259	193	50
night	.259	321	83
grass	.254	437	111
turf	.286	77	22
1st pitch	.379	29	11
ahead in count	.290	169	49
behind in count	.242	157	38
two strikes	.197	213	42
scoring position	.311	106	33
close and late	.322	90	29
none on/out	.254	142	36
batting #1	.222	221	49
batting #3	.287	289	83
other	.250	4	1
April	.253	75	19
May	.291	86	25
June	.242	95	23
July	.304	79	24
Aug	.198	96	19
Sept/Oct	.277	83	23
Pre–All-Star	.263	278	73
Post–All-Star	.254	236	60

NOTE: Data from Cramer and Dewan (1992).

hit flyball pitchers 57 points better than groundball pitchers. But in the 4-year period preceding 1992, he hit groundball pitchers 29 points higher than flyball pitchers. The 31-point 1992 home/away effect (favoring away) appears spurious, because Boggs was much better at Fenway Park for the previous 4-year period. Looking at the eight situations, only the night/day and pre- post-All-Star situational effects appear to be constant over time.

From this simple analysis, one concludes that it is difficult to interpret the significance of seasonal batting averages for a single player. In particular, it appears to be difficult to conclude that a player is a clutch or "two-strike" hitter based solely on batting statistics for a single season. But a large number of major league players bat in a particular season, and it may be easier to detect any significant situational patterns by pooling the hitting data from all of these players. The intent of this article is too look at situational effects in batting averages over the entire group of hitters for the 1992 season.

Here we look at the group of 154 regular major league players during the 1992 season. We define "regular" as a player who had at least 390 official at-bats; the number 400 was initially chosen as a cutoff, but it was lowered to 390 to accommodate Rob Deer, a hitter with unusual talents (power with a lot of strikeouts).

Using data from Cramer and Dewan (1992) for the 154 regulars, we investigate the effects of the following eight situations (with the abbreviation for the situation that we use given in parentheses):

- opposite side versus same side (OPP–SAME)
- groundball pitcher versus flyball pitcher (GBALL–FBALL)
- home versus away
- day versus night
- grass versus turf
- ahead in count versus two strikes in count (AHEAD–2 STRIKE)
- scoring position versus none on/out (SCORING–NONE ON/OUT)
- pre-All-Star game versus post-All-Star game (PRE/AS–POST/AS).

A few remarks should be made about this choice of situations. First, it is well known that batters hit better against pitchers who throw from the opposite side from the batter's hitting side. For this reason, I look at the situation "opposite side versus same side"; batters who switch-hit will be excluded from this comparison. Next, for ease of comparison of different situations, it seemed helpful to create two nonoverlapping categories for each situation. All of the situations of Table 1 are of this type except for pitch count, clutch situations, and time. Note that the pitch categories are overlapping (one can be behind in the count and have two strikes) and so it seemed simpler to just consider the nonoverlapping cases "ahead in count" and "two strikes." Likewise, one can simultaneously have runners in scoring position and the game be close and late, so I considered only the "scoring position" and "none on/out" categories. The batting data across months is interesting; however, because the primary interest is in comparing the time effect to other situational variables, I used only the pre- and post-All-Star game data.

When we look at these data across the group of 1992 regulars, there are two basic questions that we will try to answer.

Table 2. Situational Differences in Batting Averages (One Unit = .001) for Wade Boggs for 1992 and 1988–1991

Year	Right–left	Flyball–groundball	Home–away	Night–day
1992	−19	57	−31	0
1988–1991	61	−29	85	6
Year	**Turf–grass**	**Ahead–2 strikes**	**None out–scoring position**	**Pre-All Star–Post-All Star**
1992	32	93	−57	9
1988–1991	−30	137	1	−8

First, for a particular situation it is of interest to look for a general pattern across all players. Baseball people believe that most hitters perform better in various situations. In particular, managers often make decisions under the following assumptions:

- Batters hit better against pitchers throwing from the opposite side.
- Batters hit better at home.
- Batters hit better during day games (because it is harder to see the ball at night).
- Batters hit better when they are ahead in the count (instead of being behind two strikes).
- Batters hit better on artificial turf than on grass (because groundballs hit on turf move faster and have a better chance of reaching the outfield for a hit).

The other three situations in the list of eight above are not believed to be generally significant. One objective here is to measure and compare the general sizes of these situational effects across all players.

Once we understand the general situational effects, we can then focus on individuals. Although most players may display, say, a positive home effect, it is of interest to detect players who perform especially well or poorly at home. It is easy to recognize great hitters such as Wade Boggs partly because his success is measured by a well-known statistic, batting average. The baseball world is less familiar with players who bat especially well or poorly in given situations and the statistics that can be used to measure this unusual performance. So a second objective in this article is to detect these unusual situational players. These outliers are often the most interesting aspect of baseball statistics. Cramer and Dewan (1992), like many others, list the leading hitters with respect to many statistical criteria in the back of their book.

This article is outlined as follows. Section 2 sets up the basic statistical model and defines parameters that correspond to players' situational effects. The estimates of these situational effects from a single season can be unsatisfactory and often can be improved by adjusting or shrinking them towards a common value. This observation motivates the consideration of a prior distribution that reflects a belief in similarity of the set of true situational effects. Section 3 summarizes the results of fitting this Bayesian model to the group of 1992 regulars for each one of the eight situational variables. The focus, as explained earlier, is on looking for general situational patterns and then finding players that deviate significantly from the general patterns. Section 4 summarizes the analysis and contrasts it with the material presented by Cramer and Dewan (1992).

2. THE MODEL

2.1 Basic Notation

Consider one of the eight situational variables, say the home/away breakdown. From Cramer and Dewan (1992), we obtain hitting data for $N = 154$ players; for each player, we record the number of hits and official at-bats during home and away games. For the ith player, this data can be represented by a 2×2 contingency table,

	HITS	OUTS	AT-BATS
home	h_{i1}	o_{i1}	ab_{i1}
away	h_{i2}	o_{i2}	ab_{i2} ,

where h_{i1} denotes the number of hits, o_{i1} the number of outs, and ab_{i1} the number of at-bats during home games. (The variables h_{i2}, o_{i2}, and ab_{i2} are defined similarly for away games.) Let p_{i1} and p_{i2} denote the true probabilities that the ith hitter gets a hit home and away. If we assume that the batting attempts are independent Bernoulli trials with the aforementioned probabilities of success, then the number of hits h_{i1} and h_{i2} are independently distributed according to binomial distributions with parameters (ab_{i1}, p_{i1}) and (ab_{i2}, p_{i2}).

For ease of modeling and exposition, it will be convenient to transform these data to approximate normality using the well-known logit transformation. Define the observed logits

$$y_{ij} = \log\left(\frac{h_{ij}}{o_{ij}}\right), \qquad j = 1, 2.$$

Then, approximately, y_{i1} and y_{i2} are independent normal, where y_{ij} has mean $\mu_{ij} = \log(p_{ij}/(1 - p_{ij}))$ and variance $\sigma_{ij}^2 = (ab_{ij}p_{ij}(1 - p_{ij}))^{-1}$. Because the sample sizes are large, we can accurately approximate σ_{ij}^2 by an estimate where p_{ij} is replaced by the observed batting average h_{ij}/ab_{ij}. With this substitution, $\sigma_{ij}^2 \approx 1/h_{ij} + 1/o_{ij}$.

Using the foregoing logistic transformation, we represent the complete data set for a particular situation, say home/away, as a $2 \times N$ table:

	Player 1	2	. . .	N
home	y_{11}	y_{21}	. . .	y_{N1}
away	y_{12}	y_{22}	. . .	y_{N2} .

The observation in the (i, j) cell, y_{ij}, is the logit of the observed batting average of the ith player during the jth situation. We model μ_{ij}, the mean of y_{ij}, as

$$\mu_{ij} = E(y_{ij}) = \mu_i + \alpha_{ij},$$

where μ_i measures the hitting ability of player i and α_{ij} is a situational effect that measures the change in this player's hitting ability due to the jth situation. The model as stated is overparameterized, so we express the situational effects as $\alpha_{i1} = \alpha_i$ and $\alpha_{i2} = -\alpha_i$. With this change, the parameter α_i represents the change in the hitting ability of the ith player due to the first situational category.

2.2 Shrinking Toward the Mean

For a given situational variable, it is of interest to estimate the player situational effects $\alpha_1, \ldots, \alpha_N$. These parameters represent the "true" situational effects of the players if they were able to play an infinite number of games.

Is it possible to estimate accurately a particular player's situational effect based on his hitting data from one season?

To answer this question, suppose that a player has a true batting average of .300 at home and .200 away, a 100-point differential. If he has 500 at-bats during a season, half home and half away, then one can compute that his observed seasonal home batting average will be between .252 and .348 with probability .9 and his away batting average will be between .158 and .242 with the same probability. So although the player's true differential is 100 points, his seasonal batting average differential can be between 10 and 190 points. Because this is a relatively wide interval, one concludes that it is difficult to estimate a player's situational effect using only seasonal data.

How can we combine the data from all players to obtain better estimates? In the situation where one is simultaneously estimating numerous parameters of similar size, it is well known in the statistics literature that improved estimates can be obtained by shrinking the parameters toward a common value. Efron and Morris (1975) illustrated the benefit of one form of these shrinkage estimators in the prediction of final season batting averages in 1970 for 18 ballplayers based on the first 45 at-bats. (See also Steffey 1992, for a discussion of the use of shrinkage estimates in estimating a number of batting averages.) Morris (1983) used a similar model to estimate the true batting average of Ty Cobb from his batting statistics across seasons.

In his analysis of team breakdown statistics, James (1986) discussed the related problem of detecting significant effects. He observed that many team distinctions for a season (e.g., how a team plays against right- and left-handed pitching) will disappear when studied over a number of seasons. A similar pattern is likely to hold for players' breakdowns. Some players will display unusually high or low situational statistics for one season, suggesting extreme values of the parameters α_i. But if these situational data are studied over a number of seasons, the players will generally have less extreme estimates of α_i. This "regression to the mean" phenomena is displayed for the "ahead in count–two strikes" situation in Figure 1. This figure plots the 1992 difference in batting averages (batting average ahead in count–batting average 2 strikes) against the 1988–1991 batting average difference for all of the players. Note that there is an upward tilt in the graph, indicating that players who have larger batting average differences in 1992 generally had larger differences in 1989–1991. But also note that there is less variability in the 4-year numbers (standard deviation .044 versus .055 for the 1992 data). One way of understanding this difference is by the least squares line placed on top of the graph. The equation of this line is $y = .126 + (1 - .683)(x - .122)$, with the interpretation that the 4-year batting average difference generally adjusts the 1992 batting average difference 68% toward the average difference (.12) across all players.

Suppose that the underlying batting abilities of the players do not change significantly over the 5-year period. Then the 4-year batting average differences (based on a greater number of at-bats) are better estimates than the 1-year differences of the situational abilities of the players. In that case, it is clear from Figure 1 that the observed seasonal estimates of the α_i should be shrunk toward some overall value to obtain more accurate estimates. In the next section we describe a prior

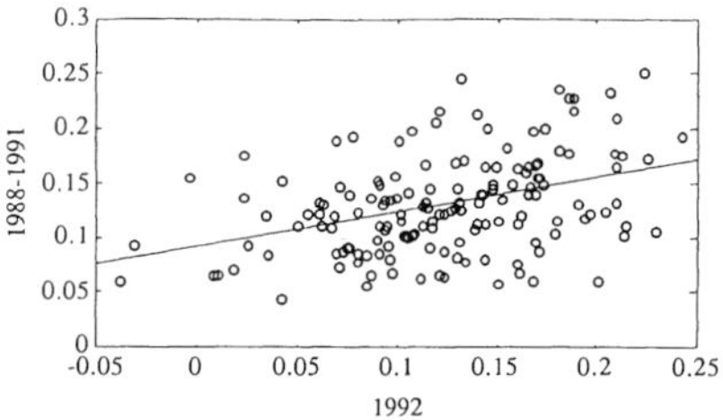

Figure 1. Scatterplot of 1992 Pitch Count Difference in Batting Averages (AHEAD–2 STRIKES) Against Previous 4-Year Difference for All 1992 Players.

distribution on the situation parameters that reflects a belief in similarity of the effects and results in sample estimates that will shrink the season values toward a pooled value.

2.3 The Prior Distribution

The model discussed in Section 2.1 contains $2N$ parameters, the hitting abilities $\mu_1, \ldots, \mu_N$, and the situational effects $\alpha_1, \ldots, \alpha_N$. Because the ability parameters in this setting are nuisance parameters, we assume for convenience that the μ_i are independently assigned flat noninformative priors.

Because the focus is on the estimation of the situational effects $\alpha_1, \ldots, \alpha_N$, we wish to assign a prior that reflects subjective beliefs about the locations of these parameters. Recall from our earlier discussion that it seems desirable for the parameter estimates to shrink the individual estimates toward some common value. This shrinkage can be accomplished by assuming a priori that the effects $\alpha_1, \ldots, \alpha_N$ are independently distributed from a common population $\pi(\alpha)$. This prior reflects the belief that the N effects are similar in size and come from one population of effects $\pi(\alpha)$.

Because the α_i are real-valued parameters, one reasonable form for this population model is a normal distribution with mean μ_α and variance σ_α^2. A slightly preferable form used in this article is a t distribution with mean μ_α, scale σ_α, and known degrees of freedom ν (here we use the relatively small value, $\nu = 4$). The parameters of this distribution are used to explain the general size of the situational effect and to identify particular players who have unusually high or low situational effects. The parameters μ_α and σ_α describe the location and spread of the distribution of effects across all players. To see how this model can identify outliers, note that a $t(\mu_\alpha, \sigma_\alpha, \nu)$ distribution can be represented as the mixture $\alpha_i \mid \lambda_i$ distributed $N(\mu_\alpha, \sigma_\alpha^2/\lambda_i)$, λ_i distributed gamma$(\nu/2, \nu/2)$. The new scale parameters $\lambda_1, \ldots, \lambda_N$ will be seen to be useful in picking out individuals who have situational effects set apart from the main group of players.

To complete our prior specification, we need to discuss what beliefs exist about the parameters μ_α and σ_α^2 that describe the t population of situational effects. First, we assume that we have little knowledge about the general size of the situational effect; to reflect this lack of knowledge, the mean μ_α is assigned a flat noninformative prior. We must be more careful about the assignment of the prior distribution on σ_α^2, because this parameter controls the size of the shrinkage

Table 3. Posterior Means of Parameters of Population Distribution of the Situation Effects and Summary Statistics of the Posterior Means of the Batting Average Differences $p_{i1} - p_{i2}$ Across All Players

Situation	$E(\mu_\alpha)$	$E(\sigma_\alpha)$	Summary statistics of $[E(p_{i1} - p_{i2})]$ (one unit = .001)			
			Q_1	M	Q_3	Q_3-Q_1
GRASS–TURF	−.002	.107	−17	−3	15	32
SCORING–NONE ON/OUT	.000	.108	−13	0	17	30
DAY–NIGHT	.004	.105	−13	2	16	29
PRE/AS–POST/AS	.007	.101	−9	3	17	26
HOME–AWAY	.016	.103	−8	8	21	29
GBALL–FBALL	.023	.109	−7	11	24	31
OPP–SAME	.048	.106	5	20	32	27
AHEAD–2 STRIKES	.320	.110	104	123	142	38

of the individual player situational estimates toward the common value. In empirical work, it appears that the use of the standard noninformative prior for σ_α^2, $1/\sigma_\alpha^2$, can lead to too much shrinkage. So to construct an informative prior for σ_α^2, we take the home/away variable as one representative situational variable among the eight and base the prior of σ_α^2 on a posterior analysis of these parameters based on home/away data from an earlier season. So we first assume that μ_α, σ_α^2 are independent, with μ_α assigned a flat prior and σ_α^2 distributed according to an inverse gamma distribution with parameters $a = 1/2$ and $b = 1/2$ (vague prior specification), and then fit this model to 1991 home/away data for all of the major league regulars. From this preliminary analysis, we obtain posterior estimates for σ_α^2 that are matched to an inverse gamma distribution with parameters $a = 53.2$ and $b = .810$.

To summarize, the prior can be written as follows:

- $\mu_1, \ldots, \mu_N$ independent with $\pi(\mu_i) = 1$.
  $\alpha_1, \ldots, \alpha_N$ independent $t(\mu_\alpha, \sigma_\alpha, \nu)$, where $\nu = 4$.
- μ_α, σ_α independent with
  $\pi(\mu_\alpha) = 1$ and $\pi(\sigma_\alpha^2) = K(1/(\sigma_\alpha^2)^{a+1})\exp(-b/\sigma_\alpha^2)$, with $a = 53.2$ and $b = .810$.

This prior is used in the posterior analysis for each of the situational variables.

3. FITTING THE MODEL

3.1 General Behavior

The description of the joint posterior distribution and the use of the Gibbs sampler in summarizing the posterior are

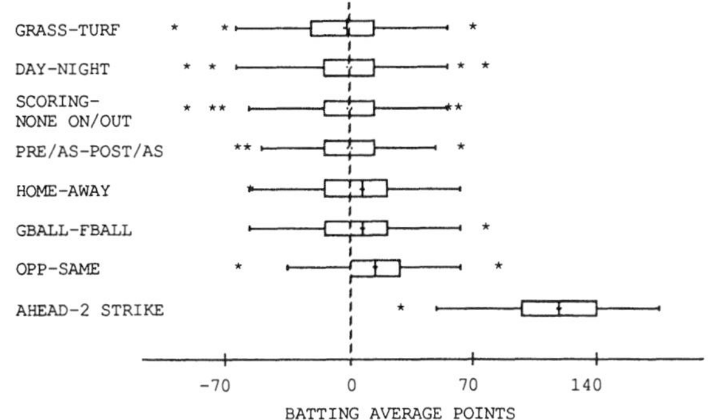

Figure 2. Boxplots of the Posterior Means of the Differences in Batting Averages for the Eight Situational Variables.

outlined in the Appendix. For each of the eight situational variables, a simulated sample of size 1,000 from the joint posterior distribution of $(\{\mu_i\}, \{\alpha_i\}, \mu_\alpha, \sigma_\alpha^2\}$ was obtained. These simulated values can be used to estimate any function of the parameters of interest. For example, if one is interested in the difference in breakdown probabilities for player i, $p_{i1} - p_{i2}$, then one can simulate values of this difference by noting that $p_{i1} - p_{i2} = \exp(\mu_i + \alpha_i)/(1 + \exp(\mu_i + \alpha_i)) - \exp(\mu_i - \alpha_i)/(1 + \exp(\mu_i - \alpha_i))$, and then performing this transformation on the simulated values of $\{\mu_i\}$ and $\{\alpha_i\}$.

Our first question concerns the existence of a general situational effect. Do certain variables generally seem more important than others in explaining the variation in hitting performance? One can answer this question by inspection of the posterior distribution of the parameters μ_α and σ_α, which describe the population distribution of the situational effects $\{\alpha_i\}$. The posterior means of these hyperparameters for each of the eight situations are given in Table 3. Another way to look at the general patterns is to consider the posterior means of the true batting average differences $p_{i1} - p_{i2}$ across all players. Table 3 gives the median, quartiles, and interquartile range for this set of 154 posterior means; Figure 2 plots parallel boxplots of these differences for the eight situations.

Note from Table 3 and Figure 2 that there are significant differences between the average situational effects. The posterior mean of μ_α is a measure of the general size of the situational effect on a logit scale. Note from Table 3 that the "ahead in count–2 strikes" effect stands out. The posterior mean of μ_α is .32 on the logit scale; the corresponding median of the posterior means of the batting average differences is 123 points. Batters generally hit much better when ahead versus behind in the pitch count. Compared to this effect, the other seven effects are relatively insignificant. Closer examination reveals that "opposite–same arm" is the next most important variable, with a median batting average difference of 20 points. The "home–away" and "groundball–flyball" effects follow in importance, with median batting average differences of 8 and 11 batting average points. The remaining four situations appear generally insignificant, as the posterior mean of μ_α is close to 0 in each case.

The posterior means of σ_α give some indication of the relative spreads of the population of effects for the eight situations. The general impression from Table 3 and Figure 2 is that all of the situational populations have roughly the same spread, with the possible exception of the pitch count situation. So the differences between the sets of two different situational effects can be described by a simple shift. For example, the "groundball–flyball" effects are approximately 10 batting average points higher than the "day–night" effects.

3.2 Individual Effects—What is an Outlier?

In the preceding section we made some observations regarding the general shape of the population of situation effects $\{\alpha_i\}$. Here we want to focus on the situation effects for individual players. Figure 3 gives stem and leaf diagrams (corresponding to the Fig. 2 boxplots) for these differences in batting averages for each of the eight situations. These plots confirm the general comments made in Section 3.1 about the comparative locations and spreads of the situational effect distributions.

Recall from Section 2 that it is desirable to shrink the batting average differences that we observe for a single season toward a common value. Figure 4 plots the posterior means of the differences $p_{i1} - p_{i2}$ against the season batting average differences for the effect "ahead in count–2 strikes." The line $y = x$ is plotted on top of the graph for comparison purposes. This illustrates that these posterior estimates shrink the seasonal estimates approximately 50% toward the average effect size of 122 points. Note that some of the seasonal batting average differences are negative; these players actually hit better in 1992 with a pitch count of 2 strikes. But the posterior means shrink these values toward positive values. Thus we have little faith that these negative differences actually relate to true negative situational effects.

We next consider the issue of outliers. For instance, for the particular effect "home–away," are there players that bat particularly well or poor at home relative to away games? Returning back to the displays of the estimates of the batting average differences in Figure 4, are there any players whose estimates deviate significantly from the main group? We observe some particular estimates, say the −95 in the "grass–turf" variable, that are set apart from the main distribution of estimates. Are these values outliers? Are they particularly noteworthy?

We answer this question by first looking at some famous outliers in baseball history. With respect to batting average, a number of Hall of Fame or future Hall of Fame players have achieved a high batting average during a single season. In particular, we consider Ted Williams .406 batting average in 1941, Rod Carew's .388 average in 1977, George Brett's .390 average in 1980 and Wade Boggs's .361 average in 1983. Each of these batting averages can be considered an outlier, because these accomplishments received much media attention and these averages were much higher than the batting averages of other players during that particular season.

These unusually high batting averages were used to calibrate our Bayesian model. For each of the four data sets—Major League batting averages in 1941 and American League batting averages in 1977, 1980, and 1983—an exchangeable model was fit similar to that described in Section 2. In each model fit we computed the posterior mean of the scale parameter λ_i corresponding to the star's high batting average. A value $\lambda_i = 1$ corresponds to an observation consistent with the main body of data; an outlier corresponds to a small positive value of λ_i. The posterior mean of this scale parameter was for Williams .50, .59 for Carew, .63 for Brett, and .75 for Boggs. Thus Williams's accomplishment was the greatest outlier in the sense that it deviated the most from American League batting averages in 1941.

This brief analysis of famous outliers is helpful in understanding which individual situational effects deviate significantly from the main population of effects. For each of the eight situational analyses, the posterior means of the scale parameters λ_i were computed for all of the players. Table 4 lists players for all situations where the posterior mean of λ_i is smaller than .75. To better understand a player's unusual accomplishment, this table gives his 1992 batting average in each category of the situation and the batting average difference ("season difference"). Next, the table gives the posterior mean of the difference in true probabilities $p_{i1} - p_{i2}$. Finally, it gives the difference in batting averages for the previous 4 years (given in Cramer and Dewan 1992).

What do we learn from Table 4? First, relatively few players are outliers using our definition—approximately one per situational variable. Note that the posterior estimates shrink the observed season batting average differences approximately halfway toward the average situational effect. The amount of shrinkage is greatest for the situational variables (such as "grass–turf") where the number of at-bats for one of the categories is small. The last column addresses the question if these nine unusual players had exhibited similar situational effects the previous 4 years. The general answer to this question appears to be negative. Seven of the nine players had 1988–1991 effects that were opposite in sign from the 1992 effect. The only player who seems to display a constant situational effect over the last 5 years is Tony Gwynn; he hits for approximately the same batting average regardless of the pitch count.

4. SUMMARY AND DISCUSSION

What have we learned from the preceding analysis of this hitting data? First, if one looks at the entire group of 154 baseball regulars, some particular situational variables stand out. The variation in batting averages by the pitch count is dramatic—batters generally hit 123 points higher when ahead in the count than with 2 strikes. But three other variables appear important. Batters on average hit 20 points higher when facing a pitcher of the opposite arm, 11 points higher when facing a groundball pitcher (as opposed to a flyball pitcher), and 8 points higher when batting at home. Because these latter effects are rather subtle, one may ask if these patterns carry over to other seasons. Yes they do. The same model (with the same prior) was fit to data from the 1990 season and the median opposite arm, groundball pitcher, and home effects were 25, 9, and 5 points respectively, which are close to the 1992 effects discussed in Section 3.

Do players have different situational effects? Bill James (personal communication) views situational variables as either "biases" or "ability splits." A bias is a variable that has the same effect on all players, such as "grass–turf," "day–

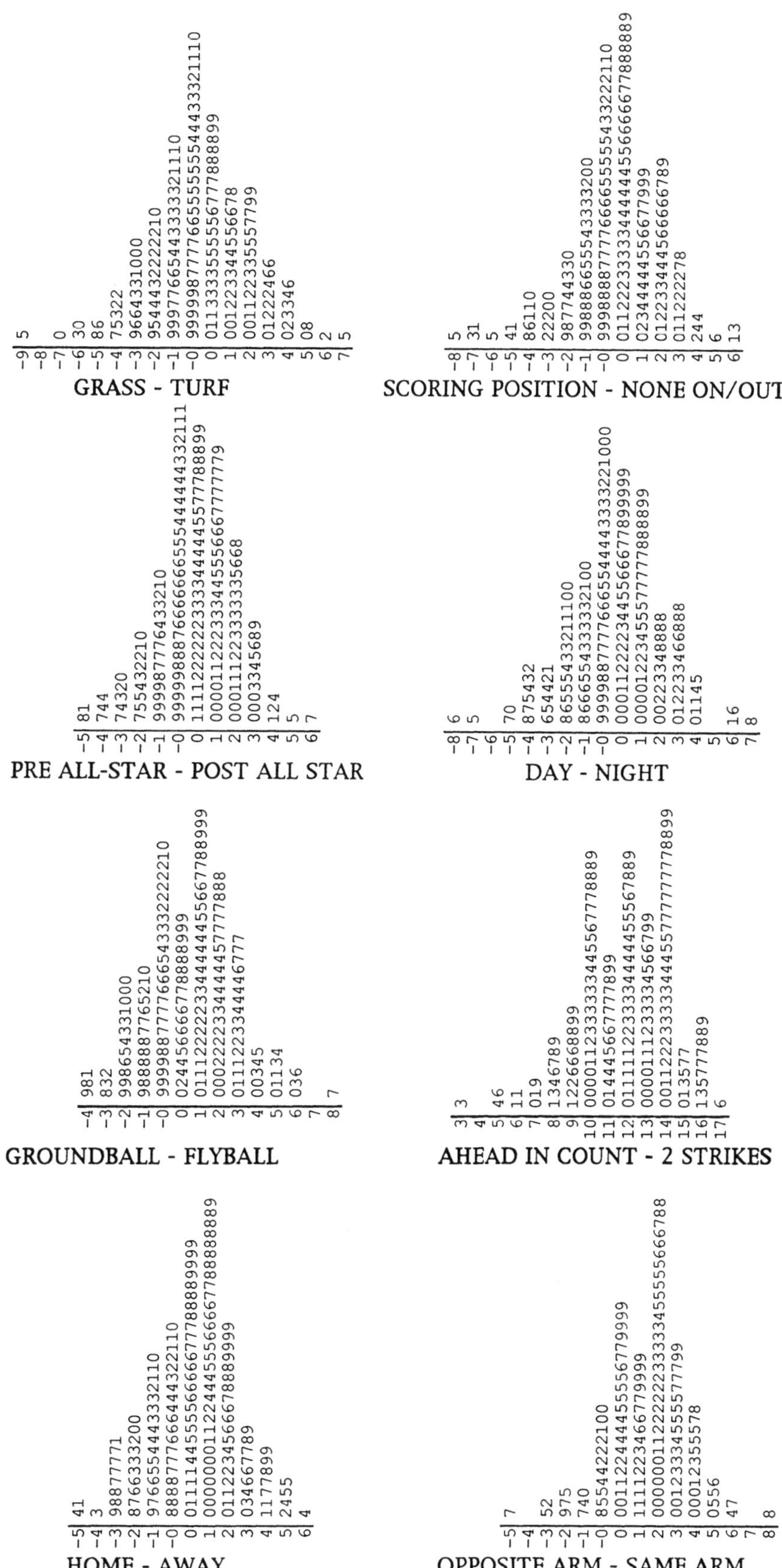

Figure 3. Stem-and-Leaf Diagrams of the Posterior Means of the Differences in Batting Averages for the Eight Situational Variables.

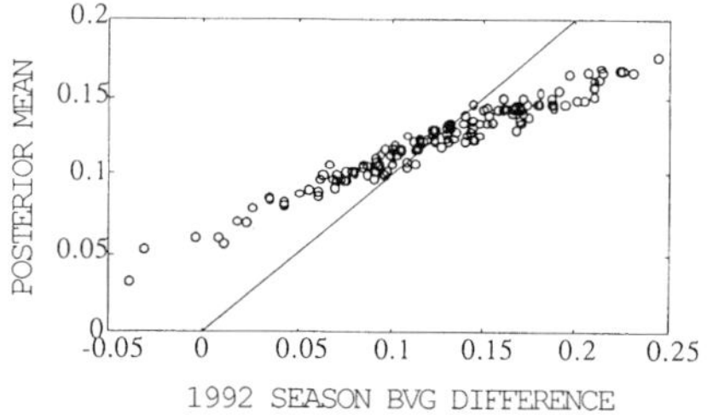

Figure 4. Posterior Means of the Differences in Batting Averages Plotted Against the Seasonal Difference in Batting Averages for the Situation "Ahead in Count/2 Strikes."

night", or "home–away." James argues that a player's ability to hit does not change just because he is playing on a different surface, or a different time of day, or at a particular ballpark, and so it is futile to look for individual differences in these situational variables. In contrast, the pitch count is an "ability split." There will exist individual differences in the "ahead in count–2 strikes" split, because one's batting average in the 2-strike setting is closely related to one's tendency to strike out. This statement is consistent with Figure 1, which indicates that players with high situational effects for pitch count during the previous 4-year period were likely to have high effects during the 1992 season. But there is much scatter in this graph, indicating that season performance is an imperfect measurement of this intrinsic ability.

Although there are clear situational patterns for the entire group of players, it is particularly difficult to see these patterns for individual players. We notice this in our brief study of nine unusual players in Section 3.2. These players had extreme estimated effects for the 1992 season, but many of them displayed effects of opposite sign for the previous 4-year period. The only player who seemed to have a clear outlying situational split was Tony Gwynn. But this does not mean that our search for players of high and low situational splits is futile. Rather, it means that we need more data to detect these patterns at an individual level.

Let us return to the book by Cramer and Dewan (1992), where the data were obtained. How does this book summarize the situational batting statistics that are listed? There is little discussion about the general size of the situational effects, so it is difficult to judge the significance of individual batting accomplishments. For example, suppose that a given hitter bats 100 points higher when ahead in the count compared with 2 strikes: Is that difference large or small? By our work, we would regard this as a relatively small difference, because it is smaller than the average of 123 batting average points for all players.

The book lists the 1992 batting leaders at the back. In the category of home games, we find that Gerry Sheffield had the highest batting average at home. But in this list, the players' hitting abilities and situational abilities are confounded; Sheffield's high average at home may reflect only the fact that Sheffield is a good hitter. In this article we have tried to isolate players' hitting abilities from their abilities to hit better or worse in different situations.

5. RELATED WORK

Because breakdown batting statistics are relatively new, there has been little statistical analysis of these data. There has been much discussion on one type of situational variable—hitting streaks or slumps during a season. Albright (1993) summarized recent literature on the detection of streaks and did his own analysis on season hitting data for 500 players. This data are notable, because a batter's hitting performance and the various situational categories are recorded for each plate appearance during this season. Albert (1993), in his discussion of Albright's paper, performed a number of stepwise regressions on this plate appearance data for 200 of the players. His results complement and extend the results described here. The "home–away" and "opposite arm–same arm" effects were found to be important for the aggregate of players. In addition, players generally hit for a higher batting average against weaker pitchers (deemed as such based on their high final season earned run averages). Other new variables that seemed to influence hitting were number of outs and runners on base, although the degree of these effects was much smaller than for the pitcher strength variable. The size of these latter effects appeared to be similar to the "home–away" effects. Players generally hit for a lower average with two outs in an inning and for a higher average with runners on base. This brief study suggests that there may be more to learn by looking at individual plate appearance data.

APPENDIX: DESCRIPTION OF THE POSTERIOR DISTRIBUTION OF THE SITUATIONAL EFFECTS AND SUMMARIZATION OF THE DISTRIBUTION USING THE GIBBS SAMPLER

The use of Bayesian hierarchical prior distributions to model structural beliefs about parameters has been described by Lindley and Smith (1972). The use of Gibbs sampling to simulate posterior distributions in hierarchical models was outlined by Gelfand, Hills, Racine-Poon, and Smith (1990). Albert (1992) used an outlier model, similar to that described in Section 2.3, to model homerun hitting data.

The complete model can be summarized as follows. For a given breakdown variable, we observe $\{(y_{i1}, y_{i2}), i = 1, \ldots, N\}$, where y_{ij} is the logit of the seasonal batting average of batter i in the jth category of the situation. We assume that the y_{ij} are independent, where y_{i1} is $N(\mu_i + \alpha_i, \sigma_{i1}^2)$ and y_{i2} is $N(\mu_i - \alpha_i, \sigma_{i2}^2)$, where the variances σ_{i1}^2 and σ_{i2}^2 are assumed known. The unknown parameters are $\alpha = (\alpha_1, \ldots, \alpha_N)$ and $\mu = (\mu_1, \ldots, \mu_N)$. Using the representation of a t density as a scale mixture of normals, the prior distribution for (α, μ) is written as the following two-stage distribution:

Stage 1. Conditional on the hyperparameters μ_α, σ_α, and $\lambda = (\lambda_1, \ldots, \lambda_N)$, α and μ are independent with μ distributed according to the vague prior $\pi(\mu) = c$ and the situational effect components $\alpha_1, \ldots, \alpha_N$ independent with α_i distributed $N(\mu_\alpha, \sigma_\alpha^2/\lambda_i)$.

Stage 2. The hyperparameters μ_α, σ_α, and λ are independent with μ_α distributed according to the vague prior $\pi(\mu_\alpha) = c$, σ_α^2 is distributed inverse gamma(a, b) density with kernel $(\sigma_\alpha^2)^{-(a+1)}\exp(-b/\sigma_\alpha^2)$, and $\lambda_1, \ldots, \lambda_N$ are iid from the gamma $(\nu/2, \nu/2)$ density with kernel $\lambda_i^{\nu/2-1}\exp(-\lambda_i\nu/2)$. The hyperparameters a, b, and ν are assumed known.

Table 4. Outlying Situational Players Where the Posterior Mean of the Scale Parameter $\lambda_i < .75$

Player	Situation	Batting average 1	Batting average 2	Season difference	Estimate of $p_1 - p_2$	Previous 4 years
Terry Steinbach	grass–turf	.251	.423	−.172	−.095	.033
Darrin Jackson	grass–turf	.278	.156	.122	.075	.020
Kevin Bass	scoring–none on/out	.205	.376	−.171	−.085	.063
Joe Oliver	scoring–none on/out	.172	.319	−.147	−.073	.079
Kent Hrbek	pre/AS–post/AS	.294	.184	.110	.074	−.007
Keith Miller	day–night	.167	.325	−.158	−.086	.022
Mike Devereaux	day–night	.193	.309	−.116	−.075	.035
Mickey Morandini	groundball–flyball	.324	.155	.169	.082	−.045
Tony Gwynn	ahead–2 strikes	.252	.291	−.039	.033	.061

Combining the likelihood and the prior, the joint posterior density of the parameters α, μ, μ_α, σ_α, and λ is given by

$$\exp\left\{-\frac{1}{2}\sum_{i=1}^{N}\left[\frac{(y_{i1}-\mu_i-\alpha_i)^2}{\sigma_{i1}^2}+\frac{(y_{i2}-\mu_i+\alpha_i)^2}{\sigma_{i2}^2}\right]\right\}$$
$$\times\left(\prod_{i=1}^{N}\lambda_i^{1/2}\right)\frac{1}{(\sigma_\alpha^2)^{N/2}}\exp\left\{-\frac{1}{2\sigma_\alpha^2}\sum_{i=1}^{N}\lambda_i(\alpha_i-\mu_\alpha)^2\right\}$$
$$\times\prod_{i=1}^{N}\left[\lambda_i^{\nu/2-1}\exp\left(-\frac{\lambda_i\nu}{2}\right)\right]\frac{1}{(\sigma_\alpha^2)^{a+1}}\exp\left(-\frac{b}{\sigma_\alpha^2}\right) \quad (A.1)$$

To implement the Gibbs sampler, we require the set of full conditional distributions; that is, the posterior distributions of each parameter conditional on all remaining parameters. From (A.1), these fully conditional distributions are given as follows:

a. $[\mu|\alpha, \mu_\alpha, \sigma_\alpha, \lambda]$. Define the variates $z_{i1} = y_{i1} - \alpha_i$ and $z_{i2} = y_{i2} + \alpha_i$. Then, conditional on all remaining parameters, the μ_i are independent normal with means $(z_{i1}/\sigma_{i1}^2 + z_{i2}/\sigma_{i2}^2)/(1/\sigma_{i1}^2 + 1/\sigma_{i2}^2)$ and variances $(1/\sigma_{i1}^2 + 1/\sigma_{i2}^2)^{-1}$.
b. $[\alpha|\mu, \mu_\alpha, \sigma_\alpha, \lambda]$. Define the variates $w_{i1} = y_{i1} - \mu_i$ and $w_{i2} = y_{i2} - \mu_i$. Then the α_i are independent normal with means $(w_{i1}/\sigma_{i1}^2 - w_{i2}/\sigma_{i2}^2 + \mu_\alpha\lambda_i/\sigma_\alpha^2)/(1/\sigma_{i1}^2 + 1/\sigma_{i2}^2)$ and variances $(1/\sigma_{i1}^2 + 1/\sigma_{i2}^2 + \lambda_i/\sigma_\alpha^2)^{-1}$.
c. $[\mu_\alpha|\mu, \alpha, \sigma_\alpha, \lambda]$ is $N(\Sigma\lambda_i\alpha_i/\Sigma\lambda_i, \sigma_\alpha^2/\Sigma\lambda_i)$.
d. $[\sigma_\alpha^2|\mu, \alpha, \mu_\alpha, \lambda]$ is inverse gamma with parameters $a_1 = N/2 + a$ and $b_1 = \Sigma_{i=1}^{N}\lambda_i(\alpha_i - \mu_\alpha)^2/2 + b$.
e. $[\lambda|\mu, \alpha, \mu_\alpha, \sigma_\alpha^2]$. The λ_i are independent from gamma distributions with parameters $a_i = (\nu + 1)/2$ and $b_i = \nu/2 + \lambda_i(\alpha_i - \mu_\alpha)^2/2\sigma_\alpha^2$.

To implement the Gibbs sampler, one starts with an initial guess at $(\mu, \alpha, \mu_\alpha, \sigma_\alpha, \lambda)$ and simulates in turn from the full conditional distributions a, b, c, d, and e, in that order. For a particular conditional simulation (say μ), one conditions on the most recent simulated values of the remaining parameters (α, μ_α, σ_α, and λ). One simulation of all of the parameters is referred to one cycle. One typically continues cycling until reaching a large number, say 1,000, of simulated values of all the parameters. Approximate convergence of the sampling to the joint posterior distribution is assessed by graphing the sequence of simulated values of each parameter and computing numerical standard errors for posterior means using the batch means method (Bratley, Fox, and Schrage 1987). For models such as these, the convergence (approximately) of this procedure to the joint posterior distribution takes only a small number of cycles, and the entire simulated sample generated can be regarded as an approximate sample from the distribution of interest.

[Received April 1993. Revised July 1993.]

REFERENCES

Albert, J. (1992), "A Bayesian analysis of a Poisson Random Effects Model for Homerun Hitters," *The American Statistician*, 46, 246–253.

—— (1993), Discussion of "A Statistical Analysis of Hitting Streaks in Baseball" by S. C. Albright, *Journal of the American Statistical Association*, 88, 1184–1188.

Albright, S. C. (1993), "A Statistical Analysis of Hitting Streaks in Baseball," *Journal of the American Statistical Association*, 88, 1175–1183.

Bratley, P., Fox, B., and Schrage, L. (1987), *A Guide to Simulation*, New York: Springer-Verlag.

Cramer, R., and Dewan, J. (1992), *STATS 1993 Player Profiles*, STATS, Inc.

Efron, B., and Morris, C. (1975), "Data Analysis Using Stein's Estimator and Its Generalizations," *Journal of the American Statistical Association*, 70, 311–319.

Gelfand, A. E., Hills, S. E., Racine-Poon, A., and Smith, A. F. M. (1990), "Illustration of Bayesian Inference in Normal Data Models Using Gibbs Sampling," *Journal of the American Statistical Association*, 85, 972–985.

James, B. (1986), *The Bill James Baseball Abstract*, New York: Ballantine.

Lindley, D. V., and Smith, A. F. M. (1972), "Bayes Estimates for the Linear Model," *Journal of the Royal Statistical Society*, Ser. B, 135, 370–384.

Morris, C. (1983), "Parametric Empirical Bayes Inference: Theory and Applications" (with discussion), *Journal of the American Statistical Association*, 78, 47–65.

Steffey, D. (1992), "Hierarchical Bayesian Modeling With Elicited Prior Information," *Communications in Statistics*, 21, 799–821.

Chapter 11

Did Shoeless Joe Jackson Throw the 1919 World Series?

Jay Bennett*

Joe Jackson and seven other White Sox were banned from major league baseball for throwing the 1919 World Series. This article examines the validity of Jackson's banishment with respect to his overall performance in the series. A hypothesis test of Jackson's clutch batting performance was performed by applying a resampling technique to Player Game Percentage, a statistic that measures a player's contribution to team victory. The test provides substantial support to Jackson's subsequent claims of innocence.

KEY WORDS: Clutch hitting; Hypothesis test; Player Game Percentage; Player Win Average; Resampling

In 1919, the Cincinnati Reds upset the Chicago White Sox in a best-of-nine World Series in eight games. A year later, two key Chicago players confessed to participating in a conspiracy to throw the 1919 World Series. Eight Chicago players (the Black Sox) were tried and found innocent of this crime. Nonetheless, all eight players were banned from major league baseball by the newly appointed commissioner Kenesaw Mountain Landis and never reinstated. The foremost player among the Black Sox was "Shoeless" Joe Jackson, who at the time of his banishment had compiled a lifetime .356 batting average (BA), the third highest average in baseball history. There is no doubt that his alleged participation in the fix is the only factor that prevents his election to the Hall of Fame.

Many questions have persisted about the 1919 World Series since the opening game of that Series. What evidence exists that the Black Sox did not play to their full ability? How well did players not accused in the scandal play in the Series? Especially mysterious is the case of Joe Jackson. He was one of the initial confessors, but later retracted his confession. Even if he had taken money for throwing the Series, his .375 batting average and record 12 hits in the Series indicate that he may have played on the level anyway. Detractors have commented that he may have compiled impressive statistics, but he didn't hit in the clutch.

This article presents a statistical analysis of the 1919 World Series. It uses the data from this Series to highlight the effectiveness of a new baseball statistic, Player Game Percentage (PGP). PGP's capability to account for game situation allows questions of clutch performance on the field to be answered. Thus PGP is uniquely capable of determining the extent of the scandal from the one true piece of available evidence: the record of play on the field. The focal point of the article is the analysis of the contention that Joe Jackson threw the 1919 World Series while batting .375.

1. PLAYER GAME PERCENTAGE

Fifty years after the Black Sox scandal, Mills and Mills (1970) introduced a remarkable baseball statistic, the Player Win Average (PWA). PWA was developed on the premise that performance of baseball players should be quantified based on the degree that he increases (or decreases) his team's chance for victory in each game. Of course, all established baseball statistics attempt to do this indirectly by counting hits, total bases, and runs. The novel part of the Mills' idea was that they estimated this contribution directly.

Consider this example which the Mills brothers estimated to be the biggest offensive play of the 1969 World Series, in which the Miracle Mets defeated the heavily favored Orioles. Al Weis of the Mets came to the plate in the top of the ninth inning of Game 2 with the scored tied. The Mets had runners on first and third with two outs. The probability of a Mets victory was 51.1%. If we define the win probability (WP) as the probability of a home team victory, then WP = 48.9%. Weis singled, which placed runners at first and second and knocked in the go-ahead run (the eventual Game Winning RBI [GWRBI]). His hit increased the Mets probability of winning to 84.9%. The difference ΔWP in the win probabilities before and after the at-bat outcome is awarded to the batter and pitcher as credits and debits. Win (Loss) Points are assigned on a play-by-play basis to players who increased (decreased) their team's chances of winning. Win and Loss Points are calculated as the product of $|\Delta WP|$ and a scale factor of 20 chosen by the Mills brothers. Thus Weis was credited $20 \times 33.8 = 676$ Win Points and the Orioles pitcher Dave McNally was awarded an equal number of Loss Points. The Win and Loss Points are accumulated for each player throughout the game or series of games. The Player Win Average for the game or series is

$$\text{PWA} = \frac{\text{Win Points}}{\text{Win Points} + \text{Loss Points}}. \quad (1)$$

The Mills' system is deceptively simple and yet solves many of the problems of standard baseball statistics:

- *Standard baseball statistics do not consider the game situation.* A walk with bases loaded and the game tied in the bottom of the ninth inning is much more important than a home run in the ninth inning of a 10–0 rout. The PWA system was expressly developed by the Mills brothers to take this "clutch" factor into account.
- *Standard baseball statistics of hitters and pitchers are not comparable.* PWA gives a single value for evaluating the performance of all players.

*Jay Bennett is Member of Technical Staff, Bellcore, Red Bank, NJ 07701. The author thanks John Flueck and John Healy for their helpful comments.

- *Standard baseball statistics do a poor job of evaluating relief pitchers.* Since PWA accounts for the game status when the relief pitcher enters the game, PWA allows meaningful comparisons among relief pitchers and between starters and relievers.
- *Standard baseball statistics do a poor job of evaluating fielding performance.* In the Mills system, the defensive player can be a fielder rather than the pitcher. Thus if a fielder makes an error, he receives the Loss Points that ordinarily would be awarded to the pitcher. Similarly, if the fielder makes a great play, the fielder would be awarded Win Points for the out rather than the pitcher. Fielding is evaluated within the context of the game like hitting and pitching.

Given all of these significant benefits, one wonders why PWA is not better known and commonly used. One major drawback is that the calculation of PWA is a greater data collection burden than standard baseball scorekeeping. A bigger obstacle is the estimation of win probabilities for every possible game situation. The Mills brothers outlined the basic estimation technique. First, data were collected on the fraction of time each standard baseball event occurred (e.g., the fraction of time that a home run was hit for each out and baserunner combination). Using these data, thousands of baseball games were simulated to obtain win probabilities for each situation. The Mills brothers provided many win probability estimates in their analysis of the 1969 World Series, but a complete set was not published.

Bennett and Flueck (1984) devised a technique for estimating these win probability values based on data collected from the 1959 and 1960 seasons (Lindsey 1961, 1963). The appendix describes the techniques used in these estimates. They demonstrated that these estimates were close to the values published by the Mills brothers in their 1969 World Series analysis. In addition, they replaced the Mills' PWA with a new statistic, Player Game Percentage (PGP). A player's PGP for a game generally is calculated as the sum of changes in the probability of winning the game for each play in which the player has participated. For most plays, half the change is credited to an offensive player (e.g., batter) and half to a defensive player (e.g., pitcher). Thus home players generally receive $\Delta WP/2$ for each play and visiting players receive $-\Delta WP/2$. In terms of Win and Loss Points, the Player Game Percentage for a game is

$$PGP = (\text{Win Points} - \text{Loss Points})/40. \qquad (2)$$

The denominator in the above equation accounts for this conversion to half credit and for the differences in scale.

To clarify the PGP analysis process, Table 1 provides excerpts from the PGP analysis for the most exciting game of the 1919 World Series, the sixth game in which the White Sox came from behind to win in extra innings by a 5–4 score. The scoring for this game (and all games in the Series) was derived from the play-by-play descriptions in Cohen, Neft, Johnson, and Deutsch (1976). The table presents the following information for each play of the game (all of which are required for PGP calculations):

Inning
Visiting team's runs
Home team's runs
Runners on base at the play's conclusion
Number of outs at the play's conclusion
Major offensive player
Major defensive player
Change in the win probability ΔWP (expressed as a percentage) for the home team from the previous play
Win probability WP (expressed as a percentage) for the home team at the play's conclusion

In plays with a major offensive player o and a major defensive player d, if the play occurs in the bottom of the inning, then

$$PGP_o = PGP_o + \Delta WP/2 \qquad (3)$$

$$PGP_d = PGP_d - \Delta WP/2, \qquad (4)$$

Table 1. Excerpts From PGP Analysis of Game 6 in the 1919 World Series

	Score				Players				
Inning	*Sox*	*Reds*	*Bases*	*Outs*	*Offense*	*Defense*	*Play*	*ΔWP*	*WP*
2h	0	0	0	1	Duncan	Kerr	go*	−2.21	52.58
2h	0	0	1	0	—	Risberg	e	5.71	58.29
2h	0	0	12	0	Kopf	Kerr	bb	6.01	64.30
2h	0	0	12	1	Neale	Kerr	go	−4.80	59.50
2h	0	0	13	2	Rariden	Kerr	go	−4.09	55.41
2h	0	0	0	3	Ruether	Kerr	go	−5.41	50.00
10v	4	4	2	0	Weaver	Ring	2b	−18.36	31.64
10v	4	4	13	0	Jackson	Ring	1b	−13.80	17.84
10v	4	4	13	1	Felsch	Ring	k	13.73	31.57
10v	5	4	13	1	Gandil	Ring	1b	−20.94	10.63
10v	5	4	0	3	Risberg	Ring	dp	7.87	18.50
10h	5	4	0	1	Roush	Kerr	go	−8.25	10.25
10h	5	4	0	2	Duncan	Kerr	fo	−5.65	4.60
10h	5	4	0	3	Kopf	Kerr	go	−4.60	.00

NOTE: 1b, single; 2b, double; bb, base on balls; dp, double play; e, error; fo, fly out; go, ground out; k, strikeout; go*, ground out if not for error.

and if the play occurs in the top of the inning, then

$$PGP_o = PGP_o - \Delta WP/2 \quad (5)$$

$$PGP_d = PGP_d + \Delta WP/2. \quad (6)$$

If no offensive player is given, ΔWP is assigned to the defensive player in its entirety. This always occurs for an error (*e*) and hit by pitcher. It is also given for a sacrifice hit and intentional walk if ΔWP is in favor of the defensive team; this is done so that the offensive player is not penalized for strategic moves made by managers. An "*" next to the play indicates that the play would have had this result if the defense had not made an error or an extraordinary play. For example, as shown in Table 1, in the bottom of the second inning with no score, Duncan of the Reds led off the inning with a ground ball to Risberg at shortstop. Risberg should have thrown out Duncan but he bobbled the ball and Duncan was safe. Risberg was charged with an error. For the purposes of PGP, this event was separated into two plays:

1. The groundout which should have occurred. This play decreases Duncan's PGP by 1.1 and increases the pitcher Kerr's PGP by 1.1. Thus, for an error, the batter is penalized (as in standard batting statistics) and the pitcher rewarded.
2. The error which actually occurred. Risberg's error turned an out with no runners on base into a no-out situation with a runner on first. His PGP was penalized 5.71.

In most plays, the major offensive player is the batter and the major defensive player is the pitcher. For stolen base and caught-stealing plays, the major offensive player is the runner and the major defensive player is the catcher. For a runner's advance or a runner thrown out, the major offensive player is the runner and the major defensive player is the fielder most involved in the play.

Using win probabilities from Table 1, the progress of a baseball game can be plotted in a manner comparable to graphs produced by Westfall (1990) for basketball. Figure 1 shows such a graph for Game 6 of the 1919 World Series. The figure plots win probability as a function of plate appearances. Innings are noted numerically along the x axis. As runs are scored, they are noted at the top of the graph for the home team and below the graph for the visiting team. Additional pointers indicate plays involving Joe Jackson. The plot dramatically describes how the Reds went out to an early lead and how the White Sox tied the game in the sixth inning and won it in the tenth. Large jumps in the graph are observed when important runs are scored (e.g., the Reds' first two runs at the end of the third inning). The importance of leadoff hits to establish scoring threats is emphasized by smaller jumps (e.g., Jackson's walk to start the eighth inning). Note how the jumps become larger in the later innings when each event becomes more important in determining the winner.

There are several advantages to PGP over PWA:

1. *PGP has a simpler interpretation than PWA.* A positive (negative) PGP value represents winning per-

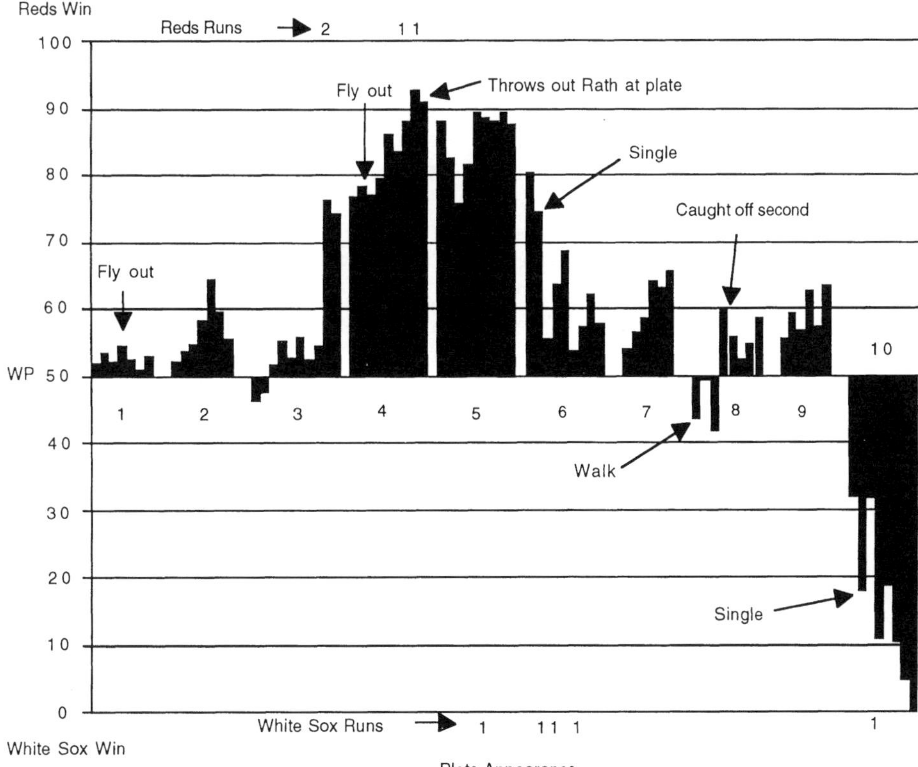

Figure 1. PGP Analysis of Game 6 of the 1919 World Series.

centage above (below) average (i.e., the .500 level). For example, if a player has 2,600 Win points and 2,400 Loss points, PGP = 5 which is easily interpreted as adding .050 to a team's winning percentage or about 8 wins to an average team's total wins for the year. The PWA value of .520 does not have such a simple interpretation.

2. *PGP is a more valid quantification of a player's contribution to victory.* Consider two sequences of plays for a relief pitcher's single inning of work:

- Sequence 1: The pitcher gives up two infield singles to the first two batters and then strikes out the side.
- Sequence 2: The pitcher faces three batters and strikes out the side.

Both sequences are the same with respect to the pitcher's contribution to team victory since the side was retired with no runs scored. PGP recognizes this and gives the same value for both sequences. PWA does not. The second sequence has no Loss Points for the pitcher; thus PWA is 1.000. The first sequence does have Loss Points from the singles; thus PWA is less than 1.000.

Questions have been posed concerning the heavy weighting that the PWA/PGP systems place in key events. For example, a PGP evaluation indicates that Kirk Gibson's only at-bat in the 1988 World Series made him the most valuable player of the Series. Gibson's game-winning home run raised the probability of a Dodger victory in Game 1 from 13.3% to 100%. This results in a 8.7 PGP average over the five-game Series, which tops that of his closest contender, teammate Orel Hershiser (the Series MVP as voted by the sportswriters) who had a 5.4 PGP Series average. While the sample size is too small for this to be a statistically significant assessment of their abilities, it is a reasonable assessment of the values of the actual performances of the two players in the Series. Gibson alone lifted his team from imminent defeat to a certain triumph. Hershiser pitched well, but unlike Gibson he received the support of other players (5.5 Dodger runs per game) to achieve his two victories. Thus, in general, baseball does have these critical points and PGP assesses them in a reasonable fashion.

Another question concerns the use of a game as the standard winning objective. For example, the win probability for the series could be used instead of summing the win probabilities for each game. In this way, events in early games would have less weight than events in the seventh game. Using this system, Bill Mazeroski's home run which won the 1960 World Series in the final play of Game 7 would probably be the most valuable play in World Series history. This system was considered at one point but was rejected primarily because the game is the basic unit of baseball achievement with respect to winning and losing. Using the game win probability allows the PGP system to be applied meaningfully and comparably to all levels of baseball play (season, championship playoffs, and World Series).

2. PGP ANALYSIS OF THE 1919 WORLD SERIES

For each Series game, Table 2 shows the score, the Most Valuable Player (MVP), and the Least Valuable Player (LVP), with respect to PGP. Typically, the MVP (LVP) is from the winning (losing) team and this holds true for each game of the 1919 World Series. All MVP's and LVP's were pitchers except for Felsch, Rath, Risberg, and Roush. Several players appear more than once. Chicago pitcher Dickie Kerr was MVP in both games he started. Pitchers Ring (Reds) and Cicotte (Sox) were MVPs and LVPs. A Black Sox player was the LVP in all five games lost by Chicago and was MVP in one of their three wins. Joe Jackson was neither MVP nor LVP in any game.

If all contributions to each player's PGP are summed and then divided by the number of games (8), the PGP results for the entire 1919 World Series are obtained as presented in Table 3. The eight Black Sox players have their names capitalized. One of these players, Swede Risberg, was the LVP of the Series. The MVP was Dickie Kerr who played on the losing side. To give some perspective to these performances, if Risberg played at this level throughout an entire season, he alone would reduce a .500 team to a .449 team. On the other hand, Kerr would raise a .500 team to a .541 winning percentage. Joe Jackson stands out clearly as the Black Sox player with the best performance.

3. JOE JACKSON'S PERFORMANCE

As shown in the previous section, Joe Jackson's contribution to team victory in the 1919 World Series was

Table 2. 1919 World Series Game Results

	Score		*Most Valuable*		*Least Valuable*	
Game	*White Sox*	*Reds*	*Player*	*PGP*	*Player*	*PGP*
1	1	9	Ruether	18.68	CICOTTE	−20.66
2	2	4	Sallee	19.30	RISBERG	−14.21
3	3	0	Kerr	17.85	Fisher	−14.33
4	0	2	Ring	29.44	CICOTTE	−10.04
5	0	5	Eller	22.33	FELSCH	−8.02
6	5	4	Kerr	15.34	Ring	−12.07
7	4	1	CICOTTE	18.28	Rath	−11.28
8	5	10	Roush	10.00	WILLIAMS	−14.58

NOTE: Black Sox players are capitalized.

Table 3. PGP per Game for the 1919 World Series Players

White Sox	PGP	Reds
Kerr, p	4.15	
	3.22	Eller, p
Schalk, c	2.86	
	2.17	Ring, p
	2.04	Wingo, c
	1.54	Sallee, p
JACKSON, lf	1.45	
	1.21	Roush, cf
	.66	Ruether, p
	.19	Duncan, lf
	.14	Luque, p
	.08	Magee, of
Lowdermilk, p	.00	
Smith, if	.00	
Mayer, p	.00	
Lynn, c	−.05	
Wilkinson, p	−.05	
MCMULLIN, 3b	−.21	
	−.27	Kopf, ss
Murphy, of	−.34	
WEAVER, 3b	−.40	
GANDIL, 1b	−.57	
S. Collins, of	−.84	
James, p	−1.06	
Leibold, rf	−1.14	
	−1.16	Rariden, c
	−1.19	Daubert, 1b
	−1.30	Neale, rf
CICOTTE, p	−1.55	
	−1.61	Fisher, p
	−2.01	Rath, 2b
FELSCH, cf	−2.79	
E. Collins, 2b	−2.89	
	−3.46	Groh, 3b
WILLIAMS, p	−3.69	
RISBERG, ss	−5.11	

NOTE: Black Sox players are capitalized.

superior not only to most players on his team, but also superior to that of most players on the opposing team. Table 4 summarizes his PGP performance in each game and traces his PGP average through the course of the Series. Clearly, Jackson's hitting was the strongest feature of his game; his batting PGP was the highest among all players in the Series. The only negatives are in stealing and base running. His most serious mistake occurred when he was doubled off second base for the third out with the score tied in the eighth inning of Game 6; however, his positive batting and fielding contributions in this game far outweighed this mistake. The fact that his key hit in the tenth inning resulted from his hustle to beat out a bunt gives further support to Jackson's case.

Jackson's batting average and slugging average (.563) were higher for the Series than during the regular season. Still, detractors state that indeed Jackson batted well, but he did not hit in the clutch (e.g., ". . . Joe Jackson's batting average only camouflaged his intentional failings in clutch situations, . . ." [Neft, Johnson, Cohen, and Deutsch 1974, p. 96]). We will examine this contention in several ways:

- By using traditional batting statistics for clutch situations;
- By using linear regression to determine if Jackson's batting PGP was low for a batter with a .563 slugging average (SA); and
- By using resampling techniques to create a PGP distribution for a .563 SA hitter batting in the situations encountered.

3.1 Traditional Statistics

Currently, the most complete analysis of clutch hitting is that performed by the Elias sports bureau in its annual publication (Siwoff, Hirdt, Hirdt, and Hirdt 1989). The analysis consists of examining standard baseball statistics (batting average, slugging average, and on-base average [OBA]) for certain clutch situations. They define a Late Inning Pressure (LIP) situation as a plate appearance occurring in the seventh inning or later when 1) the score is tied; 2) the batter's team trails by one, two, or three runs; or 3) the batter's team trails by four runs with the bases loaded. Table 5 shows the results of such an analysis applied to Jackson's plate appearances. Generally, Jackson performed well in these situations. By any yardstick, he performed especially well in LIP situations and when leading off an inning. This analysis gives exploratory-level indications that Jackson did hit in the clutch. However, it is difficult to draw a conclusion because of the multitude of clutch categories and the small sample sizes within each category.

3.2 Regression Analysis

PGP, in a sense, weights the Elias clutch categories and allows them to be pooled into an overall evaluator of clutch batting performance. We expect that a cor-

Table 4. Jackson's PGP for Each Game of the 1919 World Series

Game	Total	Batting	Stealing	Running	Fielding	Cumulative Average
1	−2.64	−2.64	.00	.00	.00	−2.64
2	6.36	6.36	.00	.00	.00	1.86
3	−.05	.76	−.81	.00	.00	1.22
4	−1.07	−1.07	.00	.00	.00	.65
5	−4.65	−4.65	.00	.00	.00	−.41
6	7.57	11.25	.00	−4.27	.59	.92
7	7.81	8.80	.00	−.99	.00	1.90
8	−1.77	−1.76	.00	.00	.00	1.45
Average	1.45	2.13	−.10	−.66	.07	

Table 5. Standard Situational Baseball Statistics for Joe Jackson's Plate Appearances in the 1919 World Series

Situation	AB	TB	H	BB	BA	SA	OBA
Leading off	9	6	4	1	.444	.667	.500
Runners on	17	7	6	0	.353	.412	.353
Scoring position	14	6	5	0	.357	.429	.357
2 Outs/runners on	6	2	2	0	.333	.333	.333
2 Outs/scoring position	5	2	2	0	.400	.400	.400
LIP	3	2	2	1	.667	.667	.750
LIP/runners on	1	1	1	0	1.000	1.000	1.000
LIP/scoring position	1	1	1	0	1.000	1.000	1.000
All	32	18	12	1	.375	.563	.394

relation should exist between batting PGP and traditional batting statistics. Since the traditional statistics do not account for the game situation, batting PGP should be higher than expected if the batter tended to hit better than expected in clutch situations and lower if the batter tended to hit worse in such situations. Thus, one method of determining if Jackson hit in clutch situations is to develop a linear regression line for the relationship between batting PGP and a standard batting statistic and see where Jackson's batting PGP stands with respect to this line. Cramer (1977) used a similar technique on PWA in an examination of the existence of clutch hitting.

Figure 2 plots batting PGP versus SA for all 17 players with a minimum of 16 at bats in the 1919 World Series. Similar results were obtained when BA and OBA were used in place of SA. The regression line shown was estimated using the data from all players except the Black Sox. Jackson's batting PGP was well above the regression line. Black Sox Risberg and Felsch not only batted poorly but also had lower batting PGP's than expected for those averages. Black Sox Weaver had a high SA but did not hit as well as expected in clutch situations. Weaver was banned from baseball not because of his participation in throwing games, but because he knew about the fix and did not reveal it to the baseball authorities. Part of his defense was his .324 BA in the Series. This analysis indicates that his high average was not as valuable as it might appear. In fact, it was no more valuable than that of the Black Sox Gandil who had a poor SA but did as well as expected in clutch situations with respect to that average. Given the limited nature of the data used to establish the regression, it is difficult to place any level of significance on the degree to which Jackson (or any other player) hit in the clutch using this analysis.

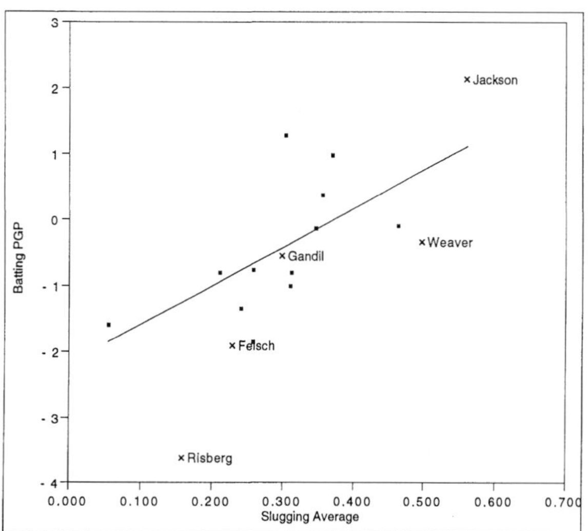

Figure 2. Regression Analysis of Batting PGP Versus Slugging Average in the 1919 World Series.

3.3 Resampling Analysis

In order to establish such a level of significance for clutch performance, a new technique using resampling was developed. Table 6 presents the situations and results for each of Joe Jackson's 33 plate appearances in the 1919 World Series. The ΔWP/2 values were calculated with the standard PGP approach using the before and after plate appearance situations. However, the resampling technique to be proposed required the classification of each plate appearance as one of the following:

- 1B/1—Single with runners advancing one base
- 1B/2—Single with runners advancing two bases
- 1B/U—Single that cannot be classified as 1B/1 or 1B/2. For example, if the single occurred with no runners on base, such a classification could not be made.
- 2B/2—Double with runners advancing two bases
- 2B/3—Double with all runners scoring
- 2B/U—Double that cannot be classified as 2B/2 or 2B/3.
- 3B—Triple
- HR—Home run
- BB—Base on balls
- K—Strikeout
- FO/NA—Fly out with no advance (e.g., foul pop-up)
- SF—Deep fly out in which a runner did score or could have scored from third
- FO/A—Fly out that advances all runners one base
- FO/U—Fly out that cannot be classified as FO/NA, FO/A, or SF

Table 6. Joe Jackson's Plate Appearances in the 1919 World Series

		Before At Bat			After At Bat				
Game	Inning	RD	Bases	Outs	RD	Bases	Outs	Play	ΔWP/2
1	2V	−1	0	0	−1	0	1	GO/U	−1.06
	4V	0	0	1	0	0	2	GO/U	−.87
	6V	−5	12	1	−5	23	2	GO/A	−.72
	9V	−8	0	0	−8	0	1	SF	.00
2	2V	0	0	0	0	2	0	2B/U	3.65
	4V	0	1	0	0	12	0	1B/1	3.53
	6V	−3	2	1	−3	2	2	K	−1.88
	8V	−2	0	2	−2	1	2	1B/U	1.06
3	2H	0	0	0	0	1	0	1B/U	1.75
	3H	2	12	0	2	12	1	FO/NA	−1.48
	6H	3	0	0	3	1	0	1B/U	.49
4	2H	0	0	0	0	2	0	2B/U	3.62
	3H	0	2	2	0	0	3	GO/U	−1.78
	6H	−2	0	0	−2	0	1	GO/U	−1.67
	8H	−2	0	1	−2	0	2	K	−1.24
5	1H	0	13	1	0	13	2	FO/NA	−2.86
	4H	0	0	1	0	0	2	GO/U	−.87
	7H	−4	0	0	−4	0	1	GO/U	−.79
	9H	−5	3	2	−5	0	3	GO/U	−.12
6	1V	0	1	2	0	0	3	FO/NA	−1.10
	4V	−2	0	1	−2	0	2	FO/NA	−.81
	6V	−3	2	0	−2	1	0	1B/2	2.95
	8V	0	0	0	0	1	0	BB	3.29
	10V	0	2	0	0	13	0	1B/1	6.90
7	1V	0	2	2	1	1	2	1B/2	4.84
	3V	1	2	2	2	1	2	1B/2	5.02
	5V	2	12	1	2	23	2	GO/A	−.80
	7V	3	0	1	3	0	2	GO/U	−.26
8	1H	−4	23	1	−4	23	2	FO/NA	−3.48
	3H	−5	0	2	−4	0	2	HR	2.21
	6H	−8	1	0	−8	1	1	SF	−.18
	8H	−9	23	1	−7	2	1	2B/U	.14
	9H	−5	23	2	−5	0	3	GO/U	−.45

NOTE: RD = (White Sox Runs) − (Reds Runs).

GO/DP—Ground ball that was or could have been turned into a double play
GO/F—Ground ball resulting in a force out
GO/A—Ground ball in which all runners advance one base
GO/U—Ground ball that cannot be classified as GO/DP, GO/F, or GO/A

The following procedure was used in the resampling technique. A random sample of 1,000 permutations of the 33 plays (listed in the "Play" column of Table 6) was created. For each permutation, a new batting PGP was calculated. This was done by calculating a new ΔWP/2 for the first play in the permutation occurring in the first situation (Game 1, top of the second inning, no outs, bases empty, White Sox trailing by a run), a new ΔWP/2 for the second play in the permutation occurring in the second situation (Game 1, top of the fourth inning, one out, bases empty, score tied), and so on through all 33 plays in the permutation for all 33 situations. The new batting PGP for the permutation is the sum of these 33 new ΔWP/2 values.

Calculation of new ΔWP/2 values for plays that had not actually occurred required the definition of reasonable rules for baserunner advancement and outs. Several plays (e.g., home run, walk) are clearly defined in their effects on baserunners, but most are less clear. For each of the play types described above (excluding the U plays such as GO/U), a table describing the motion of base runners, runs scored, and outs was constructed. Table 7 is an example of such a table for GO/

Table 7. Basic Resolution of GO/DP

Initial Base Situation	Final Base Situation	Runs Scored	Outs
0	0	0	1
1	0	0	2
2	2	0	1
12	3	0	2
3	3	0	1
13	0	1	2
23	23	0	1
123	3	1	2

DP. In some cases, the tables are overridden by special rules. For example, if a play results in the third out, no runs are scored (e.g., with bases loaded and one out, GO/DP ends the inning with no additional runs scored despite the indication in Table 7).

U plays required special procedures. The existence of U plays reflects the uncertainty of play type categorization because of the limited description of the play provided in the World Series record and because of the limited ability to predict the effect of each play in situations other than the one in which it actually occurred. The following assumptions were made here:

- GO/U—equal probability of being resolved as GO/A, GO/DP, or GO/F
- FO/U—equal probability of being resolved as FO/NA or SF
- 2B/U—equal probability of being resolved as 2B/2 or 2B/3
- 1B/U—2/3 probability of 1B/1 and 1/3 probability of 1B/2 unless there are two outs in which case the probabilities are reversed

A separate random number was selected for each U play of each permutation and the probabilities described above were used to select the appropriate play resolution. For example, Joe Jackson's batting record in the 1919 World Series had nine GO/U plays. In each permutation, nine separate random numbers were selected to determine whether each GO/U should be treated as a GO/A, GO/DP, or GO/F. Once the determination of play type was made, the normal tables and rules were used to determine ΔWP/2. Since a new random number was selected for each U play in each permutation, the variability resulting from the uncertainty of the play description was incorporated into the resulting distribution for batting PGP.

Using the resampling technique described above, both the batting situations and the batting performance (e.g., BA, SA, OBA) were kept constant. The only thing that changed the PGP was the pairing of play to situation. Using the 1,000 permutations, a sampling distribution for batting PGP/Game was established where the situations and batting performance were fixed. (Another possible approach would have been to generate random play results for each situation based on the player's season or career statistics. This was not done for three reasons. First, this would require the development of another model to generate play results from the available batting data. Second, and more importantly, the conditioning on the actual batting performance would be lost. The question to be tested here is whether Jackson "placed" his hits in a manner that was detrimental to his team. Third, given that Jackson's World Series BA and SA were higher than in the 1919 season and his career, the analysis performed provides a tougher test of his innocence.)

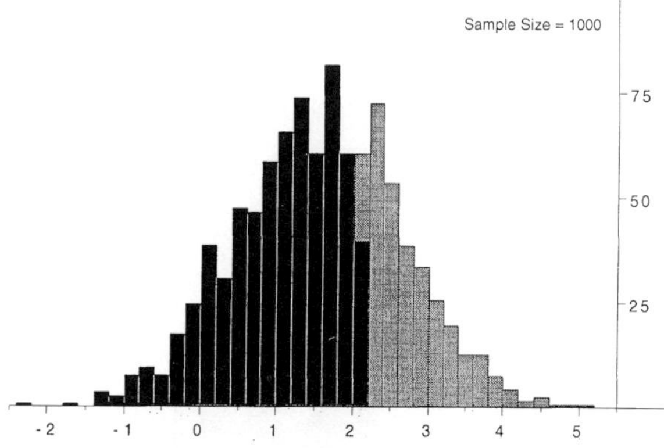

Figure 3. Sampling Distribution of Batting PGP/Game for Joe Jackson's Batting Record in the 1919 World Series. Darkened area indicates PGP/Game values less than Jackson's 2.13.

Since the distribution is a reflection only of the clutch factor of the batting performance, it can be used in a hypothesis test for clutch effect:

H_0: Joe Jackson's batting performance had no clutch effect (i.e., he batted as well as expected in clutch situations).

H_1: Joe Jackson did not hit as well as expected in the clutch.

To test at a .10 level of significance, the critical region for rejecting H_0 contains the values for the PGP sampling distribution lower than the .10 quantile.

Figure 3 shows the resulting distribution for the batting performance presented in the "Play" column of Table 6 when matched with the situations shown in that table. The mean PGP/Game value for this distribution is 1.56 and the median is 1.59. Jackson's PGP/Game value of 2.13 is the .686 quantile of this distribution. That is, Jackson's 1919 World Series batting had more clutch value than 68.6% of batters with his combination of batting results occurring in random association with his batting situations. Thus since the proposed hypothesis test has a .686 level of significance, hypothesis H_0 that Jackson's batting showed no clutch effect is accepted.

Of additional interest in this distribution is its wide variability. A player having a high BA and high SA and neutral clutch ability has an 8% chance of being a detriment to his team at bat. Thus, in one sense, Jackson's detractors were right in that it is possible to have high traditional batting statistics and yet have a negative impact on your team's chances to win. This quantitatively demonstrates that in a short series of baseball games the "when" of batting is just as (if not more) important as the quantity of hits or bases.

4. CONCLUSION

Did Shoeless Joe Jackson throw the 1919 World Series? Almost every statistical view of the game data supports the contention that Joe Jackson played to his full potential in the 1919 World Series. Not only did Jackson have higher batting statistics than he did during the regular season, but his batting PGP was also higher than expected from those statistics. An hypothesis test based on a PGP distribution developed from Jackson's World Series batting record strongly supports the null hypothesis of no clutch effect in his batting performance. This conclusion is also supported by the following analysis results:

- Jackson was the third most valuable player in the Series for his team and the seventh most valuable overall.
- As a batter, Jackson made a greater contribution to his team's chances for victory than any other batter in the Series.
- Jackson made a positive overall contribution toward White Sox victory in the Series while all other Black Sox had negative impacts.
- Jackson had high traditional batting statistics in most clutch situations (especially leading off and in late inning pressure).

The analysis also brought to light some interesting findings concerning other players. Buck Weaver appears to be the player to whom the high-BA, no-clutch contention appears to be most applicable. All of the controversy surrounding the Black Sox has obscured Eddie Collins' poor performance which was worse than six of the Black Sox. Collins had no positive contributions in any category (batting, fielding, baserunning, or stealing). His batting PGP was low even given his low SA for the Series. Only three White Sox had overall positive PGP values for the Series. Thus Chicago's loss in the Series cannot be blamed totally on the Black Sox.

This analysis has highlighted the possible applications of situational baseball statistics such as PGP. It has revealed the great degree to which game outcomes are determined not only by how many hits and of what type occur but also by when they occur. Since PGP can be directly interpreted in terms of additional games won or lost, it possesses great applicability in evaluating the value of a player to his team. Further research will focus on using the resampling techniques outlined in this article to determine batting PGP/Game distributions for different levels of BA, SA, and OBA in regular season play.

APPENDIX: ESTIMATION OF WIN PROBABILITIES

At any point in a baseball game, the probability $P(W|RD, I, H, O, B)$ of the home team winning may be defined given the following parameters:

- The run differential RD (i.e., home team runs minus visiting team runs)
- The half-inning (H and I) where $H = 1(2)$ indicates the top (bottom) half of an inning (e.g., bottom of the third inning)
- The number of outs O
- The on-base situation B (e.g., runners on first and third bases).

This appendix describes the techniques used to estimate these win probabilities as developed by Bennett and Flueck (1984).

Using data acquired from the 1959 and 1960 seasons, Lindsey (1961, 1963) estimated:

- The win probability at the end of each inning I given the score, $P(W|RD, I)$. Lindsey gave win probabilities only for cases where $|RD| < 7$. Win probabilities for $|RD| \geq 7$ were estimated as follows:

$$\begin{aligned} P(W|RD, I) &= P(W|-6, I)C_I^{-(6+RD)} && \text{if } RD < -6 \\ &= 1 - (1 - P(W|6, I))C_I^{RD-6} && \text{if } RD > 6. \end{aligned} \quad (7)$$

The values for C_I are shown in the following table. They were derived by fitting the above functional form to the win probabilities for high run differentials.

Inning I	1	2	3	4	5	6	7	8	≥9
C_I	.65	.64	.60	.57	.54	.49	.48	.30	.30

- The probability of scoring R more runs in an inning given the state of the inning, $P(R|O, B)$. Lindsey provided values only for cases in which $R < 3$. Run probabilities for $R \geq 3$ were estimated using the functional form

$$P(R|O, B) = a_{O.B} b_{O.B}^{R-3} \quad (8)$$

fitted to the expected number of runs scored $E(R|O, B)$ provided by Lindsey.

Using $P(R|O, B)$ and $P(W|RD, I)$, the probability of a home team victory at each stage of a game may be estimated. Clearly, by the definition of victory in baseball:

$$P(W|RD, I, H, O, B) = 1 \text{ if } RD > 0 \text{ and either a) } I = 9 \text{ and } H = 2 \text{ or b) } I > 9 \quad (9)$$

$$P(W|RD, I, 2, 3, B) = 0 \text{ if } RD < 0 \text{ and } I > 8. \quad (10)$$

The win probabilities at the end of each half-inning are

$$P(W|RD, I, 1, 3, B) = \sum_{R=0}^{\infty} P(R|0, 0)P(W|RD + R, I) \quad (11)$$

and

$$P(W|RD, I, 2, 3, B) = P(W|RD, I). \quad (12)$$

The win probabilities at the start of each half-inning are

$$\begin{aligned} P(W|RD, I, 1, 0, 0) &= .5 && \text{if } I = 1 \\ &= P(W|RD, I-1) && \text{if } I \geq 2 \end{aligned} \quad (13)$$

and

$$P(W|RD, I, 2, 0, 0) = P(W|RD, I, 1, 3, B). \quad (14)$$

Extra innings are assumed to be identical to the ninth inning:

$$P(W|RD, I, H, O, B) = P(W|RD, 9, H, O, B) \text{ if } I > 9 \text{ and } RD < 1. \quad (15)$$

Each win probability within a half-inning is the sum of the products of 1) the probability of scoring R more runs in the inning given the out-base situation and 2) the probability of winning given R more runs are scored:

$$P(W|RD, I, H, O, B) = \sum_{R=0}^{\infty} P(R|O, B)P(W|RD + (-1)^H R, I, H, 3, B). \quad (16)$$

The techniques described here can be applied to the appropriate baseball data from any period. The win probability estimates from the Lindsey data representing the 1959 and 1960 seasons were found to be close to the values published by the Mills brothers in their 1969 World Series analysis. In 1919 though, baseball was still in the deadball era. While batting averages and on-base percentages were quite similar, slugging averages (.388 vs. .348) and runs per game (8.6 vs. 7.7) were higher in 1960 than in 1919. Since runs were more valuable in 1919, we would expect that the $|\Delta WP/2|$ values for hits (especially those producing RBI's) are actually larger than those estimated from the 1960 data. Since probability tables such as those produced by Lindsey are not available for 1919, currently it is not possible to estimate the magnitude of these differences. However, given that the resampling analysis of Jackson's performance was performed on a relative basis and given the .686 p value for that test, it is unlikely that the use of 1919 data would lead to a different conclusion concerning Jackson's participation in the fix.

[*Received December 1991. Revised April 1993.*]

REFERENCES

Bennett, J. M., and Flueck, J. A. (1984), "Player Game Percentage," in *Proceedings of the Social Statistics Section, American Statistical Association*, pp. 378–380.

Cohen, R. M., Neft, D. S., Johnson, R. T., and Deutsch, J. A. (1976), *The World Series*, New York: Dial.

Cramer, R. D. (1977), "Do Clutch Hitters Exist?" *Baseball Research Journal*, Sixth Annual Historical and Statistical Review of the Society for American Baseball Research, 74–79.

Lindsey, G. R. (1961), "The Progress of the Score During a Baseball Game," *Journal of the American Statistical Association*, 56, 703–728.

——— (1963), "An Investigation of Strategies in Baseball," *Operations Research*, 11, 4, 477–501.

Mills, E. G., and Mills, H. D. (1970), *Player Win Averages*, New York: A. S. Barnes.

Neft, D. S., Johnson, R. T., Cohen, R. M., and Deutsch, J. A. (1974), *The Sports Encyclopedia: Baseball*, New York: Grosset & Dunlap.

Siwoff, S., Hirdt, S., Hirdt, T., and Hirdt, P. (1989), *The 1989 Elias Baseball Analyst*, New York: Collier.

Westfall, P. H. (1990), "Graphical Presentation of a Basketball Game," *The American Statistician*, 44, 305.

Chapter 12

PLAYER GAME PERCENTAGE

Jay M. Bennett, Bell Communications Research
John A. Flueck, NOAA/ERL and NCAR/FOF

1. Introduction

Player Game Percentage (PGP) is a statistical technique for evaluating a major league baseball player with respect to his contribution to winning games for his team. It is unique among baseball statistics in its ability to synthesize batting, pitching, fielding, and base running evaluations into a single value rating the player. Thus, not only may batters be compared with other batters and pitchers with other pitchers, but batters may be compared with pitchers in their ultimate contribution to team victory. PGP also simplifies the comparison of relief pitchers with other relievers as well as with starting pitchers. PGP has its genesis with the Player Win Average (PWA) concept.[1] While most baseball statistics are direct (e.g. ERA, RBI) or indirect (e.g. BA) measures of runs scored,[2] PWA goes directly to the heart of the matter by measuring changes in the probability of winning with each game event. At any point in a baseball game, the probability P of the home team winning may be estimated given the following parameters:

- The run differential RD (i.e. home team runs minus visiting team runs)
- The half-inning (H and I) (e.g. bottom of the third inning)
- The number of outs O
- The on-base situation B (e.g. runners on first and third bases).

A baseball event may then be defined as an occurrence which changes any of the above parameters. Thus, baseball events include not only hits, walks, double plays but also stolen bases, passed balls, and balks.

Before each baseball event, there is a prior probability of the home team winning $P(E-)$ where $E-$ is the state of the game prior to the event. Similarly, after the event there is an analogous probability $P(E+)$ where $E+$ is the state after the event. The net effect of the event E on the outcome of the game is

$$N(E) = P(E+) - P(E-).$$

Most baseball events have a major offensive player and a major defensive player who are most responsible for the event's occurrence. Each of these players may then be credited with half the change in probability from the event. The home team player is credited with $N(E)/2$ percentage points while the visiting player is credited with the negative of this same amount.

The Player Game Percentage (PGP) for a player is the sum of these percentage point credits over the measurement period desired. A player's PGP may be calculated over a single game, a series of games such as the World Series, a baseball season, or an entire baseball career. The PGP Average is the total PGP divided by the number of games considered.

PGP differs from PWA in several important ways:

- While PWA converts the win probabilities into Win/Loss Points on a scale of -1000 to 1000, PGP deals directly in probabilities in the form of percentages. This form enhances understanding and reduces unnecessary arithmetic conversions.
- While PWA is analogous to PGP Average, PWA does not use number of games in its denominator. Instead it uses the sum of credits (Win Points) and debits (Loss Points) in the denominator. The disadvantage of this technique is the elimination of the additive properties of the averages. It also tends to reduce the averages of players in critical situations and increase those of players in non-critical situations.
- PGP adds several new scoring concepts including an intermediate event evaluation (described in Section 4) to more precisely assign credit for events.
- PGP uses a different set of win probabilities. Their calculation is discussed in the next section.

2. Estimation of Win Probabilities

From the above discussion, it is clear that the key to the PGP concept lies in the estimation of the win probability for each possible state of a baseball game. Mills and Mills[1] obtained their estimates by recording the frequencies of each event in each state during the 1969 baseball season. They then used this information to simulate thousands of games and thus obtain the required probabilities. Unfortunately, a complete set of these probability values was not published.

For the estimation of the win probabilities for PGP, we relied on data from other published sources. Using data acquired from the 1959 and 1960 seasons, Lindsey estimated:

- The win probability at the end of each inning I given the score, $P(W|RD,I)$.[3] Since Lindsey gave win probabilities for only a limited number of cases ($|RD|<7$), we assumed that

$$P(W|RD,I) = \begin{cases} P(W|-6,I) \times 2^{6+RD} & \text{if } RD<-6 \\ P(W|6,I)+(1-P(W|6,I)) \times 2^{6-RD} & \text{if } RD>6. \end{cases}$$

- The probability of scoring R more runs in an inning given the state of the inning, $P(R|O,B)$.[4] Again, since Lindsey provides a limited number of cases ($R<3$), it was assumed for $R>2$ that

$$P(R|O,B) = (1 - \sum_{r=0}^{2} P(r|O,B)) \times 2^{2-R}.$$

With Lindsey's data, it is possible to calculate the probability of a home team victory at each stage of a game, $P(W|RD,I,H,O,B)$ where $H=1(2)$ indicates the visiting (home) team at bat. Clearly, by the definition of victory in baseball:

- $P(W|RD,I,H,O,B) = 1$ if $RD>0$ and either a) $I=9$ and $H=2$ or b) $I>9$.
- $P(W|RD,I,2,3,B) = 0$ if $RD<0$ and $I>8$.

The probabilities at the end of each half-inning are

- $P(W|RD,I,1,3,B) = \sum_{R=0}^{\infty} P(R|0,0)\, P(W|RD+R,I)$
- $P(W|RD,I,2,3,B) = P(W|RD,I)$.

The probabilities within half-innings are the sums of the products of the probability of scoring R more runs in the inning given the out-base situation and the probability of winning given R more runs are scored:

$$P(W|RD,I,H,O,B) = \sum_{R=0}^{\infty} P(R|O,B)P(W|RD+(-1)^H R,I,H,3,B).$$

To complete the calculations, we assume that all extra innings are identical in nature to the ninth inning:

$P(W|RD,I,H,O,B) = P(W|RD,9,H,O,B)$ if $I>9$ and $RD<1$.

We also assume that the game starts as an even contest:

$$P(W|0,1,1,0,0) = .5.$$

Clearly, for $1<I<10$,

$$P(W|RD,I,1,0,0) = P(W|RD,I-1,2,3,B)$$

and, for $0<I<10$,

$$P(W|RD,I,2,0,0) = P(W|RD,I,1,3,B).$$

3. Comparison of PGP and PWA Win Probabilities

A limited comparison of $P(W|RD,I,H,O,B)$ values used for PWA with those calculated using the techniques described above was made. Tables 1 and 2 show the $P(W|RD,I,H,3,0)$ used by Mills and Mills[1] in their evaluation of the 1969 World Series along with the comparable values calculated using the procedures from Section 2. Across both tables, the median difference of PWA-PGP is 0.0% and the median absolute value difference |PWA-PGP| is 0.4%. The means are the same when rounded to the nearest tenth of a percent.

Thus, although the PGP probabilities are not exactly equal to the PWA probabilities, they are reasonably close to be useful in player evaluations. These results are quite remarkable when one considers that the PGP probabilities are based on data from major league baseball a decade prior to PWA. Over this period there were many changes in baseball most notably the expansion of both leagues. In addition, PGP uses a deterministic set of equations to simply calculate the probabilities while PWA probabilities required extensive simulation for their creation. This allows PGP to calculate probabilities as needed instead of referencing a large data base of probabilities. Thus, PGP can easily fit on the smallest of personal computers (such as the TI-99/4A on which some of these calculations were made) to make it easily accessible to the general public.

4. A PGP Example

One of the great strengths of PGP is its ability to analyze defensive contributions along with offensive contributions. As an illustration of a PGP calculation, we have selected a great defensive play from the 1980 World Series. In Game Three with the score tied in the top of the tenth inning, Mike Schmidt of the Philadelphia Phillies came to bat against Dan Quisenberry of the Kansas City Royals. The Phillies had one out and runners on first and second bases. The probability $P(W|0,10,1,1,1\&2)$ of a home victory for the Royals at this point was 40.7%. Schmidt hit a low liner that appeared to be a sure single past second base. If the ball had gone through for a hit, conservatively the Phillies would have had the bases loaded and still only one out; the probability of a Royals' victory would have been reduced to $P(W|0,10,1,1,1\&2\&3) = 26.2\%$. However, the Royals' second baseman Frank White made an extraordinary play to intercept the liner and double the runner on second off the bag. The final result of the play was the end of the Phillies' turn at bat with no runs scored. The Royals' probability of victory increased to $P(W|0,10,1,3,0) = 62.6\%$.

In the player PGP evaluations for this event, Schmidt is responsible on the offensive side. His PGP total for the game is therefore debited half of the overall change in the probability

$$-10.95\% = -(62.6\% - 40.7\%)/2.$$

Evaluation on the defensive side is more complex. Quisenberry is responsible for allowing Schmidt to almost get a hit loading the bases; so, Quisenberry's PGP total is debited half of the intermediate change in probability

$$-7.25\% = (26.2\% - 40.7\%)/2.$$

White is responsible for preventing Schmidt's hit and doubling the runner off second; his PGP total is increased half of the other intermediate change

$$18.20\% = (62.6\% - 26.2\%)/2.$$

Unlike any other scoring system, PGP correctly gives credit to White for getting Quisenberry out of a jam.

5. PGP Evaluations of the 1980 World Series

Tables 3 and 4 present the PGP evaluations of the Phillies and Royals respectively on a game by game basis in addition to an overall PGP series average (i.e., PGP player total/6). Unser, Schmidt, and McBride were the outstanding players on the World Champion Phillies while Aikens and Otis were their counterparts on the Royals. All of these players were everyday players except for Unser who contributed two outstanding pinch hits for the Phillies. One of these pinch hits was the single most significant play in the 1980 World Series according to PGP. In the top of the ninth inning of Game Five, the Royals held a one run lead. Unser came to bat against the Royals' Dan Quisenberry with no outs and a runner on first base. Unser ripped a double which tied the game and put him in position to score the go-ahead run. Unser's hit raised the probability of a Phillies' victory from 32.3% to 74.2%, a change of 41.9%. Unser's performance can be contrasted with that of the Royals' key pinch hitter Cardenal, who did not perform as well in that role resulting in the second worst PGP of all participants in the series. PGP, thus, highlights the importance of pinch hitting at critical points of the game.

One of the best features of PGP is its ability to directly assess the impact of player substitution on winning. If Willie Aikens (playing at his 1980 World Series level) had replaced an average player (PGP average = 0%) on a .500 team, PGP predicts that his addition alone would have lifted that team to divisional contender status with a .580 record:

New Team Record = Old Team Record + PGP Average/100.

A similar substitution involving Quisenberry would have reduced the same team to a .403 record. However, these cases are extreme ones; it is unlikely that such high and low PGP averages would be recorded for an entire season.

The worst performance of the series was that of Quisenberry who lost several late leads for the Royals. In fact, relief pitching in general was not outstanding in this series. McGraw, widely regarded as a Phillies pitching hero, in reality had only an average series (PGP average near zero) mostly a result of his generally-neglected loss of Game Three.

One of the surprises from the PGP analysis of the series is the relatively high evaluation of Willie Wilson's performance. In spite of his record-setting 12 strikeouts, Wilson had the third highest PGP average on the Royals. Wilson's efforts in scoring the winning run of Game Three and his outstanding fielding are highlighted by PGP.

6. CONCLUSION

Assessing player performance in baseball using probabilities of victory has lain dormant since its inception in 1969. Much of this concept's relative obscurity is a result of the reticence of the concept's innovators to reveal the probabilities which are at the heart of the system. Thus, baseball researchers and fans have been unable to test and experiment with these ideas. This paper shows how the required probabilities may be calculated from published data. Although this data is nearly a quarter century old, the calculated probabilities were shown to be quite close to those used in PWA ten years later. The techniques themselves do not date and will be as applicable to 1984 baseball data as they are to 1960 baseball data. A study of how (and if) these probabilities change with time would be quite interesting.

An analysis of the 1980 World Series was shown as an example of the insights provided by PGP. The great advantages of PGP are:

- A synthesis of hitting, pitching, running, and fielding evaluations into one quantitative value
- A quantitative tool to measure the previously unmeasurable player contributions such as the advancement of base runners on outs
- A direct measure of how much a player contributes to winning
- Evaluation of pinch hitters and relief pitchers conditional on the state of the game at the moment of their entrance.

The only drawbacks to PGP are the detail of record-keeping in a game required for the proper state evaluations and the somewhat subjective assignment of responsibility for plays. However, it should be emphasized that PGP is not intended to be the final arbiter in player evaluations. Like diagnostic measures for the detection of outliers in regression, PGP is a tool to aid the baseball researcher and fan in achieving a previously unknown perspective for player evaluation.

Baseball has been relatively amenable to this probabilistic approach because of its limited quantized nature. Efforts should be made to extend these concepts to other sports such as football and basketball.

References

1. MILLS, ELDON G., and MILLS, HARLAN D. (1970), *Player Win Averages*, New York: A. S. Barnes.

2. BENNETT, JAY M., and FLUECK, JOHN A. (1983), "An Evaluation of Major League Baseball Offensive Performance Models," *The American Statistician*, 37, 1, 76-82.

3. LINDSEY, G. R. (1961), "The Progress of the Score During a Baseball Game," *American Statistical Association Journal*, 56, 703-728.

4. LINDSEY, G. R. (1963), "An Investigation of Strategies in Baseball," *Operations Research*, 11, 4, 477-501.

TABLE 1. $P(W|RD,I,1,3,0)$ Used in PWA and in PGP

I	RD	PWA	PGP	PWA-PGP
1	0	54.5	54.5	0.0
2	0	54.8	54.8	0.0
2	1	66.3	65.7	0.6
3	-3	21.0	21.9	-0.9
3	0	55.1	55.2	-0.1
3	1	67.7	67.3	0.4
3	3	85.5	85.0	0.5
4	-3	19.0	19.9	-0.9
4	-1	41.0	41.6	-0.6
4	1	69.5	69.1	0.4
4	3	87.5	87.0	0.5
5	-3	16.6	17.3	-0.7
5	-1	39.4	40.3	-0.9
5	1	71.9	71.3	0.6
5	3	89.8	89.5	0.3
5	4	94.2	94.0	0.2
6	-3	13.8	14.4	-0.6
6	-1	37.2	37.9	-0.7
6	1	75.2	74.8	0.4
6	3	92.3	92.1	0.2
6	4	95.9	95.7	0.2
7	-1	33.6	33.3	0.3
7	1	80.1	81.1	-1.0
7	3	95.0	95.1	-0.1
7	4	97.6	98.3	-0.7
8	0	59.7	60.0	-0.3
8	1	87.7	87.4	0.3
8	3	97.6	97.8	-0.2
8	4	99.0	98.9	0.1
9	-1	17.8	18.5	-0.7
9	0	62.2	62.6	-0.4
10	0	62.2	62.6	-0.4

TABLE 2. $P(W|RD,I,2,3,0)$ Used in PWA and in PGP

I	RD	PWA	PGP	PWA-PGP
1	0	50.0	50.0	0.0
1	1	61.3	61.0	0.3
2	0	50.0	50.0	0.0
2	1	62.3	61.7	0.6
2	3	81.5	80.3	1.2
3	0	50.0	50.0	0.0
3	1	63.6	63.0	0.6
3	3	83.4	82.7	0.7
4	0	50.0	50.0	0.0
4	1	65.3	64.7	0.6
4	3	85.6	84.9	0.7
4	4	91.3	90.8	0.5
5	1	67.6	66.8	0.8
5	3	88.1	87.7	0.4
5	4	93.2	93.0	0.2
6	1	71.1	70.5	0.6
6	4	95.2	94.9	0.3
7	0	50.0	50.0	0.0
7	1	76.5	77.6	-1.1
7	3	94.0	94.0	0.0
7	4	97.1	98.0	-0.9
8	0	50.0	50.0	0.0
8	1	85.2	84.6	0.6
8	2	93.4	93.6	-0.2
8	3	97.1	97.4	-0.3
8	5	99.5	99.5	0.0

TABLE 3. Phillies PGP Evaluations

		GAME					
PLAYER	PGP AVE	1	2	3	4	5	6
Unser	5.44		13.10		-1.40	20.95	
Schmidt	4.63	1.55	14.35	-5.70	-1.90	12.95	6.55
McBride	3.94	17.25	9.05	-0.95	-2.55	2.55	-1.70
Carlton	2.03		-3.55				15.75
Boone	1.45	11.80	-0.05	-3.40	-0.75	-1.95	3.05
Ruthven	1.27			7.60			
Rose	0.76	-1.05	-1.50	8.65	1.95	-10.00	6.50
Noles	0.27				1.60		
Brusstar	0.26				1.55		
McGraw	0.13	12.65		-24.40		12.60	-0.10
Reed	0.08		3.20			-2.70	
Saucier	0.02				0.10		
Bowa	-0.14	-0.65	3.40	-0.55	2.15	-3.05	-2.15
Moreland	-0.19		-1.40	-3.60		3.85	
Gross	-0.30	-0.65	-0.60	-0.55			
Luzinski	-0.50	0.10				0.75	-3.85
Smith	-1.57	-2.35	-3.85	2.35	-4.70		-0.85
Maddox	-2.15	1.55	4.20	-1.10	-2.70	-15.55	0.70
Bystrom	-2.45					-14.70	
Walk	-2.46	-14.75					
Christenson	-2.81				-16.85		
Trillo	-3.91	-0.45	-20.50	-6.50	-2.55	9.85	-3.30

TABLE 4. Royals PGP Evaluations

		GAME					
PLAYER	PGP AVE	1	2	3	4	5	6
Aikens	8.04	12.20	3.35	19.30	8.75	3.40	1.25
Otis	7.02	11.95	7.75	4.65	2.10	15.65	0.00
Wilson	1.49	-5.70	2.30	2.65	7.15	4.75	-2.20
Chalk	0.91		5.45				
Martin	0.39	0.20		-1.55			3.70
White	0.10	-4.20	-9.15	10.20	-4.10	22.05	-14.20
Pattin	0.02						0.10
Hurdle	-0.10	-0.80		-0.05	1.75	-1.50	
Splittorf	-0.54						-3.25
Gura	-0.59		1.25			-4.80	
Wathan	-1.23	-6.20	0.15				-1.35
Gale	-1.71			-2.55			-7.70
Brett	-2.05	-1.15	3.05	5.55	5.55	-22.80	-2.50
Washington	-2.09	-4.85	-3.35	-8.80	-0.95	4.05	1.35
McRae	-2.15	-0.40	2.25	3.35	0.90	-13.35	-5.65
Porter	-2.36	-3.20	-1.65	-7.95	-1.55	0.20	
Leonard	-4.13	-23.50			-1.25		
Cardenal	-4.16	-10.10			-15.65	0.80	
Quisenberry	-9.73	0.65	-35.45	-2.95	5.60	-26.45	0.25

Chapter 13

Estimation With Selected Binomial Information or Do You Really Believe That Dave Winfield is Batting .471?

George CASELLA and Roger L. BERGER*

Often sports announcers, particularly in baseball, provide the listener with exaggerated information concerning a player's performance. For example, we may be told that Dave Winfield, a popular baseball player, has hit safely in 8 of his last 17 chances (a batting average of .471). This is biased, or selected information, as the "17" was chosen to maximize the reported percentage. We model this as observing a maximum success rate of a Bernoulli process and show how to construct the likelihood function for a player's true batting ability. The likelihood function is a high-degree polynomial, but it can be computed exactly. Alternatively, the problem yields to solutions based on either the EM algorithm or Gibbs sampling. Using these techniques, we compute maximum likelihood estimators, Bayes estimators, and associated measures of error. We also show how to approximate the likelihood using a Brownian motion calculation. We find that although constructing good estimators from selected information is difficult, we seem to be able to estimate better than expected, particularly when using prior information. The estimators are illustrated with data from the 1992 Major League Baseball season.

KEY WORDS: Brownian motion; EM algorithm; Gibbs sampling; Selection bias.

1. INTRODUCTION

Sports announcers—in particular, baseball announcers—often use hyperbolical descriptions of a player's ability. For example, when Dave Winfield, a popular baseball player, is batting, rather than report his current batting average (number of hits divided by number of at-bats), it might be said that "he's really hitting well these days; he's 8 for his last 17." This is clearly selectively reported data, biased upward from the player's actual average. But with models that take this bias into account, we should be able to use the selectively reported data to recover an estimate of the player's true ability. In this article we explore various methodologies for doing this.

1.1 Background

Research in estimation and modeling from selectively reported data has always been of interest and has many applications other than analyzing baseball data. We will not attempt a thorough literature review here but will describe some general directions that such research has taken. Perhaps the most widespread use of selection-bias methodology is in the area of meta-analysis. Starting from work of Rosenthal (1979), researchers have worried about the effect of selectively reported data when combining results of different studies, where the selection is mainly through publication bias (publishing only significant studies). These concerns have been summarized and reviewed by Iyengar and Greenhouse (1988) and, more recently, in a trio of papers in *Statistical Science* (Dear and Begg 1992; Hedges 1992; Mosteller and Chalmers 1992). Cleary (1993) has used these selection models, along with likelihood theory and Gibbs sampling, to construct estimates of effects based on publication-biased data.

A Bayesian approach to inference from selected data was taken by Bayarri and DeGroot (1986a,b; 1991). A major lesson to be learned from their work is that the uncorrected maximum likelihood estimator (MLE) can be exceedingly bad. In our baseball data, it is quite obvious that the naive estimate of Winfield's batting ability ($8/17 = .471$) is vastly incorrect. In other, more complicated situations, this might not be so obvious.

Perhaps the methodology most similar to ours here is that of Dawid and Dickey (1977). They were concerned with the influence of selectively reported data on the likelihood function, and how such influence can be accounted for. In particular, they considered an example where the selectively reported data is the maximum of sums of Bernoulli random variables, an example closely related to our situation. More recently, Carlin and Gelfand (1992) have studied parametric likelihood inference in "record-breaking" data; that is, data that are a series of records, or maxima. They discussed many applications of their models, including sporting events, meteorology, and industrial stress testing. In particular, they modeled an underlying regression that attempts to explain the increasing sequence of means and illustrated their techniques using data on Olympic record high jumps. The selected data that we are concerned with here may be thought of as a special case of "record breaking" data. But our models and estimation methodologies are different from previous approaches.

1.2 Information

To make inferences from our selected data, we must make some assumptions about the data we see. For example, when we are told that Dave Winfield is 8 for his last 17, we assume that the "17" is chosen because the ratio $8/17$ is the maximum ratio of hits to at-bats. There is some hidden information in this number. For example, we know that on his 18th and 19th previous at-bats he did not get a hit; otherwise, the announcer would have reported $9/18 > 8/17$ or $9/19 > 8/17$. Moreover, our naive estimate of his ability should

* George Casella is Professor, Biometrics Unit, Cornell University, Ithaca, NY 14853. Roger L. Berger is Professor, Department of Statistics, North Carolina State University, Raleigh, NC 27695. This research was supported by National Science Foundation Grant DMS9100839 and National Security Agency Grant 90F-073. The authors thank Steve Hirdt of the Elias Sports Bureau for providing detailed data for the 1992 Major League Baseball season. They also thank Marty Wells for numerous conversations concerning the Gibbs/EM algorithm implementation and the editors and referees for many constructive comments on an earlier version of this article.

Journal of the American Statistical Association
September 1994, Vol. 89, No. 427, Statistics in Sports

not be $8/17 = .471$, but rather $8/19 = .421$, because we know that the two previous at-bats (18th and 19th) were failures. More precisely, we assume that a baseball player's sequence of at-bats is a sequence of Bernoulli(θ) random variables, $X_1, \ldots, X_n$, where the X_i's are in chronological order. After the nth at-bat, the player's batting average is $\hat{p} = \sum_{i=1}^{n} X_i/n$. This is the MLE based on observing the entire (complete) data set. But we assume that the data reported are k^* hits out of the last m^* at-bats, where k^* and m^* satisfy

$$r^* = \frac{k^*}{m^*} \geq \max_{m^* \leq i < n} \frac{X_n + X_{n-1} + \cdots + X_{n-i}}{i+1}. \quad (1)$$

Note that the quantity $(X_n + X_{n-1} + \cdots + X_{n-i})/(i+1)$ is just a player's batting average in the previous $i + 1$ at-bats. Thus we are assuming only that there is no higher hits to at-bats ratio in the unreported data than the reported ratio of $r^* = k^*/m^*$. The value of r^* also tells us the number of at-bats previous to m^* that we know to be failures, in that $r^* < 1/j$ implies failure on the previous j at-bats. (Here $r^* = 8/17 < 1/2$, which implies failure on the previous two at-bats.) There may be a higher ratio in the last m^* at-bats; for example, perhaps the batter was 1 for 1 in his last at-bat. In practice, with the exception of similar trivial cases, r^* will usually represent the maximum ratio of hits to at-bats. Also notice that the exact mechanism of choosing m^* need not be known; we only need assume that (1) is satisfied.

1.3 Data

We will illustrate the selected data information on data from the 1992 Major League Baseball season. Our first data set is the record of all of Dave Winfield's 1992 at-bats and whether the at-bat resulted in a hit or out. (Winfield actually made 670 plate appearances in 1992, but 87 of these were not official at-bats, because they resulted in a walk, sacrifice, hit-by-pitch, or other outcome that does not count as an at-bat. Thus Winfield had 583 at-bats.)

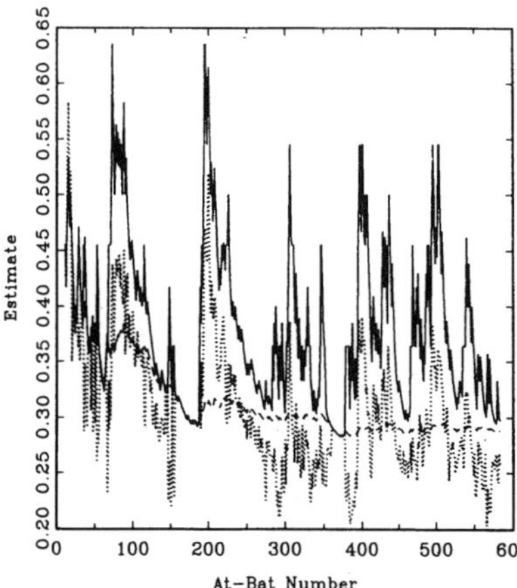

Figure 1. The 1992 Hitting Record of Dave Winfield. The complete data MLE (dashed line) is $\hat{p}$ = ratio of hits to at-bats, the selected maximum (solid line) is r^* of (1), and the selected data MLE (dotted line) is $\hat{\theta}$ of Section 3.2.

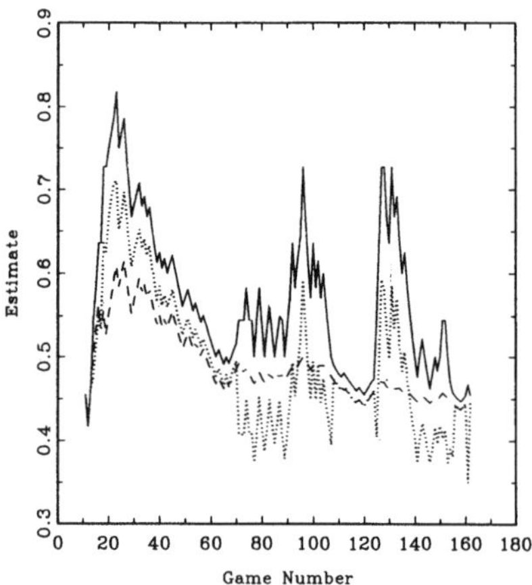

Figure 2. The 1992 Won–Loss Record of the New York Mets. The complete data MLE (dashed line) is $\hat{p}$ = ratio of wins to games, the selected maximum (solid line) is r^* of (1), and the selected data MLE (dotted line) is $\hat{\theta}$ of Section 3.2.

The data for Dave Winfield are displayed in Figure 1. The dashed line represents his batting average (i.e., ratio of hits to at-bats) for each at-bat. It can be seen that this value settles down quickly, and remains close to $169/583 = .290$, Winfield's final batting average for the 1992 season. For each at-bat, this ratio is also the MLE given that we have observed the entire sequence of all previous at-bats or, equivalently, that we know the total number of hits up to the given at-bat.

The solid line in Figure 1 is a running sequence of values of r^*. For each at-bat, this number is the maximum ratio of hits to at-bats, counting backward in time from the given at-bat. In the calculation of r^*, we required $m^* > 10$, which merely serves to eliminate trivial cases (e.g., 1 for his last 1), and smooths out the picture somewhat (eliminating multiple peaks at 2 for 3, 4 for 7, and so on). Thus Figure 1 shows 573 at-bats, starting with at-bat number 11.

Finally the dotted line in Figure 1 is a running plot of the selected data MLE, the MLE of Winfield's batting ability based on observing only k^*, m^*, and n = at-bat number. This estimator is one of the main objects of investigation in this article and is discussed in detail in later sections.

We also analyze a similar data set comprising the 1992 won–loss record of the New York Mets, pictured in Figure 2. The Mets played 162 games; we show the 152 games from game 11 to game 162. The values of r^* (solid line) are somewhat less variable than Winfield's, and again the complete data MLE (dashed line) quickly settles down to the final winning percentage of the Mets, $72/162 = .444$. The selected data MLE still remains quite variable, however.

1.4 Summary

In Section 2 we show how to calculate the exact likelihood based on observing the selected information k^*, m^*, and n. We do the calculations two ways: one method using an exact combinatoric derivation of the likelihood function and one method based on a Gibbs sampling algorithm. The combinatorial derivation and the resulting likelihood are quite complicated. But an easily implementable (albeit computer-intensive) Gibbs algorithm yields likelihood functions that are virtually identical. In Section 3 we consider maximum likelihood point estimation and estimation of standard errors, and also show how to implement Bayesian estimation via the Gibbs sampler. We also calculate maximum likelihood point estimates in a number of ways, using the combinatorial likelihood, the EM algorithm, and a Gibbs sampling-based approximation, and show how to estimate standard errors for these point estimates.

In Section 4 we adapt methodology from sequential analysis to derive a Brownian motion-based approximation to the likelihood. The approximation also yields remarkably accurate MLE values. A discussion in Section 5 relates all of this methodology back to the baseball data that originally suggested it. Finally, in the Appendix we provide some technical details.

2. LIKELIHOOD CALCULATIONS

In this section we show how to calculate the likelihood function exactly. We use two methods, one based on a combinatorial derivation and one based on Markov chain methods.

Recall that the data, $X_1, \ldots, X_n$ are in chronological order. But for selectively reported data like we are considering, it is easier to think in terms of the reversed sequence, looking backward from time n, so we now redefine the data in terms of the reversed sequence. Also, we want to distinguish between the reported data and the unreported data. We define $\mathbf{Y} = (Y_1, \ldots, Y_{m^*+1})$ by $Y_i = X_{n-i+1}$ and $\mathbf{Z} = (Z_1, \ldots, Z_m)$ by $Z_i = X_{n-m^*-i}$, where $m = n - (m^* + 1)$. Thus $\mathbf{Y}$ is the reported data (with Y_1 being the most recent at-bat, Y_2 being the next most recent at-bat, and so on), including $Y_{m^*+1} = X_{n-m^*}$, which we know to be 0. There are k^* 1s in $\mathbf{Y}$. $\mathbf{Z}$ is the unreported data. We know that the vector $\mathbf{Z}$ satisfies

$$\frac{k^* + \sum_{i=1}^{j} Z_i}{m^* + 1 + j} \leq r^* = \frac{k^*}{m^*} \quad \forall j = 1, \ldots, m. \tag{2}$$

This is assumption (1), that there is no higher ratio than the reported r^* in the unreported data.

The likelihood, given the reported data $\mathbf{Y} = \mathbf{y}$, is denoted by $L(\theta \mid \mathbf{y})$. (Generally, random variables will be denoted by upper case letters and their observed values denoted by the lower case counterparts.) It is proportional to

$$L(\theta \mid \mathbf{y}) \propto \theta^{k^*}(1-\theta)^{m^*+1-k^*} \sum_{\mathbf{z} \in Z^*} \theta^{S_{\mathbf{z}}}(1-\theta)^{m-S_{\mathbf{z}}}, \tag{3}$$

where Z^* is the set of all vectors $\mathbf{z} = (z_1, \ldots, z_m)$ that never give a higher ratio than r^* [see (2)], and $S_{\mathbf{z}} = \sum_{i=1}^{m} z_i$ = number of 1s in the unreported data. Dawid and Dickey (1977) called the factor $\theta^{k^*}(1-\theta)^{m^*-k^*}$ the "face-value likelihood" and the remainder of the expression the "correction factor." The correction factor is the correction to the likelihood that is necessary because the data were selectively reported.

2.1 Combinatorial Calculations

An exact expression for the sum over Z^* can be given in terms of constants $i^*, n_1, \ldots, n_{i^*}$, and $c_1, \ldots, c_{i^*}$, which we now define. Let i^* be the largest integer that is less than $(n - m^*)(1 - r^*)$ and define

$$n_i = \left[\frac{i}{1-r^*}\right], \qquad i = 1, \ldots, i^*,$$

where $[a]$ is the greatest integer less than or equal to a. Now define constants c_i recursively by $c_1 = 1$ and

$$c_i = \binom{n_i - 1}{i-1} - \sum_{j=1}^{i-1} c_j \binom{n_i - 1 - n_j}{i-j}, \qquad i = 2, \ldots, i^*.$$

Then (3) can be written as

$$L(\theta \mid \mathbf{y}) \propto \theta^{k^*}(1-\theta)^{m^*+1-k^*}\left\{1 - \sum_{i=1}^{i^*} c_i \theta^{n_i+1-i}(1-\theta)^{i-1}\right\}. \tag{4}$$

The equivalence of (3) and (4) is proved in Appendix A. If $i^* = 0$, which will be true if r^* is large and $n - m^*$ is small, then the sum in (4) is not present and the likelihood is just

$$L(\theta \mid \mathbf{y}) \propto \theta^{k^*}(1-\theta)^{m^*+1-k^*}.$$

The constants c_i grow very rapidly as i increases. So if i^* is even moderately large, care must be taken in their computation. They can be computed exactly with a symbolic processor, but this can be time-consuming. So we now look at alternate ways of computing the likelihood, and in Section 4 we consider an approximation of $L(\theta \mid \mathbf{y})$.

2.2 Sampling-Based Calculations

As an alternative to the combinatorial approach to calculating $L(\theta \mid \mathbf{y})$, we can implement a sampling-based approach using the Gibbs sampler. We can interpret Equation (3) as stating

$$L(\theta \mid \mathbf{y}) = \sum_{Z^*} L(\theta \mid \mathbf{y}, \mathbf{z}), \tag{5}$$

where $\mathbf{z} = (z_1, \ldots, z_m)$ are the unobserved Bernoulli outcomes, Z^* is the set of all such possible vectors, and $L(\theta \mid \mathbf{y}, \mathbf{z})$ is the likelihood based on the complete data. Equation (5) bears a striking resemblance to the assumed relationship between the "complete data" and "incomplete data" likelihoods for implementation of the EM algorithm and, in fact, can be used in that way. We will later see how to implement an EM algorithm, but first we show how to use (5) to calculate $L(\theta \mid \mathbf{y})$ using the Gibbs sampler.

We assume that $L(\theta \mid \mathbf{y})$ can be normalized in θ; that is, $\int_\Theta L(\theta \mid \mathbf{y})\, d\theta < \infty$. (This is really not a very restrictive assumption, as most likelihoods seem to have finite integrals. The likelihood $L(\theta \mid \mathbf{y})$ in (3) is the finite sum of terms that each have a finite integral, so the integrability condition is satisfied.) Denote the normalized likelihood by $L^*(\theta \mid \mathbf{y})$, so

$$L^*(\theta \mid \mathbf{y}) = \frac{L(\theta \mid \mathbf{y})}{\int_\Theta L(\theta \mid \mathbf{y})\, d\theta}. \tag{6}$$

Because $L(\theta \mid \mathbf{y})$ can be normalized, so can $L(\theta \mid \mathbf{y}, \mathbf{z})$ of (5). Denoting that normalized likelihood by $L^*(\theta \mid \mathbf{y}, \mathbf{z})$, we now can consider both $L^*(\theta \mid \mathbf{y})$ and $L^*(\theta \mid \mathbf{y}, \mathbf{z})$ as density functions in θ. Finally, from the unnormalized likelihoods, we define

$$k(\mathbf{z} \mid \mathbf{y}, \theta) = \frac{L(\theta \mid \mathbf{y}, \mathbf{z})}{L(\theta \mid \mathbf{y})}, \tag{7}$$

an equation reminiscent of the EM algorithm. The function $k(\mathbf{z} \mid \mathbf{y}, \theta)$ is a density function that defines the density of $\mathbf{Z}$ conditional on $\mathbf{y}$ and θ. If we think of $\mathbf{z}$ as the "missing data" or, equivalently, think of $\mathbf{y}$ as the "incomplete data" and $(\mathbf{y}, \mathbf{z})$ as the "complete data," we can use the unnormalized likelihoods in a straightforward implementation of the EM algorithm. But we have more. If we iteratively sample between $k(\mathbf{z} \mid \mathbf{y}, \theta)$ and $L^*(\theta \mid \mathbf{y}, \mathbf{z})$ (i.e., sample a sequence $\mathbf{z}_1, \theta_1, \mathbf{z}_2, \theta_2, \mathbf{z}_3, \theta_3, \ldots$), then we can approximate the actual normalized likelihood by

$$\hat{L}^*(\theta \mid \mathbf{y}) \approx \frac{1}{M} \sum_{i=1}^{M} L^*(\theta \mid \mathbf{y}, \mathbf{z}_i), \tag{8}$$

with the approximation improving as $M \to \infty$ (see the Appendix for details). Thus we have a sampling-based exact calculation of the true likelihood function.

Note that this sampling-based strategy for calculating the likelihood differs from some other strategies that have been used. The techniques of Geyer and Thompson (1992) for calculating likelihoods are based on a different type of Monte Carlo calculation, one not based on Gibbs sampling. That is also the technique used by Carlin and Gelfand (1992) and Gelfand and Carlin (1991). These approaches do not require an integrable likelihood, as they sample only over $\mathbf{z}$. Here we sample over both $\mathbf{z}$ and θ, which is why we require an integrable likelihood. For our extra assumption, we gain the advantage of easy estimation of the entire likelihood function. The technique used here, which closely parallels the implementation of the EM algorithm, was discussed (but not implemented) by Smith and Roberts (1993).

Implementing Equation (8) is quite easy. The likelihood $L^*(\theta \mid \mathbf{y}, \mathbf{z})$ is the normalized complete data likelihood, so

$$L^*(\theta \mid \mathbf{y}, \mathbf{z}) = \frac{\Gamma(n+2)}{\Gamma(k^* + S_{\mathbf{z}} + 1)\Gamma(n - k^* - S_{\mathbf{z}} + 1)} \times \theta^{k^* + S_{\mathbf{z}}}(1-\theta)^{n-k^*-S_{\mathbf{z}}} \tag{9}$$

(recall that $S_{\mathbf{z}} = \sum_{i=1}^{m} z_i$ and $m = n - m^* - 1$). Thus to calculate $\hat{L}^*(\theta \mid \mathbf{y})$, we use the following algorithm:

Step 0. Initialize $\theta = \theta_0$.
For $j = 1, \ldots, M$
Step 1. Generate $\mathbf{z}_j \sim k(\mathbf{z} \mid \mathbf{y}, \theta_{j-1})$
Step 2. Generate $\theta_j \sim L^*(\theta \mid \mathbf{y}, \mathbf{z}_j)$. (10)

Because $L^*(\theta \mid \mathbf{y}, \mathbf{z})$ is a beta distribution with parameters $k^* + S_{\mathbf{z}} + 1$ and $n - k^* - S_{\mathbf{z}} + 1$, it is easy to generate the θ's. To generate the $\mathbf{z}$'s, from $k(\mathbf{z} \mid \mathbf{y}, \theta)$ the following simple rejection algorithm runs very quickly:

1. Generate $\mathbf{z} = (z_1, \ldots, z_m)$, z_i iid Bernoulli(θ).
2. Calculate $S_i = (k^* + \sum_{j=1}^{i} z_j)/(m^* + 1 + i)$, $i = 1, \ldots, m$.
3. If $S_i \le r^*$ for every $i = 1, \ldots, m$, accept $\mathbf{z}$; otherwise, reject $\mathbf{z}$. (11)

Implementing this algorithm using a 486 DX2 computer with the Gauss programming language is very simple, and the running time is often quite short. The running time was increased only in situations where $n \gg m^*$, when more constrained Bernoulli sequences had to be generated. Typical acceptance rates were approximately 60% (for $r^* = 8/17$ and $n = 25$–500), which, we expect, could be improved with a more sophisticated algorithm.

Figure 3 illustrates the Gibbs-sampled likelihoods for $M = 1{,}000$, for $r^* = 8/17$ with various values of n. As can be seen, the modes and variances decrease as n increases. If we plot the likelihoods calculated from the combinatorial formula (4), the differences are imperceptible.

3. ESTIMATION

One goal of this article is to assess our ability to recover a good estimate of θ from the selectively reported data. We would be quite happy if our point estimate from the selected data was close to the MLE of the unselected data. But as we shall see, this is generally not the case. Although in some cases we can do reasonably well, only estimation with strong prior information will do well consistently.

3.1 Exact Maximum Likelihood Estimation

Based on the exact likelihood of (4), we can calculate the MLE by finding the 0s of the derivative of $L(\theta \mid \mathbf{y})$. The likelihood is a high-degree polynomial in θ, but symbolic manipulation programs can compute the constants and symbolically differentiate the polynomial. But the 0s must be solved for numerically, as the resulting expressions are too involved for analytical evaluation. In all the examples we have calculated, $L(\theta \mid \mathbf{y})$ is a unimodal function for $0 \le \theta \le 1$ and no difficulties were experienced in numerically finding the root.

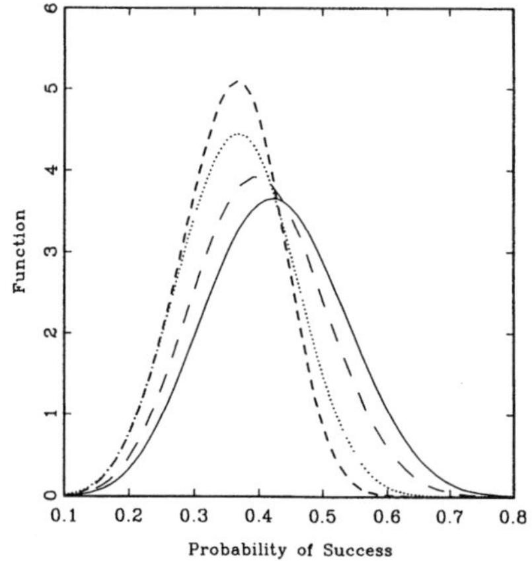

Figure 3. Likelihood Functions for $k^* = 8$ and $m^* = 17$, for $n = 19$ (Solid Line), 25 (Long-Dashed Line), 50 (Dotted Line), and 100 (Short-Dashed Line), Normalized to Have Area = 1. The value $n = 19$ gives the "naive" likelihood.

We have calculated the MLE for several different data sets; results for four values of m^* and r^* and five values of m are given in Table 1. For each value of m^*, r^*, and m, two values are given. The first is the exact MLE, computed by the method just described. The second is an approximate MLE, which is discussed in Section 4. Just consider the exact values for now.

The exact MLE's exhibit certain patterns that would be expected for this data:

1. The MLE never exceeds the naive estimate $k^*/(m^* + 1) = m^*r^*/(m^* + 1)$.
2. For fixed reported data m^* and r^*, the MLE decreases as m, the amount of unreported data, increases. It appears to approach a nonzero limit as m grows. Knowing that the ratio does not exceed r^* in a long sequence of unreported data should lead to a smaller estimate of θ than knowing only that the ratio does not exceed r^* in a short sequence.
3. For fixed r^* and m, the MLE increases to r^* as m^*, the amount of reported data increases.

This method of finding the MLE requires a symbolic manipulation program to calculate the constants c_i or else some careful programming to deal with large factorials. Also, the method can be slow if m is large. The values for $m = 200$ in Table 1 each took several minutes to calculate. Thus we are led to investigate other methods of evaluating the MLE, methods that do not use direct calculation of $L(\theta|\mathbf{y})$. Although these other methods are computationally intensive, they avoid the problem of dealing with the complicated exact likelihood function.

3.2 The EM Algorithm

As in Section 2.2, the incomplete data interpretation of the likelihood function allows for easy implementation of the EM algorithm. With $\mathbf{y}$ = incomplete data and $(\mathbf{y}, \mathbf{z})$ = complete data, we compute an EM sequence $\hat{\theta}_1, \hat{\theta}_2, \ldots$ by

$$\hat{\theta}_{i+1} = \frac{k^* + E(S_{\mathbf{z}}|\hat{\theta}_i)}{n}, \qquad (12)$$

where $E(S_{\mathbf{z}}|\hat{\theta}_i)$ is the expected number of successes in the missing data. [The E step and the M step are combined into one step in (12).] More precisely, $S_{\mathbf{z}} = \sum_{j=1}^{m} Z_j$, where the Z_j are iid Bernoulli$(\hat{\theta}_i)$ and the partial sums satisfy the restrictions in (11). Such an expected value is virtually impossible to calculate analytically, but is quite easy to approximate using Monte Carlo methods (Wei and Tanner 1990). The resulting sequence $\hat{\theta}_1, \hat{\theta}_2, \ldots$ converges to the exact complete data MLE. In all of our calculations, the value of the EM-calculated MLE is indistinguishable from the MLE resulting from (4).

3.3 Approximate MLE's from the Gibbs Sampler

Equation (8), which relates the exact likelihood to the average of the complete data likelihoods, forms a basis for a simple approximation scheme for the MLE. Although the average of the maxima is not the maximum of the average, we can use a Taylor series approximation to estimate the incomplete data MLE as a weighted average of the complete data MLE's.

Table 1. Exact (First Entry) and Approximate (Second Entry) MLE's Calculated From the Exact Likelihood (4) and the Approximation Described in (23)

		m^*							
r^*	m	5		25		45		200	
1/5	5	.109	.096	.170	.167	.183	.180	.196	.196
	20	.088	.079	.150	.145	.167	.164	.191	.190
	60	.084	.078	.139	.134	.156	.152	.185	.184
	100	.084	.078	.136	.132	.152	.149	.182	.181
	200	.084	.078	.135	.131	.149	.147	.178	.177
2/5	5	.261	.263	.359	.360	.376	.377	.394	.394
	20	.232	.232	.332	.332	.356	.356	.389	.389
	60	.227	.229	.317	.316	.341	.341	.381	.381
	100	.227	.228	.314	.313	.336	.336	.377	.377
	200	.227	.228	.312	.311	.333	.333	.372	.372
3/5	5	.437	.448	.553	.559	.572	.577	.593	.594
	20	.408	.412	.527	.531	.554	.556	.588	.589
	60	.403	.407	.511	.513	.538	.541	.580	.581
	100	.403	.407	.507	.510	.534	.535	.577	.578
	200	.403	.407	.506	.508	.530	.531	.571	.572
4/5	5	.632	.644	.754	.763	.773	.779	.794	.795
	20	.612	.615	.734	.742	.759	.765	.789	.791
	60	.609	.609	.721	.726	.746	.750	.783	.785
	100	.609	.609	.718	.722	.742	.745	.780	.782
	200	.609	.609	.717	.720	.739	.742	.776	.777

An obvious approach is to expand each complete data likelihood in (8) around its MLE, $\hat{\theta}_i = (k^* + S_{\mathbf{z}})/n$, to get

$$\begin{aligned} L^*(\theta|\mathbf{y}, \mathbf{z}_i) &\approx L^*(\hat{\theta}_i|\mathbf{y}, \mathbf{z}_i) + (\theta - \hat{\theta}_i)L^{*\prime}(\hat{\theta}_i|\mathbf{y}, \mathbf{z}_i) \\ &\quad + \frac{(\theta - \hat{\theta}_i)^2}{2} L^{*\prime\prime}(\hat{\theta}_i|\mathbf{y}, \mathbf{z}_i) \\ &= L^*(\hat{\theta}_i|\mathbf{y}, \mathbf{z}_i) + \frac{\theta - \hat{\theta}_i)^2}{2} L^{*\prime\prime}(\hat{\theta}_i|\mathbf{y}, \mathbf{z}_i), \end{aligned} \qquad (13)$$

because $L^{*\prime}(\hat{\theta}_i|\mathbf{y}, \mathbf{z}_i) = 0$. Now substituting into (8) yields

$$\hat{L}^*(\theta|\mathbf{y}) \approx \frac{1}{M}\sum_{i=1}^{M}\left[L^*(\hat{\theta}_i|\mathbf{y}, \mathbf{z}_i) + \frac{(\theta - \hat{\theta}_i)^2}{2} L^{*\prime\prime}(\hat{\theta}_i|\mathbf{y}, \mathbf{z}_i)\right],$$

and differentiating with respect to θ yields the approximate MLE

$$\hat{\theta}_A = \frac{\sum_{i=1}^{M} \hat{\theta}_i L^{*\prime\prime}(\hat{\theta}_i|\mathbf{y}, \mathbf{z}_i)}{\sum_{i=1}^{M} L^{*\prime\prime}(\hat{\theta}_i|\mathbf{y}, \mathbf{z}_i)}.$$

But it turns out that this approximation is not very accurate. A possible reason for this is the oversimplification of (13), which ignores most of the computed information. In particular, for $j \neq i$, the information in $\hat{\theta}_j$ is not used when expanding $L^*(\theta|\mathbf{y}, \mathbf{z}_i)$. Thus we modify (13) into a "double" Taylor approximation. We first calculate an average approximation for each $L^*(\theta|\mathbf{y}, \mathbf{z}_i)$, averaging over all $\hat{\theta}_j$, and then average over all $L(\theta|\mathbf{y}, \mathbf{z}_i)$. We now approximate the incomplete data likelihood $L^*(\theta|\mathbf{y})$ with

$$L^*(\theta|\mathbf{y}) = \frac{1}{M^2}\sum_{i=1}^{M}\sum_{j=1}^{M}\Big[L^*(\hat\theta_j|\mathbf{y},\mathbf{z}_i) + (\theta-\hat\theta_j)L^{*\prime}(\hat\theta_j|\mathbf{y},\mathbf{z}_i) + \frac{(\theta-\hat\theta_j)^2}{2}L^{*\prime\prime}(\hat\theta_j|\mathbf{y},\mathbf{z}_i)\Big].$$

Differentiating with respect to θ yields the approximate MLE

$$\hat{\hat\theta}_A = \frac{\sum_{i,j}\hat\theta_j L^{*\prime\prime}(\hat\theta_j|\mathbf{y},\mathbf{z}_i) - \sum_{i,j}L^{*\prime}(\hat\theta_j|\mathbf{y},\mathbf{z}_i)}{\sum_{i,j}L^{*\prime\prime}(\hat\theta_j|\mathbf{y},\mathbf{z}_i)}. \quad (14)$$

Table 2 compares the approximate MLE of (14) to the exact value found by differentiating the exact likelihood of (4). It can be seen that the Gibbs approximation is reasonable, but certainly not as accurate as the EM calculation.

3.4 Bayes Estimation

It is relatively easy to incorporate prior information into our estimation techniques, especially when using the Gibbs sampling methodology of Section 2.2. More importantly, in major league baseball there is a wealth of prior information. For any given player or team, the past record is readily available in sources such as the *Baseball Encyclopedia* (Reichler 1988).

If we assume that there is prior information available in the form of a beta(a, b) distribution, then, analogous to Section 2.2, we have the two full-conditional posterior distributions

$$\pi(\theta|\mathbf{y},\mathbf{z},a,b) = \frac{\Gamma(n+a+b)}{\Gamma(k^*+S_{\mathbf{z}}+a)\Gamma(n-k^*-S_{\mathbf{z}}+b)} \times \theta^{k^*+S_{\mathbf{z}}+a-1}(1-\theta)^{n-k^*-S_{\mathbf{z}}+b-1}$$

and

$$k(\mathbf{z}|\mathbf{y},\theta,a,b) = \text{as in (7) and (11)}, \text{ with the Bernoulli parameter of } \theta. \quad (15)$$

Running a Gibbs sampler on (15) is straightforward, and the posterior distribution of interest is given by

$$\hat\pi(\theta|\mathbf{y},a,b) = \frac{1}{M}\sum_{i=1}^{M}\pi(\theta|\mathbf{y},\mathbf{z}_i,a,b). \quad (16)$$

For point estimation, we usually use the posterior mean, given by

$$\hat E(\theta|\mathbf{y},a,b) = \frac{1}{M}\sum_{i=1}^{M}E(\theta|\mathbf{y},\mathbf{z}_i,a,b) = \frac{1}{M}\sum_{i=1}^{M}\frac{k^*+S_{\mathbf{z}_i}+a}{n+a+b}. \quad (17)$$

By using a beta(1, 1) prior, we get the likelihood function as the posterior distribution. Table 2 also shows the values of this point estimator, and we see that it is a very reasonable estimate.

Because the available prior information in baseball is so good, the Bayes posteriors are extraordinarily good, even though the data are not very informative. Figure 4 shows posterior distributions for the New York Mets using historical values for the prior parameter values. It can be seen that once the prior information is included, the selected MLE produces an excellent posterior estimate.

3.5 Variance of the Estimates

When using a maximum likelihood estimate, $\hat\theta$, a common measure of variance is $-1/l''(\hat\theta)$, where l is the log-likelihood, $l = \log L$. In our situation, where L is expressed as a sum of component likelihoods (8), taking logs is not desirable. But a few simple observations allow us to derive an approximation for the variance.

Because $l = \log L$, it follows that

$$l' = \frac{L'}{L} \quad\text{and}\quad l'' = \frac{LL'' - (L')^2}{L^2}. \quad (18)$$

We have $L'(\hat\theta) = 0$ and $-l''(\hat\theta) = -L''(\hat\theta)/L(\hat\theta)$; using (8) yields the approximation

Table 2. Comparison of Combinatoric MLE [Obtained by Differentiating the Likelihood (4)], the MLE From the EM Algorithm, the Gibbs/Likelihood Approximation of (14), the Bayes Posterior Mean Using a Beta(1, 1) Prior (the Mean Likelihood Estimate), the Brownian Motion-Based Approximation, and $\hat p$ (the Complete Data MLE)

At-bat	k^*	m^*	r^*	$\hat p$	Combinatoric MLE	EM MLE	Gibbs approximate MLE	Bayes mean	Brownian MLE
187	55	187	.294	.294	.294	.294	.294	.296	.294
188	4	11	.364	.298	.241	.240	.232	.237	.241
189	5	12	.417	.302	.292	.289	.276	.278	.293
190	5	13	.385	.300	.267	.267	.253	.258	.268
191	5	11	.455	.304	.321	.322	.303	.299	.324
339	12	39	.308	.298	.241	.240	.238	.225	.241
340	47	155	.303	.297	.273	.273	.273	.272	.273
341	13	41	.317	.299	.251	.251	.250	.243	.251
342	13	42	.310	.298	.245	.244	.242	.238	.245
343	14	43	.326	.300	.260	.261	.258	.252	.260
344	14	44	.318	.299	.254	.254	.246	.247	.254
345	4	11	.367	.301	.241	.240	.222	.234	.241
346	5	11	.456	.303	.321	.313	.300	.302	.325

NOTE: The at-bats were chosen from Dave Winfield's 1992 season.

$$\text{var}(\hat{\theta}\,|\,\mathbf{y}) \approx -\left[\frac{L^{*\prime\prime}(\hat{\theta}\,|\,\mathbf{y})}{L^{*}(\hat{\theta}\,|\,\mathbf{y})}\right]^{-1}$$

$$\approx \frac{\sum_{i=1}^{M} L^{*}(\hat{\theta}_i\,|\,\mathbf{y}, \mathbf{z}_i)}{\sum_{i=1}^{M} L^{*\prime\prime}(\hat{\theta}_i\,|\,\mathbf{y}, \mathbf{z}_i)}$$

$$= \frac{1}{n}\frac{\sum_{i=1}^{M} \hat{\theta}_i(1-\hat{\theta}_i)L^{*\prime\prime}(\hat{\theta}_i\,|\,\mathbf{y}, \mathbf{z}_i)}{\sum_{i=1}^{M} L^{*\prime\prime}(\hat{\theta}_i\,|\,\mathbf{y}, \mathbf{z}_i)}, \quad (19)$$

using (9) and (15). Of course, for the selected data the variances are much higher than they would have been had we observed the entire data set. We thus modify (19) to account for the fact that we did not observe n Bernoulli trials, but only $m^* + 1$. Because our calculations are analogous to those in the EM algorithm, we could adjust (19) as in equation (3.26) of Dempster, Laird, and Rubin (1977), where they showed that the ratio of the complete-data variance to incomplete-data variance is given by the derivative of the EM mapping. But in our case we have an even simpler answer, and assume $\text{var}(\hat{\theta}\,|\,\mathbf{y}, \mathbf{z})/\text{var}(\hat{\theta}\,|\,\mathbf{y}) = (m^* + 1)/n$. Thus we modify (19) to

$$\text{var}(\hat{\theta}\,|\,\mathbf{y}) \approx \frac{1}{m^*+1}\frac{\sum_{i=1}^{M} \hat{\theta}_i(1-\hat{\theta}_i)L^{*\prime\prime}(\hat{\theta}_i\,|\,\mathbf{y}, \mathbf{z}_i)}{\sum_{i=1}^{M} L^{*\prime\prime}(\hat{\theta}_i\,|\,\mathbf{y}, \mathbf{z}_i)}.$$

It turns out that this approximation works quite well, better than the "single" Taylor series approximation for the variance of the MLE of θ. But the double Taylor series argument of Section 3.3 results in an improved approximation. Starting from the fact that $l'' = [LL'' - (L')^2]/L^2$, we write

$$\text{var}(\hat{\theta}\,|\,\mathbf{y}) \approx \frac{-n}{m^*+1}\left[\frac{L^{*\prime\prime}}{L^*} - \left(\frac{L^{*\prime}}{L^*}\right)^2\right]^{-1}$$

$$= \frac{-n}{m^*+1}\left[\sum_{i,j}\frac{L^{*\prime\prime}(\hat{\theta}_j\,|\,\mathbf{y}, \mathbf{z}_i)}{L^{*}(\hat{\theta}_j\,|\,\mathbf{y}, \mathbf{z}_i)} - \left(\sum_{i,j}\frac{L^{*\prime}(\hat{\theta}_j\,|\,\mathbf{y}, \mathbf{z}_i)}{L^{*}(\hat{\theta}_j\,|\,\mathbf{y}, \mathbf{z}_i}\right)^2\right]^{-1}. \quad (20)$$

Table 3 compares the approximation in (20) to values obtained by calculating $l''(\hat{\theta})$ exactly (using a symbolic processor). It can be seen that the approximation is quite good for moderate to large m^* and still is acceptable for small m^*.

We can also compute Bayesian variance estimates. Proceeding analogously to Section 3.4, the Bayesian variance would be an average of beta variances,

$$\text{var}(\theta\,|\,\mathbf{y}, a, b)$$
$$\approx \frac{1}{M}\sum_{i=1}^{M}\frac{(k^* + S_{z_i} + a)(n - k^* - S_{z_i} + b)}{(n+a+b)^2(n+a+b+1)}. \quad (21)$$

But as before, we must adjust this variance to account for the fact that we observe only $m^* + 1$ trials. We do this adjustment by replacing the term $(n + a + b + 1)$ by $(m^* + 1 + a + b + 1)$. The resulting estimate behaves quite reasonably, yielding estimates close to the exact MLE values for moderate m^* and $a = b = 1$. These values are also displayed in Table 3.

As expected, the standard error is sensitive to m^*, yielding large limits when m^* is small. Figures 5 and 6 show these limits for the 1992 season of Dave Winfield and the Mets. Although the estimates and standard errors are quite variable, note that the true batting average and winning percentage are always within one standard deviation of the selected MLE.

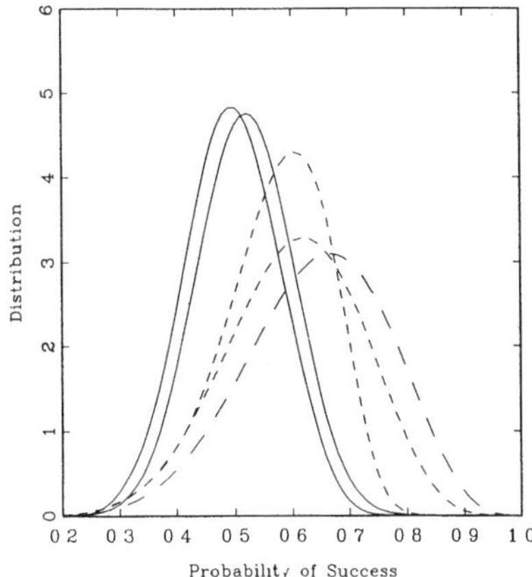

Figure 4. Posterior Distributions for the New York Mets. The value r^* = 8/11 = .727 was actually achieved at games 19 and 131. The solid lines are posterior distributions based on beta priors with parameters a = 9.951 and b = 11.967 (representing a mean of .454 and a standard deviation of .107, which are the Mets' overall past parameters), and are based on n = 19 and 131 observations with modes decreasing in n. The short-dashed lines are posteriors using a beta(1, 1) prior with n = 19 and 131 and hence are likelihood functions, again with modes decreasing in n. The long-dashed line is the naive likelihood, which assumes n = 12.

4. BROWNIAN MOTION APPROXIMATION TO THE LIKELIHOOD

The last terms in the expressions (3) and (4),

$$\sum_{\mathbf{z}\in Z^*} \theta^{S_\mathbf{z}}(1-\theta)^{m-S_\mathbf{z}} = 1 - \sum_{i=1}^{i^*} c_i\theta^{n_i+1-i}(1-\theta)^{i-1}, \quad (22)$$

are complicated to compute. But we can approximate these terms with functions derived from a consideration of Brownian motion. The resulting approximate likelihood can then be maximized to find an approximate MLE.

This was done for the data in Table 1. The second entry in each case is the approximate MLE. It can be seen that the approximate MLE's are excellent. In the 48 cases with $m \geq 60$, the approximate and exact MLE's never differ by more than .006. In fact, even for the smaller values of m, the exact and approximate MLE's are very close. In only three cases, all with $m = 5$, do the two differ by more than .01.

To develop our approximation, note from Appendix A that the expression in (22) is equal to $P_\theta(S_j/(j+1) \leq r^*$ for $j = 1, \ldots, m)$, where $Z_1, Z_2, \ldots$ are independent Bernoulli(θ) random variables and $S_j = \sum_{i=1}^{j} Z_i$. We rewrite the inequality $S_j/(j+1) \leq r^*$ as

$$S_j^* = \sum_{i=1}^{j}\frac{Z_i - \theta}{\sqrt{\theta(1-\theta)}} \leq \frac{(j+1)r^* - j\theta}{\sqrt{\theta(1-\theta)}} = b_\theta + \eta_\theta j,$$

where $\eta_\theta = (r^* - \theta)/\sqrt{\theta(1-\theta)}$ and $b_\theta = r^*/\sqrt{\theta(1-\theta)}$. Now the vector $(S_1^*, \ldots, S_m^*)$ has the same means, variances,

Table 3. Comparison of Standard Deviations Based on Exact Differentiation of the Log-Likelihood, the Gibbs/Likelihood Approximation of (20), and the Bayes Posterior Standard Deviation Using a Beta(1, 1) Prior

					Standard deviation		
At-bat	k^*	m^*	$\hat{p}$	MLE	Exact	Gibbs/approximate	Bayes
187	55	187	.294	.294	.033	.033	.033
188	4	11	.298	.241	.083	.121	.117
189	5	12	.302	.292	.086	.122	.118
190	5	13	.300	.267	.080	.115	.112
191	5	11	.304	.321	.093	.129	.125
339	12	39	.298	.241	.046	.067	.065
340	47	155	.297	.273	.029	.036	.035
341	13	41	.299	.251	.046	.067	.065
342	13	42	.298	.245	.045	.065	.064
343	14	43	.300	.260	.046	.066	.064
344	14	44	.299	.254	.045	.064	.063
345	4	11	.301	.241	.083	.119	.116
346	5	11	.303	.321	.093	.131	.126

NOTE: The at-bats were chosen from Dave Winfield's 1992 season.

and covariances as $(W(1), \ldots, W(m))$, where $W(t)$ is standard (mean 0 and variance 1) Brownian motion. So we can approximate

$$P_\theta\left(\frac{S_j}{j+1} \leq r^*, j = 1, \ldots, m\right)$$
$$= P_\theta(S_j^* \leq b_\theta + \eta_\theta j, j = 1, \ldots, m)$$

by

$$P(W(t) \leq b_\theta + \eta_\theta t, 0 \leq t \leq m).$$

If we define τ as the first passage time of $W(t)$ through the linear boundary $b_\theta + \eta_\theta t$ (i.e., $\tau = \inf\{t: W(t) > b_\theta + \eta_\theta t\}$), then

$$\begin{aligned} P(W(t) &\leq b_\theta + \eta_\theta t, 0 \leq t \leq m) \\ &= P(\tau > m) \\ &= 1 - P(\tau \leq m) \\ &= \Phi\left(\frac{b_\theta}{\sqrt{m}} + \eta_\theta\sqrt{m}\right) - e^{-2b_\theta\eta_\theta}\Phi\left(-\frac{b_\theta}{\sqrt{m}} + \eta_\theta\sqrt{m}\right), \end{aligned}$$

where Φ is the standard normal cdf and the last equality is from (3.15) of Siegmund (1985).

Because S_j^* is a discrete process, the first time that $S_j^* > b_\theta + \eta_\theta j$ it will in fact exceed the boundary by a positive amount. Also, $W(t)$ may exceed $b_\theta + \eta_\theta t$ for some $0 \leq t \leq m$, even if $(W(1), \ldots, W(m))$ does not. So the probability we want is in fact larger than the approximation. Siegmund (1985, p. 50) suggests that this approximation will be improved if b_θ is replaced by $b_\theta + \rho$, where ρ is an appropriately chosen constant. By trial and error, we found that $\rho = .85$ produced good approximate MLE's. Thus to obtain the approximate MLE's in Table 1, Expression (22) was replaced by

$$\Phi\left(\frac{b_\theta + \rho}{\sqrt{m}} + \eta_\theta\sqrt{m}\right) - e^{-2(b_\theta+\rho)\eta_\theta}\Phi\left(-\frac{b_\theta + \rho}{\sqrt{m}} + \eta_\theta\sqrt{m}\right) \tag{23}$$

in (4), and the approximate likelihood was numerically maximized. (The 0 of the derivative was found using a symbolic manipulation program, just as the exact MLE's were found.)

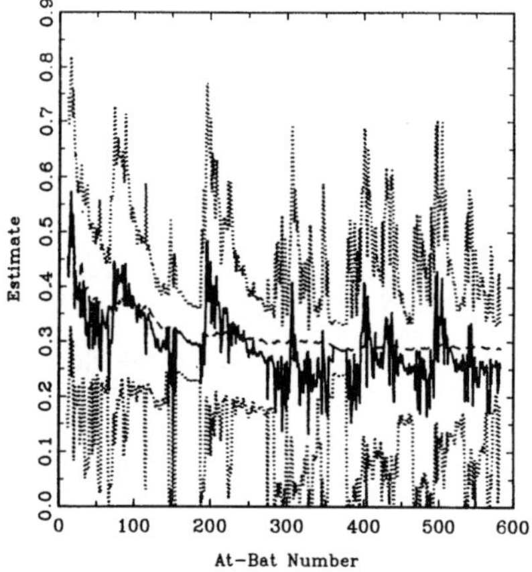

Figure 5. The 1992 Hitting Record of Dave Winfield. The dashed line is $\hat{p}$, the complete data MLE, and the solid line is $\hat{\theta}$, the selected data MLE. The standard deviation limits (dotted lines) are based on the selected data MLE.

5. DISCUSSION

If we are told that Dave Winfield is "8 for his last 17," the somewhat unhappy conclusion is that really not very much information is being given. But the somewhat surprising observation is that there is some information. Although we cannot hope to recover the complete data MLE with any degree of accuracy, we see in Figures 5 and 6 that ± 2 standard deviations of the selected data MLE always contains the complete data MLE. Indeed, in almost every case the complete data MLE is within one standard deviation of the selected data MLE. Moreover, the selected data esti-

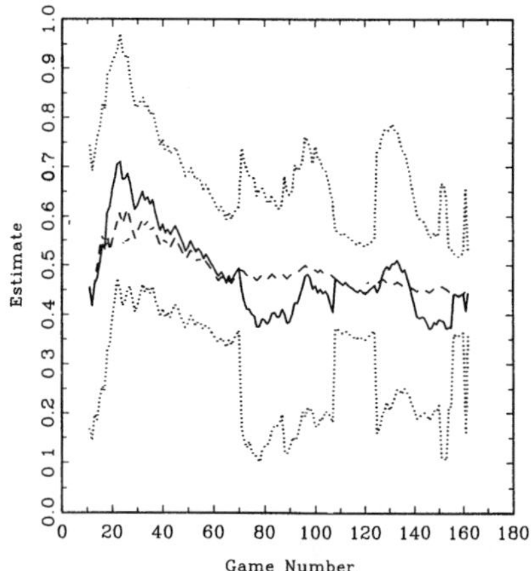

Figure 6. The 1992 Won-Loss Record of the New York Mets. The dashed line is $\hat{p}$, the complete data MLE, and the solid line is $\hat{\theta}$, the selected data MLE. The standard deviation limits (dotted lines) are based on the selected data MLE.

mates behave as expected. In particular, as n (the number of either games or at-bats) increases, the ratio 8 for 17 looks worse; that is, it results in a smaller value of the selected MLE. This is as it should be, because for a given success probability, longer strings (larger values of n) will produce larger maxima. Also, due mainly to the method of construction, the standard deviation of the selected MLE directly reflects the amount of information that it contains, through the ratio m^*/n.

Baseball is a sport well known for its accumulation of data. This readily translates into an enormous amount of prior information that can be used for estimation. In Figure 4 we saw how the New York Mets' prior information completely overwhelms the selected data (and produces very good estimates). This is in fact not an extreme case. If a picture similar to Figure 4 is constructed for Dave Winfield (with prior mean .285 and standard deviation .021), the resulting posterior is virtually a spike, no matter what data are used.

Throughout this article we have assumed that the observed selected data consist of k^*, m^*, and n. But typically the value of n is not reported, so the data are really only k^* and m^*. During the baseball season, it is quite easy to estimate n, especially for ballplayers who play regularly. Moreover, once n reaches a moderate value, its value has very little effect on that of the selected data MLE. For example, for an everyday player we expect $n \approx 100$ by May, so the value of n will have little effect on MLE's based on $m^* \leq 20$. This is evident in Table 1, in the likelihood functions of Figure 3, and also in Table 4, which explores some limiting behavior of the MLE. Although we do not know the exact expression for the limit as $n \to \infty$, two points are evident. Besides the fact that the effect of n diminishes as n grows, it is clear that $r^* = 8/17$ and $r^* = 16/34$ have different limits. Thus much more information is contained in the pair (k^*, m^*) than in the single number r^*.

Finally, we report an observation that a colleague (Chuck McCulloch) made when looking at Figure 1—an observation that may have interest for the baseball fan. When m^* and n are close together, a number of things occur. First, the selected data and complete data MLE are close; second, the selected data standard deviation is smallest. Thus the selected data MLE is a very good estimator. But McCulloch's observation is that when $\hat{\theta} \approx \hat{p}$ (i.e., the complete data MLE), which usually implies $m^* \approx n$, then a baseball player is in a batting slump (i.e., his current batting average is his maximum success ratio). This definition of a slump is based only on the player's relative performance, relative to his own "true" ability. A major drawback of our current notion of a slump is that it is usually based on some absolute measure of hitting ability, making it more likely that a .225 hitter, rather than a .300 hitter, would appear to be in a slump. (If a player is 1 for his last 10, is he in a slump? The answer depends on how good a hitter he actually is. For example, Tony Gwynn's slump could be Charlie O'Brien's hot streak!) If we examine Figure 1, Dave Winfield was in a bad slump during at-bats 156–187 (he was 5 for 31 = .161) and at-bats 360–377 (3 for 17 = .176), for in both cases his maximum hitting ability, r^*, is virtually equal to the MLE's. Similar observations can be made for Figure 2 and the Mets, particularly for games 45–70 (although many would say the New York Mets' entire 1992 season was a slump!). But the message is clear: You are in a slump if your complete data MLE is equal to your selected data MLE, for then your maximum hitting (or winning) ability is equal to your average ability.

Table 4. Limiting Behavior of the MLE for Fixed $r^* = .471$, as $n \to \infty$

	k^*/m^*			
n	8/17	16/34	64/136	128/272
25	.395			
50	.368	.418		
75	.360	.401		
100	.359	.396		
150	.358	.392	.456	
200	.358	.391	.442	
300	.358	.391	.435	.460
400	.357	.390	.431	.451
500	.357	.390	.431	.447
1000	.357	.390	.429	.442

APPENDIX A: DERIVATION OF COMBINATORIAL FORMULA FOR LIKELIHOOD

In this section we derive the exact expressions (3) and (4) for $L(\theta \mid \mathbf{y})$. The reported data are k^* successes in the last m^* trials. We have not specified exactly how this report was determined. In the baseball example, we do not know exactly how the announcer decided to report "k^* out of m^*." But what we have assumed is that the complete data $(\mathbf{y}, \mathbf{z})$ consist of a vector $\mathbf{y}$, with k^* 1s and $m^* + 1 - k^*$ 0s, and a vector $\mathbf{z} \in Z^*$, which satisfies (2). The likelihood is then

$$\sum_{\mathbf{y},\mathbf{z}} \theta^{S_{\mathbf{y},\mathbf{z}}}(1-\theta)^{n-S_{\mathbf{y},\mathbf{z}}},$$

where the sum is over all $(\mathbf{y}, \mathbf{z})$ that give the reported data and $S_{\mathbf{y},\mathbf{z}} = \sum y_i + \sum z_i = k^* + S_{\mathbf{z}}$. We have not specified exactly what all the possible $\mathbf{y}$ vectors are, but for each possible $\mathbf{y}$, $\mathbf{z}$ can be any element in Z^*. Thus if C is the number of possible $\mathbf{y}$ vectors, then the likelihood is

$$C\theta^{k^*}(1-\theta)^{m^*+1-k^*} \sum_{\mathbf{z}\in Z^*} \theta^{S_{\mathbf{z}}}(1-\theta)^{m-S_{\mathbf{z}}}. \tag{A.1}$$

Dropping the constant C, which is unimportant for likelihood analysis, yields (3).

Let Z denote the set of all sequences $(z_1, \ldots, z_m)$ of length m of 0s and 1s. Then the sum in (A.1) is

$$\sum_{\mathbf{z}\in Z^*} \theta^{S_{\mathbf{z}}}(1-\theta)^{m-S_{\mathbf{z}}} = \sum_{\mathbf{z}\in Z} \theta^{S_{\mathbf{z}}}(1-\theta)^{m-S_{\mathbf{z}}} - \sum_{\mathbf{z}\in Z^{*c}} \theta^{S_{\mathbf{z}}}(1-\theta)^{m-S_{\mathbf{z}}}$$
$$= 1 - \sum_{\mathbf{z}\in Z^{*c}} \theta^{S_{\mathbf{z}}}(1-\theta)^{m-S_{\mathbf{z}}}. \tag{A.2}$$

This sum over Z^{*c} is the sum that appears in (4), as we now explain.

The set Z^* is the set of all $\mathbf{z}$'s that satisfy (2). Let $S_j = \sum_{i=1}^{j} z_i$. Then Z^{*c} is the set of all $\mathbf{z}$'s that satisfy

$$\frac{k^* + S_j}{m^* + 1 + j} = \frac{k^* + \sum_{i=1}^{j} z_i}{m^* + 1 + j} > r^* = \frac{k^*}{m^*}, \quad \text{for some} \quad j = 1, \ldots, m.$$

But $(\mathbf{k}^* + S_j)/(m^* + 1 + j) > \mathbf{k}^*/m^*$ if and only if $S_j/(j+1) > r^*$. So Z^{*c} is the set of all $\mathbf{z}$'s that satisfy

$$\frac{S_j}{j+1} > r^*, \quad \text{for some} \quad j = 1, \ldots, m.$$

Now to complete our derivation of (4), we must show that the sums in (A.2) and (4) are equal; that is,

$$\sum_{\mathbf{z} \in Z^{*c}} \theta^{S_\mathbf{z}}(1-\theta)^{m-S_\mathbf{z}} = \sum_{i=1}^{i^*} c_i \theta^{n_i+1-i}(1-\theta)^{i-1}. \tag{A.3}$$

To show this, we must explain what the constants i^*, n_i, and c_i are.

First, the value of $i^* - 1$ is the maximum number of 0s that can occur in $(z_1, \ldots, z_j)$ if $S_j/(j+1) > r^*$. This is because $S_j/(j+1) > r^* \Leftrightarrow S_j > r^*(j+1) \Leftrightarrow j - S_j < j - r^*(j+1)$ and hence

$$\begin{aligned} j - S_j < j - r^*(j+1) &= j(1-r^*) - r^* \le m(1-r^*) - r^* \\ &= (n - m^* - 1)(1 - r^*) - r^* \\ &= (n - m^*)(1 - r^*) - 1. \end{aligned}$$

Next, suppose that when $S_j/(j+1)$ first exceeds r^*, the number of 0s in $(z_1, \ldots, z_j)$ is $j - S_j = j' - 1$. Then this must happen on trial $j = n_{j'}$, because if on trial $n_{j'}$, $n_{j'} - S_{n_{j'}} = j' - 1$, then

$$\begin{aligned} \frac{S_{n_{j'}}}{n_{j'}+1} = \frac{n_{j'} - j' + 1}{n_{j'}+1} &= 1 - \frac{j'}{n_{j'}+1} \\ &> 1 - \frac{j'}{\left(\frac{j'}{1-r^*} - 1\right) + 1} = r^*. \end{aligned}$$

But if $j - S_j = j' - 1$ and $j < n_{j'}$, then

$$\frac{S_j}{j'+1} = \frac{j - j' + 1}{j+1} = 1 - \frac{j'}{j+1} \le 1 - \frac{j'}{\left(\frac{j'}{1-r^*} - 1\right) + 1} = r^*.$$

To compute the sum in (A.3), we partition Z^{*c} into sets $Z_0, \ldots, Z_{i^*-1}$, where Z_i is the set of $\mathbf{z}$'s such that $(z_1, \ldots, z_j)$ contains exactly i 0s if $S_j/(j+1)$ is the first term to exceed r^*; that is,

$$\begin{aligned} Z_i = \{\mathbf{z}\colon j - S_j = i \text{ at the } j \text{ where } S_j/(j+1) > r^* \\ \text{and } S_{j'}/(j'+1) \le r^* \text{ for } 1 \le j' < j\}. \end{aligned}$$

If $\mathbf{z} \in Z_i$, then in fact $j = n_{i+1}$ from our foregoing argument. Let $(z_1, \ldots, z_{n_{i+1}})$ be a sequence such that $S_{n_{i+1}}/(n_{i+1}+1) > r^*$ and $S_{j'}/(j'+1) \le r^*$ for $1 \le j' < n_{i+1}$. [So the vector $(z_1, \ldots, z_{n_{i+1}})$ contains i 0s and $n_{i+1} - i$ 1s.] This initial sequence can be completed in any way to produce a $\mathbf{z} \in Z_i$. The sum of $\theta^{S_\mathbf{z}}(1-\theta)^{m-S_\mathbf{z}}$ over all $\mathbf{z}$'s with this initial sequence is $\theta^{n_{i+1}-i}(1-\theta)^i$, because the sum over all the parts that could be added to this initial sequence is 1. Note that we get the same value $\theta^{n_{i+1}-i}(1-\theta)^i$, regardless of which initial sequence we choose. So if c_{i+1} is the number of different initial sequences that could form $\mathbf{z}$'s in Z_i, then

$$\begin{aligned} \sum_{\mathbf{z} \in Z^{*c}} \theta^{S_\mathbf{z}}(1-\theta)^{m-S_\mathbf{z}} &= \sum_{i'=0}^{i^*-1} \sum_{\mathbf{z} \in Z_{i'}} \theta^{S_\mathbf{z}}(1-\theta)^{m-S_\mathbf{z}} \\ &= \sum_{i=1}^{i^*} c_i \theta^{n_i+1-i}(1-\theta)^{i-1}, \end{aligned}$$

which is Equation (A.3). It remains only to verify that the formula in Section 2.1 is the correct formula for c_i. The value of c_1 is the number of initial sequences with $1 - 1 = 0$. Of course, $c_1 = 1$, as defined. Suppose $c_1, \ldots, c_i$ are correctly defined. Then we will show that the formula

$$c_{i+1} = \binom{n_{i+1} - 1}{i} - \sum_{j=1}^{i} c_j \binom{n_{i+1} - 1 - n_j}{i + 1 - j}, \tag{A.4}$$

from Section 2.1, is correct. The value of $\binom{n_{i+1}-1}{i}$ is the number of all sequences $(z_1, \ldots, z_{n_{i+1}})$ that end in 1 and have exactly i 0s. From this we must subtract those sequences for which $S_j/(j+1) > r^*$ for some $j < n_{i+1}$. If $S_j/(j+1) > r^*$ for the first time at j', and if $j' - S_{j'} = i'$, then j' must equal $n_{i'+1}$. Among all sequences $(z_1, \ldots, z_{n_{i+1}})$ that end in 1 and have exactly i 0s, there are $c_{i'+1}\binom{n_{i+1}-1-n_{i'+1}}{i-i'}$ that first exceed r^* at $n_{i'+1}$ with i' 0s. The value of $c_{i'+1}$ is the number of initial sequences $(z_1, \ldots, z_{n_{i'+1}})$, and the combinatorial term is the number of sequences $(z_{n_{i'+1}} + 1, \ldots, z_{n_{i+1}-1})$ containing the remaining $i - i'$ 0s. Summing these terms for $i' = 0, \ldots, i-1$, changing the summation index to $j = i' + 1$ yields the sum in (A.4). Thus the formula for $c_1, \ldots, c_{i^*}$ is correct.

APPENDIX B: CALCULATING LIKELIHOODS WITH GIBBS SAMPLING

Gibbs sampling calculations for likelihood functions actually involves a mixture of some EM algorithm ideas (Dempster et al. 1977) and an implementation of successive substitution sampling (Gelfand and Smith 1990).

As in the EM algorithm, we start with $L(\theta \mid \mathbf{y})$ as the likelihood of interest, based on the "incomplete" (but observed) data $\mathbf{y}$. The augmented data is denoted by $\mathbf{z}$, yielding the complete data likelihood $L(\theta \mid \mathbf{y}, \mathbf{z})$ that satisfies

$$L(\theta \mid \mathbf{y}) = \sum_{Z^*} L(\theta \mid \mathbf{y}, \mathbf{z}). \tag{A.5}$$

The set Z^* may be quite complicated, taking into account all the restrictions imposed on the incomplete data likelihood. But it is often the case that we will be able to sample from this set.

Now normalize both likelihoods (as in Sec. 2.2) to $L^*(\theta \mid \mathbf{y})$ and $L^*(\theta \mid \mathbf{y}, \mathbf{z})$ and define $k(\mathbf{z} \mid \mathbf{y}, \theta)$ as in (7). Then

$$L^*(\theta \mid \mathbf{y}) = \int_\Theta \left(\sum_{Z^*} L^*(\theta \mid \mathbf{y}, \mathbf{z}) k(\mathbf{z} \mid \mathbf{y}, \theta') \right) L^*(\theta' \mid \mathbf{y})\, d\theta'. \tag{A.6}$$

To verify Equation (A.6), write

$$\begin{aligned} &\int_\Theta \left(\sum_{Z^*} L^*(\theta \mid \mathbf{y}, \mathbf{z}) k(\mathbf{z} \mid \mathbf{y}, \theta') L^*(\theta' \mid \mathbf{y}) \right) d\theta' \\ &\qquad = \sum_{Z^*} L^*(\theta \mid \mathbf{y}, \mathbf{z}) \left(\int_\Theta L^*(\theta' \mid \mathbf{y}, \mathbf{z})\, d\theta' \right) \end{aligned} \tag{A.7}$$

by interchanging the order of the sum and integral and noting that $k(\mathbf{z} \mid \mathbf{y}, \theta') L^*(\theta' \mid \mathbf{y}) = L^*(\theta' \mid \mathbf{y}, \mathbf{z})$. Now the integral on the right side of (A.7) is equal to 1, and the remaining sum is just (A.5). Thus from Equation (A.6) and the results of Gelfand and Smith (1990), we can calculate $L^*(\theta \mid \mathbf{y})$ by successively sampling from $L^*(\theta \mid \mathbf{y}, \mathbf{z})$ and $k(\mathbf{z} \mid \mathbf{y}, \theta)$.

Note that the implementation of the Gibbs sampler is a frequentist implementation, relying on the finiteness of the integral of $\int_\Theta L^*(\theta \mid \mathbf{y})\, d\theta$. We can, however, interpret the finiteness of this integral as using a flat prior for θ; that is, $\pi(\theta) = 1$. With the additional "parameter" $\mathbf{z}$, we then have the two full posterior distributions $\pi(\theta \mid \mathbf{y}, \mathbf{z}) (= L^*(\theta \mid \mathbf{y}, \mathbf{z}))$ and $\pi(\mathbf{z} \mid \mathbf{y}, \theta) (= k(\mathbf{z} \mid \mathbf{y}, \theta))$. Sampling from these densities will yield a sample that is (approximately) from the marginal posterior $\pi(\theta \mid \mathbf{y}) (= L^*(\theta \mid \mathbf{y}))$.

[Received April 1993. Revised November 1993.]

REFERENCES

Bayarri, M. J., and DeGroot, M. (1986a), "Bayesian Analysis of Selection Models," Technical Report 365, Carnegie-Mellon University, Dept. of Statistics.

——— (1986b), "Information in Selection Models," Technical Report 368, Carnegie-Mellon University, Dept. of Statistics.

——— (1991), "The Analysis of Published Significant Results," Technical Report 91-21, Purdue University, Dept. of Statistics.

Carlin, B. P., and Gelfand, A. E. (1992), "Parameter Likelihood Inference for Record-Breaking Problems," technical report, University of Minnesota, Division of Biostatistics.

Cleary, R. J. (1993), "Models for Selection Bias in Meta-analysis," Ph.D. thesis, Cornell University, Biometrics Unit.

Dawid, A. P., and Dickey, J. M. (1977), "Likelihood and Bayesian Inference From Selectively Reported Data," *Journal of the American Statistical Association*, 72, 845–850.

Dear, K. B. G., and Begg, C. B. (1992), "An Approach for Assessing Publication Bias Prior to Performing a Meta-Analysis," *Statistical Science*, 7, 237–245.

Dempster, A. P., Laird, N. M., and Rubin, D. B. (1977), "Maximum Likelihood From Incomplete Data Via the EM Algorithm" (with discussion), *Journal of the Royal Statistical Society*, Ser. B, 39, 1–37.

Gelfand, A. E., and Carlin, B. P. (1991), "Maximum Likelihood Estimation for Constrained or Missing Data Models," technical report, University of Minnesota, Division of Biostatistics.

Gelfand, A. E., and Smith, A. F. M. (1990), "Sampling-Based Approaches to Calculating Marginal Densities," *Journal of the American Statistical Association*, 85, 398–409.

Geyer, C. J., and Thompson, E. A. (1992), "Constrained Monte Carlo Maximum Likelihood for Dependent Data" (with discussion), *Journal of the Royal Statistical Society*, Ser. B, 54, 657–699.

Hedges, L. V. (1992), "Modeling Publication Selection Effects in Meta-Analysis," *Statistical Science*, 7, 246–255.

Iyengar, S., and Greenhouse, J. B. (1988), "Selection Models and the File Drawer Problem," *Statistical Science*, 3, 109–135.

Mosteller, F., and Chalmers, T. C. (1992), "Some Progress and Problems in Meta-Analysis of Clinical Trials," *Statistical Science*, 7, 227–236.

Reichler, J. L. (ed.) (1988), *Baseball Encyclopedia* (7th ed.), New York: Macmillan.

Rosenthal, R. (1979), "The 'File Drawer' Problem and Tolerance for Null Results," *Psychology Bulletin*, 86, 638–641.

Siegmund, D. (1985), *Sequential Analysis*, New York: Springer-Verlag.

Smith, A. F. M., and Roberts, G. O. (1993), "Bayesian Computation Via a Gibbs Sampler and Related Markov Chain Monte Carlo Methods" (with discussion), *Journal of the Royal Statistical Society*, Ser. B, 55, 3–24.

Wei, G. C. G., and Tanner, M. A. (1990), "A Monte Carlo Implementation of the EM Algorithm and the Poor Man's Data Augmentation Algorithms," *Journal of the American Statistical Association*, 85, 699–704.

Chapter 14

Good pitchers make no-hitters happen, but poor-hitting teams aren't especially vulnerable.

Baseball: Pitching No-Hitters

Cliff Frohlich

On August 11, 1991, as I watched from my seat in the third row behind the Oriole dugout at Baltimore's Memorial Stadium, Wilson Alvarez of the Chicago White Sox pitched a no-hitter, that is, in the entire nine-inning game no Oriole batter reached base except on walks and one Chicago error. In interviews after the game, Alvarez humbly gave credit for his performance to his catcher, who had chosen the type and location of pitches thrown to each batter, and to his center fielder, who had made a truly remarkable play late in the game, preventing a seemingly sure hit.

Alvarez's performance that day raises several questions about what factors are responsible for the occurrence of no-hitters. How important is the pitcher's talent and experience? Alvarez seemed an unlikely candidate for a no-hitter because this was only his second major league game, and he was not even listed in my program. In his only previous game he had failed to retire a single batter and had given up two home runs. How important is the talent of the hitting team? Alvarez's victim, the Baltimore Orioles, had the third worst winning percentage of all major league teams in 1991. How often do no-hitters come about because of questionable decisions by the official scorer, a newspaper reporter who, on each play, decides whether it was the batter's hit or a fielder's error that allowed the batter to reach base? In the seventh inning of Alvarez' game, Cal Ripken bunted and reached first safely on a close play, yet the official scorer ruled an error because the Chicago catcher had "thrown poorly." Finally, was something different about baseball in 1991 that made no-hitters especially likely? Alvarez's performance was 1 of 16 no-hitters pitched in the 1990 and 1991 seasons, the most ever pitched in any 2-year period in baseball history. There were none in 1989, two in 1992, and three in 1993. Why were there so many in 1990–1991? More generally, who pitches no-hitters, against whom, and what can we learn about why they happen?

A Probability Model for Hits/Game

Let's construct a simple probability (SP) model for the distribution of hits/game in nine-inning games. In the model we make assumptions that every student of baseball knows aren't realistic; despite this, the model provides an excellent starting point for the evaluation of no-hitters. When batters face pitchers, we assume that p is the probability that each batter makes a hit and $q = 1-p$ is the probability that each batter makes an out. We don't count walks, errors, and the like because under this simple model they affect neither the number of hits nor the number of outs. A game is just a sampling regime from which we sample batters until there are 27 outs (9 innings). This is just like drawing red balls (hits) and white balls (outs) from a hat until we have accumulated 27 white balls. Thus, the distribution of hits/game will be given by the negative binomial formula (see sidebar) and the probability a team will have no hits in a game is $(1-p)^{27}$. This model assumes that p is the same for all batters and pitchers and that serial opportunities to "draw" hits or outs are independent events (see sidebar).

Does the SP model correctly predict the incidence of no-hit games? All the available data concerning no-hit or low-hit games (Fig. 1) indicate that the SP model generally underestimates the incidence of low-hit games. For example, in 1993, 28 major league teams each started 162 games, for a total of 4,536 games. Major league pitchers averaged 9.1 hits/9 innings, which corresponds to $p = 9.1/(27+9.1) = .252$. Thus, the predicted number of no-hitters is $4{,}536(1 - .252)^{27} = 1.8$. There were three no-hitters during the 1993 season. Now, 1.8 and 3.0 aren't very different. But, if we perform this same calculation for each year for the American and National Leagues using information reported in *The Baseball Encyclopedia* about no-hitters, hits/9 innings, and games played, we predict that there should have been 135 no-hit games since 1900, but, in fact, 202 have occurred. Thus, the SP model only explains about two-thirds of the no-hit games observed. What is responsible for the 67 "extra" no-hitters?

An obvious oversimplification in the SP model is that it assumes implicitly that the performance of pitchers, batters, and official scorers is the same throughout a

Three Models for Hitting

If the probability of getting a hit or an out are p and $1-p$, respectively, the probability P_{SP} of having n hits in a game with 27 outs (9 innings) will be

$$P_{SP}(n,p) = \frac{(26+n)!}{26!\,n!}\,p^n(1-p)^{27}$$

Here the coefficient containing factorials accounts for the possible combinations of hits and outs, given that the last batter always makes an out. For the 19,385 nine-inning major league games occurring between 1989 and 1993 the distribution predicted by this simple probability (SP) model is generally similar to the observed distribution (Fig. 1). The SP model, however, predicts fewer than half of the number of no-hitters observed (Table 1). It also predicts lower-than-observed numbers for games with 9 or fewer hits and higher-than-observed numbers for games with 10 or more hits.

How does variability in the abilities of pitchers affect the distribution of hits/game? Suppose that not all pitchers are equal in their ability to get batters out or that the ability of individual pitchers varies from one game to another. In this variable pitcher (VP) model, p is not a constant but has a central value p_0 and varies such that the hits/9 innings for individual pitchers are normally distributed around $h_0 = 27p_0/(1 - p_0)$ with standard deviation σ (Fig. 2). Then the overall distribution of hits/game is

$$P_{VP}(n,h_0,\sigma) = \frac{1}{\sigma\sqrt{2\pi}} \int P_{SP}\left(n,\frac{h}{h+27}\right)\exp\left[-\frac{(h-h_0)^2}{2\sigma^2}\right]dh$$

We call this a $VP_{\sigma=1.0}$ model if σ is 1.0 hits/9 innings, a $VP_{\sigma=1.5}$ model if σ is 1.5 hits/9 innings, and so on. To find the distribution, the integral is evaluated numerically over values of h within 4σ of the central value h_0.

How does batter variability affect the incidence of no-hitters? In the variable batter (VB) model, we simulate games on the computer, assuming that in each game the values of p for individual batters are normally distributed with standard deviation. Before each "game" the computer randomly selects hitting averages p_i for batters $i = 1, 2, \ldots, 9$ from the assumed distribution. Then, each time batter i is "up" we generate a random Bernoulli trial with the probability of a hit equal to p_i and the probability of an out equal to $(1 - p_i)$. The game ends when there are 27 outs. If hitting probabilities are normally distributed with mean p_0 and standard deviation σ, then we call this a VP_σ model. Because our computer simulation is stochastic, for each p_0 and σ we simulate a large number of games N (4×10^7 for Table 1) and find the number n_k having k hits. Then, we estimate the probability of pitching a k-hitter as n_k/N.

The SP, VP, and VB models are identical if σ is 0.0. The $VP_{\sigma=1.0}$ and $VB_{\sigma=.050}$ models correctly predict more games with 0 and 1 hits than the SP model. All three models, however, predict more games than observed with about 11 and more hits. Presumably this is because baseball managers are more likely to change pitchers in games in which there are many hits.

A Test of Independence

Are batting outs independent? Or are hits bunched together so that strings of two or more hits are more common than they would be if hits were independent? To test this, between 1990 and 1993 I kept records of how often strings of 1, 2, 3, ..., hits occurred in the 119 games I attended, played by 27 major league teams at 20 different stadiums (Table 2). If there are N_{tot} hits, and outs and hits are independent, then the number N_k of strings of hits of length k should be

$$N_k = N_{tot}\,p^k(1-p)$$

where p = (number of hits)/(number of hits + outs). The data agree remarkably well with this model, supporting the assumption of event independence.

game and from game to game. We now consider how variability affects the results of the SP model.

Pitching

We can certainly attribute some of the "extra" no-hitters to the fact that not all pitchers are of equal ability and that the performance of each individual pitcher varies from day to day. Using data compiled from annual summaries of innings pitched and hits allowed, Fig. 2 demonstrates that the distribution of hits/9 innings for starting pitchers is approximately normal with a standard deviation of about 1.0 hits/9 innings. This represents the variation among the season-long performance of pitchers. Of course, the performance of each individual varies on a day-to-day basis; thus the effective standard deviation of the random quantity hits/9 innings is presumably somewhat higher than 1.0.

To evaluate the effect of variable pitching ability on the incidence of no-hitters, I construct a variable pitcher (VP) model by allowing p in the SP model to vary so that hits/9 innings has a normal distribution. I call this a $VP_{\sigma=1.0}$ model if the standard deviation of hits/9 innings is 1.0, a $VP_{\sigma=1.5}$ model if the standard deviation is 1.5, and so on. For games since 1900, the $VP_{\sigma=1.0}$ model predicts 181 no-hitters (see Table 1). Thus, apparently at least two-thirds of the 67 "extra" no-hitters occur just because all major league pitchers are not of equal ability. A $VP_{\sigma=1.5}$ model predicts 264 no-hitters, significantly more than actually occurred.

Are no-hitters pitched mostly by "average" pitchers having good days or by elite pitchers like Nolan Ryan? Nolan Ryan pitched seven no-hitters during his career; Sandy Koufax pitched four; in fact, Ryan and Koufax are the two pitchers with the lowest career average of hits allowed per nine

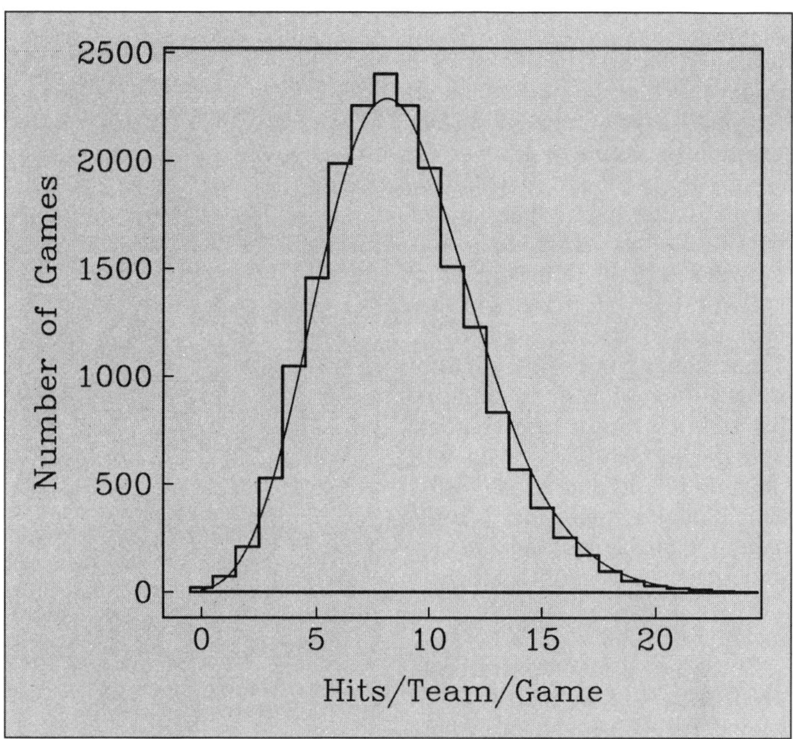

Figure 1. Distribution of hits/game in nine-inning major league games occurring between 1989 and 1993 (histogram) and distribution predicted by the simple probability (SP) model (continuous line), assuming $p = .2482$.

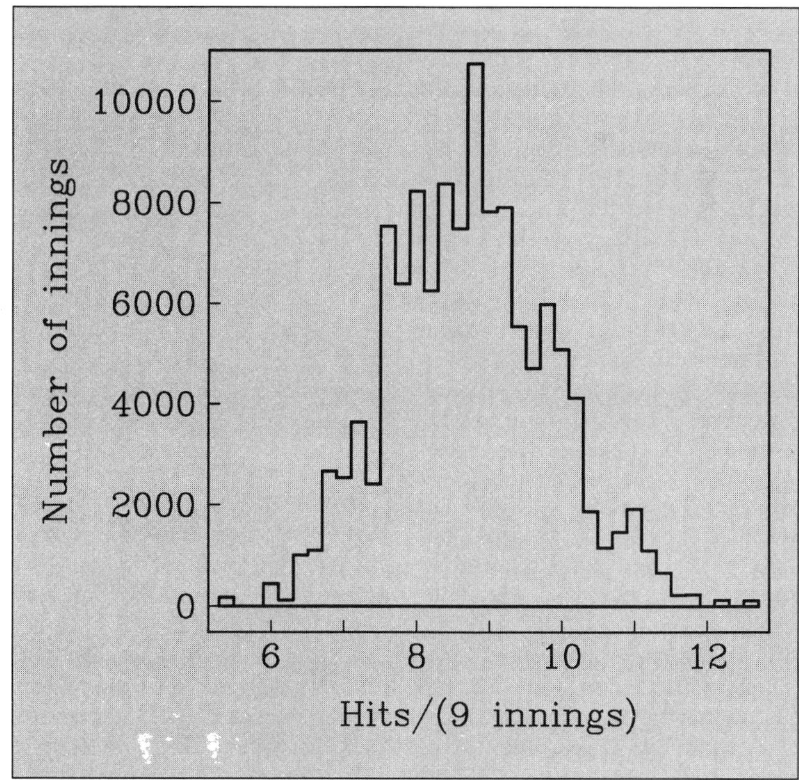

Figure 2. Distribution of hits/9 innings reported annually for all pitchers who pitched 100 or more innings between 1989 and 1993. Bin width is 0.2 hits/9 innings; mean and standard deviation for the 1989–1993 data are 8.72±1.10 hits/9 innings.

Table 1—Distribution of Hits/Team/Game for All Nine-Inning Major League Games Occurring Between 1989 and 1993 and Predictions of the SP Model, a $VP_{\sigma=1.0}$ Model, and a $VB_{\sigma=.050}$ Model

Hits/game	Observed no. of games	Expected number of games		
		SP model	$VP_{\sigma=1.0}$	$VB_{\sigma=.050}$
0	20	8.8	11.9	9.8
1	72	58.7	74.1	64.6
2	209	204.0	241.7	219.1
3	527	489.4	548.8	513.0
4	1048	911.0	975.2	938.3
5	1457	1401.7	1444.3	1418.2
6	1988	1855.3	1854.7	1853.4
7	2256	2170.6	2121.7	2148.2
8	2403	2289.4	2204.5	2248.8
9	2256	2209.5	2111.1	2164.4
10	1967	1974.0	1884.7	1934.7
11	1509	1647.8	1582.8	1620.6
12	1230	1295.0	1259.7	1282.7
13	834	964.1	955.9	964.1
14	569	683.6	695.1	693.5
15	393	463.7	486.5	478.2
16	253	302.1	328.9	318.6
17	171	189.6	215.4	204.8
18	97	115.0	137.1	128.0
19	53	67.6	85.0	77.5
20	31	38.6	51.5	45.9
21	19	21.4	30.5	26.4
22	13	11.6	17.7	14.8
23	5	6.1	10.0	8.2
24	1	3.2	5.6	4.4
25	0	1.6	3.1	2.4
26	1	0.8	1.7	1.2
27	1	0.4	0.9	0.7
28	0	0.2	0.5	0.3

Note: All three models assume a central p of .2482, consistent with the observations.

innings. Indeed, since 1900, 48 no-hitters, or 23% of the total, belong to pitchers who are among the top 60 pitchers with respect to this statistic. Thus, the top pitchers really are responsible for a substantial fraction of no-hitters.

It seems they also may get more than their fair share. Ryan averaged 6.56 hits/9 innings and started 773 games over his 27-year career; 6.56 hits/9 innings corresponds to $p = .196$. Now, $773(1 - .196)^{27} = 2.2$ games, so the SP model estimates that Ryan would have pitched two no-hitters. If we perform this calculation for each of the top 60 pitchers on the all-time hits/game list, we estimate that they should have pitched 24 no-hitters, exactly half of the number observed. A $VP_{\sigma=1.0}$ model predicts the top 60 should have pitched 33 no-hitters.

Incidentally, an analysis of seasonal hits/9 innings data since 1900 for individual no-hit pitchers indicates that they are generally having superior seasons. In particular, during the season that they perform no-hitters, their hits/9 innings rate averages 0.82 less than the league average. Only one-quarter of all no-hitters are by pitchers with seasonal hits/9 innings rates exceeding the league average.

Batting

Are no-hitters pitched mostly against poor teams or weak-hitting teams? No, definitely not. On the average, since 1900, teams that lost no-hitters actually won 49.2% of their games during those very seasons, and their seasonal batting average was only .003 lower than the league average.

Batters seem to hit better in home games than on the road; thus, we might expect that no-hitters are more likely to be pitched by the home team. For example, between 1990 and 1993, home batting averages in the major leagues were .0065 higher than away averages. If p is .2465 on the road and .2400 at home, the SP model predicts 4.80 no-hitters/10,000 games away and 6.05 no-hitters/10,000 games at home. This amounts to 56% of all no-hitters occurring at home. The data do, indeed, indicate that more no-hitters are pitched by the home team, and this was especially true before about 1980. Between 1900 and 1979, 113 of 168 (67%) no-hitters occurred at home; since 1980, 18 of 34 (53%) were at home.

How much is the incidence of no-hitters influenced by the fact that individual batters differ in their hitting ability? The SP and VP models assume that within any game there is no variation in batter hitting ability. Yet one could never pitch a no-hitter against any team if one batter had an average of 1.000, even if all the other batters were 0.000 hitters, combining to form an average p of .111. Clearly, batter variability is a factor, at least in extreme cases.

To investigate this, we conduct simulated games on the computer using a variable batter (VB) model, assuming that p for individual batters is normally distributed. To find a reasonable value for the standard deviation, we use individual batting averages and at-bats for the 1992 season reported in the *American League Red Book* and the *National League Green Book.* The results are that σ is .027 and .048 for the American League and National League, respectively, with the difference occurring because pitchers (who are notoriously weak hitters) bat in the National League. If pitchers are excluded, σ for the National and American Leagues is about the same.

As expected, the VB model does produce more no-hitters than the SP model. For a σ of .025, however, the increase is only about 2.5%, and for a σ of .050, the increase is 11.9%. I conclude that individual batter variability is less important than pitcher variability; that is, batter variability increases the no-hitter rate no more than about .5 no-hitters/10,000 games above the rate predicted by the SP model.

Scoring

Official scorers may be responsible for some no-hitters by calling close plays "errors" rather than hits. Newpaper accounts suggest that scoring bias clearly is an influence in some no-hitters. Official scorers are supplied by the home team, which may also explain why home teams pitch no-hitters more often than visiting teams. For example, in no-hitters recorded by Addie Joss in 1910, Ernie Koob in 1917, and Virgil Trucks in 1952, plays originally scored as hits were changed to errors six or more innings after they occurred. Presumably there are fewer such blatant scoring changes in the modern era, when most games are televised both at home and away and scorers have the benefit (or burden) of instant replay. Indeed, this may explain why the proportion of no-hitters pitched at home is lower since 1980 than previously.

Nevertheless, scoring a play as a hit or an error is subjective, and even today my own subjective observation is that scorers are more likely to score plays as errors if there are no hits in the game, especially after about the sixth inning. There are also clear differences among stadiums with regard to what is ruled an error. For example, there were few infield errors scored in three games I attended in 1993 at Mile High Stadium in Denver; rather, it appeared that the scorer ruled a "hit" on all ground balls if the batter reached base, regardless of how badly the infielder muffed the play.

How can we estimate how many one-hitters might be turned into no-hitters because of questionable scoring? Let's assume that scoring bias only affects a play that would ordinarily be a hit if it occurs after the sixth inning, that is, in one-third of all one-hitters. Furthermore, let's suppose that scorers only fudge on "questionable" plays and that they designate as errors exactly half of all such late-inning, questionable plays. Thus, we estimate that the fraction of one-hitters that becomes no-hitters is the product of the probability that the hit occurs after the sixth inning (1/3), the probability that such a hit occurs during a "questionable play," and the probability that the scorer rules it an error (1/2).

How common are "questionable" hits? To estimate this, in games I attended during the 1993 season I carefully noted all "questionable plays," or plays that in my judgment might have been scored as either a hit or an error. In 27 games played by 20 different major league teams at 10 different stadiums, batters reached base on 500 plays, of which official scorers ruled 481 as base hits and 19 as errors. Of these plays I scored 34 of the hits and 6 of the errors as "questionable plays." Although scoring is highly subjective, I believe that most who try this will find that under ordinary conditions (i.e., in games in which a no-hitter is no longer possible) between about 5% and 10% of all plays scored as hits occur on "questionable plays."

Of these, as we assume that half would be scored as errors if they occurred in the last third of a game with no previous hits, the fraction of one-hitters that becomes no-hitters is between 0.008 and 0.017. Moreover, the SP model predicts that in the absence of scoring bias there will be $27p$ one-hitters for each no-hitter. Thus, if p is .250, these calculations suggest that scoring bias may increase the number of no-hitters about 5–10% over that predicted by the SP model.

Table 2—Observed and Predicted Number of Strings of Length $k = 1, 2, \ldots$ Hits

k	1	2	3	4	5	6	7	Hits	Outs
Observed by author	1190	289	74	19	6	1	0	2102	6406
Predicted	1192	294	73	18	4	1	0		

Note: Predicted number calculated assuming $p = .2471$.

Variations Over Time

Are no-hitters more common during periods when the league is weaker because of expansion? Since 1960 the major leagues have expanded from 16 to 28 teams, with new teams obtaining personnel by drafting players from established teams. This dilution of the talent pool may increase the rate of occurrence of no-hitters. The expansions occurred in 1961, 1962, 1969, 1977, and 1993. To investigate whether expansion affected no-hitters, I compare the number of no-hitters pitched in 2-year periods prior to expansion with the number pitched in 2-year periods following expansion. The record shows that there were 30 no-hitters pitched in the preexpansion years (1959, 1960, 1967, 1968, 1975, 1976, 1991, and 1992) and 27 no-hitters pitched in the postexpansion years (1961, 1962, 1963, 1969, 1970, 1977, 1978, and 1993). Thus, there were actually fewer postexpansion no-hitters even though the number of games played was about 10% greater. I conclude that it is simply wrong to argue that no-hitters are pitched by pitchers feasting on leagues weakened by expansion.

Table 3—Number of and Rates of Nine-Inning No-Hitters Observed in Each Decade Since 1900 and Rates Predicted by SP and VP Models

Years	p	No. of no-hit games	Rate/10,000 games played		
			Observed	SP model	$VP_{\sigma=1.0}$
1900–1909	.242	24	10.6	5.6	7.6
1910–1919	.239	27	11.2	6.3	8.4
1920–1929	.268	9	3.7	2.2	2.9
1930–1939	.268	9	3.7	2.2	2.9
1940–1949	.249	13	5.3	4.4	5.9
1950–1959	.248	20	8.1	4.5	6.1
1960–1969	.238	35	11.0	6.5	8.7
1970–1979	.244	31	7.8	5.2	7.0
1980–1989	.247	13	3.2	4.8	6.4
Average			7.1	4.6	6.2
1990–1993	.247	21	13.5	4.7	6.3
1990–1991	.245	16	19.0	5.0	6.7

Are no-hitters more likely after September 1, when teams expand their roster from 25 to 40 to try out inexperienced players? In the 1993 season, two of the three no-hitters recorded occurred in September. The teams that were no-hit were the Cleveland Indians and the New York Mets, two of the weakest teams in the entire major leagues. Is it reasonable to suggest, as did a September 20, 1993 editorial in the *Sporting News*, that such September no-hitters occur because the weaker teams just don't care? Since 1900 there have been 48 no-hitters pitched in the month of September compared to 36, 33, and 32 in the mid-season months of May, June, and August. There have also been [illegible] in July, which has fewer games played in recent years because of the all-star break, as well as 28 and 3 in the "short" months of April and October when the season begins and ends.

Do the 48 no-hitters in September represent a significantly higher rate than the 101 no-hitters observed for the "regular" mid-season months of May, June, and August? If we presume no-hitters are distributed according to a Poisson distribution, the mean and variance are equal for samples taken over T-day periods. Thus, after scaling the rates and variances to represent occurrences reported in 30-day months, the September and mid-season since-1900 monthly rate estimates are 48/month and 33/month, respectively. Furthermore, these monthly rates are different enough so that I conclude that there probably is a "September effect," producing about one "extra" no-hitter every 6 years.

Is it fair, however, to attribute the September effect to lack of caring? Not necessarily. In the previous sections I have demonstrated that no-hitters are more common when there is an increase in variance for either pitchers or batters. When teams expand their rosters from 25 to 40 players in September it is plausible that there is an increased variance in performance due to the combination of inexperienced players competing alongside regulars whose skills are honed to late-season form. Incidentally, the season-average winning percentage of teams that are no-hit in September and October is 50.5%; thus, it is simply not true that late-season no-hitters are always pitched against weaker teams, as occurred in 1993.

Are no-hitters more common nowadays than in the past? The data indicate that the incidence of no-hitters differs during different baseball eras (Table 3). For example, during the "dead ball" era from 1900 to 1919, p was a low .241, and there were about 11 no-hitters per 10,000 games pitched. Then, during the "lively ball" period from 1920 to 1939, p rose to .268, whereas the number of no-hitters dropped to 3.7 per 10,000 games. In most decades the rate of no-hitters is approximately 1.5 times the rate predicted by the SP

Jered Weaver pitching for the 2003 USA Baseball Team. Photo courtesy of USA Baseball, © 2003.

model. An outstanding exception is the 1980–1989 decade, when there were actually fewer no-hitters than predicted by the SP model. The rate for the 1990–1993 period is unusually high; however, this is only because so many occurred in 1990 and 1991. Otherwise, there is no evidence that no-hitters are more common now than in the past.

What Happened in 1990–1991?

This is a puzzle. There were eight no-hitters in 1990, and eight in 1991. For a Poisson process with typical rates of about 7.5 no-hitters/10,000 games, the probability of observing 8 or more in a 4,200-game season is about 1%. Yet it happened 2 years in a row.

I can find nothing unusual about the 1990–1991 seasons to explain this anomaly. When compared with data from all years since 1900, the distributions for 1990–1991 of the factors that might be important are in no way unusual; these factors are hits/game/team, hits/9 innings for all pitchers in the majors, hits/9 innings for no-hit pitchers, the seasonal winning percentages of teams being no-hit, and team batting averages of teams being no-hit. There were no obvious changes in baseball in 1990–1991, such as expansion, rule changes, and the like. In addition, following 1991, the number of no-hitters has been normal, with two in 1992 and three in 1993. Thus, in the absence of any better explanation, I tentatively conclude that the high number of no-hitters in 1990–1991 is due simply to *chance*.

Discussion and Conclusions

So, what have we learned about no-hitters? The data indicate clearly that good pitchers make no-hitters happen, but, surprisingly, poor-hitting teams do not seem especially vulnerable. No-hitters occur more often late in the season and, prior to 1980, more often at home. They are not, however, more common against weak teams or in years of league expansion.

How well does the SP model explain the incidence of no-hitters? The SP model is highly successful as a reference model for evaluating how various factors affect the distribution of hits/game, even though it ignores many factors that a baseball fan knows are important; for example, pitchers tire as games progress, all hitters are not of equal ability, and so on. In this respect it is like other reference models, such as the ideal gas law in physics, or the statistical assumption that a distribution is normal—it often provides useful answers even though we know the model isn't quite true.

The variable pitcher (VP) and variable batter (VB) models are natural modifications of the simple model that evaluate the importance of two of its obvious inadequacies—variation in player ability for pitchers and hitters. All three models ignore the obvious fact that baseball is a game of strategy in which game-specific objectives such as bunting (sacrificing an out to advance a runner) or the possibility of double plays embody the fact that individual batting outs are not strictly independent.

To summarize our quantitative findings, the simple model predicts that we would have expected 135 no-hit games from 1900 to the present even if identical, average pitchers pitched against identical, average batters. We expect an additional 45 or so because pitchers aren't identical—a few are Nolan Ryans. About 15 more occur because batter ability is also variable, including several that appear to reflect a "September effect." Perhaps 10 are attributable to bias in scoring, especially prior to about 1980. Thus, the various factors that we have considered combine to explain the 202 observed no-hitters.

Statistics and baseball go hand in hand, but how much of the game is just plain luck?

Chapter 15

Answering Questions About Baseball Using Statistics

Bill James, Jim Albert, and Hal S. Stern

Is it possible for a last-place baseball team, a team with the least ability in its division, to win the World Series just because of luck? Could an average pitcher win 20 games during a season or an average player achieve a .300 batting average? The connection between baseball and statistics is a strong one. The Sunday newspaper during baseball season contains at least one full page of baseball statistics. Children buy baseball cards containing detailed summaries of players' careers. Despite the large amount of available information, however, baseball discussion is noticeably quiet on questions that involve luck.

Interpreting Baseball Statistics

Statistics are prominently featured in almost every discussion on baseball. Despite this, the way that baseball statistics are understood by the public, sportswriters, and even baseball professionals is very different from how statistics are normally used by statisticians to analyze issues. In fact, one might say that the role of statistics in baseball is unique in our culture. Baseball statistics form a kind of primitive literature that is

Illustration by John Gampert

unfolded in front of us each day in the daily newspaper. This is the way that they are understood by baseball fans and baseball professionals—not as numbers at all, but as words, telling a story box score by box score, or line by line on the back of a baseball card. That is why young children often love those baseball numbers, even though they might hate math: They love the stories that the numbers tell. People who do not analyze data on a regular basis are able to examine the career record of a player and describe, just by looking at the numbers, the trajectory of the career—the great rookie year followed by several years of unfulfilled potential, then a trade and four or five excellent years with the new team followed by the slow decline of the aging player.

Baseball has a series of standards, measures of season-long performance that are widely understood and universally accepted as marks of excellence—a .300 batting average (3 hits in every 10 attempts on average), 200 hits, 30 home runs, 20 wins by a pitcher. In fact, standards exist and have meaning in enormous detail for players at all levels of ability; a .270 hitter is viewed as being a significantly different type of player from a .260 hitter, despite the fact that the standard deviation associated with 500 attempts is about .020. These baseball standards have taken on a meaning above and beyond that suggested by the quantitative information conveyed in the statistics. To baseball fans, sportswriters, and professionals, all too often a .300 batting average does not suggest 30% of anything—.300 means excellence. Similarly, 20 wins by a pitcher is usually interpreted as great success in a way that 19 wins is not. Pitchers with 20 wins are typically described as "hardworking," and "they know how to win." The percentage of games won and other measures of pitching effectiveness are often relegated to secondary consideration.

Baseball people tend, as a consequence of how they normally understand statistics, to overestimate by a large amount the practical significance of small differences, making it very difficult to educate them about what inferences can and cannot be drawn from the data available. Take, for example, the difference between a pitcher who wins 20 games and a 15-game winner. How likely is it for an average pitcher—that is, a pitcher with average ability—to win 20 games just because he was lucky? If an average pitcher playing for an average team (we assume such a pitcher has probability .5 of winning each decision) has 30 decisions (wins or losses) in a year, then the probability of winning 20 or more games by dumb luck is .05 (actually .0494). Because these 30 decisions are influenced in a very strong way by the quality of the team for which the pitcher plays, the chance of an average pitcher on a better-than-average team winning 20 games is even larger. There are many pitchers who have about 30 decisions in a season, and, therefore, the chance that *some* pitcher somewhere would win 20 games by dumb luck is much greater than that.

For an average pitcher to win 20 games by chance is not really a true fluke; it is something that any statistical model will show *could* easily happen to the same pitcher twice in a row or even three times in a row if he is pitching for a quality team. Furthermore, real baseball provides abundant examples of seemingly average pitchers who do win 20 games in a season. For example, in 1991, Bill Gullickson of the Detroit Tigers, who had been very nearly a .500 pitcher (probability of success .5) in the surrounding seasons, won 20 games in 29 decisions. Any statistician looking at this phenomenon would probably conclude that nothing should be inferred from it, that it was simply a random event.

But to a baseball person, the suggestion that an average pitcher might win 20 games by sheer luck is anathema. Such an argument would be received by many of them as, at best, ignorance of the finer points of the game, and, at worst, as a frontal attack on their values. If it were suggested to a teammate or a coach or a manager of Bill Gullickson that he had won 20 games in 1991 simply because he was lucky, this would be taken as an insult to Mr. Gullickson. The problem is that such a suggestion would be messing with their language, trying to tell them that these particular words, "20-game winner," do not mean what they take them to mean, that excellence does not mean excellence.

The economics of baseball lock these misunderstandings into place. The difference between a pitcher who wins 20 out of 30 decisions and a pitcher who wins 15 out of 30 decisions is not statistically significant, meaning that the results are consistent with the possibility of two pitchers having equal underlying ability. But in today's market, the difference between the two pitchers' paychecks is more than a million dollars a year. A million dollars is a significant amount of money. So from the vantage point of baseball fans, players, and owners, it makes no sense to say that the difference between 20 wins and 15 wins is not significant. It is a *highly* significant difference.

This misunderstanding of the role of chance is visible throughout baseball. Is it reasonably possible, for example, that a .260 hitter (probability of success on each attempt is .26) might hit .300 in a given season simply because he is lucky? A baseball fan might not believe it, but it quite certainly is. Given 500 trials, which in baseball would be 500 at-bats, the probability that a "true" .260 hitter would hit .300 or better is .0246, or about one in 41. A .260 hitter could hit .300, .320, or even pos-

How Does One Simulate a Baseball Season?

Before considering one simulation model in detail, it is a good idea to review the basic structure of major league baseball competition. Baseball teams are divided into the National League with 12 teams and the American League with 14 teams. (This alignment will be changed in the 1993 season when two expansion teams join the National League.) The teams in each league are divided into Eastern and Western divisions, and the objective of a team during the regular season is to win more games than any other team in its division. In the National League, every team plays every other team in its division 18 times and every team in the other division 12 times for a total of 162 games. Teams in the American League also play 162 games, but they play 13 games against opponents in the same division and 12 games against teams in the other division. In post-season play in each league, the winners of the Eastern and Western divisions play in a "best-of-seven" play-off to decide the winner of the league pennant. The pennant winners of the two leagues play in a "best-of-seven" World Series to determine the major league champion.

Because a baseball season is a series of competitions between pairs of teams, one attractive probability model for the simulation is the choice model introduced by Ralph Bradley and Milton Terry in an experimental design context in 1952. Suppose that there are T teams in a league and the teams have different strengths. We represent the strengths of the T teams by positive numbers $\lambda_1, \ldots, \lambda_T$. Now suppose two teams, say Philadelphia and Houston, play one afternoon. Let λ_{phil} and λ_{hous} denote the strengths of these two teams. Then, under the Bradley–Terry choice model, the probability that Philadelphia wins the game is given by the fraction $\lambda_{phil}/(\lambda_{phil} + \lambda_{hous})$.

Can we assume that the teams have equal strengths? If this is true, $\lambda_1 = \cdots = \lambda_T$ and the result of any baseball game is analogous to the result of tossing a fair coin. The chance that a particular team wins a game against any opponent is 1/2, and the number of wins of the team during a season has a binomial distribution with sample size 162 and probability of success 1/2. If this coin-tossing model is accurate, the observed variation of winning percentages across teams and seasons should look like a binomial distribution with a probability of success equal to 1/2. If one looks at the winning percentages of major league teams from recent seasons, one observes that this binomial model is not a good fit—the winning percentages have greater spread than that predicted under this coin-tossing model. So it is necessary to allow for different strengths among the teams.

We do not know the values of the Bradley–Terry team strengths $\lambda_1, \ldots, \lambda_T$. It is reasonable to assume, however, that there is a hypothetical population of possible baseball team strengths and the strengths of the T teams in a league for a particular season is a sample that is drawn from this population. Because it is convenient to work with normal distributions on real-valued parameters, we will assume that the logarithms of the Bradley–Terry parameters, $\ln \lambda_1, \ldots, \ln \lambda_T$, are a random sample from a normal population distribution with mean 0 and known standard deviation σ.

How do we choose the spread of this strength distribution σ? Generally, we choose a value of the standard deviation so that the simulated distribution of season-winning percentages from the model is close to the observed distribution of winning percentages of major league teams from recent years. One can mathematically show that if we choose the standard deviation $\sigma = .19$, then the standard deviation of the simulated winning percentages is approximately 6.5%, which agrees with the actual observed standard deviation of season winning percentages from the past seven years. With this final assumption, the model is completely described, and we can use it to simulate one major league season. Here is how we perform one simulation.

1. Because there are 26 major league teams, we simulate a set of 26 team strengths for a particular season from the hypothetical population of strengths. In this step, we simulate values of the logarithms of the strengths from a normal distribution with mean 0 and standard deviation .19 and then exponentiate these values to obtain values for the team abilities. A team with strength λ_i has expected winning percentage p approximately described by $\ln(p / (1-p)) = 1.1 \ln \lambda_i$ (this relationship was developed empirically). In Tables 2 and 3 of the article, teams are characterized by their expected winning percentage.
2. Simulate a full season of games for each league using these values of the team strengths. In the National League, suppose Philadelphia plays Houston 12 times. If λ_{phil} and λ_{hous} denote the strength numbers for the two teams, then the number of games won by Philadelphia has a binomial distribution with 12 trials and probability of success $\lambda_{phil}/(\lambda_{phil} + \lambda_{hous})$. In a similar fashion we simulate all of the games played during the season.
3. After all the games have been played, the number of wins and losses for each team are recorded.
4. The division winners from the simulated data are determined by identifying the team with the most wins. It may happen that there are ties for the winner of a particular division and one game must be played to determine the division winner. The simulated season is completed by simulating the results of the pennant championships and the World Series using the Bradley–Terry model.

The simulation was repeated for 1000 seasons. For each team in each season, or team-season, we record the "true" Bradley–Terry strength, its simulated season winning percentage, and whether it was successful in winning its division, pennant, or World Series. Using this information, we can see how teams of various true strengths perform during 162-game seasons.

sibly as high as .340 in a given season, simply because he is lucky. Although it is true that the odds against this last event are quite long (1 in 23,000), it is also true that there are hundreds of players who participate each season, any one of whom might be the beneficiary of that luck. In discussing an unexpected event—let us say the fact that Mike Bordick of the Oakland A's, a career .229 hitter until the 1992 season, when he improved to .300—one should keep prominently in mind the possibility that he may simply have been very lucky. There are other possible explanations, such as a new stadium or a new hitting strategy. These basic probability calculations merely illustrate the large effects that can result due to chance alone.

On the other hand, an unusually poor performance by a player may also be explained by chance. The probability that a .300 hitter will have a run of 20 consecutive outs (ignoring walks and sacrifices) during a season is about .11. In fact, this calculation does not take into account that some opposing pitchers are tougher than others and that, therefore, the .300 hitter may not have the same probability of success against each pitcher. There are other explanations for a slump besides bad luck—a player does not see the ball well, a player is unsatisfied with his defensive assignment, problems at home, and so forth. All too often these alternatives are the explanation preferred by fans and sportswriters. It is rare to hear a run of successes or failure attributed to chance.

Baseball fans do not completely ignore the role of chance in baseball. Many well-hit baseballs end up right at a defensive player and many weak fly balls fall in just the right place. Conventional wisdom suggests that these events even out in the course of the season. The previous calculations suggest otherwise, that chance does not have to balance out over a season or even over several seasons.

Does the Best Team Win?

It is possible for a .260 hitter to hit .300, and it is possible for an average pitcher to win 20 games, but is it possible for a last-place team, more precisely a team with last-place ability, to win the pennant simply because they are lucky from the beginning of the season to the end? Needless to say, a baseball fan would writhe in agony even to consider such a question, and for good reason: It undermines his/her entire world.

To study this issue, a model of the modern major leagues is constructed—26 teams, 2 leagues, 14 teams in 1 league, 12 in the other. In real baseball, the exact quality of each team is unknown. If a team finishes a season with wins in 50% of their games (a .500 winning proportion), it could be that they were a .500 team, but it could also be that they were a lucky .450 team or a .550 team that fell on hard luck. In the model teams are randomly assigned abilities so that the quality of each team is known. Then the model is used to simulate baseball seasons and the probability of certain events—like the probability that the best team in baseball wins the World Series—is estimated. Some subtlety is required in creating these simulation models; it would seem natural to assign winning percentages in the model that are consistent with actual observed winning percentages, but that would create a problem. As a season is played out, there is a tendency for the empirical distribution of the simulated winning percentages to be more spread out than the randomly assigned percentages. Essentially, the differences between the records of the best teams and those of the worst teams will normally appear greater in the simulated results than the differences in assigned abilities because some of the better teams will be lucky and some of the weaker teams will be unlucky. The randomly assigned ratings were adjusted so that the simulated distribution of team records matched the observed distribution. (See the sidebar for a discussion of the simulation model used to generate the results in this article.)

One thousand seasons of baseball were simulated, a total of 26,000 team-seasons, and the answers to the following basic questions were collected:

- How often does the best team in a division (as measured by the randomly assigned ability) win the divisional title?
- How often does the best team in baseball win the World Series?
- How often does the best team in baseball fail to win even its own division?
- How often does an average team win its division?
- How often does an average team win the World Series?
- Is it possible for a last-place team (the weakest team in its division) to win the World Series simply because they are lucky?

In each simulated season it is possible to identify the "best" team in each of the four baseball divisions and the "best" team in the entire league as measured by the teams' randomly assigned abilities. The best team in the division wins its division slightly more than one-half the time. In the 1000-year simulation, 4 divisions each year, 56.4% of the 4000 divisional races were won by the best team in the division. The results are similar in the two leagues, although the larger divisions in the American League (beginning in 1993 the leagues will be the same size) lead to slightly fewer wins by the best team. The best team in baseball, the team with the highest randomly assigned ability, won the World Series in 259 out of

1000 seasons. The best team in baseball fails to win even its own division 31.6% of the time. Even if the best team in baseball does win its division, it must still win two best-of-seven play-off series to win the World Series. The probability that the best team survives the play-offs, given that it won its division, is about .38.

Table 1 defines five categories of teams—good, above average, average, below average, and poor. The categories are described in terms of the percentile of the pool of baseball teams and in terms of the expected success of the team. Good teams are the top 10% of baseball teams with expected winning proportion .567 or better (equivalent to 92 or more wins in a 162-game season). Note that in any particular season it may be that more than 10% or fewer than 10% of the teams actually achieve winning proportions of .567 or better because some average teams will be lucky and some of the good teams will be unlucky.

Table 1—Defining Five Categories of Baseball Teams

Category	Percentiles of distribution	Winning proportion
Poor	0–10	.000–.433
Below averge	10–35	.433–.480
Average	35–65	.480–.520
Above average	65–90	.520–.567
Good	90–100	.567–1.000

Table 2 describes the results of the 1000 simulated seasons. The total number of team-seasons with randomly assigned ability in each of the five categories is shown along with the actual performance of the teams during the simulated seasons. For example, of the 2491 good team-seasons in the simulations, more than half had simulated winning proportions that put them in the good category. A not insignificant number, 1%, had winning proportions below .480, the type of performance expected of below-average or poor teams.

Table 3 records the number of times that teams of different abilities achieved various levels of success. As we would expect, most of the 4000 division winners are either above-average or good teams. It is still a relatively common event, however, that an average or below-average team (defined as a team with true quality below .520) wins a divisional title. Over 1000 seasons, 871 of the 4000 division winners were average or below in true quality. Apparently, an average team wins just less than one of the four divisions per season on average. There are three reasons why this event is so common. First, an average team *can* win 92 or more games in a season by sheer luck; this is certainly not an earth-shaking event. In the simulations, 2.5% of the average or worse team-seasons had good records by chance. Second, there are a lot of average teams; average teams have good teams badly outnumbered. The third and largest reason is that it is relatively common for there to be a division that does not, in fact, *have* a good team. In real baseball, it may not be obvious that a division lacks a good team because the nature of the game is that somebody has to win. But it might be that in baseball there are five good teams with three of those teams in one division, one in a second, one in a third, but no good team in the fourth division. In another year, there might be only three good teams in baseball, or only two good teams; there is no rule of nature that says there *have* to be at least four good

Table 2—Simulated Performance of Teams in Each Category

Randomly assigned ability	No. of team-seasons	Percent of total	Performance in Simulated Season: Poor	Below average	Average	Above average	Good
Poor	2595	10.0	58.0	32.0	8.9	1.1	.0
Below avg.	6683	25.7	20.5	42.7	27.4	8.7	.7
Average	7728	29.7	5.0	26.0	37.9	26.2	4.9
Above avg.	6503	25.0	.6	8.1	26.9	42.9	21.5
Good	2491	9.6	.0	1.0	7.4	31.9	59.7

Table 3—Frequency of Winning Title for Teams in Each Category

Randomly assigned ability	No. of team-seasons	Number of times		
		Won division	Won league	Won World Series
Poor	2595	8	2	1
Below average	6683	156	42	9
Average	7728	707	287	122
Above average	6503	1702	844	403
Good	2491	1427	825	465
Total	26000	4000	2000	1000

teams every year. So, as many years as not, there simply is not a very good team in one division or another, and then a team that is in reality a .500 or .510 team has a fighting chance to win. It is fairly common.

However, most of those average teams that get to the play-offs or World Series will lose. A slight advantage in team quality is doubled in the seven-game series that constitutes baseball's play-offs and World Series. If a team has a 51% chance of winning a single game against an opponent, then they would have essentially a 52% chance of winning a seven-game series (ignoring home-field advantages and assuming independence, the actual probability is .5218). If a team has a 53% chance of winning a single game against a given opponent, that will become about 56% in a seven-game series. So if an average team is in the play-offs with three good, or at least above-average teams, their chances of coming out on top are not very good, although still non-negligible. In fact, Table 3 shows that teams of average or less ability won 132 of the 1000 simulated World Series championships.

The simulation results suggest that a more-or-less average team wins the World Series about one year in seven. But is it possible for a last-place team, the team with the least ability in its division, to win the World Championship simply because of chance? Yes, it is. In 1 of the 1000 simulated seasons, a last-place team, a team that should by rights have finished sixth in a six-team race with a record of about 75–87, did get lucky, won 90 games instead of 75, and then survived the play-offs against three better teams. The chance of a last-place team winning the World Championship can be broken down into elements, and none of those elements is, by itself, all that improbable. The combination of these elements is unlikely but clearly not impossible.

A complementary question is: How good does a team have to be so that it has a relatively large chance of winning the World Series? The simulations suggest that even teams in the top one-tenth of 1% of the distribution of baseball teams, with an expectation of 106 wins in the 162 game schedule, have only a 50-50 chance of winning the World Championship.

Conclusions

What do all these simulation results mean from the perspective of baseball fans and professionals? From their perspective, it is appalling. In baseball, every success and every failure is assumed to have a specific origin. If a team succeeds, if a team wins the World Championship, this event is considered to have not a single cause but a million causes. There has never been a member of a World Championship team who could not describe 100 reasons why his team won. It is much rarer to hear the role of chance discussed. An exception is the story of old Grover Cleveland Alexander, who scoffed at the idea that he was the hero of the 1926 World Series. Just before Alexander struck out Tony Lazzeri in the seventh and final game of that World Series, Lazzeri hit a long drive, a home-run ball that just curved foul at the last moment. Alexander always mentioned that foul ball and always pointed out that if Lazzeri's drive had stayed fair, Lazzeri would have been the hero of the World Series, and *he* would have been the goat.

The simulations suggest that, indeed, there might not be *any* real reason why a team wins a World Championship; sometimes it is just luck. That is an oversimplification, of course, for even assuming that an average team might win the World Series; it still requires an enormous effort to be a part of an average professional baseball team. Baseball teams are relatively homogeneous in ability. It is surprisingly difficult to distinguish among baseball teams during the course of the 162-game regular season and best-of-seven play-off series and World Series.

It might be argued that baseball managers and players prefer not to think of events as having a random element because this takes the control of their own fate out of their hands. The baseball player, coach, or manager *has* to believe that he can control the outcome of the game, or else what is the point of working so hard? This desire to control the outcome of the game can lead to an overreliance on small samples. There are many examples of baseball managers choosing a certain player to play against a particular pitcher because the player has had success in the past (perhaps 5 hits in 10 attempts). Or perhaps a pitcher is allowed to remain in the game despite allowing three long line drives that are caught at the outfield fence in one inning but removed after two weak popups fall in for base hits in the next inning. Can a manager afford to think that these maneuvers represent an overreaction to the results of a few lucky or unlucky trials?

In any case, this conflict between the way that statisticians see the game of baseball and the way that baseball fans and baseball professionals see the game can sometimes make communication between the groups very difficult.

Additional Reading

Bradley, R. A., and Terry, M. E. (1952), "Rank Analysis of Incomplete Block Designs. I. The Method of Paired Comparisons," *Biometrika*, 39, 324–345.

James, B. (1992), *The Baseball Book 1992*, New York: Villard Press.

Ladany, S. P., and Machol, R. E. (1977), *Optimal Strategies in Sports*, New York: North-Holland.

Lindsey, G. R. (1959), "Statistical Data Useful to the Operation of a Baseball Team," *Oper. Res.*, 7, 197–207.

Chapter 16

THE PROGRESS OF THE SCORE DURING A BASEBALL GAME

G. R. Lindsey

Defence Systems Analysis Group, Defence Research Board, Ottawa, Canada

Since a baseball game consists of a sequence of half-innings commencing in an identical manner, one is tempted to suppose that the progress of the score throughout a game would be well simulated by a sequence of random drawings from a single distribution of half-inning scores. On the other hand, the instincts of a baseball fan are offended at so simple a suggestion. The hypothesis is examined by detailed analysis of 782 professional games, and a supplementary analysis of a further 1000 games. It is shown that the scoring does vary significantly in the early innings, so that the same distribution cannot be used for each inning. But, with a few exceptions, total scores, establishments of leads, overcoming of leads, and duration of extra-inning games as observed in the actual games show good agreement with theoretical calculations based on random sampling of the half-inning scoring distributions observed.

1. INTRODUCTION

A baseball game consists of a sequence of half-innings, all commencing in an identical manner. It might therefore be supposed that a good approximation to the result of a large sample of real games could be obtained from a mathematical model consisting of random drawings from a single population of half-inning scores. To be more explicit, two simple assumptions can be postulated, which could be labelled "homogeneity" and "independence."

"Homogeneity" implies that each half-inning of a game offers the same a priori probability of scoring—i.e. the distributions of runs for each half-inning are identical. "Independence" implies that the distribution of runs scored subsequent to the completion of any half-inning is unaffected by the scoring history of the game previous to that time.

On the other hand, there are reasons to suggest that the scoring in a game may have a structure more complicated than that represented by a simple superposition of identical independent innings. Pitchers may tire and lose their dominance over batters as the game progresses. Once a lead is established the tactics may alter as the leading team tries to maintain its lead and the trailing team tries to overcome it. Reminiscences suggest that special events occur in the last half of the ninth inning. The home crowd never fails to remind the Goddess of Fortune of the arrival of the Lucky Seventh. Surely a game so replete with lore and strategy must be governed by laws deeper than those of Blind Chance!

This paper attempts to examine such questions.

The main body of observed data are taken from the results of 782 games played in the National, American, and International Leagues during the last three months of the 1958 season. All games reported during this period were recorded.

The distributions of runs are found for each inning, and compared with one another to test the postulate of homogeneity. Using the observed distribution and assuming them to be independent, the theoretical probability distributions of a number of variables such as winning margin and frequency of extra innings are calculated. These theoretical distributions are then compared with those actually observed in the sample of 782 games, in order to test the postulate of independence.

Some of the unexpected results are examined in a further sample of 1000 games chosen at random from the 1959 season of the National and American Leagues, and some of the conventional tests for correlation are applied.

2. DISTRIBUTION OF SCORES BY INNINGS

The distribution of scoring of runs in each half-inning is shown in Table 1. The columns headed 0, 1, · · · >5 show the relative frequency with which one team scored x runs in the ith inning. The column headed "N_i" shows the number of ith half-innings recorded. Games abandoned due to weather before becoming legal contests, or abandoned with the score tied, were not counted, so that the 782 games necessarily produced 1564 first, second, third and fourth half-innings. A small number of games were called in the fifth, sixth, seventh or eighth inning. Many games did not require the second half of the ninth inning, and only a few required extra innings. International League games scheduled for seven innings only were excluded.

The third last column shows the mean number of runs scored by one team in the ith inning, and the second-last column gives the standard deviation σ of the distribution. The last column, headed $\sigma/\sqrt{N_i}$, gives the standard error of the mean. It is immediately evident that the means differ considerably and significantly from inning to inning.

The means and standard deviations are also shown on the top half of Figure 1, in which the solid vertical black bars are centered at the mean number of runs, and have length $2\sigma/\sqrt{N_i}$.

TABLE 1. RELATIVE FREQUENCY WITH WHICH ONE TEAM SCORED x RUNS IN THE iTH INNING—(BASED ON 782 GAMES FROM NL, AL, IL, 1958)

Inning (i)	x = Number of Runs Scored							N_i	Mean	σ	$\sigma/\sqrt{N_i}$
	0	1	2	3	4	5	>5				
1	.709	.151	.087	.029	.012	.008	.004	1,564	.53	1.05	.03
2	.762	.142	.065	.017	.008	.004	.002	1,564	.38	.83	.02
3	.745	.119	.064	.034	.020	.010	.008	1,564	.53	1.06	.03
4	.746	.142	.063	.026	.015	.006	.002	1,564	.44	.94	.02
5	.748	.140	.060	.030	.016	.004	.002	1,564	.45	.95	.02
6	.715	.152	.077	.033	.010	.008	.005	1,558	.52	1.05	.03
7	.743	.140	.067	.026	.014	.005	.005	1,558	.46	.99	.03
8	.720	.162	.064	.027	.012	.012	.003	1,554	.50	1.04	.03
9	.737	.148	.074	.021	.011	.008	.001	1,191	.45	.87	.03
10	.72	.13	.10	.04	.01	.00	.00	134	.51	.93	.08
11	.80	.14	.03	.03	.00	.00	.00	64	.30		.08
12	.78	.11	.09	.02	.00	.00	.00	44	.4		
13	.92	.08	.00	.00	.00	.00	.00	26			
14	.72	.18	.05	.00	.00	.05	.00	22			
15	.6	.2	.1	.0	.1	.0	.0	10			
16	.0	.5	.5	.0	.0	.0	.0	2			
All Extra	.752	.132	.076	.027	.010	.003	.000	302	.42	.86	.05
All	.737	.144	.069	.027	.013	.007	.003	13,993	.475	1.00	.008

The bottom row of Table 1 shows the overall frequency distribution when all innings are combined. If the numbers in this bottom row are combined with N_i to give the frequencies expected in the other rows for the individual innings, on the assumption that all innings are merely samples of the same aggregate, and all extra innings are combined into one row, then the differences between the number of runs observed and expected produce a value of chi-square which has a probability of only about 0.005 of being exceeded by chance alone.

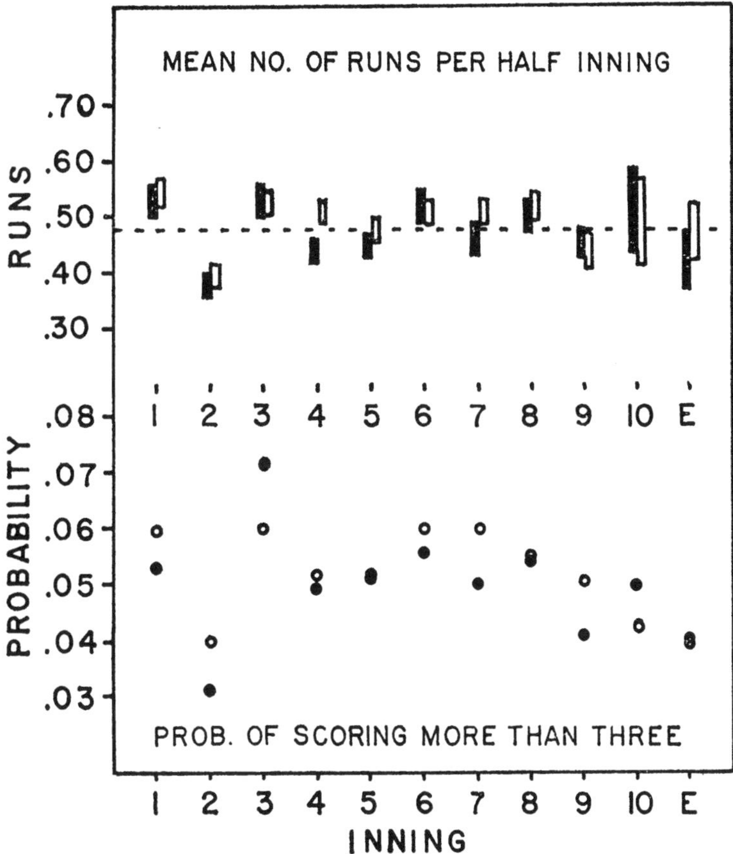

FIG. 1. Scoring by individual half-innings.
Black: 782 games of 1958
White: 1000 games of 1959

Thus it can be concluded that the distribution of runs is not the same from inning to inning, and the "postulate of homogeneity" is untrue. Examination of Table 1 or Figure 1 shows that the mean number of runs are greatest in the first, third and sixth innings, and least in the second and in the extra innings. There is nothing remarkable about the seventh or the ninth innings, both of which are very similar to the aggregate distribution.

The most notable differences occur in the first three innings, where the mean scores in the first and third innings are 0.53, while the mean score in the second inning is only 0.38. The probabilities of deviations as large as these from the overall mean of 0.475 are approximately 0.03, 0.001 and 0.03 for a homogeneous sample. Deviations of the means for all later innings, and chi-square for the later innings, are small enough to allow them to be considered as samples from the overall aggregate (at a level of $\epsilon = .05$).

A possible explanation for this peculiar pattern of scoring in the first three innings is that the batting order is designed to produce maximum effectiveness in the first inning. The weak tail of the order tends to come up in the second inning, the strong head tends to reappear in the third, and the weak tail in the fourth.

This pattern is even more evident if one plots the frequency of scoring three or more runs, obtained by adding the columns for $x = 3$, 4, 5, and >5 runs. The frequencies are shown by the solid black circles on the lower half of Figure 1. It is seen that the third inning is the most likely to produce a big score.

TABLE 2. THE RELATIVE FREQUENCY WITH WHICH ONE TEAM SCORED x RUNS IN THE iTH INNING—(BASED ON 1000 GAMES FROM NL, AL, 1959)

Inning (i)	x = Number of Runs Scored							N_i	Mean	σ	$\sigma/\sqrt{N_i}$
	0	1	2	3	4	5	>5				
1	.700	.159	.081	.031	.018	.010	.001	2,000	.54	1.03	.02
2	.768	.139	.053	.023	.009	.003	.005	2,000	.40	.92	.02
3	.730	.131	.079	.029	.016	.009	.006	2,000	.53	1.10	.03
4	.719	.151	.078	.027	.013	.007	.005	2,000	.51	1.03	.02
5	.730	.145	.074	.033	.011	.005	.002	2,000	.48	.98	.02
6	.721	.157	.062	.036	.012	.008	.004	2,000	.51	1.03	.02
7	.731	.138	.071	.029	.020	.008	.003	2,000	.51	1.04	.02
8	.710	.162	.073	.027	.019	.006	.003	2,000	.52	1.03	.02
9 C	.770	.129	.060	.027	.010	.004	.000	1,576	.40	.87	.02
9 I	.000	.46	.26	.23	.03	.02	.00	61	1.88	.98	.13
EC	.825	.112	.033	.012	.003	.006	.009	331	.32	.95	.05
EI	.00	.73	.17	.06	.04	.00	.00	53	1.40	.76	.10
All E	.711	.198	.052	.018	.008	.005	.008	384	.47	1.00	.05
All C	.730	.146	.070	.029	.014	.007	.004	17,907	.487	1.01	.008
All I	.00	.59	.22	.15	.03	.01	.00	114	1.66	.92	.09
All	.726	.148	.071	.030	.014	.007	.004	18,021	.493	1.05	.008

The main arguments in this paper are based on the data from the 782 games played in 1958. However, in order to obtain additional evidence regarding results that appeared to deserve further examination, data were also collected from a sample of 1000 games selected at random from the 1959 seasons of the National and American Leagues. No abandoned games were included in this sample. Table 2 shows the frequency distributions of inning-by-inning scores, and the means and frequencies of scoring three or more runs are plotted on Figure 1, using hollow white bars and circles.

Comparison of the two tables show very close agreement. A chi-square test shows no significant difference between the two samples. The low scoring in the second inning as compared to the first and third is evident again.

At this point we must abandon our postulate of homogeneity, and treat the inning-by-inning distributions individually.

In the subsequent analysis, the scoring probabilities are deduced directly from the observed data of Table 1, without any attempt to replace the observations by a mathematical law. However, examination of the shape of the distributions is interesting for its own sake, and is described in Appendix A.

Since baseball games are terminated as soon as the decision is certain, the last half of the ninth (tenth, eleventh, $\cdots$) inning is incomplete for games won by the home team in that particular half-inning. The results in Table 2 show separate rows of complete and incomplete ninth and extra half-innings. The frequency distributions are very different for complete and incomplete half-innings. A winning (and therefore incomplete) last half must have at least one run. A large proportion of winning last halves of the ninth show two or three runs, and the mean is 1.88. One run is usually enough to win in an extra inning, and the mean score for winning last halves is 1.40. Completed halves include all the visiting team's record, but exclude all winning home halves, so that the mean scores are low: 0.40 for completed ninth half-innings, and lower still, 0.32, for completed extra halves.

The basic probability function required for the calculation is $f_i(x)$, the a priori probability that a team will score x runs in its half of the ith inning, provided that the half-inning is played, and assuming it to be played to completion. If half-innings were always played to completion, the distributions observed in Tables 1 and 2 would provide direct estimates of $f_i(x)$. However, for $i \geq 9$ the distributions for complete half-innings recorded in the tables cannot be used directly, since they have excluded the decisive winning home halves, while the distributions for incomplete halves are ineligible just because they are incomplete.

The problem of estimating $f_i(x)$ from the observed frequencies for $i \geq 9$ is discussed in Appendix B. There it is concluded that $f_9(x)$ is found to be very nearly the same as $\overline{f_{18}}(x)$, the mean of the first eight innings, so that the ninth inning does not present any unusual scoring pattern. There are too few extra innings to allow calculation of the tenth, eleventh, $\cdots$ innings separately, so a grouped distribution $f_E(x)$ applying to all extra innings is sought. $f_E(x)$ as calculated from Table 1 (the 782 games of 1958) does not agree with $\overline{f_{18}}(x)$, showing a substantial deficit in ones and an excess of twos, threes, and fours. But the 1000 games of 1959 (Table 2) produce distributions consistent with the hypothesis that $f_E(x) \approx \overline{f_{18}}(x)$.

3. INHOMOGENEITY INHERENT IN THE SAMPLE

The 1782 games from which all of the data in this paper have been obtained include two seasons, three leagues, twenty-four teams, day and night games, and single games and doubleheaders. If the sample were subdivided according to these, or other categories, it is possible that significant differences in the populations might be discernible. However, as was pointed out in an earlier study of batting averages [4, p. 200], subdivision of the sample soon reduces the numbers to the point where the sampling error exceeds the magnitude of any small effects being examined, while extension of the period of time covered will introduce new sources of inhomogeneity such as changes in the personnel of the teams. Therefore, many small effects will be measurable only if they appear consistently in large samples necessarily compounded from many categories of games. These are the most interesting effects for the general spectator, although a manager would prefer to know their application to the games of his own team exclusively.

In any case, it is the contention of the author that professional baseball games offer a very homogeneous sample. Teams differ from one another so little that it is very unusual for a team to win less than one third or more than two-thirds of their games in a season. Of the sixteen major league teams, the lowest mean total score for the 1958 season was 3.40, the highest 4.89, the mean 4.39, and the standard deviation between mean scores was 0.375. Therefore the difference between teams, if they represented sixteen separate populations with means as measured over the season, would show a standard deviation of approximately one-third of 0.375, or 0.12, for each of the nine innings, which is a small magnitude as compared to the standard deviation of approximately 1.0 for the game-to-game variation in the pooled sample.

One hundred Chicago and Washington games from the 1959 American League were extracted from the 1000-game sample, and their inning-by-inning distributions analyzed separately (as in Table 1). (These teams finished first and last, respectively.) The mean scores and standard deviations for the first eight innings combined were $\mu=0.40$, $\sigma=0.94$ for Chicago, and $\mu=0.47$, $\sigma=0.85$ for Washington. They do show slightly smaller standard deviations than the pooled distribution, but σ is still about 2μ.

A small inhomogeneity probably does exist between the visiting and home halves of any inning. The home team won 55.0% of the major league games of 1958, and 54.7% of the 1000 games of 1959. In the latter sample the total score of the home teams showed $\mu=4.50$, $\sigma=3.03$, while for visitors the result was $\mu=4.38$, $\sigma=3.13$. The differences between these are negligible as compared to the inning-to-inning differences of the pooled distribution.

While it might be interesting to extract separate sources of variances due to various inhomogeneities, it seems much more profitable in this exploratory study to use pooled data and seek effects which appear in spite of whatever inhomogeneity may be present.

4. INDEPENDENCE BETWEEN HALF-INNINGS

The postulate of independence could be tested by conventional methods, such as computation of coefficients of correlation and comparison of conditional distributions. This is in fact done in Appendix C, for certain pairs of half-innings, and shows no significant correlation. The method is applied only to pairs of half-innings, whereas the interesting questions pertain to tendencies noticeable over the whole game.

A different approach to the testing of independence is followed in the main body of this paper, which seems more closely related to the question at the forefront of the minds of the participants and spectators during the course of a baseball game, and which is likely to be more sensitive than the methods of correlation coefficients or conditional distributions. This is to examine the probability that the team which is behind at a particular stage of the game will be able to overcome the lead and win. It is possible to deduce this by probability theory, if independence is assumed and the $f_i(x)$ derived from the data of Table 1 used, and the predicted result can be compared with that actually observed in practice. Also, in addition to the probability of the lead being overcome, several other distributions of interest, such as the winning margin, the total score of each team, and the length of the game, can be computed and compared with observed results.

5. LENGTH OF GAME

It is shown in Appendix B that if we know the distribution of runs scored by a team in each half-inning (whether homogeneous or not), and if the scores obtained in successive half-innings are independent, then it is possible to calculate the probability that the score will be tied at the end of nine innings, and the probability that the two teams make equal scores during any single extra inning. For the data collected here, these probabilities are 0.106 and 0.566, which implies that, of all the games which had a ninth inning, a fraction 0.106 would require a tenth, and that of all the games which required a tenth (elev-

TABLE 3. THE STAGES AT WHICH VARIOUS GAMES CONCLUDED

<table>
<tr><th>Sample</th><th></th><th>$i=9$</th><th>10</th><th>11</th><th>12</th><th>13</th><th>14</th><th>15</th><th>16</th><th>17</th><th>18</th></tr>
<tr><td>1958</td><td>Prob. $(=i)$</td><td>.894</td><td>.046</td><td>.026</td><td>.015</td><td>.008</td><td>.005</td><td>.003</td><td>.002</td><td>.001</td><td>.001</td></tr>
<tr><td>1958</td><td>$N_{\geq i}$ (obs)</td><td>777</td><td>67</td><td>32</td><td>22</td><td>13</td><td>11</td><td>5</td><td>1</td><td>0</td><td>0</td></tr>
<tr><td></td><td>$N_{\geq i}$ (exp)</td><td></td><td>82</td><td>38</td><td>22</td><td>12</td><td>7</td><td>4</td><td>2</td><td>1</td><td>0</td></tr>
<tr><td>1959</td><td>$N_{\geq i}$ (obs)</td><td>1000</td><td>95</td><td>44</td><td>20</td><td>15</td><td>8</td><td>4</td><td>4</td><td>1</td><td>0</td></tr>
<tr><td></td><td>$N_{\geq i}$ (exp)</td><td></td><td>106</td><td>54</td><td>25</td><td>11</td><td>8</td><td>5</td><td>2</td><td>2</td><td>1</td></tr>
<tr><td></td><td>Game finished</td><td colspan="2">At end of $V9$</td><td colspan="2">During $H9$</td><td colspan="2">At end of $H9$</td><td colspan="2">During E</td><td colspan="2">Total</td></tr>
<tr><td>1959</td><td>N (obs)</td><td colspan="2">424</td><td colspan="2">61</td><td colspan="2">420</td><td colspan="2">95</td><td colspan="2">1000</td></tr>
<tr><td></td><td>N (exp)</td><td colspan="2">400</td><td colspan="2">47</td><td colspan="2">447</td><td colspan="2">106</td><td colspan="2">1000</td></tr>
</table>

enth, twelfth, $\cdots$) inning a fraction 0.566 would require an eleventh (twelfth, thirteenth, $\cdots$) inning.

The row labelled "Prob$(=i)$" of Table 3 shows the calculated a priori probability that a game will require exactly i innings. The third row shows the number of games (out of the 782 in 1958) which actually did require an ith inning. The fourth row shows the number that would be expected, based not on the a priori probability but on the formula

$$N_{10}\,(\text{exp}) = 0.106N_9\,(\text{obs})$$

$$N_i\,(\text{exp}) = 0.566N_{i-1}\,(\text{obs}) \qquad \text{for } i \geq 11$$

which is based on the number of games actually requiring the inning previous to the ith.

Similar results are shown for the 1000 games of 1959.

For both samples, the number of games actually requiring a tenth and eleventh inning is rather less than predicted, but the difference is not significant to to the 10% level of chi-square.

Another examination of the lengths of games can be made by observing the number that end without requiring the home half of the ninth inning (designated as $H9$), the number that are concluded during $H9$, the number ending with the conclusion of $H9$, and the number requiring one or more extra innings. These are listed in Table 3 for the 1000 games of 1959. Also listed are the number that would be predicted by the independent model, using the theory of Appendix B. The number of games decided in the last half of the ninth exceed the predicted number (significantly at the level $\epsilon \approx 0.04$).

Of all the 531 extra innings observed in both samples, a fraction 0.538 were tied (as compared to the expected fraction 0.566: the difference is not significant).

It seems fair to conclude that the length of games actually observed is consistent with the prediction made on the hypothesis of independence between innings, although there is a slight tendency for games to be concluded in the last of the ninth, or in the tenth, more often than predicted.

6. TOTAL SCORE

The distribution of the total number of runs scored by one team in an entire game is shown in Table 4, in the rows labelled "Observed." The frequencies are compared with those predicted on the model of independence, by calculations described in Appendix B, and employing estimates of $f_i(x)$ obtained from the data of Table 1. The mean score is 4.25, and the standard deviation is 3.11.

Figure 2 shows the same results in graphical form, expressed as normalized probabilities. The continuous line shows $F(x)$, the probability of one team scoring x runs in a complete game as predicted by the theory. The midpoints of the vertical black bars represent the frequencies observed in the 782 games of 1958. The length of each bar is $2\sqrt{F(x)[1-F(x)]/N}$, where N is the number of observations, 1564. About two-thirds of the bars would be expected to overlap the predicted values on the continuous line.

The vertical white bars show the results observed for the 1000 games of 1959.

TABLE 4. DISTRIBUTION OF TOTAL SCORE BY ONE TEAM

Sample	Total Runs:	0	1	2	3	4	5	6	7	8	9	10
	Frequency											
1958	Observed	100	193	225	240	193	138	136	115	70	51	34
	Expected	95	167	218	233	219	184	145	108	77	52	31
	(O–E)		26			−26	−46*					
1959	Observed	134	196	249	276	286	241	171	140	100	76	41
	Expected	122	214	278	298	280	236	186	138	98	66	40
	(O–E)			−29								
	Total Runs:	11	12	13	14	15	16	17	18	19	20	Sum
	Frequency											
1958	Observed	24	18	13	7	1	2	1	2	1	0	1564
	Expected	17	9	5	2	2	0	0	0	0	0	1564
	(O–E)	7	←				27*				→	
1959	Observed	42	14	13	10	3	3	3	1	0	1	2000
	Expected	22	12	6	2	2	0	0	0	0	0	2000
	(O–E)	20*	←				26*				→	

The results for the larger scores are also shown with an expanded vertical scale on the right of the diagram.

The rows labelled (O-E) in Table 4 show the difference (observed-expected) only when chi-square is larger than would be expected on 10% of occasions. When the value has an asterisk this signifies that chi-square exceeds the value for $\epsilon = 0.001$. There are two anomalous points—a very small number of 5-run games in 1958 and a very large number of 11-run games in 1959—and a general tendency to observe more large scores (over 10 runs) than predicted. With these exceptions, the agreement is good.

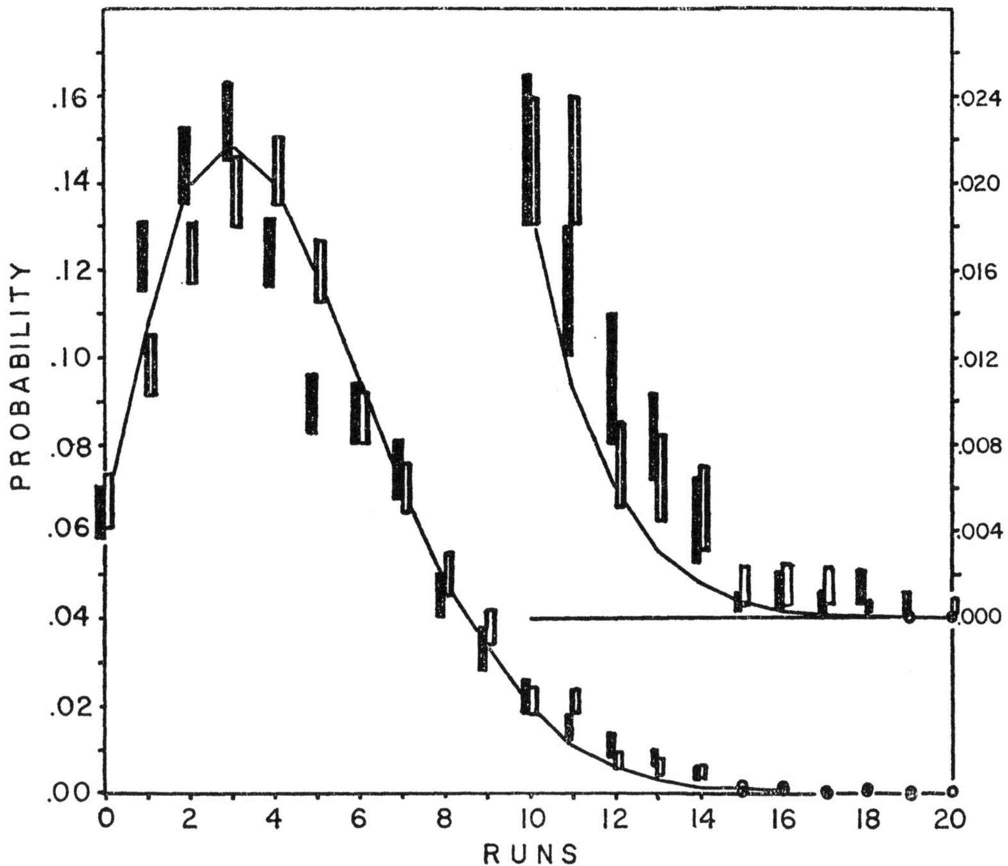

FIG. 2. Distribution of total scores by one team.
Black: 782 games of 1958
White: 1000 games of 1959

The excess number of large scores suggests the possibility that a team against which a large score is being made might regard the game as hopelessly lost, and decide to leave the serving but ineffective pitcher to absorb additional runs, instead of tiring out more relief pitchers in a continuing effort to halt their opponents. However, this would imply that games in which a large total score was obtained by one team would show a disproportionate share of their runs coming in the later innings, but an examination of those games in which the winner scored more than 10 runs showed no such effect.

It may be that there are a small number of occasions on which the batters are in abnormally good form, and for which the statistics do not belong to the general population.

7. ESTABLISHMENT OF A LEAD

As well as the number of runs scored in each half-inning, the difference between the scores of the two teams totalled up to the end of each full inning was recorded. Table 5 shows the number of games in which a lead of l runs had been established by the end of the ith inning, and at the end of the game, for the 782 games of 1958.

TABLE 5. ESTABLISHMENT OF A LEAD

Lead	Inning:	1	2	3	4	5	6	7	8	Final	Final
	Total:	782	782	782	782	782	779	779	777	782	1000
0	Established	420	286	203	147	128	118	109	82		
	Expected	418	281	196	152	126	106	96	87		
	(O-E)	—	—	—	—	—	—	—	—		
1	Established	192	254	239	247	211	187	164	160	239	316
	Expected	192	242	231	222	206	188	176	163	256	327
	(O-E)	—	—	—	25	—	—	—	—	—	—
2	Established	105	135	131	138	167	154	144	146	145	178
	Expected	105	139	152	158	158	152	148	142	147	188
	(O-E)	—	—	−21	—	—	—	—	—	—	—
3	Established	31	50	83	105	87	100	113	101	113	146
	Expected	36	60	88	100	109	114	114	114	113	145
	(O-E)	—	—	—	—	−22	—	—	—	—	—
4	Established	18	28	59	62	76	85	79	90	79	103
	Expected	16	30	52	64	72	81	84	87	85	109
	(O-E)	—	—	—	—	—	—	—	—	—	—
5–9	Established	16	28	64	78	106	118	150	168	178	226
	Expected	16	31	63	84	104	126	148	168	163	208
	(O-E)	—	—	—	—	—	—	—	—	—	—
≥10	Established	0	1	3	5	7	17	20	30	28	31
	Expected	0	0	2	3	5	9	12	16	18	23
	(O-E)	—	—	—	—	—	8	8	14*	10*	8
$\epsilon<0.1$		—	—	—	—	—	—	—	.02	—	—

In Appendix B, a formula is derived for the probability that one team will have established a lead of l at the end of i innings, on the assumption that the probability of scoring x runs in the ith inning is $f_i(x)$, and that there is no correlation between innings or with the score being made by the opposing team. The lines in Table 5 labelled "Expected" show the number predicted by this formula.

The differences between the numbers observed and expected are entered in the lines of Table 5 labelled (O-E) only when the difference has a chi-square corresponding to $\epsilon \leq 0.1$. When $\epsilon \leq 0.02$ an asterisk is attached to the number.

In the bottom line, the value of ϵ appropriate to chi-square for all observations in the inning is shown only if $\epsilon < 0.1$.

Figures 3A and 3B show the results in graphical form, with the observed fre-

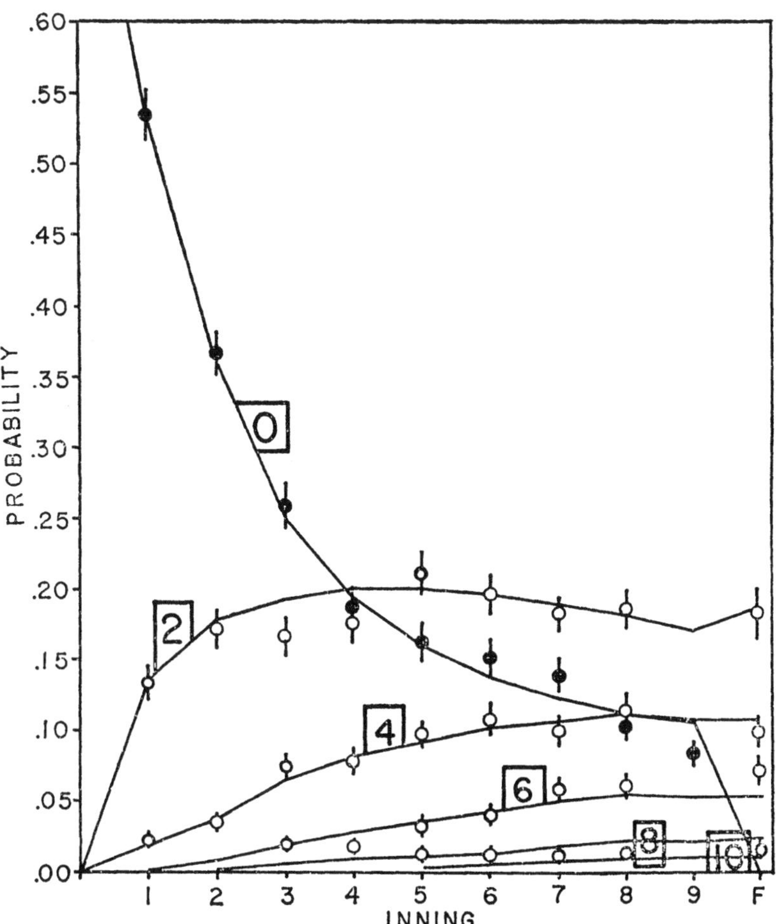

FIG. 3A. Probability of establishing a lead.
782 games of 1958
(Even Numbers Only)

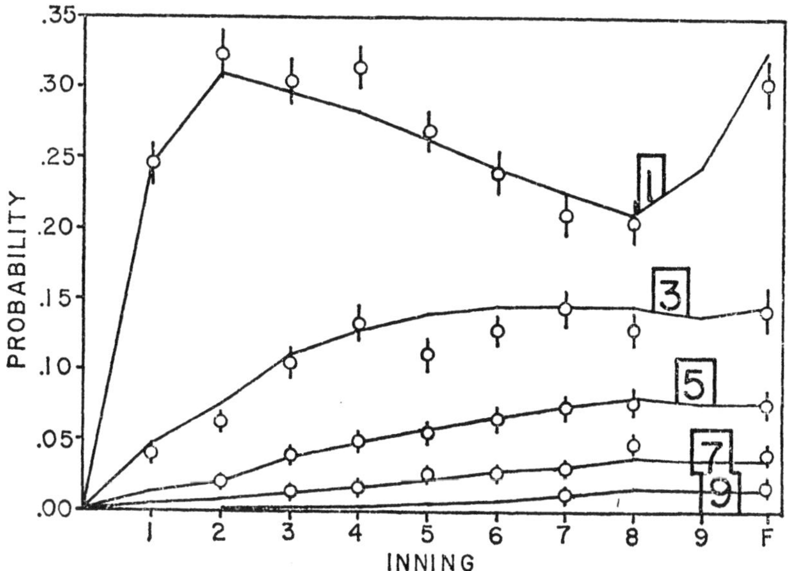

FIG. 3B. Probability of establishing a lead.
782 games of 1958
(Odd Numbers Only)

quencies of leads of 0, 2, 4, 6, and 8 runs being shown by the circles on Figure 3A and leads of odd value on Figure 3B. The observed frequencies for $l=0$ are shown as solid black circles, to distinguish them from the other hollow circles. Circles are shown only where the number has been observed in at least 10 games.

The continuous lines on Figures 3A and 3B represent the theoretical frequencies based on the postulate of independence between innings. The length of the vertical bars on the circles representing observed frequencies indicate twice the standard deviation, $2\sqrt{pq/N}$, where Np is the theoretical frequency.

All curves except $l=0$ start from 0, since the lead is 0 when the game begins. The probability of the score being tied (i.e. $l=0$) decreases as the game proceeds. dropping to 0 for the final result, since extra innings must be played until $l\neq 0$. The probability of a small lead ($l=1$ or 2) increases for a few innings, but then decreases in the later innings because of the steady rise in the probability of larger leads ($l>2$).

The practice of stopping the game as soon as the decision is certain causes sudden deviation in these probabilities for the scores at the end of the ninth inning and at the end of the game. If the home team leads by $l>1$ at the end of the eighth, and the visitors reduce the lead but do not erase it in their half of the ninth, then the home half of the ninth is not played and there is no opportunity to increase the lead once more. The home half of the ninth, and of extra innings, always commences with the visitors ahead or the score tied, and ceases if and when the home team achieves a lead of 1 (except in the circumstances that the play that scores the winning run also scores additional runs, as might occur if a tie were broken by a home run with men on base). Thus, almost all games won in the lower half of the ninth or extra innings will show a final margin of one. The formulae of Appendix B predict that 82% of games decided in extra innings will be by a margin of one run.

The consequence of these two factors (i.e. not starting the home half of the ninth if the home team is ahead, and not finishing the home half of the ninth or extra inning if the home team gets ahead) is to raise the probability for a lead of $l=1$ at the end of the ninth, and especially at the end of the game, with a corresponding adjustment to the other probabilities.

The final winning margin at the end of the game, indicated in the last columns of Table 5, and above the label "F" on Figures 3A and 3B, is given for the 1000 games of 1959 as well and is also displayed separately on Figure 4. The continuous line shows the margin calculated on the hypothesis of independence, expressed as a normalized probability. The vertical black bars show the results observed for the 782 games of 1958, and the vertical white bars for the 1000 games of 1959. As before, the centre of the bar marks the observed frequency, and the total length of the bar is $2\sqrt{pq/N}$ where the predicted frequency is Np.

It is evident from inspection of Figure 4 that the fit is extremely satisfactory, except for the excess of margins of ten or more. If we combine the results for 1958 and 1959 the excess for $l\geq 10$ is significant at the level of $\epsilon=.005$.

The slight excess of large final margins may be caused by the excess number of games with high scores already noted in Table 4 and Figure 2, since games with one team making a large total score will tend to show larger than average

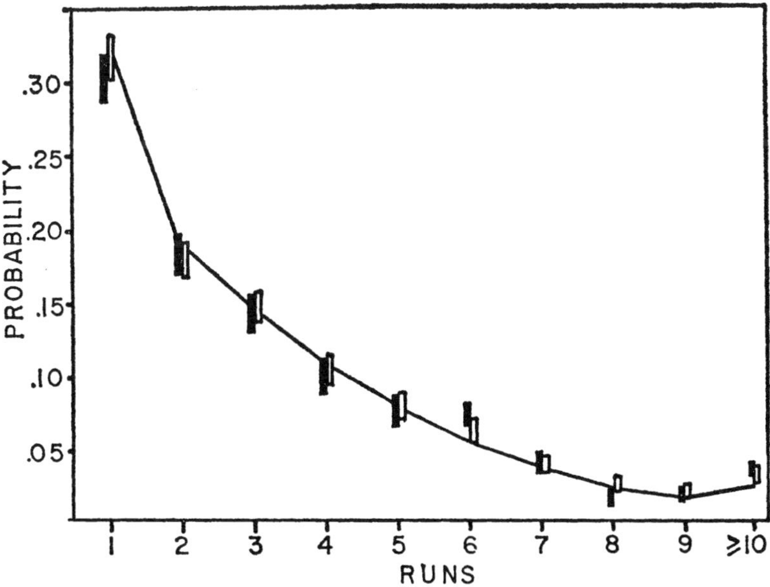

FIG. 4. Distribution of final winning margins.
Black: 782 games of 1958
White: 1000 games of 1959

winning margins. (The mean winning margin for the games of 1959 in which the winner scored more than 10 runs was 7.6 runs).

The observed mean winning margin for all games is 3.3 runs.

8. OVERCOMING A LEAD AND WINNING

When the frequency of establishment of a lead was recorded, it was also noted whether the leading team eventually won or lost the game.

Table 6 gives the number of occasions on which the lead of l (established at the end of the ith inning with frequency given in Table 5) was overcome, and the game won by the team which had been behind. Appendix B shows how this frequency can be predicted on the postulate of independence. Table 7 shows

TABLE 6. OVERCOMING OF A LEAD AND WINNING

Lead	Inning:	1	2	3	4	5	6	7	8
	Total:	782	782	782	782	782	779	779	777
1	Overcome:	59	79	87	80	72	53	36	19
	Expected:	75	97	88	87	70	55	40	24
	(O–E)	−16	−18	—	—	—	—	—	—
2	Overcome:	31	30	32	42	39	26	15	10
	Expected:	30	38	34	33	35	26	18	9
	(O–E)	—	—	—	—	—	—	—	—
≥3	Overcome:	9	16	22	16	12	11	6	1
	Expected:	10	15	24	24	18	15	10	4
	(O–E)	—	—	—	—	—	—	—	—

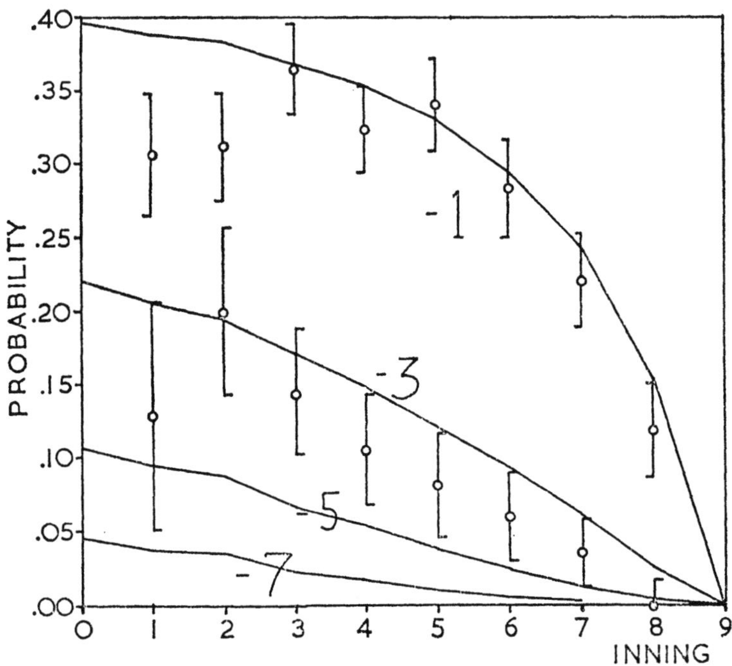

FIG. 5A. Probability of overcoming a lead and winning.
782 games of 1958
(Odd Numbers Only)

the computed frequencies expressed as probabilities, while Table 6 shows them as number of games (out of 782) in the lines labelled "Expected." The lines labelled (O-E) show the difference between observed and expected only when chi-square would attain such a magnitude 10% or less of the time. Chi-square for the combined totals of each inning is less than the value for $\epsilon = 0.1$ for every inning.

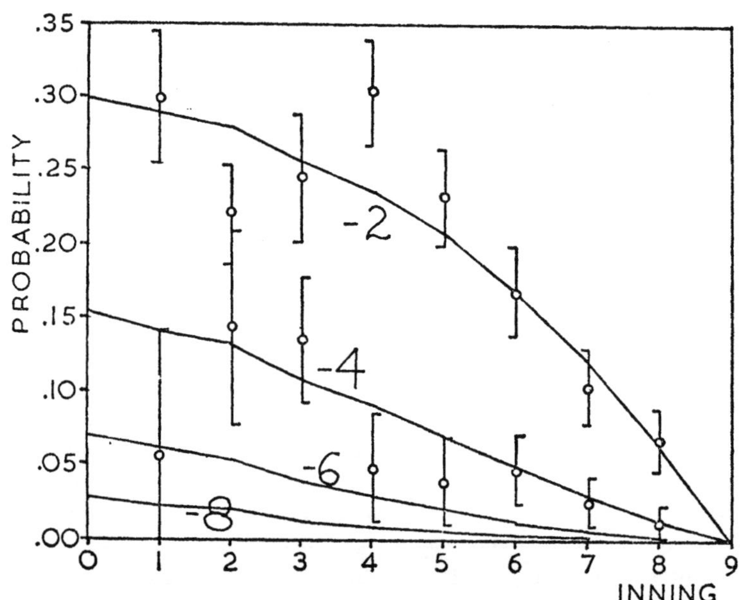

FIG. 5B. Probability of overcoming a lead and winning.
782 games of 1958
(Even Numbers Only)

TABLE 7. CALCULATED PROBABILITY THAT TEAM WITH LEAD l AT END OF iTH INNING WILL WIN THE GAME

Inning i	1	2	3	4	5	6	7	8
Lead l								
6	.939	.944	.959	.970	.979	.988	.994	.998
5	.905	.913	.932	.947	.961	.974	.987	.995
4	.857	.867	.889	.908	.930	.949	.980	.987
3	.793	.803	.827	.849	.877	.907	.940	.974
2	.700	.720	.741	.764	.793	.831	.878	.936
1	.610	.617	.630	.647	.668	.705	.776	.846
0	.500	.500	.500	.500	.500	.500	.500	.500
−1	.389	.383	.368	.353	.331	.295	.244	.153
−2	.290	.280	.257	.236	.207	.168	.122	.063
−3	.207	.196	.171	.150	.122	.093	.060	.025
−4	.142	.133	.109	.091	.070	.050	.029	.011
−5	.095	.087	.067	.053	.038	.025	.013	.004
−6	.061	.054	.039	.029	.020	.012	.005	.000

Figures 5A and 5B show the observed and expected frequencies of overcoming a lead of l (i.e. of winning after having established a lead of $-l$) at the end of the ith inning. Figure 5A shows observed frequencies for margins of $l=-1$ and -3 indicated by the circles with standard deviation bars, and calculated frequencies for $l=-1$, -3, -5 and -7, indicated by the continuous lines. Figure 5B shows observed values of $l=-2$, and -4, and calculated values for $l=-2$, -4, -6 and -8. In all cases, as might be expected, the probability of overcoming a given deficit decreases as the game proceeds. The theoretical values at $i=0$ have no significance in an ordinary game, since it begins with a score of 0-0. They would apply only if there were a handicap, or in the last game of a series played with the total run score to count, a practice not adopted in baseball. The curve for $l=0$ would be a straight line at $P=0.5$, since with the score tied, either team is equally likely to win.

To illustrate, a team that is two runs behind at the end of the fourth inning has a theoretical probability of 0.236 of winning the game, as given by Table 7. Table 5 shows that a lead of 2 was established in 138 games (out of 782). Table 6 shows that the team that was 2 runs behind eventually won 42 of them. The expected number is $0.236 \times 138 = 33$. Figure 5B shows the observed point at $42/138 = 0.303$, with a standard error of $\sigma = \sqrt{pq/N} = \sqrt{(.236)(.764)/128} = .036$. Chi-square is $(42-33)^2/33 = 2.5$, which for 1 degree of freedom has $\epsilon > 0.1$. Therefore, the lead was overcome rather more often than predicted by theory, but the difference is not statistically significant at the level of $\epsilon = 0.1$.

It may be concluded from these data that the predicted frequency of overcoming a lead is confirmed quite well by observation, so that the numbers in Table 7 (based on the postulate of independence) provide a measure of the probability that a team can overcome a lead and win the game.

9. CONCLUSIONS

1. The distribution of runs scored per half-inning is not the same for all innings. The expected number of runs scored is significantly less in the second inning than in the first or third. The mean expectation per half-inning is 0.48 runs.

2. The total score achieved by one team is consistent with a model based on random sampling from separate and independent half-inning samples, except that the number of cases of large scores of more than ten runs is somewhat in excess of the prediction based on the theory of independence. The mean total score for one team is 4.2 runs.

3. The establishment of a lead appears to follow the postulate of independence quite well except for a slight tendency toward very large final winning margins (of 10 runs or more).

4. The frequency with which a lead is overcome and the game won agrees very well with the postulate of independence. It does not appear that there is any significant tendency for the trailing team to overcome the lead either more or less frequently than would be predicted by the model of random drawings from successive half-innings. From Table 7 it is possible to estimate the probability that the game will eventually be won by the team which is ahead.

5. The number and length of extra-inning games is not inconsistent with the postulate of independence, although the number of games ended in the ninth and tenth innings is somewhat greater than would be expected.

6. It seems logical to attribute the inhomogeneity of the first three innings to the structure of the batting order.

10. REMARKS

Perhaps these findings will be considered disappointing. There is no observable tendency for the underdog to reverse the position. Although the game is never over until the last man is out in the ninth inning, it is lost at the $2\frac{1}{2}\%$ level of probability if the team is more than 2 runs behind when they start the ninth inning. There is nothing unusual about the seventh inning. The peculiar innings from the point of view of scoring are the first three. The number of runs scored in the ninth inning is normal, but the 6% of games won during the last half of the ninth exceed the expected proportion.

One way in which these results could be useful to a manager is when he is considering changing personnel during the game. If he wishes to rest a winning pitcher when the game appears to be won, or to risk the use of a good but injured player in order to overcome a deficit, he will want an estimate of the probability that the present status will be reversed. He would use the information in conjunction with other statistical data [4] and his evaluation of the immediate form and special skills of his players.

Another application is to test certain scoring records believed to be unusual. For instance, the Chicago White Sox won 35 games by one run during the 1959 American League season, which was generally considered to be remarkable. From Figure 4, the expected number would be $\frac{1}{2}\times0.327\times154\approx25$, (assuming that .327 of their games would be decided by one run, and that they would win half of these). Taking $\sigma=\sqrt{Npq}=4.6$ the difference $(\text{O-E})=10=2.2\sigma$ would be expected to occur with probability 0.028. In other words their accomplishment had a probability of about 3%, and would be expected to be repeated or exceeded by one of sixteen teams about once in every two seasons.

In a full season of two eight-team major leagues we would expect the number of games lasting eighteen innings or more to be $154\times8\times0.106\times0.566^8\approx1$.

In the World Series of 1960, the three games won by New York were by scores of 16-3, 10-0, and 12-0. Pittsburgh won the last game by 10-9.

From Figure 4, the observed probability of the final margin exceeding nine runs in 0.033. Therefore the probability of three or more games out of seven having margins of ten or more is

$$\sum_{r=3}^{7} \binom{7}{r} .033^r \ .967^{7-r} = 0.0014.$$

From Table 4, the observed probability of one team scoring more than nine runs in a game is 0.0655. Therefore the probability that there will be four or more scores of 10 or more in seven games is approximately

$$\sum_{r=4}^{14} \binom{14}{r} .0655^r \ .9345^{14-r} = 0.015.$$

Therefore we can conclude that the three one-sided games did constitute a very unusual combination, to be expected only once or twice in 1000 seven-game series. The four large scores are less extraordinary, and would be expected to occur on about $1\frac{1}{2}\%$ of seven-game series.

Thus even if this investigation has committed the sin of exploding some cherished beliefs, it does permit estimates to be made of the probability of occurrence of certain rare accomplishments.

APPENDIX A

DISTRIBUTION OF RUNS WITHIN AN INNING

The shape of the frequency distributions of Table 1 could form the subject of an interesting analysis. Since the standard deviations are about twice as great as the means, the distributions do not follow a Poisson law.

If the negative binomial distribution [3, p. 155; 5, p. 179]

$$\phi(y; a, p) = \binom{y + a - 1}{y} p^a(1 - p)^y$$

is fitted to the bottom line of Table 1 by choosing $p = \mu/\sigma = 0.475$ and $a = \mu^2/(\sigma - \mu) = 0.43$, so that it has the same mean and standard deviation as the experimental distribution, the values listed in the line of Table 8 labelled $\phi(x; a, p)$ are obtained. Use of such a law might allow mathematical manipulation of the distributions, but it is difficult to suggest any mathematical model of baseball which would indicate a law of this precise form.

TABLE 8

Function	a	p	Number of Runs						
			0	1	2	3	4	5	>5
Observed $f(x)$	—	—	.737	.144	.069	.027	.013	.007	.003
$\phi(x; a, p)$	0.43	0.475	.726	.164	.062	.026	.012	.006	.008
$\phi(x; a, p)$	3	0.685	.321	.304	.191	.101	.047	.021	.015
$.35\phi_1 + .65\phi_2$	3	0.685	.749	.133	.065	.030	.013	.006	.004

A model representing an extension of the negative binomial distribution is suggested by the fact that, if a set of trials for which the individual probability of success is p is repeated until a successes have been obtained, then the probability that y failures will have occurred before the ath success is $\phi(y;\, a,\, p)$. If a trial consists of the fielding team dealing with a batter, and putting him out is considered a success, then the probability that y men will bat but not be put out before the inning is over will be $\phi(y;\, 3,\, p)$ where p is the probability that a player who appears at bat is put out during the inning. A player appearing at bat but not put out must either score or be left on base. The number left on base, L, can be 0, 1, 2 or 3. The number of runs scored is $x = y - L$.

To be accurate, the mathematical model would need to account for all possible plays and for their relative frequencies of occurrence. Models somewhat simpler than this have been tested [1, 2, 7] with success.

A crude approximation would be to assume that in a fraction λ of the innings in which runs were scored, one man was left on base, while two were left on in the remaining $(1-\lambda)$ scoring innings. The distribution of runs would then be given by:

$$f(x) = \lambda\phi_1(x) + (1 - \lambda)\phi_2(x)$$

where

$$\begin{aligned}
\phi_1(0) &= \phi(0; 3, p) + \phi(1; 3, p) \\
\phi_1(x) &= \phi(x + 1; 3, p) \qquad \text{for } x = 1, 2, 3, \cdots \\
\phi_2(0) &= \phi(0; 3, p) + \phi(1; 3, p) + \phi(2; 3, p) \\
\phi_2(x) &= \phi(x + 2; 3, p) \qquad \text{for } x = 1, 2, 3, \cdots
\end{aligned}$$

TABLE 9. DATA FOR AMERICAN AND NATIONAL LEAGUES (1958)

Event	A.L.	N.L.	Total
(1) Official Times at Bat	41,684	42,143	83,827
(2) Bases on Balls	4,062	4,065	8,127
(3) Hit by Pitcher	252	247	499
(4) Sacrifice Hits	531	515	1,046
(5) Sacrifice Flies	322	322	644
(6) (1)+(2)+(3)+(4)+(5)	46,851	47,292	94,143
(7) Safe Hits	10,595	11,026	21,621
(8) Errors	1,002	1,083	2,085
(9) Double Plays	1,313	1,287	2,600
(10) (2)+(3)+(7)+(8)−(9)	14,598	15,134	29,732

To estimate the constant p, the probability that a man appearing at bat will be put out during the inning, we could note that in many nine-inning games some players make four and some five appearances at bat, so that out of $9\times4\frac{1}{2}\approx40$ batters, 27 are put out, and $p\approx0.68$. Or, to be more methodical, we could use data from the major league season of 1958 [6] listed in Table 9. These show 94,143 batters appearing, of which 29,732 were not put out, so that $p=1-.315=.685$.

The last line of Table 8 shows the values of $f(x)$ calculated for $\lambda=0.35$, for which reasonable agreement is found with the distribution actually observed.

APPENDIX B

THE FIRST EIGHT INNINGS

Let $f_i(x)$ be the probability that team A will score exactly x runs in the ith inning, where $1\leq i\leq 8$.

We assume throughout that both teams have the same $f_i(x)$.

Let $F_{ij}(x)$ be the probability that team A will score exactly x runs between the ith and jth innings (inclusive), where $i<j\leq 8$

$$F_i(x) = f_i(x).$$

Tables of $F_{ij}(x)$ can be accumulated by successive summations such as

$$F_{1,2}(x) = \sum_{r=0}^{r=x} F_1(r)f_2(x-r)$$

and

$$F_{ij}(x) = \sum_{r=0}^{r=x} F_{i,j-1}(r)f_j(x-r)$$

$$= \sum_{r=0}^{r=x} f_i(r)F_{i+1,j}(x-r) \qquad i<j\leq 8.$$

The probability that A will increase their lead by exactly x during the ith inning is

$$G_i(x) = \sum_{r=0}^{\infty} f_i(r)f_i(x+r) \qquad \text{for } x = 0, 1, 2, \cdots.$$

$G_i(-x)=G_i(x)=$probability that A's lead will decrease by x during the ith inning. A negative lead is, of course, the same as a deficit.

The probability that A will increase their lead by exactly x between the ith and jth innings (inclusive) is

$$G_{ij}(x) = \sum_{r=0}^{\infty} F_{ij}(r)F_{ij}(x+r).$$

INNINGS WHICH MAY BE INCOMPLETE

For $i=9$, the home half-inning ($H9$) will not be started if the home team is ahead at the end of the visitor's half ($V9$). For $i\geq 9$, the home half, if started at all, will be terminated if and when the home team achieves a lead.

Define $f_i(x)$ to be the probability that team A would score x runs in their half of the ith inning if it were played through to completion (even after the results were assured). Define $S_{Vi}(x)$ and $S_{Hi}(x)$ as the probabilities that the visitor's and home halves of the ith inning, if started at all, will produce x runs as baseball is actually played.

$$S_{Vi}(x) = f_i(x)$$

since the visitor's half is always completed. Thus we could use the measured distribution of visitor's scores to estimate $f_i(x)$. However, the assumption that the visitor's and home team have the same distribution is open to question, and it seems preferable to use the data from the home halves as well.

$$S_{Hi}(x) = f_i(x) \qquad \text{for } i < 9, \text{ but not for } i \geq 9.$$

Define $g(l)$ as the probability that the visitors have a lead of l at the end of $V9$.

$$g(l) = \sum_{r=0}^{\infty} G_{18}(l - r)f_9(r).$$

The probability that $H9$ is required is then

$$g(l > 0) = \sum_{l=0}^{\infty} g(l) = a \text{ (say)}.$$

Out of a large sample of N games we would observe N $V9$'s and Na $H9$'s.

Of the Na $H9$'s, $NG_{19}(0)$ would leave the score tied and require extra innings, and $N[1-G_{19}(0)]/2$ would leave the game won by the visitors. These $N[1+G_{19}(0)]/2$ $H9$'s would all be complete. In them the home team would score x runs only when $x \leq l$.

If we recorded only completed ninth half-innings (as in Table 2) and computed the frequency $S_{9C}(x)$ of scoring x runs

$$\frac{N}{2}[3 + G_{19}(0)]S_{9C}(x) = Nf_9(x)\left[1 + \sum_{l=x}^{\infty} g(l)\right].$$

The remaining $Na-N[1+G_{19}(0)]/2$ $H9$'s would result in victory for the home team, and be terminated with less than three out. The winning margin will almost always be one run, but can be greater on occasion when the final play that scores the decisive run produces additional runs. For example, if the home team were one behind in $H9$, but produced a home run with the bases full, the final margin would be recorded as three runs.

If we neglect the infrequent cases when the winning play produces a margin in excess of one, then all scores of x runs in incomplete $H9$'s will be associated with leads of $l=x-1$ for the visitors at the end of $V9$. But the probability of H scoring x under these circumstances is now $f_9(\geq x)$ rather than $f_9(x)$. We could imagine that the inning is played to completion, but a score of $(l+1)$ only recorded.

If we recorded only incomplete halves of the ninth inning (as in Table 2) and computed $S_{H9I}(x)$ the result would be

$$S_{H9I}(0) = 0$$

$$S_{H9I}(x) = \frac{g(x-1)}{a - 1/2[1 + G_{19}(0)]} \sum_{r=x}^{\infty} f_9(r) \qquad \text{for } x > 0.$$

If we combine all halves of the ninth (as in Table 1), and compute $S_9(x)$, the result is

$$S_9(0) = f_9(0)$$

$$S_9(x) = \frac{1}{1+a}\left\{f_9(x)\left[1 + \sum_{l=x}^{\infty} g(l)\right] + g(x-1)\sum_{r=x}^{\infty} f(r)\right\} \qquad \text{for } x > 0.$$

For extra innings, paucity of data will make it advisable to group the results and seek $f_E(x)$, the probability that a team starting its half of any extra inning would score x runs if it played the inning to completion.

Reasoning similar to that applied to the ninth inning allows us to deduce $S_{EC}(x)$, the distribution of runs for completed extra half-innings, $S_{EI}(x)$ for incomplete extra half-innings, and $S_E(x)$ for all extra innings.

$$S_{EC}(x) = \frac{2}{3 + G_E(0)} f_E(x)\left[1 + \sum_{r=x}^{\infty} f_E(r)\right]$$

$$S_{EI}(0) = 0$$

$$S_{EI}(x) = \frac{2}{1 - G_E(0)} f_E(x-1) \sum_{r=x}^{\infty} f_E(r) \qquad \text{for } x > 0$$

$$S_E(0) = f_E(0)$$

$$S_E(x) = 1/2\left\{f_E(x) + [f_E(x-1) + f_E(x)] \sum_{r=x}^{\infty} f_E(r)\right\} \qquad \text{for } x > 0.$$

NUMERICAL RESULTS

Starting with $S_9(x)$ as tabulated in Table 1 (T1), we can compute $f_9(x)$. The resulting distribution is shown in the first line of Table 10, and beneath it is $\overline{f_{18}}(x)$, the mean of the distribution $f_i(x)$ for $1 \leq i \leq 8$. They are obviously very

TABLE 10

Function	Source	x: 0	1	2	3	4	5	N	ϵ <0.1
$f_9(x)$	Calc. fm. $S_{H9}(x)$ of T 1	.737	.140	.076	.023	.013	.010	1,191	
$\overline{f_{18}}(x)$	Obs. in T 1	.736	.143	.069	.028	.013	.007	12,500	
$\overline{f_{18}}(x)$	Obs. in T 2	.726	.148	.071	.029	.013	.007	16,000	
$S_{9C}(x)$	Calc. fm. $\overline{f_{18}}(x)$ of T 1	.757	.137	.062	.023	.010	.005		
$S_{9C}(x)$	Obs. in T 2	.770	.129	.060	.027	.010	.004	1,576	
$S_{H9I}(x)$	Calc. fm. $\overline{f_{18}}(x)$ of T 1	.00	.58	.26	.10	.04	.02		.04
$S_{H9I}(x)$	Obs. in T 2	.00	.46	.26	.23	.03	.02	61	
$S_E(x)$	Calc. fm. $\overline{f_{18}}(x)$ of T 1	.736	.187	.047	.016	.007	.004		.001
$S_E(x)$	Obs. in T 1	.752	.132	.076	.027	.010	.003	302	
$S_E(x)$	Obs. in T 2	.711	.198	.052	.018	.008	.005	384	
$S_{EC}(x)$	Calc. fm $\overline{f_{18}}(x)$ of T 1	.827	.102	.039	.017	.007	.004		
$S_{EC}(x)$	Obs. in T 2	.825	.112	.033	.012	.003	.006	331	
$S_{EI}(x)$	Calc. fm $\overline{f_{18}}(x)$ of T 1	.00	.90	.08	.02	.00	.00		<.001
$S_{EI}(x)$	Obs. in T 2	.00	.73	.17	.06	.04	.00	53	

close, as may be confirmed by chi-square, so that the ninth inning is not an unusual one.

In subsequent calculations $\overline{f_{18}}(x)$ from T1 was used as the best estimate of $f_9(x)$. $\overline{f_{18}}(x)$ calculated from T2 is not significantly different.

The observed results of Table 2, as separated between complete and incomplete ninth innings, are compared with the predicted functions. The agreement is close for the 1576 complete half-innings, but less satisfactory for the 61 incomplete half-innings.

For extra innings, the distributions are calculated on the assumption that $f_E(x) = \overline{f_{18}}(x)$, and compared with those observed. There is a significant difference from the 302 observations in Table 1, for which less ones and more twos and threes were found than predicted. If the calculation is reversed, and $f_E(x)$ derived from the observed $S_E(x)$, the result shows $f_E(1) = .061$, $f_E(2) = .119$, $f_E(3) = .043$, and $f_E(4) = .019$, which is radically different from $\overline{f_{18}}(x)$ or any other $f_i(x)$. However, $S_E(x)$ and $S_{EC}(x)$ as calculated from $\overline{f_{18}}(x)$ agree quite well with the distributions observed in the 384 extra innings and 331 completed extra innings of Table 2. The agreement is poor for the 53 incomplete extra innings.

All cases of disagreement involve incomplete innings and show a deficit of one-run scores, suggesting that the neglect of scoring plays resulting in margins of more than one run may be partially to blame.

These results for extra innings are contradictory. However, in view of the excellent agreement with the 331 completed extra innings of Table 2, it would seem unjustifiable to conclude that extra innings offer any substantially different probability of scoring from the first nine innings.

LENGTH OF GAME

The probability that a tenth inning will be necessary is the same as the probability that A's lead will be 0 after nine innings, i.e. $G_{1,9}(0)$.

The probability that an eleventh inning will be necessary is $G_{1,9}(0)G_{10}(0)$.

If we group all extra innings together, so that

$$\left.\begin{aligned} f_i(x) &= f_E(x) \\ G_i(x) &= G_E(x) \end{aligned}\right\} \quad \text{for } i > 9$$

and

then the probability that an ith inning will be required is

$$G_{1,9}(0)G_E(0)^{i-10} \qquad \text{for } i = 10, 11, 12, \cdots.$$

To state the probability in a manner more suitable for comparison with observed results, if N_9 games are observed to require a ninth inning, we would expect $N_9G_{1,9}(0)$ to require a tenth. If N_i games are observed to require an ith inning (for $i>9$), then we would expect $N_iG_E(0)$ to require an $(i+1)$th inning.

The probability that a game will require *exactly* i innings is

$$1 - G_{1,9}(0) \qquad \text{for } i = 9$$

and

$$G_{1,9}(0)[1 - G_E(0)][G_E(0)]^{i-10} \qquad \text{for } i \geq 10.$$

DISTRIBUTION OF RUNS SCORED IN ENTIRE GAME

The distribution of runs accumulated over nine complete innings is $F_{1,9}(x)$. However, the distribution of total game scores must include the possibilities that the home half of the ninth ($H9$) is not played, or is started but not completed, that the game lasts more than nine innings, and that the home half of the final extra inning is not completed.

During the first nine innings, the visiting team has probability $F_{1,9}(x)$ of scoring x runs. The probability that the home team will score y consists of four terms:

$$(a) = F_{1,9}(y) \sum_{r=y+1}^{\infty} F_{1,9}(r) \qquad (V \text{ win})$$

$$(b) = F_{1,8}(y) \sum_{r=0}^{y-1} F_{1,9}(r) \qquad (H \text{ win without needing } H9)$$

$$(c) = F_{1,9}(y-1) \sum_{r=0}^{y-1} F_{1,8}(r) \sum_{s=y-r}^{\infty} f_9(s) \ (H \text{ win by one run during an incomplete } H9)$$

$$(d) = [F_{1,9}(y)]^2 \qquad (\text{Extra innings are needed}).$$

If we set $F_{1,9D}(x)=\frac{1}{2}[F_{1,9}(x)+(a)+(b)+(c)]$, this will represent the distribution of scores for games decided in nine innings, for both visiting and home team.

The distribution of runs scored in each indecisive extra inning is $[f_E(x)]^2$.

In the decisive extra inning, the visitor's score has distribution $f_E(x)$, while the home score consists of two terms:

$$(e) \ f_E(x) \sum_{r=x+1}^{\infty} f_E(r) \qquad (V \text{ win})$$

$$(f) \ f_E(x-1) \sum_{r=x}^{\infty} f_E(x) \qquad (H \text{ win in an incomplete half-inning}) \quad (x \geq 1).$$

If we set $F_{ED}(x)=\frac{1}{2}[f_E(x)+(e)+(f)]$, this will represent the distribution of scores for the decisive extra inning.

The distribution of scores for games decided in 10 innings is then

$$F_{1,10D}(x) = \sum_{r=0}^{x} [F_{1,9}(r)]^2 F_{ED}(x-r)$$

and for games decided in $i \geq 11$ innings it is

$$F_{1,iD}(x) = \sum_{r=0}^{x} F_{1,i-1D}(r)[f_E(x-r)]^2.$$

When these are calculated we can compute the probability that a team will score a total of x runs in a whole game, whatever its length, as

$$F(x) = F_{1,9D}(x) + \sum_{i=10}^{\infty} F_{1,iD}(x).$$

$F(x)$ as computed in this way from the 1958 data is shown in Table 4 and Figures 3A and 3B.

ESTABLISHMENT AND OVERCOMING OF LEAD

The probability that team A will have established a lead of exactly l by the end of the ith inning is $G_{1,i}(l)$, for $i \leq 8$. The "expected" figures of Table 5 and the curves of Figures 3A and 3B are calculated from this function for $i \leq 8$.

The probability that team A will increase its lead by l or more over the ith to jth innings, inclusive, is

$$H_{ij}(l) = \sum_{r=l}^{\infty} G_{ij}(r), \qquad \text{for } i < j \leq 8.$$

If team A is exactly l runs behind at the end of the ith inning, the probability that they will win is

$$H_{i+1,9}(l+1) + 1/2 G_{i+1,9}(l),$$

the first term representing a win in nine innings, the second a win in extra innings, where the assumption is made that when a game enters extra innings each team has probability $\frac{1}{2}$ of winning. The numbers of Tables 6 and 7, and the curves of Figures 5A and 5B are computed from this function.

To calculate the distribution of the winning margin at the end of the game it is necessary to allow for the possibilities that $H9$ will not be needed, and that the game will be won in $H9$ or an HE with the decisive half-inning being terminated as soon as a lead of one run is obtained.

As in the preceding section, the margin at the end of nine innings has a probability composed of four terms:

(a) $G_{19}(l)$ (V win by l) $l>0$

(b) $g(-l)$ (H win by l without needing $H9$)

(c) a probability of $a - \frac{1}{2}[1 + G_{19}(0)]$ that H win by 1 in the decisive incomplete $H9$

(d) a probability of $G_{19}(0)$ that the score is tied.

For those games which are decided in an extra inning, the probability of a final margin of l is composed of two terms:

(e) a term proportional to $G_E(l)$ (V win by l) $l>0$

(f) a probability of $\frac{1}{2}$ that H wins by $l=1$.

Combining these, and normalizing where necessary, we obtain the probability $G(l)$ that the winning margin will be l runs.

$$G(0) = 0$$

$$G(1) = G_{19}(1) + g(-1) + a - 1/2 + \frac{G_{19}(0)G_E(1)}{1 - G_E(0)}$$

$$G(l) = G_{19}(l) + g(-l) + \frac{G_{19}(0)G_E(l)}{1 - G_E(0)} \qquad \text{for } l \geq 2.$$

This is the function used to calculate the last columns of Table 5, the points above "F" on Figures 3A and 3B, and the curve of Figure 4.

TABLE 11. CORRELATIONS BETWEEN HALF-INNINGS

(1000 games of 1959)

Pair of Half-innings	r	N	σ_r
$V1$–$V2$	−0.001	1000	.032
$H1$–$H2$	+0.05	500	.047
$V1$–$H1$	+0.045	500	.047
$H1$–$V2$	−0.024	500	.047
$V7$–$V8$	−0.014	500	.047
$H7$–$H8$	+0.009	500	.047
7–8	−0.002	1000	.032

APPENDIX C

CORRELATIONS BETWEEN HALF-INNINGS AND BETWEEN TOTAL SCORES

A common method of testing for independence between two distributions is to compute the correlation coefficient.

The pair of scores for the visiting team's first and second innings ($V1$ and $V2$) for the 1000 games of 1959 show a correlation coefficient of −0.001. The standard deviation for small coefficients and large numbers of readings is very nearly $N^{-1/2}=0.032$. Table 11 shows several other correlation coefficients, and it is evident that none of them are significant. Thus, there is no evidence here of any linear correlation between the scores in half-innings of the same or of the opposing team. Absence of linear correlation does not, however, constitute proof of independence [3, p. 222].

Another method of testing for independence is to draw up conditional distributions for particular half-innings in which the scores in other half-innings have been below, or above, average. Some of these are shown in Table 12. $S_{V2}(x|V1>0)$ represents the measured distribution of runs in the visitors half of the second inning for all cases (out of 1000 games) in which the visitors scored

TABLE 12. CONDITIONAL DISTRIBUTIONS FOR CERTAIN INNINGS

(1000 games of 1959)

	x: 0	1	2	≥ 3	ϵ
$S_{V2}(x\|V1>0)$	246	31	14	14	0.4
Exp.	237	40	16	12	
$S_{H2}(x\|H1>0)$	111	20	12	9	0.5
Exp.	112	23	10	7	
$S_{H1}(x\|V1>0)$	108	15	11	9	0.3
Exp.	101	21	14	8	
$S_{V2}(x\|H1>0)$	119	21	3	9	0.5
Exp.	119	20	7	6	
$S_{V8}(x\|V7>0)+S_{H8}(x\|H7>0)$	191	42	25	20	0.5
Exp.	199	41	23	15	

one or more runs in their half of the first inning. "Exp" shows the number that would be expected on the null hypothesis that $S_{V2}(x|V1>0)=S_{V2}(x|V1=0)$. Five such conditional distributions are shown in Table 12, with ϵ giving the probability of occurrence of a chi-square as large as observed. No single difference between "observed" and "expected" for which "expected" was 10 or more showed a chi-square with $\epsilon<.05$. Thus there is no evidence of dependence between the half-innings investigated.

The coefficient of correlation between the total scores of the home and visiting team for the 1000 games is $+0.090\pm0.032$. This is significant at the 1% level, and indicates a slight tendency for the two teams to score together. Such a tendency would reduce the number of games with large winning margins below the predictions of the independent model. Since the number actually observed is greater than predicted (see Table 5 and Figure 4 for margins of 10 or more) the small correlation must be overcome by the excess number of games with a large total score and an associated large margin.

REFERENCES

[1] Briggs, Hexner, Meyers and Stewart, "A Simulation of a Baseball Game," *Bulletin of the Operations Research Society of America*, 8 Supplement 2 (1960), B-99.

[2] D'Esopo, Donato and Lefkowitz, Benjamin, "The Distribution of Runs in the Game of Baseball." Stanford Research Institute, August 1960.

[3] Feller, William, *An Introduction to Probability Theory and Its Application*, Volume I, Second Edition. New York: John Wiley and Sons, Inc., 1957.

[4] Lindsey, George, "Statistical Data Useful for the Operation of a Baseball Team," *Operations Research*, 7 (1959), 197.

[5] Parzen, Emanuel, *Modern Probability Theory and Its Applications*. New York: John Wiley and Sons, Inc., 1960.

[6] Spink, J. G. Taylor, *Baseball Guide and Record Book, 1959*. St. Louis, Missouri: Charles C. Spink and Son, 1959.

[7] Trueman, Richard, "A Monte Carlo Approach to the Analysis of Baseball Strategy," *Bulletin of the Operations Research Society of America*, 7, Supplement 2 (1959), B-98.

Part III
Statistics in Basketball

Chapter 17

Introduction to the Basketball Articles

Robert L. Wardrop

17.1 Background

Basketball was invented in 1891 by James Naismith, a physical education instructor at the YMCA Training School in Springfield, Massachusetts, USA. The game achieved almost immediate acceptance and popularity, and the first collegiate game, with five players on each team, was played in 1896 in Iowa City, Iowa, USA. Professional basketball in the United States dates from the formulation of the National Basketball League in 1898, which survived for six years. A later NBL was formed in 1937 and existed until 1949 when it merged with the three-year-old Basketball Association of America to become the National Basketball Association (NBA). Currently, there is one women's professional basketball league in the United States and a number of men's and women's professional leagues around the world. Basketball is one of the core sports played at high schools and colleges in the United States.

Considering the popularity of basketball, the amount of statistical research on the sport has been small compared with other sports. The topics of the chapters in this section are representative of the basketball research topics in various statistical journals. Two chapters of this section consider modeling the National Collegiate Athletic Association (NCAA) basketball tournament. The remaining three chapters investigate modeling the outcomes of individual shots. For a more in-depth analysis of basketball statistics research, including a discussion of these five papers, see the book *Statistics in Sport*, edited by Jay Bennett.

17.2 Modeling Basketball Tournaments

The NCAA basketball tournament consists of four regional tournaments, the winners of which advance to the "Final Four" to determine the U.S. collegiate champion. Dating back to 1985, a regional tournament has consisted of 16 teams, seeded $1, 2, \ldots, 16$, by a panel of experts, with the 1 seed going to the team perceived to be the best, the 2 seed going to the team perceived to be second best, and so on. Let $P(i; j)$ denote the probability that seed i will defeat seed j and let W_i denote the probability that seed i will win a regional tournament.

In 1991, Schwertman, McCready, and Howard proposed three models for the $P(i; j)$. None of their models has any unknown parameters. The motivation for the models was to provide classroom examples that illustrate how individual probabilities can be combined to obtain the probability of an event of interest, in this case, the W_i's.

In Chapter 20 of this section, Schwertman, Schenk, and Holbrook propose and examine eight regression models that use data from 10 years of tournaments to estimate the values of $P(i; j)$. Once the $P(i; j)$'s have been estimated, they can be combined, as in Schwertman, McCready, and Howard (1991), to provide estimates of the W_i's. The chapter discusses the performances of the eight models at estimating the $P(i; j)$'s and the performances of the 11 models (eight regression plus three from Schwertman, McCready, and Howard (1991)) at estimating the W_i's.

Chapter 18 by Carlin provides an alternative approach. The earlier works use seed position as the only predictor of outcome. Carlin suggests that one might get improved models by using a computer ranking of teams, in particular the Sagarin ratings published in *USA Today*, or casino point spreads, as predictors. Carlin's suggestions have two new features. First, the probability of seed i winning is allowed to vary from region to region and from year to year. Second, the probability of winning the region need not decrease with seed number.

17.3 The Hot Hand Phenomena

Chapters 19, 21, and 22 explore individual shot attempts. Note that each chapter title includes the phrase "hot hand." The basic question addressed in these chapters is simple. Is the model of Bernoulli trials adequate for describing the outcomes of a player's shots? If the answer is no, then one explanation is that there is nonstationarity in the sequence—that is, the probability that the player makes a shot changes over time. Alternatively, there may be an autocorrelation structure in the shooting sequence, where the probability of making a shot will depend on a player's previous performance.

The authors of Chapter 21 believe that the Bernoulli trials model is adequate for modeling shooting data. Their chapter begins with a discussion of difficulties researchers have with predicting random sequences. For example, researchers tend to predict too many switches—from success to failure or failure to success—in a random sequence. Next, the authors analyze data from three sources: Professional (NBA) game shooting from the floor; NBA game free-throw shooting; and collegiate practice shooting from the floor. Their analyses of these data certainly support their conclusion, but they have several serious problems. First, they search for only one type of departure from Bernoulli trials, which is autocorrelation. They make no attempt to search for nonstationarity, which may be a reasonable description of the shooting patterns. Second, the sample sizes in the collegiate study are so small that there is little power for any reasonable autocorrelation alternative. Third, when the authors find significant results, they discount them by distorting the size of the estimated effect, for example, for a player who has a 60% success rate after a success compared to only a 40% success rate after a miss, the authors describe this difference as an autocorrelation of 0.20, which they say is "quite low." No serious basketball fan would argue that the practical difference between 60% and 40% shooters (from a fixed location) is unimportant!

In Chapter 19 Larkey, Smith, and Kadane examine data from 39 NBA telecasts from the 1987–88 NBA season. They analyze 18 players, including Detroit's Vinnie Johnson, who had a reputation for being an extremely streaky shooter, and many of the top stars in the NBA. They do not attempt statistical inference; their work is purely descriptive. They want to see whether they can find any justification for the widespread belief that Vinnie Johnson is a particularly streaky shooter. They conjecture that whatever Johnson is doing, it must be noticeable and memorable, and perhaps unlikely. They present many clever ideas. First, they focus on nonstationarity rather than autocorrelation and do this by searching the data for streaks of successes. Second, they incorporate game-time into their analysis in the following way. They argue that if a player makes five successive shots in a short period of time, say three minutes of game-time, this feat is more noticeable and memorable than making five successive shots equally spaced (in time) during a 48-minute game. From their analysis, they conclude that Vinnie Johnson's reputation as a streak shooter is apparently well deserved; he is different from other players in the data in terms of noticeable, memorable field goal shooting accomplishments.

In the final chapter in this section, Wardrop reexamines Tversky and Gilovich's free-throw data from Chapter 21, which introduced this set of data with the following question asked of a sample of basketball fans: When shooting free-throws, does a player have a better chance of making his second shot after making his first shot than after missing his first shot? 68% of the fans answered yes. Tversky and Gilovich analyze data on nine members of the Boston Celtics for the seasons of 1980–81 and 1981–82 and conclude that the correct answer is no, and, hence, that a majority of the fans are incorrect. Wardrop suggests a different interpretation of these free-throw data. He shows that if the data are aggregated over players, then the correct answer to the question becomes yes. He points out that aggregation might be appropriate for the purpose of trying to understand why the fans believe what they do. In order for a fan to adapt Tversky and Gilovich's analysis of the Celtics to his/her experiences, it must be assumed that the fan has separate two-by-two contingency tables for each of the hundreds, if not thousands, of players he/she has seen play. This is a big assumption to make! Wardrop suggests that it might be more reasonable to assume that the fan has a table only for the aggregated data. Thus, instead of concluding that, as Tversky and Gilovich write, "People ... 'detect' patterns even where none exist," it would be more productive to teach people the dangers of aggregation. Wardrop also shows that the Celtics players, as a group, were statistically significantly better free-throw shooters on their second attempts than on their first.

Reference

Schwertman, N. C., McCready, T. A., and Howard, L. (1991), "Probability models for the NCAA regional basketball tournaments," The American Statistician, 45, 35–38.

Chapter 18

Improved NCAA Basketball Tournament Modeling via Point Spread and Team Strength Information

Bradley P. CARLIN

Several models for estimating the probability that a given team in an NCAA basketball tournament emerges as the regional champion were presented by Schwertman, McCready, and Howard. In this article we improve these probability models by taking advantage of external information concerning the relative strengths of the teams and the point spreads available at the start of the tournament for the first round games. The result is a collection of regional championship probabilities that are specific to a given region and tournament year. The approach is illustrated using data from the 1994 NCAA basketball tournament.

KEY WORDS: Computer ratings; Oddsmaking; Sports outcome prediction.

1. INTRODUCTION

For many years the analysis of sports statistics was limited to huge tables of sums, means, and the occasional descriptive display. More recently, however, statisticians have begun to undertake more sophisticated analyses of these data. In teaching they can provide interesting and more easily grasped illustrations of important concepts (see, e.g., Albert 1993). In research they often enable testing of new approaches for handling difficult modeling scenarios, such as nonrandomly missing data (Casella and Berger 1994) and time-dependent selection and ranking (Barry and Hartigan 1993). And, of course, in many cases the data are interesting in and of themselves; recent papers have investigated the existence of the "hot hand" in basketball (Tversky and Gilovich 1989; Larkey, Smith, and Kadane 1989) and the likelihood of "Shoeless" Joe Jackson's complicity in the famous "Black Sox" scandal (Bennett 1993). The recent creation of a new section of the American Statistical Association devoted to sports statistics provides further testimony to their increasing popularity.

Perhaps the oldest inferential problem related to sports statistics is that of predicting the ultimate winner of some event, based on whatever information is available concerning the various competitors. In the realm of college basketball the most talked-about such event is the NCAA men's tournament, held every year in March and early April. In this tournament 64 teams (some invited by a selection committee, others receiving automatic bids thanks to their having won their own conference tournaments) are divided into four regional tournaments (West, Midwest, East, and Southeast) of 16 teams each. Some effort is made by the committee to balance the overall team strength in each region, while at the same time to place teams in the appropriate geographical region. The teams in each region are then "seeded" (ranked) based on their relative strengths as perceived by the committee. In a given region the tournament begins by having the strongest team (seed 1) play the weakest team (seed 16), the second-strongest (seed 2) play the second-weakest (seed 15), and so on. The winners of these eight first round games then play off in a predetermined order (e.g., the 1–16 winner plays the 8–9 winner) in four second round games, and so on until a single regional champion is determined after four rounds of play. Finally, the four regional champions face each other in fifth and sixth round games to determine a single national champion. For the time being we focus only on prediction of the regional champions (the "Final Four").

In a recent paper, Schwertman, McCready, and Howard (1991) consider three alternatives for specifying a 16×16 matrix P of regional win probabilities. That is, $P(i,j)$ is the probability that seed i defeats seed j in a contest between the two on a neutral court where, of course, $i \neq j$ and $P(j,i) = 1 - P(i,j)$. Together with the assumption that the games are independent, they derive the probability that seed i wins the region for $i = 1, \ldots, 16$ using elementary (although fairly tedious) calculations implemented in a Fortran program. Their models for $P(i,j)$ are somewhat ad hoc, although the most sophisticated (and best fitting) plausibly assumes a normal distribution of national team strengths, with the 64 tournament teams comprising the upper tail of this distribution. Subsequent work by Schwertman, Schenk, and Holbrook (1993) refines the approach by using past NCAA tournament data to fit linear and logistic regression models for $P(i,j)$ as a function of the difference in either team seeds or normal scores of the seeds.

In this article, we extend this approach by taking advantage of valuable external information available at the tournament's outset. Specifically, we may employ any of the various computer rankings of the teams, such as RPI index, Sagarin ratings, and so on, which typically arise as a linear function of several variables (team record, opponents' records, strength of conference, etc.) monitored over the course of the season preceding the tournament. These rankings provide more refined information concerning relative team strengths than is captured by the regional seedings. Such rankings also enable differentiation between identically seeded teams in different regions.

A second source of information for the first round games is the collection of point spreads offered by casinos and sports wagering services in states that allow gambling on

Bradley P. Carlin is Associate Professor, Division of Biostatistics, University of Minnesota, Minneapolis, MN 55455-0392. The author thanks Prof. Neil Schwertman for supplying the Fortran code to convert the P matrix into the regional championship probabilities, Prof. Jim Albert for supplying the 1994 pretournament Sagarin ratings, Prof. Hal Stern for invaluable advice and for suggesting the scoring rule used in Table 4, and Profs. Lance Waller and Alan Gelfand for helpful discussions.

Table 1. Data from Round 1 of the 1994 NCAA Tournament

Region	*j − i*	*S(i) − S(j)*	Y_{ij}	R_{ij}	*Region*	*j − i*	*S(i) − S(j)*	Y_{ij}	R_{ij}
West	15	20.28	20	23	East	15	20.37	25	20
West	13	18.43	24	26	East	13	15.78	18.5	18
West	11	11.95	18	9	East	11	10.98	10.5	2
West	9	9.81	11	14	East	9	9.31	11.5	22
West	7	4.78	6	−4	East	7	2.00	4.5	12
West	5	6.39	8.5	14	East	5	5.53	5	−10
West	3	1.20	4	3	East	3	2.53	4	−5
West	1	1.64	4	−8	East	1	−.19	−3.5	−3
Midwest	15	23.59	28	15	Southeast	15	20.87	23	31
Midwest	13	13.32	16	18	Southeast	13	18.32	20.5	12
Midwest	11	8.97	11.5	4	Southeast	11	17.31	18	13
Midwest	9	8.58	12	10	Southeast	9	9.90	10.5	29
Midwest	7	4.45	4	−10	Southeast	7	5.71	9	10
Midwest	5	5.71	7	14	Southeast	5	4.96	7	22
Midwest	3	1.04	−1.5	−8	Southeast	3	2.60	2	11
Midwest	1	1.43	2	−7	Southeast	1	5.63	5	−6

college basketball. A point spread is a predicted amount by which one team (the "favorite") will defeat the other (the "underdog"); gamblers may bet on whether or not the favorite's actual margin of victory will exceed the point spread ("cover the spread"). Point spreads are potentially even more valuable as pregame data than computer rankings because, besides team strengths, they account for game- and time-specific information, such as injuries to key players. Previous work by Harville (1980) and Stern (1992) shows that point spread information is the "gold standard" against which all other pregame information as to outcome must be judged.

Unfortunately, point spreads for potential games in rounds 2–4 will be unavailable at the tournament's outset. To remedy this, in Section 2 we describe an approach for imputing point spreads for these later games, and subsequently converting the resulting point spread matrix into the win probability matrix P. Section 3 applies our approach to data from the 1994 NCAA men's basketball tournament. Finally, Section 4 summarizes our findings and comments briefly on the prediction of the ultimate NCAA basketball national champion.

2. DETERMINATION OF WIN PROBABILITIES

Working with data from three seasons of professional football, Stern (1991) showed that the favored team's actual margin of victory, R, was reasonably approximated by a normal distribution with mean equal to the point spread, Y, and standard deviation $\sigma = 13.86$. That is,

$$\text{Pr(favorite defeats underdog)} = \Pr(R > 0) \approx \Phi(Y/\sigma) \quad (1)$$

where $\Phi(\cdot)$ denotes the cumulative distribution function of the standard normal distribution. This rather surprising result indicates that the group of bettors who determine the

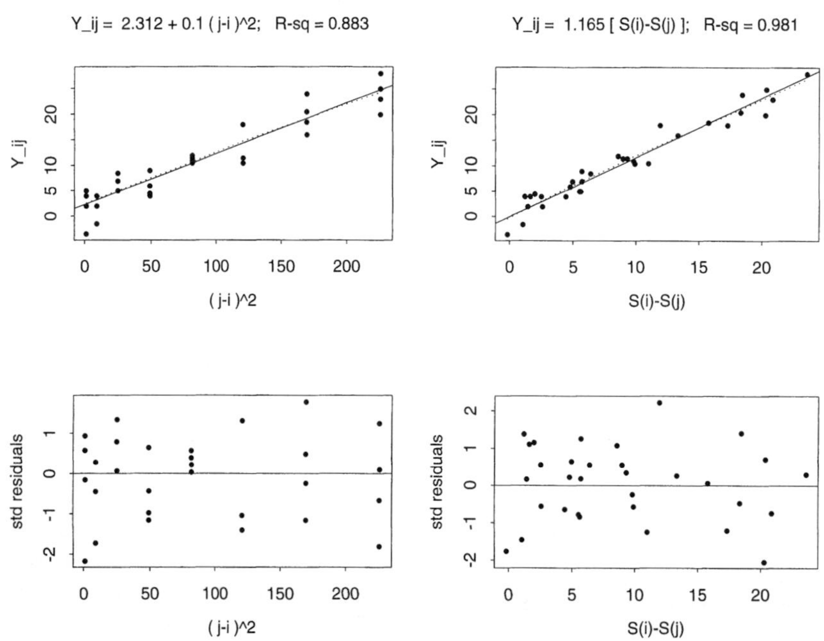

Figure 1. Regression of Round 1 Point Spreads on Differences in Seeding and Sagarin Rating.

Table 2. Comparison of Pr(Seed Wins Southeast Region) Across Models

Seed	Team	Schwertman method	Seed regression	Sagarin differences	Sagarin regression	Sagarin regression with R1 spreads
1	Purdue	.459	.326	.316	.343	.349
2	Duke	.188	.235	.151	.148	.150
3	Kentucky	.110	.155	.245	.260	.255
4	Kansas	.068	.110	.111	.108	.103
5	Wake Forest	.047	.064	.032	.024	.027
6	Marquette	.036	.045	.026	.019	.021
7	Michigan State	.026	.028	.033	.026	.025
8	Providence	.015	.018	.067	.061	.058
9	Alabama	.011	.008	.005	.003	.003
10	Seton Hall	.011	.005	.010	.006	.007
11	SW Louisiana	.009	.003	.002	.001	.001
12	Charleston	.006	.001	.002	.001	.000
13	TN–Chattanooga	.005	.001	.001	.000	.000
14	Tennessee State	.004	.000	.000	.000	.000
15	Texas Southern	.003	.000	.000	.000	.000
16	Central Florida	.001	.000	.000	.000	.000
	Sum	1.000	1.000	1.000	1.000	1.000

point spreads are, on the average, correct in their predictions of game outcome. (Note that it would be wrong to give too much credit for this accuracy of the point spread to the bookies, who merely set an initial spread and subsequently raise or lower it so that roughly the same total amount is bet on both the favorite and the underdog.)

Subsequent unpublished analysis by Stern of two seasons of professional basketball data indicates that (1) again holds, this time with $\sigma = 11.5$. Intuition suggests that this σ value may be a bit large for our purposes because college basketball is generally a lower scoring game, and we would expect the variability in the victory margin to increase with the total points scored. Indeed, data from the first four rounds of the 1994 NCAA tournament produce a value of $\hat{\sigma} = 8.83$, although using this precise value in our analysis would of course be unfair because it was unavailable at the tournament's outset. In what follows we take $\sigma = 10$ as a reasonable and somewhat conservative compromise.

Hence for a given region, we may use Equation (1) with the point spreads for the first round games to determine $P(i, 17-i), i = 1, \ldots, 16$. But this fills in only the antidiagonal (lower left to upper right) of P; how should we determine the remaining entries? A natural solution would be to obtain a general prediction equation for the point spread y in terms of the difference in team seeding (perhaps after some suitable transformation), and then again use (1) to complete the P matrix. Relevant data for this calculation from the 32 first round games in the 1994 tournament are displayed in Table 1. Besides the seeding differences $(j-i)$ and point spreads Y_{ij} for each game matching seeds i and j where $i < j$, the table also shows the actual victory margin R_{ij} for comparison. Our Y_{ij} values were obtained immediately prior to the beginning of tournament play from the Thursday morning edition of a local newspaper, which in turn got them from a prominent Las Vegas oddsmaker. A negative value of Y_{ij} implies that the team with the poorer seeding (team j) was favored by the bettors to win the game; a negative value of R_{ij} indicates that the result was, in fact, a victory by the poorer seed (an "upset").

Table 3. Comparison of Pr(Seed Wins Region) Across Regions

Seed	West	Midwest	East	Southeast
1	.182	.310	.349	.349
2	.379	.232	.306	.150
3	.147	.176	.089	.255
4	.103	.093	.109	.103
5	.053	.046	.043	.027
6	.072	.047	.041	.021
7	.009	.013	.020	.025
8	.037	.044	.009	.058
9	.011	.020	.017	.003
10	.003	.011	.004	.007
11	.002	.002	.002	.001
12	.003	.005	.010	.000
13	.000	.000	.000	.000
14	.000	.001	.000	.000
15	.000	.000	.000	.000
16	.000	.000	.000	.000
Sum	1.000	1.000	1.000	1.000

The first column of plots in Figure 1 shows the results of regressing point spread on squared seeding difference. The fitted regression line obtained is $\hat{Y}_{ij} = 2.312 + .100(j-i)^2$, where $i < j$. The extremely close agreement between this fitted regression line (solid line in upper panel) and the lowess smoothing line (dotted line) indicates a high degree of linearity on this scale, and the standardized residual plot in the lower panel does not indicate any failure in the usual regression assumptions. Further, the R^2 value of .883 indicates reasonably good fit. Note that the data line up in eight vertical columns because there are four games pitting seed i against seed $(17-i)$ for $i = 1, \ldots, 8$.

In the second column of plots in Figure 1, we replace squared seed difference as the predictor variable with the difference in Sagarin rating, a numerical measure of team strength that we denote by $S(k)$ for seed k. These ratings, which account for the won–lost record in Division I games and strength of schedule, are published every Monday during the season in the newspaper *USA Today*. Table 1 gives the relevant differences from the collection of rankings published on the Monday immediately prior to

Table 4. Information Scores for the Five Tournament Probability Estimation Methods

Region	*Schwertman method*	*Seed regression*	*Sagarin differences*	*Sagarin regression*	*Sagarin regression with R1 spreads*
West	−.116	−.111	−.106	−.101	−.102
Midwest	−.134	−.147	−.134	−.134	−.127
East	−.154	−.148	−.149	−.152	−.145
Southeast	−.114	−.103	−.116	−.114	−.111
All	−.517	−.508	−.505	−.502	−.485

the 1994 first round games. The ratings are designed to produce hypothetical point spreads when differenced, and our results bear this out: the data seem to require no transformation to achieve linearity, and the intercept in the full regression model is not significantly different from zero. Forcing the line through the origin, we obtain the fitted model $\hat{Y}_{ij} = 1.165[S(i) - S(j)]$, where $i < j$. The fitted slope coefficient suggests that the rating difference tends to slightly underestimate the point spread in matches between opponents of widely differing strengths. The improved R^2 value of .981 confirms the visual impression from the figure that the Sagarin rating difference is superior to seed difference as a predictor of point spread. We remark that while the best 39 of the 301 Division I teams (as measured by Sagarin rating) were included in the 1994 tournament, the four #16 seeds had Sagarin rankings 166, 173, 196, and 216, calling into question the assumption of Schwertman et al. (1991) that the 64 tournament teams may be safely thought of as the best 64 teams in the country.

3. APPLICATION TO THE 1994 NCAA TOURNAMENT

We begin by comparing the results of several approaches suggested by Figure 1 using the 1994 Southeast regional tournament data because differences among the approaches are most apparent using data from this region. Table 2 gives the estimated probability of emerging as the regional champion for each of the 16 teams. The first method listed is the one recommended by Schwertman et al. (1991). This method assigns nearly 50% of the mass to the #1 seed, while giving only 5% to the entire lower half of the bracket (seeds 9–16). The next column provides results obtained by using the regression of point spread on squared seed difference to obtain the win probabilities. This method gives slightly more mass to the upper seeds other than #1, but even less mass to the lower division teams (total probability less than 2%). Like the Schwertman method, it uses only seed information to determine win probabilities, so the results in these columns would apply to any of the four regional tournaments. The seed regression results *are* specific to this *year,* however, because 1994 spread data were used to fit the model.

Listed next are results obtained by simply taking the unadjusted difference of the Sagarin ratings as the hypothetical point spread between the two teams. Note that the probability of a triumph by the #1 seed has dropped again, and total support for the lower division teams remains a mere 2%. Note also that these probabilities are specific to *both* year and region because teams with identical seeds in different regions need not have identical Sagarin ratings. Indeed, the probabilities are no longer strictly decreasing from seed 1 down to seed 16, due to ordering conflicts between the seedings and the Sagarin ratings (e.g., #2 Duke rated 89.90, #3 Kentucky rated 91.59). The next column in the table adjusts the Sagarin differences using the regression model obtained in the previous section before converting them to win probabilities via Equation (1). This adjustment amounts to giving a boost to the two most highly rated teams in the region (Purdue and Kentucky) at the expense of the others; notice that the total probability allocated to the lower half of the bracket is now barely 1%. Finally, the last column replaces the imputed first round point spreads used in the previous method with the actual point spreads. This results in subtle changes to only 16 entries in the P matrix, but because they are the entries corresponding to the first games played, the effect on the regional championship probabilities is apparent, occasionally visible in the second decimal place.

Table 3 applies this final method (using the Sagarin regression model plus actual first round spreads) to data from each of the four 1994 regional tournaments. There are several reversals of the seeding order, the most interesting of which is the prediction of #2 Arizona, not #1 Missouri, as the team most likely to win the West regional. (To the model's credit this was indeed the outcome in this region.) Support for seeds in the lower half of each region is again quite low. The largest single probability in the table is .379 (Arizona), substantially lower than that given to any #1 seed by the Schwertman model, and perhaps indicative of the increased "parity" in college basketball often mentioned by sportswriters and coaches during the 1994 season. Again, the tournament results bear this out somewhat, as the actual regional champions were seeded 2 (Arizona), 1 (Arkansas), 3 (Florida), and 2 (Duke).

Schwertmann et al. (1991) explore model fit more formally by comparing observed and expected numbers of teams with a given seeding to become regional champions over the first six years of 64-team NCAA tournament play (i.e., 24 regional champions). This method of model validation is unavailable to us, however, because the probabilities in Table 3 are specific to both team and year, not just seeding. Instead, we judge a method that produces the win probability matrix P using the scoring rule

$$I(P, Z) = \frac{1}{K}\sum_{k=1}^{K} \log[p_k Z_k + (1 - p_k)(1 - Z_k)] \qquad (2)$$

where $Z = \{Z_1, \ldots, Z_K\}$, $Z_k = 1$ if the team with the more favorable seeding wins game k, and $Z_k = 0$ otherwise.

Equation (2) is simply the average log-win probability predicted by the model for those teams that actually won, so that models are rewarded for assigning high probability to these teams. $I(P, Z)$ can also be thought of more formally as the average Shannon information in the likelihood (see, e.g., Lindley 1956).

Table 4 gives the information scores for each of the five methods compared in Table 2. The bottom line sums over all $K = 60$ games in Rounds 1–4 of the 1994 tournament. To provide a reference point for the scores on this line, the completely naive method that assumes every game is an even toss-up (all $P(i, j) = .5$) would have a score of $\log(.5) = -.693$. Notice the monotonic improvement in score as we move from left to right, with our final method (the one used in Table 3) emerging as noticeably better than the rest. From the component scores within each region we see that this final method scores higher in every region than the Schwertman method, which does not intend to be team- or region-specific. Note however that there is little difference in score among the methods for the Southeast region, where Kentucky's early exit and Duke's ultimate win contradicted the Sagarin ratings. But the methods based on these ratings perform quite well in the West region, where the ratings correctly predicted that #2 Arizona would beat #1 Missouri.

4. CONCLUSION

In this article we have developed a method for improved probability modeling of NCAA regional basketball tournaments. The method requires only elementary ideas in probability theory, statistical graphics, and linear regression analysis, and as such should provide an interesting and instructive exercise for students. Implementation for a given year requires only the Sagarin ratings for the appropriate 64 teams, perhaps the collection of first round point spreads (if refitting the regression model is desired), and the Fortran program for computing the P matrix and reducing it to the collection of regional championship probabilities (extended from the original program by Schwertman, and available from the author upon request via electronic mail).

We have argued on behalf of the use of (actual or imputed) point spreads in determining win probabilities, on the grounds that true point spreads are superior to computer rankings, which are in turn superior to the crude summaries provided by tournament seedings. Our regression analyses in Section 2 and the information scores in Table 4 support this belief, as does other, more anecdotal evidence from the 1994 tournament. For example, Table 1 shows that, in two first round games, the lower seed was actually favored by the bettors: East #9 (Boston College) 3.5 points over East #8 (Washington State), and Midwest #10 (Maryland) 1.5 points over Midwest #7 (St. Louis). The Sagarin ratings corrected the seeding error in the former case, but not in the latter (Maryland rating 83.43, St. Louis rating 84.47, a difference of 1.04). Perhaps the bettors were influenced in this case by their additional knowledge that St. Louis had lost three of their last four games prior to the tournament, or that the team's tallest player was injured, suggesting that they would have a hard time guarding Maryland's 6-foot, 10-inch center, Joe Smith. As it turned out, Smith had 29 points and 15 rebounds in the game, and Maryland won by 8 points.

As a final comment we note that in some cases interest may focus not on the prediction of the Final Four, but on the prediction of the ultimate national champion. While our ideas could, of course, be extended to the case of a single 64×64 P matrix, the programming involved in reducing this matrix to the vector of championship probabilities would be almost unbearably tedious. As a simple alternative we might assume that the Final Four teams are more or less evenly matched, and thus select as our national champion the team most likely to make it this far in the tournament (in our case, W#2 Arizona). It is worth pointing out, however, that this logic, along with Table 3, suggests that no single team would be likely to have even a 10% chance at the tournament's outset of winning the required six consecutive games, so that any such prediction is almost certain to be incorrect.

[Received April 1994. Revised October 1994.]

REFERENCES

Albert, J. H. (1993), "Teaching Bayesian Statistics using Sampling Methods and MINITAB," *The American Statistician*, 47, 182–191.

Barry, D., and Hartigan, J. A. (1993), "Choice Models for Predicting Divisional Winners in Major League Baseball," *Journal of the American Statistical Association*, 88, 766–774.

Bennett, J. (1993), "Did Shoeless Joe Jackson Throw the 1919 World Series?," *The American Statistician*, 47, 241–250.

Casella, G., and Berger, R. L. (1994), "Estimation with Selected Binomial Information or Do You Really Believe that Dave Winfield is Batting .471?," *Journal of the American Statistical Association*, 89, 1080–1090.

Harville, D. (1980), "Predictions for National Football League Games via Linear-Model Methodology," *Journal of the American Statistical Association*, 75, 516–524.

Larkey, P. D., Smith, R. A., and Kadane, J. B. (1989), "It's Okay to Believe in the 'Hot Hand,'" *Chance*, 2(4), 22–30.

Lindley, D. V. (1956), "On the Measure of Information Provided by an Experiment," *Annals of Statistics*, 27, 986–1005.

Schwertman, N. C., McCready, T. A., and Howard, L. (1991), "Probability Models for the NCAA Regional Basketball Tournaments," *The American Statistician*, 45, 35–38.

Schwertman, N. C., Schenk, K. L., and Holbrook, B. C. (1993), "More Probability Models for the NCAA Regional Basketball Tournaments," Technical Report, Department of Mathematics and Statistics, California State University, Chico.

Stern, H. (1991), "On the Probability of Winning a Football Game," *The American Statistician*, 45, 179–183.

——— (1992), "Who's Number One?—Rating Football Teams," in *Proceedings of the Section on Sports Statistics* (Vol. 1), Alexandria, VA: American Statistical Association, pp. 1–6.

Tversky, A., and Gilovich, T. (1989), "The Cold Facts about the 'Hot Hand' in Basketball," *Chance*, 2(1), 16–21.

Basketball players who do the improbable and memorable do *have shooting streaks according to a new data set that rebuts the Tversky and Gilovich case against them.*

Chapter 19

It's Okay to Believe in the "Hot Hand"

Patrick D. Larkey, Richard A. Smith, Joseph B. Kadane

If economics is known as the "dismal science" for its temerity to insist that resources are always scarce and allocation decisions always difficult, perhaps psychology should be known as the "debilitating science" for its continuing fascination with our gross inadequacies as thinking and acting organisms. Psychologists of an earlier era told us that much of our adult behavior was ultimately caused by unresolved emotional conflicts in childhood that arrest our development. Psychologists more recently have experimentally confirmed a host of limitations on our ability to remember and reason. Now two psychologists, Amos Tversky and Thomas Gilovich, tell us that if we are basketball fans and believe in the "hot hand" and "streak shooting," we are misconceiving the laws of chance and suffering from, of all things, a "cognitive illusion" (see "The Cold Facts About the 'Hot Hand' in Basketball," *Chance*, Winter 1989).

Tversky and Gilovich boldly claim that "this misconception of chance has direct consequences for the conduct of the game. Passing the ball to the hot player, who is guarded closely by the opposing team, may be a non-optimal strategy if other players who do not appear hot have a better chance of scoring. Like other cognitive illusions, the belief in the hot hand could be costly." While it is not entirely clear why offensive coaches would be more vulnerable than defensive coaches to holding and acting upon such a fallacious belief, Tversky and Gilovich's claim raises the possibility that legendary coaches such as Red Auerbach and Johnny Wooden suffered from cognitive illusions throughout their careers; if they had only conferred with the right psychologists and understood the error of their ways, they might have improved on their strategies and won even more games and championships.

Is nothing sacred? Are fans and coaches really wrong to believe that all basketball players occasionally get hot in their shooting and that a few players have a greater propensity than others to shoot in streaks? Perhaps not. This paper briefly reviews Tversky and Gilovich's conceptualization of the hot hand and streak shooting, proposes a different conception of how observer's beliefs in streak shooting are based on National Basketball Association (NBA) player shooting performances, and tests this alternative conception on data from the 1987–1988 NBA season.

For the Existence of Streak Shooting

Tversky and Gilovich argue that fan beliefs in the hot hand or streak shooting imply two specific departures in player shooting sequences from the simple binomial with a constant hit rate: (1) ". . . the probability of a hit should be greater following a hit than following a miss (i.e., positive association); and (2) . . . the number of streaks of successive hits or misses should exceed the number produced by a chance process with a constant hit rate (i.e., nonstationarity)."

While Tversky and Gilovich report the results of questionnaire studies and shooting experiments, the empirical centerpiece of their argument is an analysis of field goal shooting data from the 48 home games of the Philadelphia 76ers and their opponents during the 1980–1981 NBA season contrasted with the expectations of a simple binomial process. They check for dependence and nonstationarity using a variety of tests including conditional probabilities, first order serial correlations, Wald-Wolfowitz, Lexis ratios, and a comparison of player results on four-shot sequences with expectations from binomials based on the players' sample shooting percentages.

Their data analysis yields no evidence of either dependence between shots or of nonstationarity. Tversky and Gilovich "attribute the discrepancy between the observed basketball statistics and the intuitions of highly interested and informed observers to a general misconception of the laws of chance that induces the expectation that random sequences will be far more balanced than they generally are, and creates the illusion that there are patterns or streaks in independent sequences."

There is, however, a serious problem with both their conceptualization and their data analyses for analyzing the origination, maintenance, and validity of beliefs about the streakiness of particular players. The shooting data that they analyze are in a very different form than the data usually available to observers qua believers in streak shooting. The data analyzed by Tversky and Gilovich consisted of isolated individual player shooting sequences by game. The data available to observers including fans, players, and coaches for analysis are individual players' shooting efforts in the very complicated context of an actual game.

To the extent that beliefs in relative player propensities for streak shooting are connected to shooting performance data, they are almost certainly connected to the data as presented to the believers. In order to do any analysis at all, of course, it is necessary to abstract radically from the enormous complexity of the observations of an actual basketball game. Tversky and Gilovich go beyond mere abstraction. Extracting individual player shooting sequences from an actual game is a very complex task. This task is clearly beyond the abilities of the average observer without external aids to memory. Even team and media "statisticians" do not routinely preserve shot sequence information on individual players.

Observers of NBA games see a succession of field goal shooting opportunities, interrupted by a lot of nonshooting activity, from the beginning to the end of a game. Each of these opportunities can be taken by any one of the ten players on the floor with the result of a hit or a miss. Most observers of real basketball games attend to the unfolding sequence of shooting opportunities in the game, not to the efforts of an individual player. Observers note individual players primarily when they do notable things in a cognitively manageable chunk of shooting opportunities. While we have no systematic basis for knowing precisely how large a "manageable chunk of opportunities" is—it would vary with both the observers and the overall content of the chunk—we estimate that it is not longer than 20 field goal opportunities or a little less than half a quarter in the average NBA game in our data set. Longer chunks increase the probability of failures in the observers' attention and memory.

The observers' focus on the unfolding sequence of shot opportunities rather than the activities of an individual player suggests a very different model of player shooting activities than the one used by Tversky and Gilovich. It also suggests that their model, the separate binomial for each player, is wrong both descriptively and normatively. It is almost certainly not the model *used* by any observers of a real NBA game as the basis of their expectations about a player's shooting performance (i.e., whether that player is "hot" or "cold") because of the extreme difficulty in sorting out and remembering the individual shooting performances of ten to twenty players over the course of a real game. It is not a model that observers *should use* as the basis of their expectations about a player's shooting performance on the next shot or for an entire game because the model ignores game context, how that player's shooting activities interact with the activities of the other players.

Consider two players A and V who each have a run of 5 consecutive field goal successes in the same game in which 220 field goals were attempted. Beginning with the 70th shot, V makes 5 consecutive field goals for the entire floor; he takes and makes the 70th shot, the 71st shot, . . . , and the 74th shot. A's 5 shots are maximally dispersed across the 220 shots in the game; he takes and makes the 1st shot, the 55th shot, the 110th shot, the 165th shot, and the 220th shot.

Suppose that both players have the same percentage of success, $P_i = .5$, in shooting field goals, that shots are independent, and that P_i is stationary across samples of shots. In the Tversky and Gilovich analysis each streak that is identical in length and content (hits and misses) for a particular player is equally probable. Yet, few of us would believe that the probabilities of A and V accomplishing their respective streak shooting feats are equal. We also do not believe that the probabilities of an observer seeing, noticing, and remembering the streaks of A and V are equal. The way a streak falls in the context of a game matters a lot to observers. Some streaks are much less probable and much more noticeable and memorable to observers than other streaks of the same length and identical hit and miss content.

The sequence for V is much less likely than the sequence for A. V's accomplishment is so unlikely that we did not see even one consecutive streak of 5 hits by one player with no intervening field goal activity by other players in our data on 39 NBA games with over 9000 field goal attempts by 139 different players. Indeed, we did not see even one 4-shot streak of this type. There were only two 3-shot streaks of this type in all the data. As for A's accomplishment, there were approximately 100 instances in our data where players made 5 consecutive shots with some intervening activity by other players.

In the example, any fan watching would notice and remember V's performance in the 70th to 74th shot in the game context. The noticeability and memorability of this sequence might be lessened if the shots were all tip-ins or if the break for the end of the first quarter occurred between the 71st and 72nd shot. It is hard to imagine plausible circumstances that leave this event as something other than the most memorable shooting sequence in the entire game. If all of the shots were 30-foot jumpers at the end of a close seventh game in the NBA championship series, the sequence and V would be instant legends.

On the other hand, A's sequence would probably be missed by everyone but his agent, relatives, and close friends watching the game. Player A would never be accused by anyone, except perhaps psychologists analyzing isolated player shooting sequences or A's agent renegotiating his contract, of having had the hot hand in this hypothetical game.

These differences in the proba-

Photo courtesy of Joe Labolito, Temple University Photography Department, © 2004.

bility, noticeability, and memorability of sequences in an unfolding context leads us to the following hypothesis: The field goal shooting patterns of players with reputations for streakiness will differ from the patterns of reputationless players; a streak shooter will accomplish low-probability, highly noticeable and memorable events with greater frequency than reputationless players in the data set and with greater frequency than would be expected of him in the context of a game.

All that remains is to test this hypothesis.

The Hot Hand In and Out of Context

Our data consist of 39 games telecast in the Pittsburgh market during the 1987–1988 season. The games were videotaped and all shots were coded by the authors in sequence from the beginning to the end of the game. The data include teams with varying frequency:

1. Detroit Pistons—20 games
2. Boston Celtics—17 games
3. Los Angeles Lakers—16 games
4. Atlanta Hawks—6 games
5. Dallas Mavericks—5 games
6. Chicago Bulls—4 games
7. New York Knicks—3 games
8. Utah Jazz—2 games

and one game each for the Philadelphia 76ers, the Washington Bullets, the Denver Nuggets, the Milwaukee Bucks, and the Cleveland Cavaliers.

Twenty-four of the games were playoff games with the balance in the regular season. The data set was driven primarily by the scheduling habits of Pittsburgh television stations, WTBS (Atlanta), and WOR (New York) from January to the conclusion of the 1987–1988 playoffs. While every effort was made to tape the games of the Celtics, Lakers, Hawks, Bulls, and Mavericks, taping Piston games had the highest priority. One of the Piston's players, Vincent ("Vinnie") Johnson, "the Microwave," has arguably the greatest current reputation in the NBA for streak shooting. His entry in *The Complete Handbook of Pro Basketball* (15th Edition, 1989, Signet, pp. 244–245) is:

> The Microwave was only lukewarm. . . . Had worst shooting season (.443) since rookie year. . . . And his 12.2 ppg was lowest in 6 1/2. . . . Still the most lethal streak shooter in game who can pour in line-drive jumpers by the bushel. . . . When he's on, nothing known to man can stop him. . . . Best reserved for reserve role due to his streakiness. And when he hits his first two shots, opponents can light novena candles . . .

In contrast, the same publication does not mention the hot hand or streak shooting in describing any of the other players analyzed. For our analysis Johnson's reputation as a streak shooter is fairly current and distinctively different from the reputations of the other players. While data from Vinnie's best years, the years in which he was building his reputation as the "most lethal streak shooter" in the game, would afford the best chance of differentiating him from other players, such data are not available.

One hundred and thirty-nine different players took at least one shot in the 39 games. We reduced the set of players for analysis to 18, including those who attempted more than 100 field goals or who attempted 10 or more field goals in 4 or more games. The Pistons, Celtics, and Lakers are disproportionately represented, but these were 3 of the best teams in the NBA. Our data include at least a few games for a number of the most skilled offensive players from other teams in the NBA such as Michael Jordan, Dominique Wilkins, and Mark Aguirre.

We hypothesize that Vinnie's field goal shooting patterns will differ from the patterns of reputationless players. Further, these differences will take the form of Vinnie accomplishing low-probability, highly noticeable and memorable events with greater

Table 1. Conditional Probabilities of a Hit and Autocorrelation

(1) Player	(2) $P(H/3M)$	(3) $P(H/2M)$	(4) $P(H/1M)$	(5) $P(H)$	(6) $P(H/1H)$	(7) $P(H/2H)$	(8) $P(H/3H)$	(9) ACF(1)
Jordan	.57(7)	.47(17)	.53(43)	.55(104)	.56(57)	.47(32)	.47(15)	.027
Bird	.49(57)	.40(103)	.38(177)	.44(338)	.49(145)	.51(69)	.47(34)	.141
McHale	.58(12)	.62(37)	.61(108)	.57(270)	.53(146)	.60(72)	.55(40)	−.068
Parish	.56(9)	.59(27)	.48(66)	.52(163)	.54(80)	.61(41)	.65(23)	.107
D. Johnson	.54(26)	.45(56)	.44(109)	.41(201)	.35(75)	.42(24)	.50(8)	−.107
Ainge	.40(25)	.41(49)	.43(96)	.42(184)	.44(71)	.46(28)	.36(11)	.014
D. Wilkins	.62(21)	.53(45)	.51(92)	.47(176)	.42(78)	.33(33)	.27(11)	−.088
E. Johnson	.61(28)	.45(60)	.46(123)	.43(230)	.40(91)	.50(34)	.33(15)	−.045
A-Jabbar	.38(24)	.48(50)	.49(103)	.47(209)	.49(90)	.61(41)	.50(22)	.015
Worthy	.73(22)	.54(59)	.48(124)	.47(259)	.48(119)	.48(54)	.64(25)	.025
Scott	.60(20)	.56(48)	.50(109)	.52(246)	.54(121)	.55(60)	.55(31)	.037
Aguirre	.70(10)	.54(24)	.47(47)	.46(93)	.39(41)	.40(15)	.33(6)	−.077
Dantley	.33(21)	.43(42)	.50(101)	.50(224)	.51(104)	.50(50)	.50(24)	.014
Laimbeer	.41(17)	.47(45)	.47(103)	.47(219)	.43(96)	.45(40)	.39(18)	−.084
Dumars	.60(25)	.46(52)	.47(115)	.45(234)	.41(99)	.40(40)	.40(15)	−.035
Thomas	.49(45)	.47(93)	.44(187)	.44(361)	.44(154)	.41(66)	.50(26)	−.001
V. Johnson	.47(17)	.45(42)	.44(97)	.46(213)	.49(96)	.51(45)	.52(23)	.036
Rodman	1.00(3)	.69(13)	.63(38)	.62(112)	.55(55)	.78(23)	.92(12)	−.057
Wt. Mean	.5248	.4809	.4727	.4752	.4883	.5034	.5402	−.0016

Table 2. Perfect Runs of Hits—Acontextual "Occurrences to Opportunities"

Player	8/8	7/7	6/6	5/5	4/4	3/3
Jordan	—	—	.02	.05	.08	.16
Bird	—	—	.01	.02	.06	.11
McHale	—	—	.02	.05	.10	.18
Parish	.04	.06	.07	.09	.13	.19
D. Johnson	—	—	—	.01	.03	.06
Ainge	—	—	—	.01	.03	.09
D. Wilkins	—	—	—	—	.02	.07
E. Johnson	—	—	—	.01	.03	.09
A-Jabbar	—	—	.02	.04	.07	.14
Worthy	.01	.02	.03	.05	.08	.11
Scott	—	.01	.03	.05	.09	.15
Aguirre	—	—	—	—	.03	.07
Dantley	—	.01	.02	.04	.07	.13
Laimbeer	—	—	—	.01	.04	.10
Dumars	—	—	.01	.01	.03	.08
Thomas	—	—	.01	.02	.04	.08
V. Johnson	—	.01	.03	.04	.08	.13
Rodman	.17	.15	.13	.16	.18	.24

Table 3. One Miss in Hit Run—Acontextual "Occurrences to Opportunities"

Player	7/8	6/7	5/6	4/5	3/4	2/3
Jordan	.11	.16	.19	.24	.38	.49
Bird	.03	.05	.06	.12	.18	.29
McHale	.08	.14	.19	.25	.32	.42
Parish	.04	.03	.06	.14	.23	.32
D. Johnson	—	.01	.03	.09	.12	.26
Ainge	—	.01	.04	.11	.17	.31
D. Wilkins	—	.01	.03	.11	.22	.40
E. Johnson	.01	.04	.09	.13	.21	.28
A-Jabbar	.03	.08	.12	.16	.24	.31
Worthy	.04	.06	.09	.12	.18	.35
Scott	.10	.10	.13	.16	.23	.35
Aguirre	—	—	.01	.10	.17	.33
Dantley	.04	.06	.09	.16	.26	.39
Laimbeer	—	.01	.08	.17	.25	.36
Dumars	—	.01	.03	.10	.19	.36
Thomas	.04	.06	.08	.11	.19	.35
V. Johnson	.11	.14	.17	.22	.26	.34
Rodman	—	.05	.10	.11	.22	.36

frequency than other players in the data set and with greater frequency than would be expected of him in the context of a game. We do not expect Vinnie to look different from the other players in the data set from acontextual analyses of his shooting patterns much like those of Tversky and Gilovich. If he does look different on this portion of our analysis, it implies that Tversky and Gilovich lacked a streak shooter, at least one of Vinnie's distinction, in their data set.

We began with an acontextual player shooting sequences and repeated two of the analyses utilized by Tversky and Gilovich—serial correlations between the outcomes of successive shots and the probabilities of a hit conditioned on 1, 2, and 3 hits or misses. The results are shown in Table 1.

The results from this data set are not as clearly against the existence of streak shooting as Tversky and Gilovich's results. Half of the players have positive serial correlations indicating some positive dependence between pairs of shots with like results. Larry Bird has the highest positive serial correlation at .141; this is the only one significantly different from a zero expectation of the simple binomial assuming independence. At .036, Vinnie has the third largest positive serial correlation.

Vinnie is one of six players whose conditional probabilities increase monotonically with the number of prior hits. He is not, however, the most interesting case. Robert Parish of the Boston Celtics and Dennis Rodman of the Detroit Pistons show considerable evidence that for them "success breeds success" in shooting field goals. For Rodman, it also looks as if "failure breeds success."

The entries in Table 2 are simple proportions: the number of times that each player accomplished a hit sequence of a given length from 8 to 3 divided by the number of sequence opportunities of that length that the player had. The numerator is the number of times that a player went 8 for 8, 7 for 7, and so on. The denominator is the number of times the player took a sequence of shots of length L (L = 8,7,6 . . . , 3). If a player takes 9 shots in a game, this game contributes two 8-shot opportunities, three 7-shot opportunities, four 6-shot opportunities, and so forth.

In 4% of all the 8-shot sequences that Robert Parish took in our data set, he made all 8 shots. Dennis Rodman stands out in this table as a candidate streak shooter. He clearly dominates all of the other players on this simple proportional measure. A look at the data reveals that Rodman had 1 game in which he made 9 consecutive shots, the longest isolated player streak of hits in the entire data set. Because both occurrences and opportunities are counted completely, the single long sequence of hits contributes substantially to all sequence lengths. Rodman also had the highest shooting percentage in our data set (see Table 1). Vinnie Johnson, purportedly the most lethal streak shooter in the NBA, disappears completely relative to other players by this measure.

Table 3 differs from Table 2 only in that the runs are imperfect; there was one miss in each sequence of given length. Here we see Michael Jordan and Kevin McHale, who were not notable at all in Table 2 on perfect runs,

Table 4. Perfect Runs of Hits—Acontextual "Occurrences to Expectations"

Player	8/8	7/7	6/6	5/5	4/4	3/3
Jordan	—	—	.86	.90	.83	.94
Bird	—	—	1.60	1.33	1.47	1.34
McHale	—	—	.61	.81	.94	.98
Parish	6.56	5.48	3.57	2.37	1.78	1.37
D. Johnson	—	—	—	.64	.94	.87
Ainge	—	—	—	.66	.97	1.17
D. Wilkins	—	—	—	—	.39	.65
E. Johnson	—	—	—	.82	.80	1.08
A-Jabbar	—	—	1.44	1.80	1.40	1.36
Worthy	2.86	3.63	3.11	2.24	1.55	1.10
Scott	—	1.30	1.52	1.30	1.17	1.10
Aguirre	—	—	—	—	.57	.74
Dantley	—	1.15	1.49	1.30	1.15	1.08
Laimbeer	—	—	—	.31	.90	.97
Dumars	—	—	.90	.70	.84	.91
Thomas	—	—	1.06	1.29	1.15	.99
V. Johnson	—	2.29	2.71	1.79	1.73	1.36
Rodman	7.63	4.26	2.27	1.70	1.24	.99

Table 5. One Miss in Hit Run—Acontextual "Occurrences to Expectations"

Player	7/8	6/7	5/6	4/5	3/4	2/3
Jordan	1.92	1.86	1.40	1.16	1.27	1.20
Bird	1.86	1.60	1.05	1.18	.96	.88
McHale	1.13	1.34	1.22	1.11	1.00	.99
Parish	.89	.42	.54	.80	.84	.83
D. Johnson	—	.49	.83	1.07	.74	.89
Ainge	—	.53	.88	1.24	1.01	1.02
D. Wilkins	—	.36	.47	.86	.98	1.15
E. Johnson	.68	1.78	1.72	1.30	1.15	.89
A-Jabbar	1.43	1.99	1.59	1.23	1.10	.88
Worthy	1.90	1.53	1.23	.95	.84	1.00
Scott	2.46	1.51	1.16	.94	.84	.89
Aguirre	—	—	.22	.79	.79	.95
Dantley	1.38	1.15	.91	.99	1.05	1.05
Laimbeer	—	.25	1.14	1.34	1.11	1.03
Dumars	—	.27	.49	.86	.95	1.06
Thomas	2.53	2.19	1.45	1.05	.98	1.07
V. Johnson	5.69	3.91	2.56	1.82	1.23	.99
Rodman	—	.33	.46	.40	.60	.81

emerge as the most interesting imperfect streak shooters. Jordan is tied with Vinnie in the 7 for 8 category and either Jordan or McHale have the highest proportion in all other categories. Vinnie is more apparent here with the second highest proportions in the 6 for 7 and 5 for 6 categories and with the third highest in the 4 for 5 and 3 for 4 categories. It would be difficult, however, to argue that his reputation for streak shooting is warranted based on the data in the first three tables.

Tables 4 and 5 parallel Tables 2 and 3 with one important difference. While the numerators are identical, the denominators in Tables 4 and 5 are expectations computed as the simple binomial for the player for the given type of sequence times the number of opportunities in the data set that the player had to accomplish the sequence. For example, the expectations in the denominators for Table 4 were calculated as

$$(P_i^L)\left(\sum_{g=1}^{G}(T_{ig} - L + 1)\right)$$

where

P_i = Probability of a hit given a shot by player i.
T_{ig} = Shots taken by ith player in gth game.
L = Length of run.
G = Number of games.

The expectation for each entry in Tables 4 and 5 is 1.0. If a player is below 1.0 for any sequence length, it means that he accomplished that sequence less often than expected in this data set. If the player is above 1.0, he accomplished the sequence more often than expected. 2.0 means that the player accomplished the sequence twice as often as expected. For example, a player with a .5 shooting percentage has by the binomial assuming independence a .0039 probability of making 8 field goals in a row. If this player took 100 8-shot sequences in our data set, we expect to see .39 successes. If the player had one success in the data set, the table entry would be 1/.39 = 2.5641.

Dennis Rodman and Robert Parish are the streak shooters by Table 4's measure. Vinnie Johnson does not stand out as a streak shooter on perfect sequences.

In Table 5, the ratio of occurrences to expectations on imperfect sequences, Vinnie emerges as different from other players including several great players. He accomplishes the 7 for 8 sequence 5½ times more often than expected; the closest players on 7 for 8 are Isiah Thomas of the Pistons and Byron Scott of the Lakers at about 2½ times the expectation. Vinnie is above all other players in the 4 longest sequences and a close second in 3 for 4 sequences.

To this point in the analysis we have, like Tversky and Gilovich, looked only at isolated player shooting sequences. While our streak shooter looks somewhat different by at least one measure from the other 17 players, we have not yet considered context which we hypothesize is what really enables observers of NBA basketball to differentiate streak shooters from the other players.

Tables 6 and 7 explore the same measures as Tables 4 and 5 with a contextual restriction. Context is defined as a sequence of 20 consecutive field goal attempts taken by all players in a game. In a game in which all players attempted N

field goals, the total number of 20 shot contextual sequences is $N - 20 + 1$. The numerator for each entry is the number of times that a player accomplishes the sequence of a given length in context. The denominator for each entry in Tables 6 and 7 is an expectation: the number of shot opportunities times the probability of a player taking T or more shots (where T is greater than or equal to the sequence length, L) in a 20 field goal context and of making r of L shots regardless of position in the T shots.

For example, the expectation in the denominators of entries for Table 6 is:

$$P_i^L\left\{\sum_{j=L}^{m}\binom{m}{j}\gamma_i^j(1-\gamma_i)^{m-j}(j+1-L)\right\}S$$

where

P_i = Probability of a hit given a shot by player i.

L = Length of run.

m = Number of possible shots (size of content) = 20.

γ_i = Probability of player i taking a shot.

G = Number of games.

$$S = \sum_{g=1}^{G}(A_g - m + 1)$$

where

A_g = All field goal attempts in game g.

The probability of any player taking the next shot is a rough approximation using averages across games in our data. The estimate is the proportion of total field goals the player's team takes on average times the proportion of his team's field goals that the player takes on average. While it would be better to build the model conditional on information as to which specific players are on the floor for each particular shot opportunity, this approach is much more complicated and requires data that was not coded. An examination of season data on minutes played and shots taken by the players analyzed here indicates no substantial bias from this simplification.

As hypothesized, Vinnie Johnson does the highly unlikely, highly noticeable, and memorable in shooting field goals much more frequently than the rest of the players examined. We also explored shooting streaks in spans of 10 and 15 field goals of context. In general, the results strengthen further: The purer the sequence (fewer misses), the longer the sequence of hits, and the briefer the context, the more distinctive Vinnie is from other players in this data set.

We will leave it to the reader to apply Berkson's Interocular Traumatic Test in examining Tables 6 and 7. We conclude by this test that Vinnie is different from the other players, particularly on the most improbable and most memorable feats. Players who looked interesting as potential streak shooters in the acontextual analysis such as Dennis Rodman disappear almost completely when context is considered. In examining the data, the reason for the disappearance is obvious. Rodman's 9 for 9 streak fell in the following positions of field goal attempts in that game: 72, 86, 120, 126, 136, 146, 150, 156, 160 (i.e., Rodman took and made the 72nd field goal, the 86th field goal. . .).

Table 6. Perfect Shot Sequences in 20 FG Context—"Occurrences to Expectations"

Player	8/8	7/7	6/6	5/5	4/4	3/3
Jordan	—	—	—	1.11	.67	.49
Bird	—	—	—	5.50	2.52	1.09
McHale	—	—	—	—	1.31	.93
Parish	—	—	—	—	6.93	3.51
D. Johnson	—	—	—	—	2.43	1.35
Ainge	—	—	—	—	3.26	3.08
D. Wilkins	—	—	—	—	.22	.29
E. Johnson	—	—	—	—	.84	.83
A-Jabbar	—	—	82.93	15.33	2.63	1.50
Worthy	—	—	—	2.52	2.12	1.27
Scott	—	—	16.40	3.94	1.74	1.11
Aguirre	—	—	—	—	1.69	.92
Dantley	—	—	—	—	5.48	2.01
Laimbeer	—	—	—	—	7.74	2.62
Dumars	—	—	180.80	28.37	4.14	1.52
Thomas	—	—	14.86	5.21	1.98	.86
V. Johnson	—	4645.41	566.01	81.69	18.67	4.32
Rodman	—	—	—	—	33.62	10.42

Table 7. Imperfect Shot Sequences in 20 FG Context—"Occurrences to Expectations"

Player	7/8	6/7	5/6	4/5	3/4	2/3
Jordan	—	—	1.12	1.09	.90	.68
Bird	—	23.34	5.70	3.67	1.16	.67
McHale	—	—	3.90	3.08	1.80	1.19
Parish	—	—	—	10.91	6.88	1.95
D. Johnson	—	—	—	—	.42	1.00
Ainge	—	—	—	5.70	4.13	2.06
D. Wilkins	—	—	.76	.87	.69	.51
E. Johnson	—	—	—	2.21	1.42	1.04
A-Jabbar	—	—	—	1.36	1.36	.89
Worthy	—	—	—	2.24	1.49	.97
Scott	—	—	5.92	2.56	1.50	.90
Aguirre	—	—	—	2.88	1.80	1.01
Dantley	—	—	—	6.33	3.29	1.82
Laimbeer	—	—	62.43	25.89	5.83	2.10
Dumars	—	—	24.65	2.32	1.98	1.62
Thomas	—	17.26	9.73	2.73	1.50	.90
V. Johnson	10050.27	1131.63	118.41	41.75	11.13	3.01
Rodman	—	—	—	—	13.71	5.15

Photo courtesy of Joe Labolito, Temple University Photography Department, © 2004.

Conclusion

At least one *streak shooter* with an occasional *hot hand* was alive and well and living in Detroit during the 1987–1988 season. While 1987–1988 was not his best year and we cannot from this research support him as "the most lethal streak shooter in basketball," Vinnie Johnson's reputation as a streak shooter is apparently well deserved; he is different from other players in the data in terms of noticeable, memorable field goal shooting accomplishments.

The coaches, players, and fans accused of "misconceiving chance processes" stand somewhat vindicated. At least in the case of the Microwave and our other players for this data set, observers are able to notice and remember improbable shooting feats. They can also apparently make proper reputational attributions to those players who do the improbable and memorable more regularly than other players.

Attributing error in reasoning about chance processes requires at the outset that you know the correct model for the observations about which subjects are reasoning. Before you can identify errors in reasoning and explain those errors as the product of a particular style of erroneous reasoning, you must first know the correct reasoning. It is much easier to know the correct model in an experimental setting than in a natural setting. In the experimental setting you can choose it. In a natural setting such as professional basketball you must first discover it.

Tversky and Gilovich employed a model, the set of simple binomials for isolated individual player shooting sequences, in their data analysis that could not and, indeed, should not be used by observers of NBA games in understanding sequences of shots in real games. Their model uses shooting data in a form that knowledgeable fans who believe in hot hands and streak shooting never encounter.

The binomials for isolated player sequences are not very useful models for formulating or critiquing game strategies. The model cannot be used to reproduce the sequence of shots in a basketball game; there is nothing in the model to indicate which player shoots next. This model, which assumes that defensive responses are irrelevant to a player's shooting success, has some strange strategic implications. For example, the model suggests the optimal strategy for allocating field goal shots is to have the player on a team with the highest shooting percentage take all shots. Coaches such as Red Auerbach and Johnny Wooden probably never even considered such a strategy. If they had considered and adopted such a strategy there would be many fewer championship banners hanging in the rafters of the Boston Garden and Pauly Pavilion.

Basketball fans and coaches who once believed in the hot hand and streak shooting and who have been worried about the adequacy of their cognitive apparatus since the publication of Tversky and Gilovich's original work can relax and once again enjoy watching the game. It is even okay to admire the feats of a Vinnie Johnson and to think about him as fundamentally different from other shooters, even great shooters like Larry Bird, Michael Jordan, and Isiah Thomas.

Additional Reading

Gilovich, T., Vallone, R., and Tversky, A. (1985), "The Hot Hand in Basketball: On the Misperception of Random Sequences," *Cognitive Psychology*, 17, 295–314.

Tversky, A. and Gilovich, T. (1989), "The Cold Facts about the 'Hot Hand' in Basketball," *Chance*, 2(1), 16–21.

Chapter 20

More Probability Models for the NCAA Regional Basketball Tournaments

Neil C. SCHWERTMAN, Kathryn L. SCHENK, and Brett C. HOLBROOK

Sports events and tournament competitions provide excellent opportunities for model building and using basic statistical methodology in an interesting way. In this article, National Collegiate Athletic Association (NCAA) regional basketball tournament data are used to develop simple linear regression and logistic regression models using seed position for predicting the probability of each of the 16 seeds winning the regional tournament. The accuracy of these models is assessed by comparing the empirical probabilities not only to the predicted probabilities of winning the regional tournament but also the predicted probabilities of each seed winning each contest.

KEY WORDS: Basketball; Logistic regression; Regression.

1. INTRODUCTION

Enthusiasm for the study of probability is enhanced when the concepts are illustrated by real examples of interest to students. Athletic competitions afford many such opportunities to demonstrate the concepts of probability and have been extensively studied in the literature; see, for example, Mosteller (1952), Searls (1963), Moser (1982), Monahan and Berger (1977), David (1959), Glenn (1960), Schwertman, McCready, and Howard (1991), and Ladwig and Schwertman (1992). One excellent probability analysis opportunity for use in the classroom occurs each spring when "March Madness," as the media calls it, occurs. "March Madness" is the National Collegiate Athletic Association (NCAA) regional and Final Four basketball tournaments that culminate in a National Collegiate Championship game. The NCAA selects (actually, certain conference champions or tournament winners are included automatically) 64 teams, 16 for each of 4 regions, to compete for the national championship. The NCAA committee of experts not only selects the 64 teams from 292 teams in Division 1-A, but assigns a seed position to each team in the four regions based on their consensus of team strengths. The format for each regional tournament is predetermined following the pattern in Figure 1, where the number one seed (strongest team) plays the sixteenth seed (weakest team), the number two seed (next strongest team) plays the fifteenth seed (second weakest), etc. The experts attempt to evenly distribute the teams to the regional tournament to achieve parity in the quality of each region.

Schwertman et al. (1991) suggested three rather ad hoc probability models that predicted remarkably well the empirical probability of each seed winning its regional tournament and advancing to the "final four." The validity of the three models was measured only by each seed's probability of winning its regional tournament. In this article we use the NCAA regional basketball tournament data as an example to illustrate ordinary least squares and logistic regression in developing prediction models. The parameter estimates for the simple models considered are based on the 600 games played (1985–1994) during the first ten years using the 64-team format. Validity of the eight new empirical and the three previous models in Schwertman et al. (1991) are assessed by comparing the empirical probabilities not only to the predicted probabilities for each seed winning the regional tournament but also to the predicted probabilities of each seed winning each contest.

2. TOURNAMENT ANALYSIS

Predicting the probability of each seed winning the regional tournament (and advancing to the final four) requires the consideration of all possible paths and opponents. Even though there are 16 teams in each region, the single elimination format (only the winning team survives in the tournament, i.e., one loss and the team is eliminated) is relatively easy to analyze compared to a double-elimination format. [See, for example, the analysis of the college baseball world series by Ladwig and Schwertman (1992).] In the first game each seed has only one possible opponent, but in the second game there are two possible opponents, four possible in the third game and eight possible in the regional finals. Hence there are $1 \cdot 2 \cdot 4 \cdot 8 = 64$ potential sets of opponents for each seed to play in order to eventually win the regional tournament. For example, for the number 2 seed to win, it must defeat seed 15 in game 5, either 7 or 10 in game 11, either 3, 14, 6, or 11 in game 14, and either 1, 16, 8, 9, 4, 13, 5, or 12 in game 15. The probability analysis for the regional championship must include not only the probability of defeating each potential opponent, but also the probability of each potential opponent advancing to that particular game. To illustrate, suppose the second seed wins the regional tournament by defeating seeds 15, 7, 6, and 1 in games 5, 11, 14, and 15, respectively. Then the probability that this occurs is $P(2,15) \cdot P(2,7) \cdot P(7$ plays in game 11$) \cdot P(2,6) \cdot P(6$ plays in game 14$) \cdot P(2,1) \cdot P(1$ plays in game 15$)$ where $P(i,j)$ is the probability that an ith seed defeats a jth seed and $P(j,i) = 1 - P(i,j)$. A more detailed explanation of the various paths and the associated probability analysis is contained in Schwertman et al. (1991). As in that article and most all such probability analyses of athletic competitions, we assume that the games are

Neil C. Schwertman is Chairman and Professor of Statistics, Department of Mathematics and Statistics, California State University, Chico, CA 95929-0525. Kathryn L. Schenk is Instructional Support Coordinator, Computer Center, California State University, Chico, CA 95929-0525. Brett C. Holbrook is Student, Department of Experimental Statistics, New Mexico State University, Las Cruces, NM 88003-0003.

independent and the probabilities remain constant throughout the tournament. To complete the analysis we now must find probability models for determining $P(i, j)$.

3. PROBABILITY MODELS

The purpose of the probability models is to incorporate the relative strength of the teams in estimating the probability of each team winning in each game. It seems reasonable to use some function of seed positions because these were determined by a consensus of experts. In order to have the broadest possible use in the classroom we use the simplest linear straight line model $E(Y) = \beta_0 + \beta_1(S(i) - S(j))$ where $S(i)$ is some function of i, the team's seed position. Clearly, multiple regression models could be used for more advanced classes, but our basic model is appropriate even for most introductory classes. In addition to the three models used in Schwertman et al. (1991), eight other models for assigning probabilities of success in each individual game are considered. The eight models are defined by the 2^3 possible combinations of three factors: (1) type of regression (ordinary or logistic), (2) type of intercept β_0 (estimated or specified constant), and (3) type of independent variable (linear or nonlinear function of seed positions).

There are obviously many functions of seed position that could be used in the models. The choice of $S(i)$ and $S(j)$ is quite arbitrary. We have chosen two rather simple functions for our investigation that fortunately provide excellent preditor models. The first is $S_1(i) = -i$ for all i, which is simply using the difference in seed position as a single independent variable. This function of seed position suggests a linearity in team strengths, for example, the difference in strength between seeds 1 and 3 is the same as between seeds 14 and 16. Intuitively it seems likely that there is a greater difference in quality between a number 1 and 3 seed than between a 14 and 16 seed. Thus this linearity may not be appropriate,

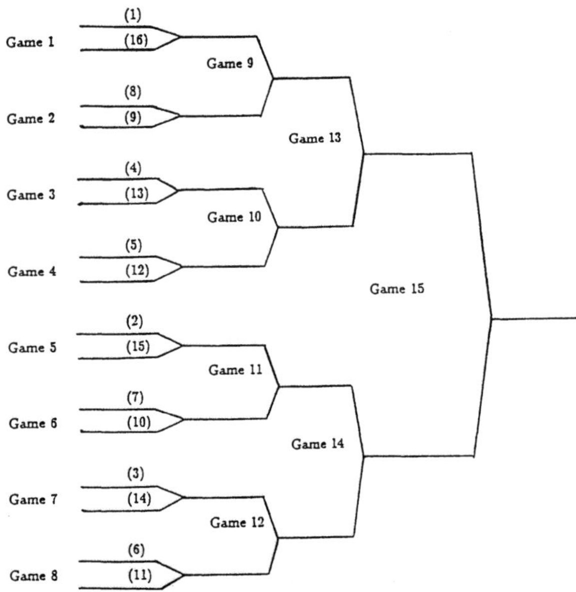

Figure 1. NCAA Regional Basketball Tournament Pairings.

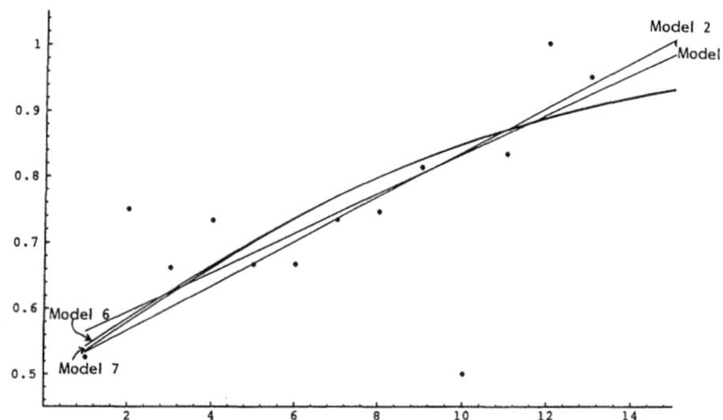

Figure 2. Points in Scatterplot May Represent as Few as 1 or as Many as 97 Games.

and therefore we consider a nonlinear function $S_2(i)$ for incorporating team strength. Since the normal distribution occurs naturally in describing many random variables and is included in most introductory classes, it seems reasonable to use the normal distribution to describe a nonlinear relationship in team strengths. If we assume that the strength of the 292 teams is normally distributed and that the experts properly ordered the top 64 teams for the tournament from 229 to 292 with 229 the weakest and 292 the strongest, we can then determine a percentile and a corresponding z score for that percentile. That is, adding a correction for continuity to 292, $S_2(i) = \Phi^{-1}((294.5 - 4i)/292.5)$ where Φ is the cumulative distribution function of the standard normal. For example, the transformed seed position z score for the number 1 seeds (289, 290, 291, 292 when ordered from weakest to strongest) was calculated from the percentile of this group's median, that is, $290.5/292.5 = 99.316$ percentile, which corresponds to a z score of 2.466. Similarly, the number 2 seed's (285, 286, 287, 288) z score is calculated from the $286.5/292.5 = 97.9487$ percentile, corresponding to a z score of 2.044. For seeds 3–16 the corresponding z scores are: 1.823, 1.666, 1.542, 1.438, 1.348, 1.267, 1.194, 1.127, 1.064, 1.006, .951, .898, .848, and .800, respectively.

The dependent variable is 1 if the lower seed (stronger team) defeats the higher seed and 0 otherwise. It should be noted that the logistic regression models (models 5–

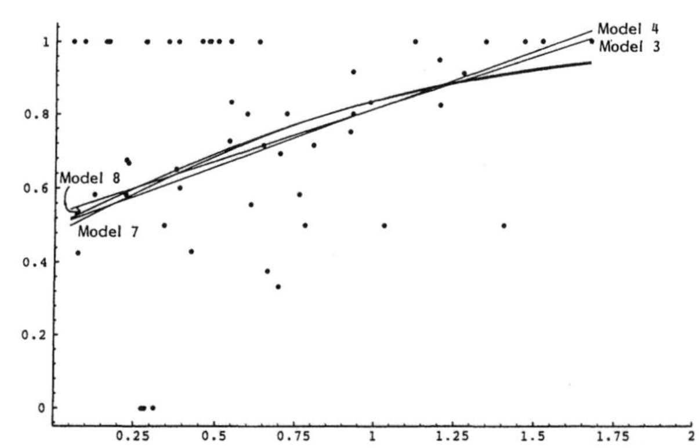

Figure 3. Points in Scatterplot Represent as Few as 1 or as Many as 40 Games.

Table 1. Empirical and Estimated Probabilities of Seed i Defeating Seed j in the NCAA Regional Basketball Tournament Games

Seed *i*	Seed *j*	Games played	Wins by lower seed no	empirical probability	$P_1(i,j)$	$P_2(i,j)$	$P_3(i,j)$	$P_4(i,j)$	$P_5(i,j)$	$P_6(i,j)$	$P_7(i,j)$	$P_8(i,j)$	$P_9(i,j)$	$P_{10}(i,j)$	$P_{11}(i,j)$
1	2	14	6	.429	.565	.534	.651	.634	.536	.543	.657	.666	.667	.531	.619
1	3	7	5	.714	.595	.567	.715	.704	.580	.586	.738	.742	.750	.563	.681
1	4	14	10	.714	.625	.601	.760	.754	.622	.627	.788	.788	.800	.594	.725
1	5	12	11	.917	.655	.635	.795	.793	.663	.666	.822	.820	.833	.625	.760
1	6	4	2	.500	.685	.669	.825	.826	.701	.703	.847	.844	.857	.656	.789
1	7	1	1	1.000	.715	.702	.851	.855	.737	.738	.866	.862	.875	.688	.815
1	8	17	14	.824	.745	.736	.874	.880	.770	.770	.882	.877	.889	.719	.837
1	9	23	21	.913	.775	.770	.895	.904	.800	.799	.894	.890	.900	.750	.858
1	10	2	2	1.000	.805	.804	.914	.925	.826	.826	.905	.900	.909	.781	.877
1	11	2	1	.500	.835	.837	.932	.945	.850	.849	.914	.909	.917	.813	.894
1	12	8	8	1.000	.864	.871	.949	.963	.871	.870	.921	.916	.923	.844	.911
1	13	1	1	1.000	.894	.905	.964	.981	.890	.888	.928	.923	.929	.875	.926
1	16	40	40	1.000	.984	1.000	1.000	1.000	.932	.930	.944	.939	.941	.969	.969
2	3	12	7	.583	.565	.534	.594	.570	.536	.543	.575	.590	.600	.531	.562
2	4	1	1	1.000	.595	.567	.639	.620	.580	.586	.640	.650	.667	.563	.606
2	6	9	5	.556	.655	.635	.704	.692	.663	.666	.726	.730	.750	.625	.671
2	7	26	18	.692	.685	.669	.730	.721	.701	.703	.756	.758	.778	.656	.696
2	8	2	1	.500	.715	.702	.753	.747	.737	.738	.781	.782	.800	.688	.719
2	10	12	9	.750	.775	.770	.793	.791	.800	.799	.820	.818	.833	.750	.758
2	11	6	5	.833	.805	.804	.811	.811	.826	.826	.836	.833	.846	.781	.776
2	15	40	38	.950	.924	.939	.873	.879	.906	.904	.881	.877	.882	.906	.837
3	4	1	1	1.000	.565	.534	.575	.550	.536	.543	.547	.564	.571	.531	.544
3	5	1	1	1.000	.595	.567	.611	.589	.580	.586	.600	.613	.625	.563	.579
3	6	20	12	.600	.625	.601	.641	.622	.622	.627	.643	.653	.667	.594	.608
3	7	4	4	1.000	.655	.635	.666	.651	.663	.666	.678	.686	.700	.625	.634
3	9	1	1	1.000	.715	.702	.711	.700	.737	.738	.734	.737	.750	.688	.677
3	10	3	1	.333	.745	.736	.730	.721	.770	.770	.756	.758	.769	.719	.696
3	11	12	7	.583	.775	.770	.748	.741	.800	.799	.776	.776	.786	.750	.714
3	14	40	32	.800	.864	.871	.795	.793	.871	.870	.822	.820	.824	.844	.760
4	5	24	14	.583	.565	.534	.566	.539	.536	.543	.533	.551	.556	.531	.535
4	6	3	2	.667	.595	.567	.596	.572	.580	.586	.578	.592	.600	.563	.564
4	8	2	1	.500	.655	.635	.645	.627	.663	.666	.648	.658	.667	.625	.612
4	9	1	1	1.000	.685	.669	.666	.650	.701	.703	.677	.684	.692	.656	.633
4	10	1	1	1.000	.715	.702	.685	.671	.737	.738	.702	.708	.714	.688	.652
4	12	8	3	.375	.775	.770	.719	.709	.800	.799	.744	.747	.750	.750	.686
4	13	40	32	.800	.805	.804	.735	.727	.826	.826	.762	.764	.765	.781	.701
5	8	1	0	.000	.625	.601	.609	.587	.622	.627	.598	.611	.615	.594	.577
5	9	1	1	1.000	.655	.635	.630	.610	.663	.666	.628	.639	.643	.625	.598
5	12	40	29	.725	.745	.736	.684	.670	.770	.770	.701	.707	.706	.719	.651
5	13	5	4	.800	.775	.770	.700	.688	.800	.799	.720	.725	.722	.750	.666
6	7	3	3	1.000	.565	.534	.556	.529	.536	.543	.518	.537	.538	.531	.525
6	10	1	0	.000	.655	.635	.619	.599	.663	.666	.613	.625	.625	.625	.588
6	11	40	26	.650	.685	.669	.638	.619	.701	.703	.638	.649	.647	.656	.605
6	14	6	5	.833	.775	.770	.685	.671	.800	.799	.702	.708	.700	.750	.652
7	10	40	27	.675	.625	.601	.594	.570	.622	.627	.575	.590	.588	.594	.562
7	11	1	0	.000	.655	.635	.612	.590	.663	.666	.601	.614	.611	.625	.580
7	14	1	1	1.000	.745	.736	.659	.643	.770	.770	.668	.677	.667	.719	.627
7	15	1	1	1.000	.775	.770	.674	.659	.800	.799	.687	.694	.682	.750	.641
8	9	40	17	.425	.565	.534	.551	.523	.536	.543	.511	.530	.529	.531	.521
10	15	1	1	1.000	.685	.669	.610	.589	.701	.703	.599	.612	.600	.656	.579
11	14	2	2	1.000	.625	.601	.578	.553	.622	.627	.551	.568	.560	.594	.547
12	13	3	3	1.000	.565	.534	.546	.517	.536	.543	.503	.523	.520	.531	.515
		χ_1^2			51.899	51.388	52.168	53.443	57.925	57.924	56.986	56.963	60.103	52.022	52.551
		χ_2^2			26.849	25.688	20.971	19.818	31.235	31.406	24.980	25.905	27.651	26.779	24.051

NOTE: χ_1^2 based on all 52 pairings of seeds that have occurred. χ_2^2 based only on the 26 pairings that have had at least 5 games.

8) are equivalent to linear regression models using log odds, for example, model 5 is equivalent to $\log[P(i,j)/(1-P(i,j))] = .0328 - .177(j-i)$. It is intuitively appealing to use .5 for the y intercept in the ordinary regression models (models 2 and 4) since, if $i = j$ (two clones play), we would want the probability of a win by either to be a half. Similarly, for the logistic regression we would want the y intercept to be zero (models 6 and 8) since, if $i = j$, then the ratio of $p/q = 1$ implies that the probability is the intuitive .5.

The eight new models with the estimated parameters are

$$P_1(i,j) = .535329 + .029922(j-i)$$
$$P_2(i,j) = .5 + .033746(j-i)$$

Table 2. Predicted Probabilities of Each Seed Winning the NCAA Regional Basketball Tournament

	Model										
Seed	1	2	3	4	5	6	7	8	9	10	11
P(1) =	.309	.293	.524	.526	.295	.298	.508	.504	.519	.275	.459
P(2) =	.224	.219	.195	.194	.225	.226	.207	.208	.216	.208	.188
P(3) =	.154	.157	.104	.104	.164	.163	.109	.110	.107	.154	.110
P(4) =	.107	.110	.060	.059	.112	.112	.060	.061	.057	.111	.068
P(5) =	.071	.077	.037	.038	.076	.075	.038	.038	.034	.080	.047
P(6) =	.050	.053	.028	.028	.050	.049	.025	.026	.022	.058	.036
P(7) =	.033	.035	.018	.018	.032	.031	.016	.017	.014	.040	.026
P(8) =	.020	.021	.010	.009	.019	.019	.010	.010	.009	.026	.015
P(9) =	.012	.013	.006	.006	.012	.011	.007	.007	.006	.017	.011
P(10) =	.008	.009	.006	.006	.007	.007	.006	.006	.005	.012	.011
P(11) =	.005	.006	.005	.005	.004	.004	.004	.004	.004	.008	.009
P(12) =	.003	.003	.003	.002	.002	.002	.003	.003	.003	.005	.006
P(13) =	.002	.002	.002	.002	.001	.001	.002	.002	.002	.003	.005
P(14) =	.001	.001	.002	.001	.001	.001	.002	.002	.002	.002	.004
P(15) =	.000	.000	.001	.001	.000	.000	.001	.001	.001	.001	.003
P(16) =	.000	.000	.000	.000	.000	.000	.001	.001	.001	.000	.001

$$P_3(i,j) = .530385 + .286507(S_2(i) - S_2(j))$$

$$P_4(i,j) = .5 + .317258(S_2(i) - S_2(j))$$

$$P_5(i,j) = 1/(1 + e^{.0328 - .1770(j-i)})$$

$$P_6(i,j) = 1/(1 + e^{-.1727(j-i)})$$

$$P_7(i,j) = 1/(1 + e^{.0847 - 1.7449(S_2(i) - S_2(j))})$$

$$P_8(i,j) = 1/(1 + e^{-1.6405(S_2(i) - S_2(j))}).$$

Figures 2 and 3 display the graphs of models 1–8 and a scatterplot of the data.

The other three models, 9–11, used by Schwertman et al. (1991), consist of one nonlinear type (model 9), $P_9(i,j) = j/(i+j)$; a linear type (model 10), $P_{10}(i,j) = .5 + .03125(S_1(i) - S_1(j))$; and one based on normal scoring using the $S_2(i)$ (model 11), $P_{11}(i,j) = .5 + .2813625(S_2(i) - S_2(j))$. For details see Schwertman et al. (1991).

Estimates of the unspecified parameter(s) in the first four models were obtained by ordinary least squares, while the next four models (5–8) were determined using the SAS logistic procedure. (See *SAS/STAT User's Guide,* Vol. 2, Version 6, 4th ed., pp. 1069–1126 for details.)

4. COMPARISON OF MODELS

The 11 different models for assigning probabilities of winning for each seed in each individual game were compared in three ways by using a chi-square statistic as a measure of the relative fit of the models. Of the possible 120 pairings of seeds $(16 \cdot 15/2)$ only 52 have occurred. Table 1 lists the pairs that have occurred, the number of games played between these seeds, the number of wins by the lower seed number (stronger team), and the empirical and estimated probabilities of the lower seed number winning from the 11 models. Using the empirical data for the seed pairings, a chi-square goodness-of-fit

$$\left(\chi^2 = \sum_i \sum_j ((X_{ij} - n_{ij}\hat{P}_{ij})^2 / n_{ij}\hat{P}_{ij}(1 - \hat{P}_{ij})\right)$$

for each of the 52 seed pairs was computed, and the sum of these chi-squares is given for each model. Twenty-six of the seed pairs had fewer than five games played, and the small expected numbers in these cells, being used as a divisor, may place too much emphasis on these cells and distort the chi-square values. Hence a second set of chi-square statistics based on just the 26 seed pairings with at least 5 games was computed and is also given in Table 1. Models 1–8 use the data to estimate the model parameters, and consequently these chi-square values are not entirely independent. Nevertheless, the chi-square values do provide a measure of the relative accuracy of the various models.

Table 3. Goodness of Fit Analysis

		Expected numbers Probability model number										
Group (seed no.)	Obs.	1	2	3	4	5	6	7	8	9	10	11
1	16	11.22	10.50	18.51	18.76	10.76	10.59	17.94	17.88	18.69	9.89	16.52
2	8	8.07	7.80	7.06	6.96	8.12	8.06	7.47	7.46	7.78	7.50	6.77
3	4	5.51	5.64	3.80	3.77	5.80	5.85	3.98	3.98	3.84	5.55	3.97
4	3	3.86	3.95	2.22	2.15	4.01	4.04	2.24	2.25	2.04	4.00	2.45
5 or more	5	7.34	8.11	4.40	4.36	7.31	7.47	4.38	4.42	3.65	9.06	6.29
	$\chi^2_{(4)}$	3.3837	4.7865	.8276	1.0070	4.0921	4.4287	.5937	.5652	1.3545	6.3057	.6283
p value		.4958	.3099	.9347	.9087	.3937	.3511	.9638	.9669	.8521	.1775	.9599

Models 2, 3, and 4 produced $P(1, 16)$ that were greater than 1.0. When this occurred the probabilities were set to .99999 in order to compute the chi-square statistics.

The third comparison of the models was done by using $P(i, j)$ to compute the probabilities of each seed winning the regional tournament. These probabilities are displayed in Table 2, and a chi-square goodness-of-fit to the empirical probabilities, used for evaluating the 11 models, is given in Table 3.

5. CONCLUSIONS

For the set of 26 pairings with 5 games or more, both the ordinary and logistic models using the z scoring of seed number $S_2(i)$ had smaller chi-square values than the corresponding models using just the seed position. In many cases the z scoring substantially improved the fit of the predicted value to the empirical data, and seems to be a worthwhile technique.

Two unexpected results of the analysis occurred. The first is that when predicting the probability of success in each game, $P(i, j)$, the regressions with the y intercept specified (.5 for ordinary least squares and zero for logistic regression) occasionally provided somewhat smaller chi-square values than the unrestricted models. Because least squares minimizes the squared deviations between predicted and observed values it was anticipated that the unrestricted models (1, 3, 5, 7) would have slightly smaller chi-square values than the corresponding restricted models (2, 4, 6, 8). The chi-square statistic, however, is a weighted sum of these squared deviations, and hence is not necessarily a minimum when the unweighted sum is minimized.

The second unexpected result was that the models that were best (the smaller chi-square values) at predicting $P(i, j)$ for each game did not do as well as some of the other models at predicting the overall regional tournament champion. Models 7 and 8 (logistic, with and without intercept, z-scored seeds) were the best at predicting the regional winner but were about in the middle (when ranked) of the models for predicting individual games. On the other hand, models 3 and 4 (ordinary least squares, with or without specified intercept, z-scored seeds) were the best prediction models for the individual games, but only fourth or fifth best for predicting the regional champion.

If the objective is to develop a model for predicting the winner between various seed pairs, then model 3 (ordinary least squares, no specified intercept, z-scored seeds) seems to be the most satisfactory, whereas if the objective is to predict the regional winner, then the logistic models, 7 and 8 (logistic regression, with and without intercept, z-scored seeds), are the most satisfactory models. Interestingly, the ad hoc model 11 seemed to be very adequate at predicting both.

Obviously there are numerous models that could be used. We have focused on the simplest straight line models and elementary methods of incorporating team strength in order to make the methodology accessible to a broad spectrum of students. We have attempted to present an application of ordinary least squares, logistic regression, and probability that should be of interest to many students. The ever-increasing interest in "March Madness" can be used to motivate and stimulate this instructive, timely application of several principles and methods of probability and statistics. Students seem to learn better when they can see application of the subject to something of interest to them. We believe that this analysis of the regional basketball tournaments can promote student learning and enthusiasm for studying probability and statistics.

[Received August 1993. Revised October 1994.]

REFERENCES

David, H. A. (1959), "Tournaments and Paired Comparisons," *Biometrika*, 46, 139–149.

Glenn, W. A. (1960), "A Comparison of the Effectiveness of Tournaments," *Biometrika*, 47, 253–262.

Ladwig, J. A., and Schwertman, N. C. (1992), "Using Probability and Statistics to Analyze Tournament Competitions, *Chance*, 5, 49–53.

Monahan, J. P., and Berger, P. D. (1977), "Playoff Structures in the National Hockey League," in *Optimal Strategies in Sports*, eds. S. P. Ladany and R. E. Machol, Amsterdam: North-Holland, pp. 123–128.

Moser, L. E. (1982), "A Mathematical Analysis of the Game of Jai Alai," *The American Mathematical Monthly*, 89, 292–300.

Mosteller, F. (1952), "The World Series Competition," *Journal of the American Statistical Association*, 47, 355–380.

Schwertman, N. C., McCready, T. A., and Howard, L. (1991), "Probability Models for the NCAA Regional Basketball Tournaments," *The American Statistician*, 45, 35–38.

Searls, D. T. (1963), "On the Probability of Winning with Different Tournament Procedures," *Journal of the American Statistical Association*, 58, 1064–1081.

Chapter 21

The Cold Facts About the "Hot Hand" in Basketball

Do basketball players tend to shoot in streaks? Contrary to the belief of fans and commentators, analysis shows that the chances of hitting a shot are as good after a miss as after a hit.

Amos Tversky and Thomas Gilovich

You're in a world all your own. It's hard to describe. But the basket seems to be so wide. No matter what you do, you know the ball is going to go in.

—Purvis Short, of the NBA's Golden State Warriors

This statement describes a phenomenon known to everyone who plays or watches the game of basketball, a phenomenon known as the "hot hand." The term refers to the putative tendency for success (and failure) in basketball to be self-promoting or self-sustaining. After making a couple of shots, players are thought to become relaxed, to feel confident, and to "get in a groove" such that subsequent success becomes more likely. The belief in the hot hand, then, is really one version of a wider conviction that "success breeds success" and "failure breeds failure" in many walks of life. In certain domains it surely does—particularly those in which a person's reputation can play a decisive role. However, there are other areas, such as most gambling games, in which the belief can be just as strongly held, but where the phenomenon clearly does not exist.

What about the game of basketball? Does success in this sport tend to be self-promoting? Do players occasionally get a "hot hand"?

Misconceptions of Chance Processes

One reason for questioning the widespread belief in the hot hand comes from research indicating that people's intuitive conceptions of randomness do not conform to the laws of chance. People commonly believe that the essential characteristics of a chance process are represented not only globally in a large sample, but also locally in each of its parts. For example, people expect even short sequences of heads and tails to reflect the fairness of a coin and to contain roughly 50% heads and 50% tails. Such a locally representative sequence, however, contains too many alternations and not enough long runs.

This misconception produces two systematic errors. First, it leads many people to believe that the probability of heads is greater after a long sequence of tails than after a long sequence of heads; this is the notorious gamblers' fallacy. Second, it leads people to question the randomness of sequences that contain the expected number of runs because even the occurrence of, say, four heads in a row—which is quite likely in even relatively small samples—makes the sequence appear nonrepresentative. Random sequences just do not look random.

Perhaps, then, the belief in the hot hand is merely one manifestation of this fundamental misconception of the laws of chance. Maybe the streaks of consecutive hits that lead players and fans to believe in the hot hand do not exceed, in length or frequency, those expected in any random sequence.

To examine this possibility, we first asked a group of 100 knowl-

Photos courtesy of Joe Labolito, Temple University Photography Department, © 2004.

Photo by Marcy Dubroff © 1986

Does Cornell senior Mike Pascal have a "hot hand"?

edgeable basketball fans to classify sequences of 21 hits and misses (supposedly taken from a basketball player's performance record) as *streak shooting*, *chance shooting*, or *alternating shooting*. Chance shooting was defined as runs of hits and misses that are just like those generated by coin tossing. Streak shooting and alternating shooting were defined as runs of hits and misses that are longer or shorter, respectively, than those observed in coin tossing. All sequences contained 11 hits and 10 misses, but differed in the probability of alternation, $p(a)$, or the probability that the outcome of a given shot would be different from the outcome of the previous shot. In a random (i.e., independent) sequence, $p(a) = .5$; streak shooting and alternating shooting arise when $p(a)$ is less than or greater than .5, respectively. Each respondent evaluated six sequences, with $p(a)$ ranging from

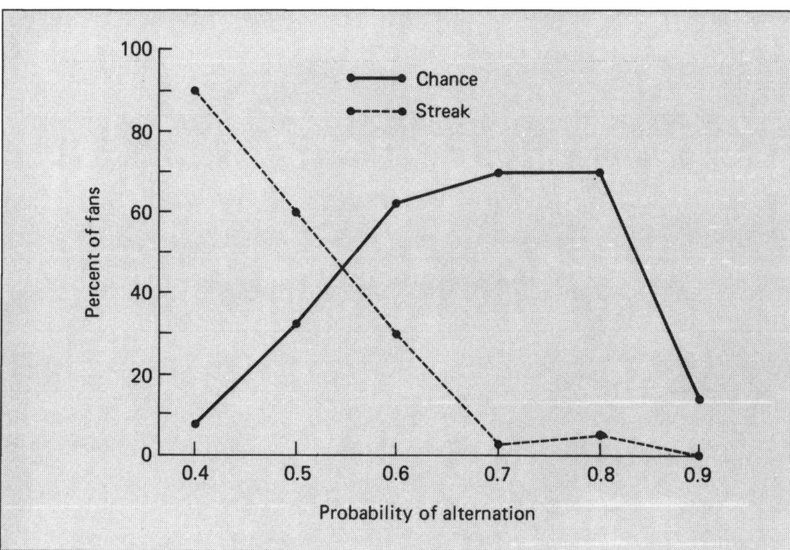

Figure 1. Percentage of basketball fans classifying sequences of hits and misses as examples of streak shooting or chance shooting, as a function of the probability of alternation within the sequences.

.4 to .9. Two (mirror image) sequences were used for each level of $p(a)$ and presented to different respondents.

The percentage of respondents who classified each sequence as "streak shooting" or "chance shooting" is presented in Figure 1 as a function of $p(a)$. (The percentage of "alternating shooting" is the complement of these values.) As expected, people perceive streak shooting where it does not exist. The sequence of $p(a) = .5$, representing a perfectly random sequence, was classified as streak shooting by 65% of the respondents. Moreover, the perception of chance shooting was strongly biased against long runs: The sequences selected as the best examples of chance shooting were those with probabilities of alternation of .7 and .8 instead of .5.

It is clear, then, that a common misconception about the laws of chance can distort people's observations of the game of basketball: Basketball fans "detect" evidence of the hot hand in perfectly random sequences. But is this the main determinant of the widespread conviction that basketball players shoot in streaks? The answer to this question requires an analysis of shooting statistics in real basketball games.

Cold Facts from the NBA

Although the precise meaning of terms like "the hot hand" and "streak shooting" is unclear, their common use implies a shooting record that departs from coin tossing in two essential respects (see accompanying box). First, the frequency of streaks (i.e., moderate or long runs of successive hits) must exceed what is expected by a chance process with a constant hit rate. Second, the probability of a hit should be greater following a hit than following a miss, yielding a positive serial correlation between the outcomes of successive shots.

To examine whether these patterns accurately describe the performance of players in the NBA, the field-goal records of individual players were obtained for 48 home games of the Philadelphia 76ers during the 1980–81 season. Table 1 presents, for the nine major players of the 76ers, the probability of a hit conditioned on 1, 2, and 3 hits and misses. The overall hit rate for each player, and the number of shots he took, are presented in column 5. A comparison of columns 4 and 6 indicates that for eight of the nine players the probability of a hit is actually higher following a miss (mean = .54) than following a hit (mean = .51), contrary to the stated beliefs of both players and fans. Column 9 presents the (serial) correlations between the outcomes of successive shots. These correlations are not significantly different than zero except for one player (Dawkins) whose correlation is negative. Comparisons of the other matching columns (7 vs. 3, and 8 vs. 2) provide further evidence against streak shooting. Additional analyses show that the probability of a hit (mean = .57) following a "cold" period (0 or 1 hits in the last 4 shots) is higher than the probability of a hit (mean = .50) following a "hot" period (3 or 4 hits in the last 4 shots). Finally, a series of Wald-Wolfowitz runs tests revealed that the observed number of runs in the players' shooting records does not depart from chance expectation except for one player (Dawkins) whose data, again, run counter to the streak-shooting hypothesis. Parallel analyses of data from two other teams, the New Jersey Nets and the New York Knicks, yielded similar results.

Although streak shooting entails a positive dependence between the outcomes of successive shots, it could be argued that both the runs test and the test for a positive correlation are not sufficiently powerful to detect occasional "hot" stretches embedded in longer stretches of normal performance. To obtain a more sensitive test of stationarity (suggested by David Freedman) we partitioned the entire record of each player into non-overlapping series of four consecutive shots. We then counted the number of series in which the player's performance was high (3 or 4 hits), moderate (2

Table 1. Probability of making a shot conditioned on the outcome of previous shots for nine members of the Philadelphia 76ers; hits are denoted *H*, misses are *M*.

Player	$P(H/3M)$	$P(H/2M)$	$P(H/1M)$	$P(H)$	$P(H/1H)$	$P(H/2H)$	$P(H/3H)$	Serial Correlation r
Clint Richardson	.50	.47	.56	.50 (248)	.49	.50	.48	−.020
Julius Erving	.52	.51	.51	.52 (884)	.53	.52	.48	.016
Lionel Hollins	.50	.49	.46	.46 (419)	.46	.46	.32	−.004
Maurice Cheeks	.77	.60	.60	.56 (339)	.55	.54	.59	−.038
Caldwell Jones	.50	.48	.47	.47 (272)	.45	.43	.27	−.016
Andrew Toney	.52	.53	.51	.46 (451)	.43	.40	.34	−.083
Bobby Jones	.61	.58	.58	.54 (433)	.53	.47	.53	−.049
Steve Mix	.70	.56	.52	.52 (351)	.51	.48	.36	−.015
Darryl Dawkins	.88	.73	.71	.62 (403)	.57	.58	.51	−.142*
Weighted Mean =	.56	.53	.54	.52	.51	.50	.46	−.039

NOTE: The number of shots taken by each player is given in parentheses in Column 5.

*$p<.01$

hits) or low (0 or 1 hits). If a player is occasionally "hot," his record must include more high-performance series than expected by chance. The numbers of high, moderate, and low series for each of the nine Philadelphia 76ers were compared to the expected values, assuming independent shots with a constant hit rate (taken from column 5 of Table 1). For example, the expected percentages of high-, moderate-, and low-performance series for a player with a hit rate of .50 are 31.25%, 37.5%, and 31.25%, respectively. The results provided no evidence for non-stationarity or streak shooting as none of the nine chi-squares approached statistical significance. The analysis was repeated four times (starting the partition into quadruples at the first, second, third, and fourth shot of each player), but the results were the same. Combining the four analyses, the overall observed percentages of high, medium, and low series are 33.5%, 39.4%, and 27.1%, respectively, whereas the expected percentages are 34.4%, 36.8%, and 28.8%. The aggregate data yield slightly fewer high and low series than expected by independence, which is the exact opposite of the pattern implied by the presence of hot and cold streaks.

At this point, the lack of evidence for streak shooting could be attributed to the contaminating effects of shot selection and defensive strategy. Streak shooting may exist, the argument goes, but it may be masked by a hot player's tendency to take more difficult shots and to receive more attention from the defensive team. Indeed, the best shooters on the team (e.g., Andrew Toney) do not have the highest hit rate, presumably because they take more difficult shots. This argument however, does not explain why players and fans erroneously believe that the probability of a hit is greater following a hit than following a miss, nor can it account for the tendency of knowledgeable observers to classify random sequences as instances of streak shooting. Nevertheless, it is instructive to examine the performance of players when the difficulty of the shot and the defensive pressure are held constant. Free-throw records provide such data. Free throws are shot, usually in pairs, from the same location and without defensive pressure. If players shoot in streaks, their shooting percentage on the second free throws should be higher after having made their first shot than after having missed their first

Table 2. Probability of hitting a second free throw (H_2) conditioned on the outcome of the first free throw (H_1 or M_1) for nine members of the Boston Celtics.

Player	$P(H_2/M_1)$	$P(H_2/H_1)$	Serial Correlation r
Larry Bird	.91 (53)	.88 (285)	−.032
Cedric Maxwell	.76 (128)	.81 (302)	.061
Robert Parish	.72(105)	.77 (213)	.056
Nate Archibald	.82 (76)	.83 (245)	.014
Chris Ford	.77 (22)	.71 (51)	−.069
Kevin McHale	.59 (49)	.73 (128)	.130
M.L. Carr	.81 (26)	.68 (57)	−.128
Rick Robey	.61 (80)	.59 (91)	−.019
Gerald Henderson	.78 (37)	.76 (101)	−.022

NOTE: The number of shots on which each probability is based is given in parentheses.

shot. Table 2 presents the probability of hitting the second free throw conditioned on the outcome of the first free throw for nine Boston Celtics players during the 1980–81 and the 1981–82 seasons.

These data provide no evidence that the outcome of the second shot depends on the outcome of the first. The correlation is negative for five players and positive for the remaining four, and in no case does it approach statistical significance.

The Cold Facts from Controlled Experiments

To test the hot hand hypothesis, under controlled conditions, we recruited 14 members of the men's varsity team and 12 members of the women's varsity team at Cornell University to participate in a shooting experiment. For each player, we determined a distance from which his or her shooting percentage was roughly 50%, and we drew two 15-foot arcs at this distance from which the player took 100 shots, 50 from each arc. When shooting baskets, the players were required to move along the arc so that consecutive shots were never taken from exactly the same spot.

The analysis of the Cornell data parallels that of the 76ers. The overall probability of a hit following a hit was .47, and the probability of a hit following a miss was .48. The serial correlation was positive for 12 players and negative for 14 (mean r = .02). With the exception of one player (r = .37) who produced a significant positive correlation (and we might expect one significant result out of 26 just by chance), both the serial correlations and the distribution of runs indicated that the outcomes of successive shots are statistically independent.

We also asked the Cornell players to predict their hits and misses

What People Mean by the "Hot Hand" and "Streak Shooting"

Although all that people mean by streak shooting and the hot hand can be rather complex, there is a strong consensus among those close to the game about the core features of non-stationarity and serial dependence. To document this consensus, we interviewed a sample of 100 avid basketball fans from Cornell and Stanford. A summary of their responses are given below. We asked similar questions of the players whose data we analyzed—members of the Philadelphia 76ers—and their responses matched those we report here.

Does a player have a better chance of making a shot after having just made his last two or three shots than he does after having just missed his last two or three shots?

Yes 91%
No 9%

When shooting free throws, does a player have a better chance of making his second shot after making his first shot than after missing his first shot?

Yes 68%
No 32%

Is it important to pass the ball to someone who has just made several (2, 3, or 4) shots in a row?

Yes 84%
No 16%

Consider a hypothetical player who shoots 50% from the field.

What is your estimate of his field goal percentage for those shots that he takes after having just made a shot?

Mean = 61%

What is your estimate of his field goal percentage for those shots that he takes after having just missed a shot?

Mean = 42%

by betting on the outcome of each upcoming shot. Before every shot, each player chose whether to bet *high*, in which case he or she would win 5 cents for a hit and lose 4 cents for a miss, or to bet *low*, in which case he or she would win 2 cents for a hit and lose 1 cent for a miss. The players were advised to bet high when they felt confident in their shooting ability and to bet low when they did not. We also obtained betting data from another player who observed the shooter and decided, independently, whether to bet high or low on each trial. The players' payoffs included the amount of money won or lost on the bets made as shooters and as observers.

The players were generally unsuccessful in predicting their performance. The average correlation between the shooters' bets and their performance was .02, and the highest positive correlation was .22. The observers were also unsuccessful in predicting the shooter's performance (mean r = .04). However, the bets made by both shooters and observers *were* correlated with the outcome of the shooters' previous shot (mean r = .40 for the shooters and .42 for the observers). Evidently, both shooters and observers relied on the outcome of the previous shot in making their predictions, in accord with the hot-hand hypothesis. Because the correlation between successive shots was negligible (again, mean r = .02), this betting strategy was not superior to chance, although it did produce moderate agreement between the bets of the shooters and the observers (mean r = .22).

The Hot Hand as Cognitive Illusion

To summarize what we have found, we think it may be helpful to clarify what we have *not* found. Most importantly, our research does not indicate that basketball shooting is a purely chance process, like coin tossing. Obviously, it requires a great deal of talent and skill. What we have found is that, contrary to common belief, a player's chances of hitting are largely independent of the outcome of his or her previous shots. Naturally, every now and then, a player may make, say, nine of ten shots, and one may wish to claim—after the fact—that he was hot. Such use, however, is misleading if the length and frequency of such streaks do not exceed chance expectation.

Our research likewise does not imply that the number of points that a player scores in different games or in different periods within a game is roughly the same. The data merely indicate that the probability of making a given shot (i.e., a player's shooting *percentage*) is unaffected by the player's prior performance. However, players' willingness to shoot may well be affected by the outcomes of previous shots. As a result, a player may score more points in one period than in another not because he shoots better, but simply because he shoots more often. The absence of streak shooting does not rule out the possibility that other aspects of a player's performance, such as defense, rebounding, shots attempted, or points scored, could be subject to hot and cold periods. Furthermore, the present analysis of basketball data does not say whether baseball or tennis players, for example, go through hot and cold periods. Our research does not tell us anything general about sports, but it does suggest a generalization about people, namely that they tend to "detect" patterns even where none exist, and to overestimate the degree of clustering in sports events, as in other sequential data. We attribute the discrepancy between the observed basketball statistics and the intuitions of highly interested and informed observers to a general misconception of the laws of chance that induces the expectation that random sequences will be far more balanced than they generally are, and creates the illusion that there are patterns or streaks in independent sequences.

This account explains both the formation and maintenance of the belief in the hot hand. If independent sequences are perceived as streak shooting, no amount of exposure to such sequences will convince the player, the coach, or the fan that the sequences are actually independent. In fact, the more basketball one watches, the more one encounters what appears to be streak shooting. This misconception of chance has direct consequences for the conduct of the game. Passing the ball to the hot player, who is guarded closely by the opposing team, may be a non-optimal strategy if other players who do not appear hot have a better chance of scoring. Like other cognitive illusions, the belief in the hot hand could be costly.

Additional Reading

Gilovich, T., Vallone, R., and Tversky, A. (1985). "The hot hand in basketball: On the misperception of random sequences." *Cognitive Psychology*, 17, 295–314.

Kahneman, D., Slovic, P., and Tversky, A. (1982). "Judgment under uncertainty: Heuristics and biases." New York: Cambridge University Press.

Tversky, A. and Kahneman, D. (1971). "Belief in the law of small numbers." *Psychological Bulletin*, 76, 105–110.

Tversky, A. and Kahneman, D. (1974). "Judgment under uncertainty: Heuristics and biases." *Science*, 185, 1124–1131.

Wagenaar, W. A. (1972). "Generation of random sequences by human subjects: A critical survey of literature." *Psychological Bulletin*, 77, 65–72.

Chapter 22

Simpson's Paradox and the Hot Hand in Basketball

Robert L. Wardrop

A number of psychologists and statisticians are interested in how laypersons make judgments in the face of uncertainties, assess the likelihood of coincidences, and draw conclusions from observation. This is an important and exciting area that has produced a number of interesting articles. This article uses an extended example to demonstrate that researchers need to use care when examining what laypersons believe. In particular, it is argued that the data available to laypersons may be very different from the data available to professional researchers. In addition, laypersons unfamiliar with a counterintuitive result, such as Simpson's paradox, may give the wrong interpretation to the pattern in their data. This paper gives two recommendations to researchers and teachers. First, take care to consider what data are available to laypersons. Second, it is important to make the public aware of Simpson's paradox and other counterintuitive results.

KEY WORDS: Hot hand phenomenon; McNemar's test; Multiple analyses; Simpson's paradox.

1. INTRODUCTION

Schoolchildren routinely learn to identify optical illusions. It is arguably as important that the general public learn to identify statistical illusions. Many outstanding researchers have addressed this issue. As examples, Diaconis and Mosteller (1989) investigate computing the probabilities of coincidences; Kahneman, Slovic, and Tversky (1983) consider judgments made in the presence of uncertainty; and Tversky and Gilovich (1989) investigate the popular belief in the hot hand phenomenon in basketball. This article examines some of the data presented by Tversky and Gilovich.

Suppose that a basketball player plans to attempt 20 shots, with each shot resulting in a hit or a miss. A statistician might assume tentatively that the assumptions of Bernoulli trials are appropriate for this experiment. Suppose next that the experiment is performed and the player obtains the following data:

HMHMM MHHHM HHHMM HMHHH

Do these data provide convincing evidence against the tentative assumption of Bernoulli trials? Are the three occurrences of three successive hits convincing evidence of the player having a "hot hand"? These are difficult questions to answer because of the myriad of possible alternatives to Bernoulli trials that exist. It is mathematically and conceptually convenient to restrict attention to alternatives that allow the probability of success on any trial to depend on the outcome of the previous trial or, perhaps, the outcomes of some small number of previous trials. (This restriction may be unrealistic, but that issue will not be addressed in this article.) With the restrictive class of alternatives described here, Tversky and Gilovich devised a clever experiment to obtain convincing evidence that knowledgeable basketball fans are much too ready to detect occurrences of streak shooting—the hot hand—in sequences that are, in fact, the outcomes of Bernoulli trials.

Having established that basketball fans detect the hot hand in simulated random data, Tversky and Gilovich next examined three sets of real data. The data sets are: shots from the field during National Basketball Association (NBA) games; pairs of free throws shot during NBA games; and a controlled experiment using college varsity men and women basketball players. Using the restrictive alternatives described above, Tversky and Gilovich found no evidence of the hot hand phenomenon in any of their data sets. In addition, using a test statistic that is sensitive to certain time trends in the probability of success, they again found no evidence of the hot hand phenomenon.

This article examines the free throw data presented by Tversky and Gilovich. Tversky and Gilovich began by asking a sample of 100 "avid basketball fans" from Cornell and Stanford: "When shooting free throws, does a player have a better chance of making his second shot after making his first shot than after missing his first shot?" A "Yes" response was interpreted as indicating belief in the existence of the hot hand phenomenon, and a "No" as indicating disbelief. (Actually, a "No" response combines persons who believe in independence with those who believe in a negative association between shots; but the researchers apparently were not interested in separating these groups.) Sixty-eight of the fans responded "Yes" and the other 32 "No." Thus, a large majority of those questioned believed in the hot hand phenomenon for free throw shooting. Tversky and Gilovich investigated the above question empirically by examining data they obtained on a small group of well-known and widely viewed basketball players, namely, nine regulars on the 1980–1981 and 1981–1982 Boston Celtics basketball team.

After their analysis of the Celtics data, Tversky and Gilovich concluded that "These data provide no evidence that the outcome of the second shot depends on the outcome of the first." Section 2 of this article will examine the Celtics data with the goal of reconciling what Tversky and Gilovich found and what their basketball fans believed. In particular, it will be shown that, in a certain sense, the prevalent fan belief in the hot hand is not necessarily at odds with Tversky and Gilovich's conclusion.

The analysis presented in Section 3 of this paper indicates that several Celtics players were better at their second shots than at their first.

2. INDEPENDENCE

It is instructive to begin by considering just two of the nine Boston Celtics players who are represented in the free throw data, namely, Larry Bird and Rick Robey. During

Robert L. Wardrop is Associate Professor, Department of Statistics, University of Wisconsin—Madison, Madison, WI 53706. The author thanks the referees and associate editor for helpful comments.

Table 1. Observed Frequencies for Pairs of Free Throws by Larry Bird and Rick Robey, and the Collapsed Table

Larry Bird				Rick Robey				Collapsed Table			
	Second:				*Second:*				*Second:*		
First:	*Hit*	*Miss*	*Total*	*First:*	*Hit*	*Miss*	*Total*	*First:*	*Hit*	*Miss*	*Total*
Hit	251	34	285	Hit	54	37	91	Hit	305	71	376
Miss	48	5	53	Miss	49	31	80	Miss	97	36	133
Total	299	39	338	Total	103	68	171	Total	402	107	509

the 1980–1981 and 1981–1982 seasons, Larry Bird shot a pair of free throws on 338 occasions. Five times he missed both shots, 251 times he made both shots, 34 times he made only the first shot, and 48 times he made only the second shot. These data are presented in Table 1, as are the same data for Rick Robey. Let $\widehat{p}_{\text{hit}}$ and $\widehat{p}_{\text{miss}}$ denote the proportion of first shot hits that are followed by a hit and the proportion of first shot misses that are followed by a hit, respectively. For Bird, $\widehat{p}_{\text{hit}} = 251/285 = .881$ and $\widehat{p}_{\text{miss}} = 48/53 = .906$. For Robey, these numbers are .593 and .612, respectively. Note that, contrary to the hot hand theory, each player shot slightly better after a miss than after a hit, although, as shown below, the differences are not statistically significant.

It is possible, of course, to ignore the identity of the player attempting the shots and examine the data in the collapsed table in Table 1. For example, on 509 occasions either Bird or Robey attempted two free throws, on 305 of those occasions both shots were hit, and so on. For the collapsed table, $\widehat{p}_{\text{hit}} = .811$ and $\widehat{p}_{\text{miss}} = .729$. These values support the hot hand theory—a hit was much more likely than a miss to be followed by a hit.

The data from Bird and Robey illustrate Simpson's paradox (Simpson 1951), namely, $\widehat{p}_{\text{hit}} < \widehat{p}_{\text{miss}}$ in each component table, but $\widehat{p}_{\text{hit}} > \widehat{p}_{\text{miss}}$ in the collapsed table. For further examples and discussion of Simpson's paradox, see Shapiro (1982), Wagner (1982), the essay by Alan Agresti in Kotz and Johnson (1983), and their references.

Figure 1 provides a visual explanation of Simpson's paradox. The top picture in the figure presents the proportion of second-shot successes after a hit for Bird, Robey and the collapsed table. The bottom picture in the figure presents the same three proportions for second shots attempted after a miss. It is easy to verify algebraically that the proportion of successes for a collapsed table equals the weighted average of the individual player's proportions, with weights equal to the proportion of data in the collapsed table that comes from the player. For the after-a-hit condition, for example, the weight for Bird is $285/376 = .758$, the weight for Robey is $91/376 = .242$, and the proportion of successes for the collapsed table, $305/376 = .811$, is

$$\frac{285}{376} \times \frac{251}{285} + \frac{91}{376} \times \frac{54}{91}.$$

In Figure 1, the heights of the four rectangles above the Bird and Robey proportions equal the weights associated with the relevant player–condition pair. For example, the height of the rectangle for Bird in the after-a-hit condition equals .758, in agreement with the computation of the previous paragraph. Thus, the proportion of successes for each collapsed table in the figure is located at the center of gravity of the two rectangles. As a result, even though both Bird and Robey shot better after a miss than after a hit, the collapsed values show the reverse pattern due to the huge variation in weights associated with each player. In short, Simpson's paradox has occurred because the after-a-miss condition, when compared to the after-a-hit condition, has a disproportionately large share of its data originating from the far inferior shooter Robey.

When I first examined the Bird and Robey data several years ago, my immediate reactions were that this is an interesting example of Simpson's paradox, the analysis of individual tables is "correct," and the analysis of the collapsed table is "incorrect." Now I believe these labels were applied too hastily. The reasons I changed my mind are discussed below after the entire data set is examined.

Table 2 introduces symbols to represent the various numbers in a 2×2 table. The values n_1, n_2, m_1, and m_2 denote the marginal totals, and the values of a, b, c, and d denote the cell counts. The null hypothesis states that the outcome of the second shot is statistically independent of the outcome of the first shot. If the null hypothesis is true, then conditional on the values of the marginal totals, the cell count a has a hypergeometric distribution with

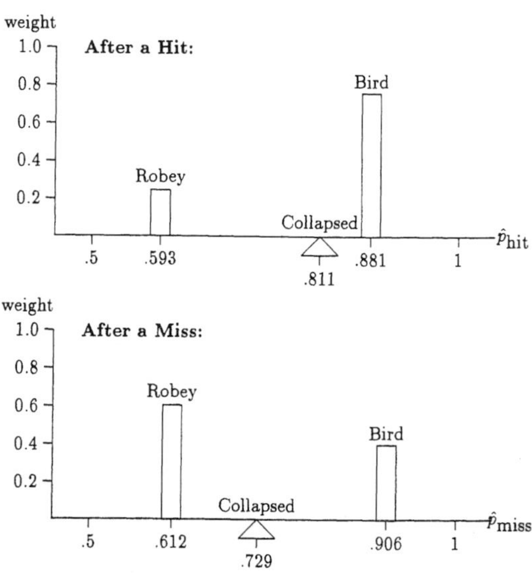

Figure 1. A Visual Explanation of Simpson's Paradox for the Free Throw Study.

Table 2. Standard Notation for a 2 × 2 Table

First:	Second: Hit	Miss	Total
Hit	a	b	n_1
Miss	c	d	n_2
Total	m_1	m_2	n

expectation and variance:

$$E(a) = \frac{n_1 m_1}{n} \tag{1}$$

and

$$\text{var}(a) = \frac{n_1 n_2 m_1 m_2}{(n-1)n^2}. \tag{2}$$

The null distribution of

$$Z = \frac{a - E(a)}{\sqrt{\text{var(a)}}} \tag{3}$$

can be approximated by the standard normal curve. For Larry Bird, $a = 251, E(a) = 252.12$, and $\text{var}(a) = 4.575$. Substituting these values into Equation (3) gives

$$z = \frac{251 - 252.12}{\sqrt{4.575}} = -.52.$$

Thus, as stated earlier, the results are not statistically significant. For Robey, $z = -.25$, and for the collapsed table, $z = 1.99$. Thus, an analysis of the collapsed table alone would lead one to conclude that there is statistically significant evidence in support of the hot hand theory.

Tversky and Gilovich report data for all nine men who played regularly for the Celtics during 1980–1982. The summaries needed for analysis are given in Table 3. The first column of the table lists the players' names. The second and third columns list the values of $\widehat{p}_{\text{hit}}$ and $\widehat{p}_{\text{miss}}$ defined above. The fourth, fifth, and sixth columns list the values of $a, E(a)$, and $\text{var}(a)$ which are obtained from their data and Equations (1) and (2). The seventh column lists the value of z from Equation (3) for each player. The men are listed in the table by decreasing values of $\widehat{p}_{\text{hit}} - \widehat{p}_{\text{miss}}$ which, not too surprisingly, also lists them by decreasing values of z. Thus, McHale, with a difference of 73 − 59 = 14 percentage points, is listed first and Carr, with a difference of 68 − 81 = −13 percentage points, is listed last. In terms of either the point estimates or the test statistic value, McHale provides the strongest evidence in support of the hot hand theory, and Carr provides the strongest evidence in support of an inverse relationship between the outcomes of the two shots. Note that four players—McHale, Maxwell, Parish, and Archibald—shot better after a hit, while the remaining five players shot better after a miss.

The data for McHale give a one-sided approximate P value of .0418. This is not particularly noteworthy for two reasons:

(1) It is difficult to justify the use of a one-sided alternative, especially given that five players shot better after a miss and four shot better after a hit.

Table 3. Selected Statistics for the Investigation of Independence of Shots for Nine Members of the Boston Celtics

Player	$\widehat{p}_{hit}$	$\widehat{p}_{miss}$	a	$E(a)$	var(a)	z
Kevin McHale	.73	.59	93	88.23	7.633	1.73
Cedric Maxwell	.81	.76	245	240.20	14.667	1.25
Robert Parish	.77	.72	164	160.75	13.061	.90
Nate Archibald	.83	.82	203	202.26	8.380	.26
Rick Robey	.59	.61	54	54.81	10.257	−.25
Gerald Henderson	.76	.78	77	77.58	4.858	−.26
Larry Bird	.88	.91	251	252.12	4.575	−.52
Chris Ford	.71	.77	36	37.03	3.100	−.58
M. L. Carr	.68	.81	39	41.20	3.620	−1.16

(2) Even if one believes a one-sided alternative is appropriate, on the assumption that all nine players have independence between shots, the approximate probability is $1 - (1 - .0418)^9 = .32$, or about one-third, that at least one of the nine P values would be as small or smaller than McHale's.

Table 4 presents the observed frequencies and row proportions for the free throw data collapsed over the nine Celtics under investigation. For the collapsed table, the relative frequency of a hit after a hit is 78.9 − 74.3 = 4.6 percentage points higher than the relative frequency of a hit after a miss. Moreover, for the collapsed table, it can be shown that $a = 1{,}162, E(a) = 1{,}143.03$, and $\text{var}(a) = 72.015$, yielding $z = 2.24$, which is statistically significant.

To summarize, separate analyses of individual players indicate that four players shot better after a hit and five players shot better after a miss, but none of the individual player patterns is convincing. By contrast, the analysis of the collapsed table gives statistically significant evidence in support of the hot hand phenomenon.

In view of the Celtics data, what, if anything, are we to make of the fact that 68 out of 100 of Tversky and Gilovich's avid basketball fans believe in the hot hand phenomenon for free throw shooting? Perhaps these fans have been watching players who do exhibit the hot hand. Perhaps these fans see patterns in data where no patterns exist. I prefer the following explanation.

I am an avid basketball fan. Over the past 30 years, I have observed several thousand different players shooting free throws. It is difficult to imagine that I (or any other basketball fan) could remember the equivalent of thousands of 2 × 2 tables. Yet these individual tables are exactly what I would need in order to investigate properly the question of the hot hand phenomenon. It is much more reasonable to assume that I have a single 2 × 2 table

Table 4. Observed Frequencies and Row Proportions for Free Throw Data Collapsed Over Nine Celtics

First:	Second: Hit	Miss	Total	First:	Second: Hit	Miss	Total
Hit	1,162	311	1,473	Hit	.789	.211	1.000
Miss	428	148	576	Miss	.743	.257	1.000
Total	1,590	459	2,049				

in my mind, namely, the collapsed table for all players I have seen. Just like the Celtics data, my collapsed table indicates that a success is more likely than a failure to be followed by a success. Thus, there *is* a pattern in the data that are reasonably available to me and, I conjecture, in the data that are reasonably available to Gilovich and Tversky's 100 basketball fans. It seems reasonable to suggest to basketball fans that the mental equivalent of Simpson's paradox could lead to a cognitive statistical illusion that results in their "seeing patterns in the data that do not exist."

3. STATIONARITY

Tversky and Gilovich correctly concluded that there is no evidence of the hot hand phenomenon in the free throw data. In this section, it is demonstrated, however, that the simple model of Bernoulli trials is also inappropriate. In particular, it is shown that several of the Celtics players shot significantly better on their second free throw, perhaps as a result of the practice afforded by the first shot.

Look at Table 1 again. Larry Bird made 84.3% (285 of 338) of his first shots compared to 88.5% (299 of 338) of his second shots. Thus, there is evidence that he improved on his second shot. The null hypothesis that his probability of success was constant can be investigated with McNemar's test, which uses the fact that the null distribution of

$$Z_1 = \frac{b-c}{\sqrt{b+c}} \tag{4}$$

can be approximated by the standard normal curve. (Recall that b and c are defined in Table 2.) For Larry Bird, $b = 34$ and $c = 48$, giving

$$z_1 = \frac{34-48}{\sqrt{34+48}} = -1.55.$$

The same analysis can be performed for the other eight Celtics; the results are given in Table 5. The first column of the table lists the player's names. The second and third columns list, respectively, the relative frequencies of successes on the first and second shots. The remaining columns list the values of b and c from each player's 2×2 table and the value of z_1 computed from Equation (4). The players are listed according to the difference in relative frequencies between the first and second shots.

Table 5. Selected Statistics for Comparing the Success Rates on the First and Second Free Throws for Nine Members of Boston Celtics

Player	$\widehat{p}(S_1)$	$\widehat{p}(S_2)$	b	c	z_1
Cedric Maxwell	.70	.80	57	97	− 3.22
Robert Parish	.67	.75	49	76	− 2.41
Nate Archibald	.76	.83	42	62	− 1.96
Rick Robey	.53	.60	37	49	− 1.29
Larry Bird	.84	.88	34	48	− 1.55
Gerald Henderson	.73	.77	24	29	−.69
M. L. Carr	.69	.72	18	21	−.48
Chris Ford	.70	.73	15	17	−.35
Kevin McHale	.72	.69	35	29	.75
Total	—	—	311	428	$z_2 = -4.30$

Thus, Maxwell, who shot ten percentage points better on the second shot than on the first, is listed first, and McHale, who shot three percentage points better on the first shot, is listed last. Note the following features of the data.

(1) Eight of nine players had a higher success rate on their second shots.

(2) Three players had one-sided approximate P values below .05: Maxwell (.0006), Parish (.0080), and Archibald (.0250). The interpretation of these P values should take into account that nine tests were performed. If, in fact, each player had a constant success rate on his two shots, the approximate probability of obtaining at least one P value equal to or smaller than .0006 is: $1 - (1 - .0006)^9 = .0054$. Similarly, the approximate probability of obtaining at least two P values equal to or smaller than .0080 is .0022. Finally, the approximate probability of obtaining at least three P values equal to or smaller than .0250 is .0012. Thus, the three statistically significant results do not seem to be attributable to the execution of many tests.

(3) McNemar's test can be viewed as testing that a Bernoulli trial success probability equals .5 based on a sample of size $b + c$. Thus, several of the analyses of individual players presented in Table 5 are based on very little data and, hence, have very low power. To combat this difficulty, it is instructive to combine the data across the nine players. In particular, if the null hypothesis of constant success probability is true for all nine players, then the observed value of

$$Z_2 = \frac{\sum(b-c)}{\sqrt{\sum(b+c)}},$$

where the sum is taken over the nine tables, can be viewed as an observation from a distribution that is approximately the standard normal curve. The observed value of Z_2 is −4.30, given in the bottom row of Table 5. This value indicates that there is overwhelming evidence against the assumption that all nine null hypotheses are true.

4. SUMMARY

This article puts forth an argument to reconcile what avid basketball fans believe and what Tversky and Gilovich found. It is argued that the fans and the researchers were analyzing different sets of data. While the researcher's data had no pattern, the fan's data had a pattern. This pattern, however, was due to the effects of aggregation and not the hot hand phenomenon. This finding indicates that researchers should take care to consider what data are available to laypersons. In addition, this finding underscores the importance of increasing the awareness of statistical fallacies among the general public.

This article also demonstrates that several Celtics players showed a significant improvement in their shooting ability on the second free throw. Thus, while the hot hand phenomenon is not supported by these free throw data, neither is the simple model of Bernoulli trials.

[*Received March 1992. Revised November 1993.*]

REFERENCES

Diaconis, P., and Mosteller, F. (1989), "Methods for Studying Coincidences," *Journal of the American Statistical Association*, 84, 853–861.

Kahneman, D., Slovic, P., and Tversky, A. (1983), *Judgement Under Uncertainty: Heuristics and Biases*, Cambridge, U.K.: Cambridge University Press.

Kotz, S., and Johnson, N. L. (eds.) (1983), *Encyclopedia of Statistical Science* (Vol. 3), New York: John Wiley, pp. 24–28.

Shapiro, S. H. (1982), "Collapsing a Contingency Table—A Geometric Approach," *The American Statistician*, 36, 43–46.

Simpson, E. H. (1951), "The Interpretation of Interaction in Contingency Tables," *Journal of the Royal Statistical Society*, Ser. B, 13, 238–241.

Tversky, A., and Gilovich, T. (1989), "The Cold Facts About the 'Hot Hand' in Basketball," *CHANCE: New Directions for Statistics and Computing*, 2, 16–21.

Wagner, C. H. (1982), "Simpson's Paradox in Real Life," *The American Statistician*, 36, 46–47.

Part IV
Statistics in Ice Hockey

Chapter 23

Introduction to the Ice Hockey Articles

Robin H. Lock

We provide a short description of ice hockey and its history in this introduction. We also briefly discuss the application of statistical methods in hockey and we identify particular research areas. We use the articles selected for this part of the volume to give the reader a sense of the history of statistical research in hockey.

23.1 Background

Ice hockey is a fast-paced winter sport that naturally enjoys its greatest popularity among sports enthusiasts in the upper northern hemisphere. The sport is believed to have been first played in the early nineteenth century in Windsor, Nova Scotia; Kingston, Ontario; or Montreal, Quebec; the first known rules were published in 1877 by the *Montreal Gazette*. The first U.S. collegiate ice hockey game was played between Yale and Johns Hopkins in 1896. Professional hockey was established in the early twentieth century; the National Hockey League (NHL) was founded in 1917 and currently includes 30 professional teams across North America. After ice hockey was introduced as an Olympic sport at the 1920 Summer Olympics, its international popularity and stature gradually grew, culminating in the introduction of women's ice hockey as an Olympic sport in 1994.

In an ice hockey game, two opposing teams of skaters use long, curved sticks to try to drive a puck (a hard rubber disk) into each other's goal net. The team that scores the most goals by hitting the puck into its opponent's goal net with their sticks wins the game. If the two opponents have scored the same number of goals at the conclusion of regulation play, some leagues will declare the game a tie. In other leagues, this leads to an overtime period that is played under "sudden death" (i.e., the first team to score in overtime is declared the winner). If neither team scores during the overtime period, some leagues then declare the game a tie, while others may use some other means of determining the winner. Additional overtime periods can be played until one team scores, or the teams can have a shoot-out, where a series of players from each team alternate in taking shots on goal.

The sport may be played outdoors or indoors on a structure called a rink. A rink consists of an oval ice surface, surrounded by a wall (usually referred to as the boards), with goals on both ends. Parallel lines are painted across the ice to divide the rink into zones. The rink is divided in half by a red centerline, and blue lines between the centerline and the goals divide the rink further. The ice between the two blue lines is referred to as the neutral zone; the ice outside of the neutral zone that contains a team's goal is called their defending zone, and the ice outside of the neutral zone that contains their opponent's goal is called their attacking zone. Collectively, the defending and attacking zones are referred to as the end zones. The rink also has a blue circle in its center and eight red circles placed strategically around the perimeter of the rink. Play often begins or resumes after a stoppage at one of these circles.

Red lines (referred to as goal lines) also exist at each end of the rink, where the boards begin to curve. The goals (also called nets) are placed in the middle of these lines. A half-circle, called the crease, is painted in front of each goal. Attacking players may not enter the crease unless the puck is already there, and they may not make contact with the goalie (the opponent's player who is designated with the responsibility of defending the goal) while in the crease.

Each team may have up to six players on the ice at any

time in a regulation ice hockey game. These players occupy three positions: forward, defense, and goalie. The three forwards—the center, left wing, and right wing—form a unit (called a line) that is primarily responsible for their team's offense. Centers, who are usually their team's best "passers," generally skate between and feed the puck to the wings, who are usually their team's best shooters. The two defenders comprise the last line of defense before the goalie; they attempt to disrupt their opponent's offense before they are able to shoot (attempt to score). Finally, the goalie (or goaltender) is responsible for guarding his team's goal and preventing the opponent's shots from entering his net. The goalie may stop a shot on his goal with any part of his body (which is usually protected with heavy padding, a helmet, a collar, and a mask), the small glove on his stick hand (the blocker), his stick, or the large leather glove on his free hand (the trapper). Although the names of the positions imply certain duties and responsibilities, the rules do not prevent players (other than the goalie) from skating to any part of the rink; forwards help with defense, defenders sometimes score, and goalies will even risk leaving the net unmanned late in a game their team is losing to allow a substitute player to generate additional offense.

At the highest competitive levels (professional, international, and college), games are comprised of three 20-minute periods with 15-minute intermissions after the first and second periods. The ice is usually resurfaced during the intermissions. At lower levels of competition, the three periods generally last 10 or 15 minutes. Each period begins with a face-off at the blue circle at center ice. One player from each team lines up at the blue circle with his stick blade on the ice. The referee drops the puck onto the ice and the two players attempt to gain control of it. Play stops when a goal is scored, the puck leaves the rink, a player is injured, a penalty or other infraction is called, or a goalie covers the puck. Substitute players may enter the game during any stoppage of play or even jump in "on the fly" when a teammate steps off the ice as play continues.

A player who violates a rule may be charged with a penalty. A penalized player must leave the ice and spend time in the penalty box. Minor penalties last two minutes and are usually charged to players who illegally impede an opponent's progress by tripping or holding. A player can also be charged with a minor penalty for delaying the game or risking injury to an opponent. Major penalties lasts five minutes and are charged to players who commit more serious fouls.

A misconduct penalty may be assessed against a player or team for a variety of infractions. A player charged with a misconduct penalty must leave the game for 10 minutes, but his team may immediately place a substitute on the ice. Game misconduct penalties are charged for dangerous play, often when the official deems that the player's intent is to injure an opposing player. A game misconduct penalty results in ejection of the penalized player and assessment of a minor penalty against his team. While goalies do not serve minor, major, or misconduct penalties (their penalties are served by a teammate who was on the ice at the time of the infraction) they may be ejected for game misconduct penalties.

When penalties leave one team with more skaters on the ice than its opponent, the team with more players on the ice is said to have a power play and this advantage lasts for the duration of the penalty or until they score a goal (unless it's a major penalty). Only two minor penalties can be served simultaneously—if a team accumulates more than two overlapping minor penalties, the additional penalty is served immediately after one of the two current penalties expires. If a goal is scored on a team that is serving two penalties, only one of the penalized players may return to the game, and the team remains shorthanded. If a player is ejected from the game, his team plays shorthanded for five minutes and may then substitute for the ejected player.

Other rule violations do not result in the assessment of penalty time. An offsides penalty occurs when an offensive player crosses into the attacking zone ahead of the puck. At this point a face-off is held in the neutral zone, giving the opponent an opportunity to regain control of the puck. Some leagues also use a two-line offsides rule that prohibits an offensive player from making a pass across the centerline from the defensive zone. At this point, a face-off is held in the red circle nearest the point from where the illegal pass was made.

Icing occurs when the puck moves from behind the centerline to a point beyond the opponent's goal line without being touched by another player. In some leagues, this infraction is automatically called, while other leagues call icing only after a defender has touched the puck behind the goal line. Once icing is called, a face-off is held at the red circle closest to the penalized team's net.

Available data for the line and defenders include position played (center, left wing, right wing, defense, goalie), games played, goals scored, assists, points (the sum of goals and assists), plus-minus (difference between goals scored and goals allowed while the player is on the ice), penalty minutes, power play goals, shorthanded goals, game-winning goals (goals that gives the winning team just enough goals to win), game-tying goals (the final goal in a tie game), shots on goal, shooting percentage, average number of shifts per game, average time on ice per game, face-offs won, face-offs lost, and percentage of face-offs won. Available data for goalies include games played, wins (a goalie is credited with a win if he is on the ice when his team scores the game-winning goal), losses (a goalie is charged with a loss if he is on the ice when the opposing team scores the game-winning goal), ties (a goaltender is credited with a tie if he is on the ice when the game-tying goal is scored), goals against, shots against, goals-against average (goals allowed per full game played), saves, save percentage, shutouts (a goalie must play the entire game to receive credit for a shutout), and penalty minutes.

Team standings in the National Hockey League are determined by a point system. A team is awarded two points for a win, one point for a tie, and one point for an overtime loss. Other available team data include aggregated individual player statistics.

23.2 Determining the Superior Team

Although the papers in this section appear here in chronological order, they also follow a natural order as a typical season reaches its conclusion. In Chapter 24, Danehy and Lock look at ways to compare teams from different leagues who play a regular season schedule with varying degrees of overlapping interleague play. One goal of the methods they develop is to provide statistically reasonable procedures for selecting teams to compete in a postseason tournament. Once the tournament begins, each game must be played to a definite conclusion so that only one team moves on after each contest. In Chapter 25 Hurley investigates several schemes for reaching that decision when two teams are tied at the end of regulation game-time. In Chapter 26 Morrison and Schmittlein complete the playoff cycle by examining the factors that play a role in determining the winning team in the "best of seven playoff" format used by the National Hockey League to award the Stanley Cup.

Harville (1977) considered the use of linear regression methods to produce comparative ratings of teams based on individual game scores in football. Danehy and Lock looked for similar models to apply to ice hockey and introduced two significant revisions. Whereas Harville and other authors (e.g., Stern, 1995) use a single rating value for each team (based on the winning and losing margins of its games), Danehy and Lock suggest using separate offensive and defensive ratings to capture both important aspects of a team's performance. This also allows them to predict individual game scores and distributions based on the ratings given to the participating teams. When analyzing this model with actual college hockey scores, they found that it was susceptible to large discrepancies between the ratings given to different leagues when there was little interleague play. This motivated the introduction of a method for parameterizing the model to move smoothly from one that gives heavy weight to the quality of an opponent to one that ignores this factor completely. Thus the model can be fine-tuned to reflect quality of opponents when producing the ratings while remaining robust to a few extreme scores.

As the earliest paper in this group (Chapter 24), Danehy and Lock's work on rating methods for college hockey has generated the most follow-up work to date. They published a second paper (Danehy and Lock, 1995) that extends the method from an ordinary least squares model to consider scores to be generated by a pair of Poisson processes with the team ratings determining the values of the Poisson scoring rates for any particular contest. This was consistent with work done by Reep, Pollard, and Benjamin (1971), who found that game scores in football (soccer), cricket, baseball, and ice hockey followed a negative binomial distribution that reflected mixtures of Poisson distributions with parameters varying with the teams in each match. In later work, Danehy and Lock (1997) modified the model further to allow the offensive, defensive, and home ice parameters to interact multiplicatively, rather than use the additive model introduced in the original paper. Details of the current methods can be found on the College Hockey Offensive and Defensive Ratings (CHODR) website at http://it.stlawu.edu/~chodr. This site also contains historical ratings and predicative results collected since 1996 and was expanded in 1999 to include ratings for NCAA Division I Women's Ice Hockey (WCHODR).

In the 2000 NCAA Division I Men's Ice Hockey playoffs, St. Lawrence University (SLU) played Boston University (BU) in a second round game at the Knickerbocker Arena in Albany, New York. The score was tied 2–2 after 60 minutes of regulation play and the teams proceeded to play three full 20-minute overtime periods before SLU's Robin Carruthers scored a goal at 3:53 of the fourth overtime to end the longest-ever NCAA tournament game. The two goalies (both freshmen) each smashed the previous record for most saves in an NCAA playoff game. The two other teams (Michigan and Maine) that were scheduled to play in the second game that day had to wait until well after the time their game was scheduled to finish before even taking the ice; this created havoc with pregame meal schedules, travel arrangements, broadcasters' voices, and fans' nerves. Perhaps these disruptions could have been avoided if NCAA officials had adopted some proposals made in Hurley's 1995 article, presented in Chapter 25.

Hurley compares overtime and shootout as methods for determining the victor in a tied contest. His work was motivated, in part, by a series of high-profile hockey and soccer tournaments the previous year in which the championship games were decided by shootout rather than a goal scored by traditional team play. The shootouts had been instituted to prevent the sort of protracted overtime affairs typified by the SLU–BU contest. But some players, fans, and sportswriters have heavily criticized the shootout as an inappropriate way to determine a very important outcome for a sport that emphasizes team play. Liu and Schultz (1994) looked at previous overtime methods employed by the National Hockey League to examine how often the tie was broken and the chances of the better team winning. Hurley suggests a novel alternative to the two extremes. First, hold the shootout and then play the overtime period. The team that wins the shootout starts the overtime with a one-goal advantage so that if neither team scores during the overtime they are declared the winner, but the shootout loser still has a chance to recover through traditional play to retie the match. Hurley provides analysis of the likely time needed to implement this procedure and considers the probability that the better team will prevail.

Mosteller (1952) performed an early analysis of the results of a best 4 out of 7 playoff system by analyzing World Series (baseball) data to estimate the chances that the better team will come out on top in the series. Maisel (1966) follows this up with an extensive analysis of a best k of $2k - 1$ competitions that, ironically, looks at properties both with and without the possibility of some contests ending in ties. But Stanley Cup games are always played to conclusion and we could always resort to Hurley's tie-breaking scheme if the prospect of televising unreasonably long overtime games becomes a problem!

Morrison and Schmittlein (Chapter 26 in this volume) examine the past history of Stanley Cup series and compare the number of games played to what one would expect if equally matched teams played a best 4 out of 7 series with independent game results. While theory would suggest that 6- and 7-game series are the most likely, the actual results show far more sweeps (4-game series) and fewer 7-game series than one would expect. What might cause this discrepancy? The authors suggest and examine three possibilities. Perhaps the teams are not well matched and one is actually significantly better than its opponent so that the probability of winning each game moves away from the theoretical 0.5. Or the discrepancy might be due to a home ice advantage since an odd number of games allows one team an extra contest at its home rink. Finally, the independence assumption might be faulty. Could one team ride a "hot" goalie to several quick wins and an early series victory? This chapter examines each of these possibilities to see which might be consistent with previous Stanley Cup results.

23.3 Summary

A great deal of game-level information is collected about individual hockey players and teams. While much of the focus of statistical research in hockey has been devoted to ranking or comparing teams, a great deal of opportunity exists for analyzing performances in a manner similar to baseball.

While the three papers included in this section are based on the structure and results of ice hockey competitions, the methods they introduce could be generalized to other sports, particularly soccer, lacrosse, field hockey, and water polo, which share the same basic format of teams trying to get an object past the opposing team's players into a confined space that is guarded by a goalie as the final line of defense.

References

Danehy, T. J. and Lock, R. H. (1995), "CHODR—Using statistics to predict college hockey," STATS, 13, 10–14.

Danehy, T. J. and Lock, R. H. (1997), "Using a Poisson model to rate teams and predict scores in ice hockey," in 1997 *Proceedings of the Section on Statistics in Sports*, Alexandria, VA: American Statistical Association, 25–30.

Harville, D. (1977), "The use of linear-model methodology to rate high school or college football teams," Journal of the American Statistical Association, 72, 278–289.

Liu, Y. and Schultz, R. (1994), "Overtime in the National Hockey League: Is it a valid tie-breaking procedure?" in 1994 *Proceedings of the Section on Statistics in Sports*, Alexandria, VA: American Statistical Association, 55–60.

Maisel, H. (1966), "Best k of $2k - 1$ comparisons," Journal of the American Statistical Association, 61, 329–344.

Mosteller, F. (1952), "The World Series competition," Journal of the American Statistical Association, 47, 355–380.

Reep, C., Pollard, R., and Benjamin, B. (1971), "Skill and chance in ball games," Journal of the Royal Statistical Society, Series A (General), 134, 623–629.

Stern, H. (1995), "Who's number 1 in college football?... And how might we decide?" Chance, 8, 7–14.

Chapter 24

Statistical Methods for Rating College Hockey Teams

Timothy J. Danehy, Clarkson University; Robin H. Lock, St. Lawrence University
Robin H. Lock, Mathematics Department, St. Lawrence University, Canton, NY 13617

KEY WORDS: Sports rating models; Poisson regression; Schedule graphs; Sports rankings

ABSTRACT: *We investigate methods for rating sports teams based solely on past game results. Techniques are illustrated using data from NCAA Division I Men's Ice Hockey competition, although the methods can easily be applied to other sports and levels of play. The proposed systems produce offensive, defensive, and overall ratings based on past performance and the quality of opponents. These ratings can then be used to compare teams and forecast outcomes of future games. A significant challenge in such rating schemes, particularly in the college environment, is the lack of connectivity in the schedule graph. We demonstrate how an insufficient amount of inter-league play can cause traditional regression methods to break down and produce clearly inappropriate ratings. We suggest a modified procedure designed to avert such pitfalls and examine the effectiveness of various models in predicting future game outcomes.*

1. INTRODUCTION

In this paper we investigate statistical models for rating sports teams, with specific applications to college hockey. Although the methods we develop are designed for college hockey, the general principles could be easily applied in other settings. In the next section we describe some overall goals which might be addressed by a rating system. Section 3 gives an overview of the schedule structure in college hockey, describes some characteristics of the game which make it amenable to ratings, and presents two existing ratings systems. A traditional additive model and least squares approach to estimating offensive and defensive ratings as well as a home ice advantage is presented in Section 4. We then demonstrate a deficiency in this model, using an example from early in the 1992–93 hockey season. A mechanism for dealing with problems arising from the lack of many inter-league connections in the schedule graph is described in Section 6. Finally we examine the performance of these ratings methods when predicting game outcomes in the 1992–93 regular season and playoffs.

2. GOALS OF A RATING SYSTEM

We must be clear to distinguish between *rating* systems and *ranking* systems. Most sports "Top 10" type polls are rankings - they provide a relative ordering of the teams by (perceived) ability, but do not give any numerical measure (other than number of votes) of that ability. A rating system should go further to produce direct estimates of a team's strength on some interpretable scale. Some functions of a rating system might include:

a. To rank order all teams (or individuals).
b. To compare specific teams (or individuals).
c. To adjust ratings for the quality of opponents.
d. To predict game outcomes (winner/loser).
e. To predict specific game differentials.
f. To predict specific game scores.

One obvious rating system is a simple won-lost percentage which could be used to accomplish tasks (a), (b), and (d). Another common rating system would be to take each team's average points scored and subtract its average points given up. Such a rating would allow for the prediction of game differentials or could be used to predict individual game scores. However, neither of these simplistic methods would address the issue of adjusting ratings to account for strengths of opponents.

3. COLLEGE HOCKEY

3.1 Structure of Division I Ice Hockey

In the 1992–93 season forty-four schools competed in Men's Ice Hockey at the NCAA Division I level. Most of these teams were organized into four leagues - Hockey East (HE), ECAC, CCHA, and WCHA. League members played from 22 (ECAC) to 32 (WCHA) of their games within their own league, with as few as zero (St. Cloud) to as many as 11 (Lowell and Maine) against non-league opponents.

We consider a *schedule graph* consisting of 44 vertices, representing individual teams, and a set of edges connecting teams involved in each game. The graph consists of four main blocks, corresponding to the leagues, with lots of edges (in fact multiple complete subgraphs) within each league block, but relatively few connections between the blocks. For example, in 1992–93 there were no regular season games between ECAC and WCHA teams and only two between the WCHA and HE. Overall, 537 of 683 matchups (79%) in the 1992–93 regular season were league contests. Within each league we might be confident in using standings based on a balanced league schedule as basis for ranking teams. However, we need an alternate mechanism to reliably compare teams between different leagues.

3.2 Why College Hockey?

Several aspects of college hockey are particularly well-suited for investigating ratings methods. It is generally well-recognized that a team's performance can be measured primarily by its ability to score goals and to prevent its opponent from scoring. Scores directly reflect the number of goals scored, as compared to football or basketball where each "score" can yield a different number of "points" Each goal is an individual event without a clustering of scores as might be found in baseball. The scoring rate is relatively low (averaging around four goals per game), but considerably higher than in soccer. Finally, there is considerable fan interest in comparisons between the leagues, particularly as the regular season winds down and 12 teams are selected and seeded to play for the national championship in the NCAA post-season tournament.

3.3 Previous Hockey Rating Systems

One rating system which is used in the NCAA tournament selection process (for basketball as well as hockey) is the Ratings Percentage Index (RPI). The system considers a teams own strength as well as strength of schedule via a weighted average of three parts: winning percentage (20%), opponents' winning percentage (40%), and opponents' opponents' winning percentage (40%). Another model is TCHCR (The College Hockey Computer Rating) (Instone 1992). This system uses a least squares optimization (Leake 1976) to match the ratings differences between two teams as closely as possible with actual game outcomes, as measured by a performance function (see Stern 1992 for a discussion of performance functions in football). The function used in TCHCR depends only on the game outcome (win, loss, or tie) so it primarily refelcts winning percentage and strength of schedule, just as the RPI. Both systems allow us to compare specific teams while adjusting for the quality of opponents and predict game outcomes (winner/loser) with the differences in ratings providing a vague indication of the disparity between two teams. However, neither of these systems is able to predict specific game differentials or actual scores.

3.4 Assumptions

As a database for computing ratings we use only games played between NCAA Division I schools, excluding any contests with other divisions or Canadian schools. Since we model regulation time scoring rates, we ignore all goals scored in overtime. We assume a "home ice" advantage which is estimated as part of the ratings. A team's expected scoring rate in a particular game is assumed to depend on its offensive ability, the defensive ability of its opponent, and the site of the contest. When computing ratings, data from past games is limited to game scores and indicators for home ice and overtime.

4. AN ADDITIVE MODEL

Let S_{ijk} denote the goals scored by Team i against opposing Team j, with the index k reflecting a game counter. If n games are played, each producing two scores, we have $k = 1, 2, \ldots 2n$. Our additive model produces an expected scoring rate according to

$$E(S_{ijk}) = O_i + D_j - \mu + HI_k \qquad (1)$$

where

O_i = offensive rating for team i.
D_j = defensive rating for team j.
H = home ice adjustment.
I_k = (+1 if i home, 0 if neutral, -1 if road).
μ = mean scoring rate (goals/game).

The key rating quantities here are O_i and D_j. One may think of O_i as the expected scoring rate for Team i against a hypothetical "average" team on neutral ice. Similarly, D_j represents the expected number of goals which Team j will give up when playing the "average" team. Since the object of the game is to score more goals than you give up, we define an *overall rating* of a team's strength by

$$R_i = O_i - D_i \qquad (2)$$

Thus the "average" team should have an overall rating of zero. Direct comparisons are easily managed since $R_i - R_j$ gives the expected goal differential when Team i plays Team j on neutral ice.

Actual ratings ($\hat{O}_i$ and $\hat{D}_j$) and other parameters ($\hat{\mu}$ and $\hat{H}$) can be estimated based on past game data using a least squares fit. To obtain unique estimates we add the condition that the (weighted) average offensive and defensive ratings be the same and, consequently, equal to $\hat{\mu}$. Thus

$$\hat{\mu} = (1/2n) \sum_i n_i \hat{O}_i = (1/2n) \sum_i n_i \hat{D}_i \qquad (3)$$

where n_i counts the games played by Team i.

The least squares solution provides the following relations among scores and estimates:

$$\hat{\mu} = \overline{S} \tag{4}$$

$$\hat{H} = \frac{1}{2}((\overline{S}_h - \overline{S}_r) - (\overline{O}_h - \overline{O}_r) + (\overline{D}_h - \overline{D}_r)) \tag{5}$$

$$\hat{O}_i = \frac{1}{n_i}\sum_{\mathcal{B}_i}(S_{ijk} - (\hat{D}_j + \hat{H}I_k)) + \hat{\mu} \tag{6}$$

$$\hat{D}_j = \frac{1}{n_j}\sum_{\mathcal{A}_j}(S_{ijk} - (\hat{O}_i + \hat{H}I_k)) + \hat{\mu} \tag{7}$$

where

$\overline{S}$ = average goals scored

$\overline{S}_h, \overline{S}_r$ = average goals scored at home/on road

$\overline{O}_h$ = average offensive rating for home teams

$\overline{O}_r, \overline{D}_h, \overline{D}_r$ are defined analogously

$\mathcal{A}_j = \{k : k^{th}$ score was AGAINST team $j\}$

$\mathcal{B}_i = \{k : k^{th}$ score was BY team $i\}$

Although one could use a standard package to calculate the least squares estimates, we have chosen an iterative method which can be adapted more easily to the modified situations which we discuss shortly. We start with initial estimates $\hat{O}_i$ = average goals scored by team i, $\hat{D}_j$ = average goals scored against team j, $\hat{\mu} = \overline{S}$, and $\hat{H} = \frac{1}{2}(\overline{S}_h - \overline{S}_r)$. We then adjust $\hat{H}$ using (5) and the initial offensive and defensive estimates, recompute new offensive estimates with (6), and finally calculate new defensive ratings using (7). This cycle is continued until the estimates converge to fixed values.

5. SPARSE CONNECTIONS

The lack of frequent connections between league blocks in the schedule graph can lead to some serious difficulties in computing comparative ratings. A vivid example can be seen in the fifth week of the 1992–93 season when Air Force (an independent team) traveled to Colorado College (of the WCHA) and lost by a score of 12–3. At that early point in the season there was only one other connection between the WCHA and the rest of the schedule graph. The subsequent ratings accounted for the extreme CC–AF score by raising the offensive ratings of *all* the WCHA teams by a considerable margin and compensating by significantly lowering their defensive ratings. Predicted scores between two WCHA teams were still reasonable, but comparisons with the rest of college hockey made little sense (see Table 1 - WCHA teams in bold). Although this dramatic lack of stability is magnified by its occurrence early in the season (only 129 games and few interleague connections), more subtle problems can be attributed to sensitivity to sparse connections throughout the season.

6. MODIFIED ADDITIVE MODEL

As an alternate model, one might consider predicting the number of goals a team should score based only on its offensive ability, totally disregarding the defensive strength of its opponent. Using the notation introduced in Section 4 we would have

$$E_O(S_{ijk}) = 2O_i - \mu + HI_k \tag{8}$$

Thus the ratings estimates as well as each predicted score would depend only on each teams average goals scored with an adjustment for home ice.

But why should the offense be so important? An equally plausible model might place the burden of prediction solely on the opponent's defensive ability.

$$E_D(S_{ijk}) = 2D_j - \mu + HI_k \tag{9}$$

We could even combine both viewpoints to calculate an expected score by averaging

$$E(S_{ijk}) = \frac{E_O(S_{ijk}) + E_D(S_{ijk})}{2} \tag{10}$$

Although this gives the original model (1), we might consider using (8) and (9) in the iterative procedure for computing the estimates. We would eliminate the difficulty of one unusual game affecting estimates throughout a league, however we would also lose any ability to adjust ratings based on the quality of opponents. Thus a team could enhance its ratings by scheduling weak opponents and building up attractive goals for and goals against averages.

To minimize the liabilities of both approaches, we include an additional parameter to allows us to vary smoothly between them

$$E_O(S_{ijk}) = (1+\alpha)O_i + (1-\alpha)D_j - \mu + HI_k \tag{11}$$

$$E_D(S_{ijk}) = (1-\alpha)O_i + (1+\alpha)D_j - \mu + HI_k \tag{12}$$

Clearly $\alpha = 0$ gives our original model, while $\alpha = 1$ produces equations (8) and (9). We still use (1) to obtain score predictions, but the updating equations in the iterative procedure become

$$\hat{O}_i = \frac{\left(\frac{1}{n_i}\sum_{\mathcal{B}_i}(S_{ijk} - ((1-\alpha)\hat{D}_j + \hat{H}I_k)) + \hat{\mu}\right)}{1+\alpha} \tag{13}$$

$$\hat{D}_j = \frac{\left(\frac{1}{n_j}\sum_{\mathcal{A}_j}(S_{ijk} - ((1-\alpha)\hat{O}_i + \hat{H}I_k)) + \hat{\mu}\right)}{1+\alpha} \tag{14}$$

Table 2 shows how the introduction of an α value as small as 0.05 or 0.10 can alleviate many of the problems caused by extreme instances such as the CC–AF game, while still giving significant weight to quality of opponents.

7. RATING THE 1992–93 SEASON

7.1 Comparison of Models

Data from the 1992–93 season were used to evaluate the effectiveness of these models in predicting college hockey scores and game outcomes. The first seven weeks of the season (200 games) were applied to develop initial ratings. Thereafter (553 games), we forecast each week's games using ratings based only on data available prior to that week. We compare results for the modified additive model using $\alpha =$ 0, 0.05, 0.10, and 1.

Measures for evaluating the accuracy of the ratings are presented in Table 3. Several of these were suggested by Stern (1992) in his analysis of football rating systems. The most basic quantity is the percentage of games for which the predicted scores accurately forecast the winner of the contest. The percentages in Table 3 are based on the 473 games after Week 7 in which a winner was determined in regulation time. To test a model's ability to forecast the margin of victory we use the mean absolute deviation between the predicted and actual goal differentials. An information statistic is computed as the average negative logarithm of the joint probablity of the observed game scores, using a Poisson probability model based on predicted scoring rates. We also include the square root of the average squared prediction errors and the median absolute deviation between predicted and actual team scores.

Although many of the differences in Table 3 are small, the trends are quite consistent. The α=0.10 case is invariably better than the two extremes, while the models allowing no adjustment for opponent's strength (α=1) are inevitably the least effective. To help assess the percentage correct data, one can check that the home team won 256 of the 421 (60.8%) non-tie, non-neutral ice games during this period. The additive least squares model (with α=0) forecast 319 (75.8%) of those outcomes correctly.

7.2 Forecasting the 1992–93 NCAA Playoffs

As another means of assessing the effectiveness of our ratings systems, we examine performance in predicting the outcomes in postseason tournaments. Despite the fact that only one regular season leader won its conference tournament (Maine in HE), the modified additive ratings (α=0.10) correctly predicted the winners for 41 of the 52 (78.8%) non-tie games in the league playoffs. Predictions in the NCAA tournament should be more challenging since the opponents are frequently from different leagues.

Table 4 gives the ratings (α=0.10) for all 44 Division I teams following conference playoffs. At that point in the season an NCAA selection committee (using RPI ratings as one of the criteria) chose 12 teams and seeded them into its tournament. The participants in 1993 included the top 9 teams in Table 4, plus Northern Michigan (#13), Brown (#14) and Minnesota (#18) which received and automatic bid for winning the WCHA tournament. Maine, Michigan, Lake Superior, and Boston University received the top seeds and first round byes. The four first round games featured three "upsets" according to the seedings and our ratings (see Table 5). After that point, the ratings correctly predicted winners of the final seven games in the playoffs, including Maine's victory over Lake Superior State for the National Championship.

8. CONCLUSION

We have proposed statistical models for rating systems to satisfy the criteria set forth in Section 2. By the end of a full season, when most teams have played a balanced league schedule and some (but never enough) connections exist between the leagues, the ratings and predictions produced for various values of α are not terribly different. However, it is desirable to account for strength of opponents in producing reasonable ratings for comparing teams between conferences at earlier points in the season. Section 5 clearly demonstrates that traditional regression methods may not be optimal for this task. Thus we have suggested a method for moving smoothly between regression methods (which can be overly sensitive to sparse conections) and simple averages (which completely disregard opponent's strength). The results from the 1992–93 season indicate that the modified system is likely to be superior to either extreme. These methods have been used to produce a weekly rating service called CHODR (College Hockey Offensive and Defensive Ratings) which is disseminated through the HOCKEY-L electronic discussion list.

REFERENCES

Instone, K. (1992), "Inside the College Hockey Computer Rating," unpublished manuscript available through e-mail at *instone@bullwinkle.bgsu.edu.*

Leake, R.J. (1976), "A Method for Ranking Teams with an Application to College Football," in *Management Science in Sports*, North Holland, eds. R.E. Machol, S.P. Ladany, D.G. Morrison, 27–46.

Stern, H. (1992), "Who's Number One - Probability and Statistics in Sports," *1992 Proceedings of the ASA Section on Statistics in Sports.*

Table 1. Additive Model - Least Squares Regression ($\alpha = 0.00$) (Through November 14, 1993 – Week 5)

Without CC - AF					With CC - AF					
Rank	Team	Off	Def	Overall	Rank	Team	Off	Def	Overall	ΔOverall
1	Maine	8.4	1.9	6.5	1	**Denver**	8.1	0.2	7.9	2.6
2	**Denver**	7.0	1.6	5.4	2	**Wisconsin**	7.7	1.0	6.7	2.3
3	Yale	8.7	3.5	5.3	3	**Michigan Tech**	6.5	0.0	6.5	2.4
4	**Wisconsin**	6.5	2.1	4.4	4	**North Dakota**	8.2	2.1	6.1	2.6
5	**Michigan Tech**	5.5	1.4	4.1	5	**Minnesota-Duluth**	7.2	1.1	6.1	2.6
6	Princeton	6.2	2.5	3.8	6	**Colorado College**	7.9	2.1	5.8	2.8
7	**North Dakota**	6.7	3.2	3.5	7	**St Cloud**	6.7	0.9	5.8	2.4
8	**Minnesota-Duluth**	5.9	2.4	3.5	8	Maine	7.7	2.3	5.4	-1.1
9	**St Cloud**	5.4	2.1	3.3	9	**Minnesota**	6.4	1.0	5.4	2.4
10	Cornell	3.8	0.7	3.2	10	Yale	8.2	3.7	4.5	-0.8
11	**Colorado College**	6.0	3.0	3.0	11	Princeton	5.7	2.7	3.0	-0.8
12	**Minnesota**	5.4	2.5	2.9	12	**Northern Michigan**	6.0	3.2	2.8	2.0
18	Air Force	5.0	3.1	1.9	13	Cornell	3.6	1.3	2.3	-0.9
21	**Northern Michigan**	4.9	4.1	0.8	20	Air Force	5.1	4.6	0.5	-1.4

Table 2. Additive Model - Least Squares Regression (With CC - AF) (Through November 14, 1993 – Week 5)

$\alpha = 0.05$				$\alpha = 0.10$				$\alpha = 1.00$		
Rank	Team	Overall	Δ	Rank	Team	Overall	Δ	Rank	Team	Overall
1	Maine	5.6	-0.3	1	Maine	5.1	-0.2	1	Maine	2.8
2	Yale	3.6	-0.2	2	Yale	3.1	0.0	2	Yale	1.8
3	**Denver**	2.6	0.7	3	**Denver**	2.1	0.4	3	Lake Superior	1.3
4	Princeton	2.1	-0.2	4	Princeton	1.8	0.0	4	Clarkson	1.2
5	**Wisconsin**	1.8	0.5	5	Lake Superior	1.7	0.0	5	Princeton	1.1
6	**Michigan Tech**	1.6	0.7	6	**Wisconsin**	1.4	0.3	6	**Denver**	1.0
7	Lake Superior	1.4	0.0	7	Clarkson	1.3	-0.3	7	St Lawrence	0.9
8	Cornell	1.3	-0.3	8	**Michigan Tech**	1.2	0.4	8	**Wisconsin**	0.8
9	Clarkson	1.2	-0.5	9	Miami (Ohio)	1.0	0.0	9	Harvard	0.8
10	**Colorado College**	1.2	1.5	10	Michigan	1.0	0.0	10	Boston University	0.7
14	**North Dakota**	0.8	0.7	14	**Colorado College**	0.7	1.2	12	**Michigan Tech**	0.6
17	**Minnesota-Duluth**	0.7	0.6	20	**North Dakota**	0.3	0.4	19	**Colorado College**	0.1
20	**Minnesota**	0.5	0.7	21	**Minnesota-Duluth**	0.2	0.3	24	**Minnesota**	0.0
21	**St Cloud**	0.5	0.5	22	**Minnesota**	0.2	0.4	26	**North Dakota**	-0.3
31	Air Force	-1.1	-1.8	24	**St Cloud**	0.0	0.2	27	**Minnesota-Duluth**	-0.3
35	**Northern Michigan**	-1.5	0.5	34	Air Force	-1.3	-1.6	29	**St Cloud**	-0.5
				38	**Northern Michigan**	-1.7	0.2	39	Air Force	-1.1
								42	**Northern Michigan**	-1.2

Δ = Change in overall rating as a result of including CC - AF game.

Table 3. Comparison of Model Effectiveness (Weeks 8–25)

Method	α	Percent Correct	MAD Goal Differential	Average -log (prob)	Root MSE Goals	Med Abs Pred Err
Additive-LSQ	0.00	74.0%	2.22	2.08	2.02	1.34
Additive-LSQ	0.05	74.8%	2.21	2.07	2.00	1.34
Additive-LSQ	0.10	74.8%	2.20	2.07	2.00	1.33
Additive-LSQ	1.00	70.4%	2.34	2.09	2.04	1.38
RPI		68.9%				
TCHCR		71.0%				

Table 4. Additive Model - Least Squares Regression $(\alpha = 0.10)$
(Through March 21, 1993 – Week 23)

Overall Rank	Team	Record			Offense Rating	Offense Rank	Defense Rating	Defense Rank	Overall Rating
1	Maine	37	1	2	6.1	1	2.5	1	3.6
2	Michigan	28	6	2	5.8	2	2.7	2	3.2
3	Lake Superior	27	7	5	4.7	5	2.9	3	1.8
4	Boston University	28	8	2	4.8	4	3.1	5	1.7
5	Clarkson	20	9	5	4.5	7	2.9	4	1.6
6	Miami (Ohio)	27	8	5	4.6	6	3.1	6	1.5
7	Minnesota-Duluth	26	10	2	4.9	3	3.6	13	1.3
8	Harvard	22	5	3	4.3	10	3.1	7	1.2
9	Wisconsin	23	14	3	4.4	8	3.4	10	1.0
10	Michigan State	22	14	2	4.0	24	3.3	8	0.6
11	RPI	19	11	3	4.0	22	3.4	9	0.6
12	Michigan Tech	17	15	5	4.1	17	3.5	11	0.6
13	Northern Michigan	20	17	4	4.3	11	3.7	17	0.5
14	Brown	16	11	3	4.4	9	4.0	24	0.4
15	New Hampshire	18	17	3	4.1	19	3.7	14	0.4
16	UMass-Lowell	20	17	2	4.2	13	3.8	19	0.4
17	St Lawrence	17	12	3	4.1	14	3.8	20	0.3
18	Minnesota	21	11	8	4.0	23	3.7	16	0.3
19	Providence	16	16	4	4.0	20	3.9	22	0.1
20	St Cloud	14	18	2	3.8	25	3.7	15	0.1
21	Ferris State	18	16	4	3.7	30	3.7	18	0.0
22	Yale	15	12	4	4.1	18	4.1	26	-0.1
23	Western Michigan	20	16	2	3.8	26	3.9	21	-0.1
24	Alaska-Fairbanks	10	11	2	4.2	12	4.3	28	-0.1
25	Bowling Green	18	21	1	4.1	15	4.2	27	-0.1
26	Denver	19	17	2	3.7	28	4.1	25	-0.4
27	Kent	12	21	3	4.0	21	4.6	36	-0.6
28	Vermont	11	16	3	2.9	38	3.5	12	-0.6
29	North Dakota	12	25	1	3.7	29	4.4	29	-0.6
30	Northeastern	10	24	1	4.1	16	5.0	40	-0.9
31	Colgate	11	18	2	3.6	32	4.6	35	-1.0
32	Dartmouth	11	16	0	3.5	33	4.6	34	-1.1
33	Illinois-Chicago	9	25	2	3.3	35	4.4	30	-1.1
34	Boston College	9	24	5	3.4	34	4.5	32	-1.1
35	Princeton	8	17	3	3.1	36	4.4	31	-1.3
36	Merrimack	12	20	2	3.6	31	5.0	39	-1.4
37	Colorado College	8	28	0	3.8	27	5.2	41	-1.4
38	Alaska-Anchorage	9	11	3	2.5	42	4.0	23	-1.4
39	Notre Dame	6	27	2	3.0	37	4.7	37	-1.7
40	Cornell	6	19	1	2.8	39	4.5	33	-1.7
41	Union	3	22	0	2.3	44	4.9	38	-2.6
42	Ohio State	4	30	2	2.8	40	5.6	43	-2.8
43	Army	2	5	0	2.5	41	5.6	42	-3.0
44	Air Force	4	17	1	2.3	43	5.6	44	-3.3

$\hat{\mu} = 3.95$ $\hat{H} = \pm 0.44$ goals per game.

Table 5. 1993 NCAA Division I Men's Ice Hockey Tournament

Date	Prediction				Score
3/26/93	4.2–2.9	Clarkson	vs	Minnesota	1–2
	4.1–3.4	Harvard	vs	Northern Michigan	2–3 (OT)
	4.9–4.1	Minnesota-Duluth	vs	Brown	7–3
	4.0–3.6	Miami (Ohio)	vs	Wisconsin	1–3
3/27/93	5.8–2.5	Maine	vs	Minnesota	6–2
	4.5–3.4	Boston University	vs	Northern Michigan	4–1
	4.3–3.9	Lake Superior	vs	Minnesota-Duluth	4–3
	5.3–3.1	Michigan	vs	Wisconsin	4–3 (OT)
4/01/93	3.8–3.7	Lake Superior	vs	Boston University	6–1
	4.8–4.3	Maine	vs	Michigan	4–3 (OT)
4/03/93	5.0–3.2	Maine	vs	Lake Superior	5–4

Shootouts have been criticized as unfair methods for deciding tie games. A statistical analysis compares the shootout and two alternative methods.

Chapter 25

Overtime or Shootout: Deciding Ties in Hockey

William Hurley

Introduction

The year 1994 appears to have been the Year of the Shootout. A shootout, in which five players from each team attempt to score in a series of one-on-one contests with the opposing team's goaltender, decided the Olympic Gold Medal hockey game, the deciding game in the World Hockey Championship, and, in soccer, the World Cup. Most fans see shootouts as exciting. The players, however, argue to a man that they are a poor way to decide an important championship, largely because a shootout results in a more random outcome.

There is a great deal to be said for the players' position. A championship hockey game ought to be decided in the traditional way—five skaters against five skaters until a winner is determined. However, there is a need for some mechanism to shorten the tournament playoff games that precede the championship game. For instance, a semifinal game that went to four overtime periods before one of the teams finally scored would seriously damage that team's chances in a championship game played two days later.

To my knowledge, there is minimal literature on tie-breaking mechanisms. Liu and Schutz (1994) examine the National Hockey League's regular season 5-minute overtime period and find that the better team has a 65% chance of scoring first in overtime play.

This article compares three formats for deciding tie games. In traditional overtime (OT), the teams play until one of the teams scores. In a shootout format (SH), teams play overtime for up to y minutes and if no team has scored, there is a shootout. A third format is a variation of the existing shootout format. The only way to have shootouts *and* soothe the players is to have a period of regular play *following* a shootout. Here is one possibility if two teams are tied at the end of regulation time. Step 1: Immediately run a shootout. The team winning the shootout would be awarded a goal. Step 2: Then play an overtime period, termed a *recourse period*, having a *design length* of y minutes. If the team winning the shootout scores first, the game is over. If there is no goal scored in the overtime, the game is over. If the team that lost the shootout scores, however, stop

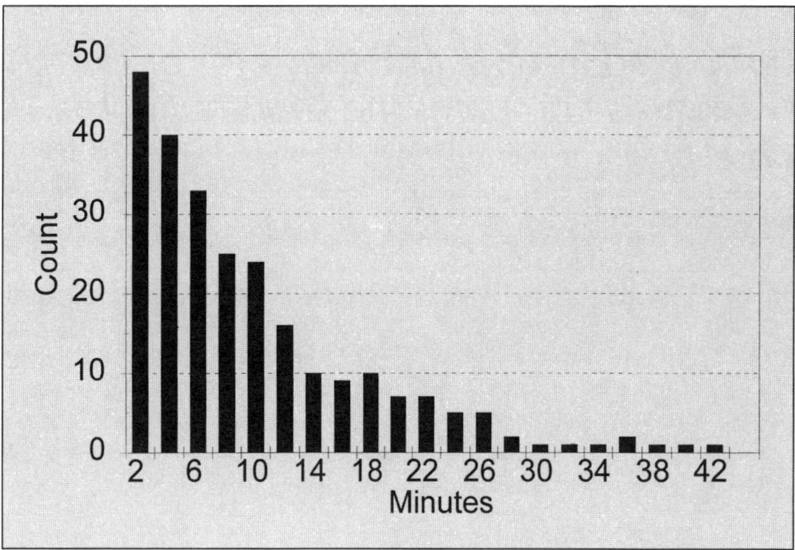

Figure 1. Frequency plot of the time to the winning goal in overtime.

play and repeat steps 1 and 2. I term this overtime format a *shootout with recourse* (SR).

In this article, we compare the three formats with regard to the expected length of the overtime, the variance of this length, and the stronger team's probability of winning.

The Length of Overtime

An important factor in this analysis is the distribution of the time between goals in a hockey game. To estimate this distribution, I examined the 251 National Hockey League playoff overtime games between 1970 and 1993. The data are taken from *The NHL Official Guide and Record Book, 1993-94* (The National Hockey League, New York, 1993). A frequency plot of the time of the winning goal is shown in Fig. 1. These data are consistent with an exponential distribution having a mean of 9.15 minutes (see Box).

Using the exponential distribution, and assuming that the two teams are of similar strength (quality), the expected length of an overtime period and its variance can be calculated for each overtime format. Traditional overtime is over at the point of the first goal, so the expected length is just the mean of the exponential distribution, 9.15 minutes. The standard deviation of the length is also 9.15.

To obtain the expected length of a traditional shootout, we must consider two possibilities—either the game ends on a goal during the sudden-death period or it ends with a shootout. Given these possibilities and a sudden-death period of length y, the expected length of overtime can be shown to be $(1-e^{-\lambda y})/\lambda$, where $1/\lambda$ is the mean time to a goal (9.15 minutes). The standard deviation of the length of overtime is $\sqrt{1-2\lambda e^{-\lambda y}-e^{-2\lambda y}}/\lambda$. The first two columns of Table 1 show the expected length and standard deviation for various overtime lengths. Note that as the length of the sudden-death period gets larger, a shootout is less likely, and, thus, for large y, this overtime format is equivalent to the traditional format.

Computing the expected length and standard deviation of the shootout with recourse requires

The Exponential Distribution

The exponential distribution is often used to approximate the distribution of a random variable that is the waiting time until some event, for example, a goal in a hockey game. The probability density function is $p(x) = \lambda e^{-\lambda x}$ and the mean waiting time is $1/\lambda$, as is the standard deviation. The probability of waiting beyond some time T for the event is $e^{-\lambda T}$.

The exponential distribution has a certain "memoryless" property. For instance, given that we have waited T time units with no event, the expected waiting time is $1/\lambda$, just as it was at the start. In many applications, like waiting for parts to fail, this memoryless property is not realistic. The exponential distribution seems to provide a good approximation to the waiting time for hockey scores, however.

Table 1—Mean Length and Standard Deviations for the SH and SR Formats for Various Design Lengths

	SH		SR	
y	Mean length	S.D. length	Mean length	S.D. length
5	3.9	6.7	4.9	6.8
10	6.1	8.3	9.1	10.8
15	7.4	8.8	12.4	13.2
20	8.1	9.0	14.6	14.7
25	8.6	9.1	16.1	15.8

an assumption about the probability of winning the initial shootout. We take this to be .5 for each team. The expected length is found by averaging over three possibilities:

1. The team that won the shootout scores first in the recourse period and, hence, wins the game.
2. Neither team scores in the recourse period and, hence, the shootout winner wins the game.
3. The shootout loser scores first in the recourse period, thus giving rise to another shootout.

This last possibility makes the shootout with recourse longer than the traditional shootout format if the recourse period has the same length as the sudden-death period of the traditional shootout.

Expressions for the expected length and standard deviation of the SR format are developed in the sidebar. Values of the expected length and standard deviation for various recourse period design lengths are presented in Table 1. Note that a recourse period having a design length of 5 minutes has an expected length of 4.9 minutes, or about half the length of the standard overtime format (9.15 minutes).

For the 5-minute SR, the probability that there is a second shootout is just the probability that the team losing the shootout scores in the first recourse period, or $p = .5 \int_0^5 \lambda\, e^{-\lambda t}\, dt = .21$, and the expected number of shootouts is $1(1-p) + 2p(1-p) + 3p^2(1-p) + \ldots = 1/(1-p) = 1.27$.

Computing Expected Length of Overtime and the Probability That the Stronger Team Will Win

The calculations to produce Tables 1 and 2 are averages over the possible outcomes. We illustrate two of them here.

The expected length of the standard shootout is either the time of the first goal (exponential distribution with parameter $1/\lambda$ or y minutes, whichever comes first). Denoting the length of this format by L_{SH}, the average length is

$$E(L_{SH}) = \int_0^y t\lambda e^{-\lambda t}\, dt + y\int_y^\infty \lambda e^{-\lambda t}\, dt$$
$$= (1 - e^{-\lambda t})/\lambda$$

where the first term is the case in which a goal is scored and the second term is the case in which no goal is scored.

The shootout with recourse is a little trickier because of the possibility of multiple periods. Letting L_{SR} denote the length of this format, the expected length is now computed in three pieces:

$$E(L_{SR}) = 0.5\int_0^y [t + E(L_{SR})]\, \lambda e^{-\lambda t}\, dt$$
$$+ 0.5\int_0^y t\lambda e^{-\lambda t}\, dt$$
$$+ y\int_y^\infty \lambda e^{-\lambda t}\, dt$$

The first integral covers the case in which the loser of the shootout scores at time t with resulting expected length $t + E(L_{SR})$, the second integral covers the case in which the team winning the shootout scores in the recourse period, and the third the case in which there is no goal scored in the recourse period. Standard deviations are calculated using the same approach to determine the expected value of the length squared.

If two teams score goals independently and exponentially with parameters λ_S and λ_W, then the probability that the stronger team will win a traditional overtime is $\lambda_S/(\lambda_S + \lambda_W)$. For shootouts, the probability that the stronger team will win is computed by listing all of the possible outcomes. In a shootout with recourse, the stronger team will win if

- the stronger team wins the shootout, and no goals are scored in overtime
- the stronger team wins the shootout, and the stronger team scores first in overtime
- the stronger team wins the shootout, the weak team scores first in overtime, and the stronger team wins given the remainder of the process
- the weak team wins the shootout, the stronger team scores first in overtime, and the stronger team wins given the remainder of the process

If we take p_S to be the probability that the stronger team wins a shootout, p to be the stronger team's overall probability of winning, $f_S(y)$ to be the probability that the stronger team scores first during a recourse period of length y minutes, and $f_W(y)$ to be the probability that the weak team will score first during a recourse period of length y minutes, then

$$p = p_S\,[1 - (1-e^{-\lambda_S y})\,(1-e^{-\lambda_W y})$$
$$+ p_S f_S(y)] + p\, p_S f_W(y)$$
$$+ p(1-p_S) f_S(y)$$

which can be solved for p. The probabilities that the strong or weak team scores first, f_S and f_W, are computed from two independent exponential distributions.

Probability That the Stronger Team Wins

A typical objection to the standard shootout is that it results in outcomes determined mainly by luck or by an arbitrarily chosen skill. In this section, we compare the degree to which the three overtime formats reward the stronger team.

Suppose that two teams of unequal strength are tied at the end of regulation time. Label the teams S (strong) and W (weak). Suppose that the time between goals for team S has an exponential distribution with mean $1/\lambda_S$, and the time between goals for team W has exponential distribution with mean $1/\lambda_W$. We assume that the two goal-scoring processes operate independently. No doubt this last assumption is not completely realistic. For instance, the distri-

Table 2—Probabilities That the Stronger Team Wins for Various Parameter Values (the probability that the stronger team wins in traditional overtime is .6055)

	SH			SR		
y	p_S=.5	p_S=.6	p_S=.7	p_S=.5	p_S=.6	p_S=.7
5	.5006	.5949	.6892	.5003	.6003	.7002
10	.5034	.5859	.6683	.5019	.6018	.7016
15	.5089	.5780	.6472	.5053	.6050	.7044
20	.5164	.5729	.6294	.5105	.6100	.7087
25	.5251	.5706	.6161	.5173	.6165	.7143

bution of time between goals for one team clearly depends on the quality of the opposing team. Nevertheless, in thinking about a game between two teams of fixed quality, the independence assumption may not be far off because we can incorporate the independence into the choice of approximate exponential parameters.

For traditional overtime, the probability that the stronger team wins is $\lambda_S(\lambda_S+\lambda_W)$. What are plausible values for λ_S and λ_W? I examined NHL Stanley Cup Final Series data for the years between 1980 and 1991. Using the scores for each of the 62 games, I calculated the total goals scored by Stanley Cup winning teams (252) and the total goals scored by Stanley Cup losing teams (164). The average time between goals for winning teams is 14.76 minutes (λ_S = .0677) and for losing teams 22.68 minutes (λ_W = .0441). These estimates ought to be consistent with the average time to the first goal of 9.15 minutes computed earlier in the article. Given our assumptions, the time to the first goal has an exponential distribution with parameter $\lambda_S + \lambda_W$, and using λ_S and λ_W, the average time to the first goal is 8.94 minutes, which is not significantly different from 9.15 minutes. Hence, based on these estimates for λ_S and λ_W, the probability that the stronger team wins a traditional overtime is .6055.

The probability that the stronger team wins under the standard shootout or the shootout with recourse depends on the length of the associated overtime, y, and the probability that the stronger team wins the shootout. It would seem that the probability that the stronger team wins a shootout should be greater than .5, but it may actually be close to .5 if the weaker team has five relatively good shooters. Table 2 presents probabilities that the stronger team wins for each shootout format and for various values of the length of overtime. A brief explanation of the computation of these probabilities is given in Box 2.

The interesting aspect of the probabilities in Table 2 is that *the probability that the stronger team wins is dominated by its probability of winning the shootout.* This is especially true for short overtime periods. As the overtime period gets longer, however, this probability gets closer to the traditional overtime result. The exception appears to be the column for p_S = .6, where the probabilities appear to be moving away from .6055. However, this is not the case. If y is sufficiently large, we again reach .6055.

Summary

The purpose of this article has been to compare three overtime formats on the basis of expected length of the game and the probability that the stronger team wins the overtime.

Based on National Hockey League data, the shootout with recourse format, having a design length of 5 minutes, would have a much smaller expected length than a standard international shootout format in which there is a shootout after 20 minutes. Even with this short design length, there is no guarantee that the SR format will always be shorter. One solution would be to limit the number of recourse periods.

Finally, the probability that a stronger team wins any overtime format having a shootout is dominated by the stronger team's probability of success in a shootout. Assuming that a stronger team's probability of success in a shootout is less than it is in regular play, overtime formats with shootouts give the weaker team a better chance of winning.

Additional Reading

Liu, Y., and Schutz, R. W. (1994), "Overtime in the National Hockey League: Is It a Valid Tie-Breaking Procedure?" School of Human Kinetics, University of British Columbia, Canada.

The role of team ability, home ice, and the "hot hand" in the Stanley Cup finals.

Chapter 26

It Takes a Hot Goalie to Raise the Stanley Cup

Donald G. Morrison and David C. Schmittlein

In May of 1997, Mike Vernon skated around the Philadelphia Spectrum with the National Hockey League's Stanley Cup raised high in his hands. Vernon played spectacularly well during the playoffs, including the finals against the Philadelphia Flyers. He was voted the Most Valuable Player (MVP) of the playoffs. Vernon's Detroit Red Wings ended a 42-year quest to regain the Stanley Cup. Was Vernon the latest in a long line of Hot Goalies to lead his team to victory? We think so—and the statistical evidence is quite compelling.

In ice hockey as in most professional sports, winning the regular season "title" is much less important than winning the postseason playoffs. Many years ago, there were only six teams in the National Hockey League and the top four made the playoffs. They played two rounds of best-of-seven series. Now there are 26 teams and 16 teams make the four-round (again, best-of-seven) playoff series. At one point in the 1970s, there were 21 teams and all but five made the playoffs. This caused many to feel that the 80-game regular season was almost meaningless. Quotes such as "they play 80

Goalie Mike Vernon's experience in the net, and his "hot hand" made him the logical choice during the Stanley Cup finals.

Courtesy of Jim Leary/Bruce Bennett Studios

games merely to determine home ice advantage" appeared frequently. That is, when Teams A and B meet in the playoffs, the seven games (if necessary) are played on A and B's home arenas in the following order: A A B B A B A. When one team wins four games, the series is terminated. Thus, if a crucial seventh game is played, it is on the home ice of Team A, the team with the better regular season record. We will return to the home ice factor, but first we give a related anecdote.

Mike Vernon, the Hot Goalie

After a 42-year hiatus, the treasured Stanley Cup returned to Detroit as the Red Wings swept the Philadelphia Flyers four games to none. The hero and MVP of the playoffs was Mike Vernon, the Red Wings' veteran goalie. In fact Vernon played every one of the games in the four rounds of the playoffs. During the regular season, Vernon played about one-third of the games as Chris Osgood's back-up. Osgood had the better winning percentage and goals against average during the regular season. Legendary coach Scotty Bowman felt Vernon was the "hot" goalie as the playoffs started, however, and went with his hot goalie for all of the playoff games. (This is not uncommon because all coaches now use two or more goalies during the regular season and many stick with just one goalie in the playoffs.)

In fact, two years earlier the Red Wings lost the Stanley Cup final series—in another 4–0 sweep—to New Jersey. In *that* series New Jersey's goaltending was seen as superior, and Detroit's was much criticized. It was this disappointing final series that led Detroit to go out and acquire the services of Mike Vernon. We thank an alert reviewer for this note on the acquisition of Vernon by Detroit.

Hot Goalie Versus Home Ice

If goalies did not get hot and home ice had no effect, and if the two teams were of equal ability, the outcome of the Stanley Cup final series games would be a Bernoulli process, with each team having a .5 probability of winning each game. The duration of the series would be four games if Team A wins the first four or Team B wins the first four. This would happen with probability $(1/2)^4 + (1/2)^4 = 1/8 = .125$. Similar reasoning gives the full probability distribution:

Duration of series	Probability
4	.1250
5	.2500
6	.3125
7	.3125

If instead the teams were of equal ability, but the home ice advantage was so strong that the home team *always* won, then every series would go to seven games because each team plays three home games out of the first six. On the other hand, if the team with the hot goalie always won, then each series would last the minimum of four games. In the less extreme cases, home ice advantage will push the preceding distribution more to 6- and 7-game series. For example, if the home team won each game with probability .7 (approximately the league-wide figure for the regular season), the probability of a four-game sweep would drop from the value of .1250 to .0882. Conversely, the impact of a "hot goalie" will cause more 4- and 5-game series.

Despite being the goalie of choice for the Detroit Red Wings during the 1997 Stanley Cup Finals, Mike Vernon found himself playing for the San José Sharks the following season.

Stanley Cup Final Duration

The Stanley Cup final series has been contested 59 times in a best-of-seven-games series format. The frequency table of duration for these series is:

Duration of series	Number
4	19
5	15
6	15
7	10
	Total 59

Table 1—Success of the Home Team Versus That of the Away Team in Stanley Cup Final Series of Varying Duration

Series record	Frequency	Number of series won by the home team	Number of series won by the away team
4–0	19	14	5
4–1	15	12	3
4–2	15	12	3
4–3	10	8	2
Total	59	46	13

Clearly this shows a lot more sweeps (4-game series) than one would expect *even with no home-ice advantage* and equal ability.

We also calculated the maximum likelihood estimate of the game-victory probability p for the better of the two teams assuming no home-ice advantage—that is, still a Bernoulli process, but allowing p to be different from .5. This resulted in an estimated value of $p = .73$. The observed frequency table of duration for the final series and the two sets of expected values are:

Series duration	Observed # of series	Expected # of series: Bernoulli $p = .5$	Expected # of series: Bernoulli $p = .73$
4	19	7.4	17.1
5	15	14.7	19.0
6	15	18.4	13.9
7	10	18.4	9.0

The Bernoulli $p = .5$ model is strongly rejected because $\chi^2 = 22.65$, compared to a critical value of $\chi^2(.01) = 11.30$ with 3 df. The Bernoulli $p = .73$ model is not rejected; that is, $\chi^2 = 1.25$, while $\chi^2(.05) = 5.99$ with 2 df.

Allowing heterogeneity across years in the game-victory probability p (for the $p = .73$ model) does not help much—it would boost the expected number of 7-game series (a help), but reduce the number of 4-game series (a hurt).

Explaining the Data: Hot Goalie, Home Ice, and/or Dominant-Team Effects

The hot-goalie theory clearly can explain the far too many 4-game series and the almost as dramatic dearth of 7-game series. But so does the Bernoulli (no home-ice advantage) model with the better team being "much better" with $p = .73$— that is, a dominant-team hypothesis. Which of these competing models is more plausible? And where does a home-ice advantage fit in?

During the regular season the elite teams do not win three-fourths of their games. Of course, some regular season games end in ties (postseason games do not), but during the regular season, the elite teams play many very poor teams. The Stanley Cup finals usually pit two elite teams against each other. The better team may be a little better, but $p = .73$ implies the better team would win over the long run three times as often as they lose.

To put this in perspective, consider baseball, which with its 162-game regular season schedule comes the closest to "the long run." No team in this century has won three-fourths of its games. The inaugural 1962 Amazin' Mets went 40 and 120 (mercifully two rainouts were never played). Thus, the worst baseball team in 100 years lost three-quarters of its games. For the $p = .73$ model to be plausible means that the two teams in the Stanley Cup finals are as disparate in their abilities as the 1962 Mets versus the average remaining National League teams of 1962. This just doesn't seem reasonable. Actually, examining our series-duration histogram in slightly more detail will show that such a $p = .73$ dominant-team hypothesis does not adequately explain the data. Specifically, the inclination of the home team to win Stanley Cup final series of varying duration will enable us to sort out much more clearly the dominant-team, home-ice, and hot-goalie effects. We will be able to rule out one of these three effects, and see that the other two together (but neither alone) suffice to explain the data.

Of the 59 Stanley Cup finals in a best-of-seven game format, the distribution for winner's record in the series, and whether the series was won by the home or away team (i.e. "home team" is the team with most regular-season points and so the home-ice advantage if the series goes seven games) is indicated in Table 1. If the two teams' point totals were equal, the number of games won in the regular season is the tiebreaker to determine the team with the home-ice advantage. So, overall, the home team (i.e., better regular season record) won 46/59 = 78% of these series.

This seems a reasonably large fraction. It is also remarkably close to our game-victory probability for the "better" team, estimated for a homogeneous Bernoulli process (i.e., .73). Maybe having home-ice is indeed worth fighting for.

Is the Home-Ice Effect Real?

As suggested earlier, the home team would be expected to prevail in Stanley Cup finals as a result of two distinct phenomena:

1. Dominant team effect: Because home ice goes by design to the team with the better regular-season record, that team should generally be better even in the absence of a "real" home-ice effect (i.e., increased propensity to win when playing at home).
2. "Real" home-ice effect: The "home" team gets to play the "odd" game number seven (if game seven is required) at home.

Table 1 shows clearly that the high 78% series-win probability by the home team has everything to do with factor 1 above (dominant team) and probably nothing at all to do with factor 2— that is, a real home-ice effect. This is seen in two ways:

1. Only 10 of the 59 series went seven games—that is, enabling a real home-ice effect to arise. About twice as many series (19) went only four games—and were therefore balanced with respect to home ice (two home games, two away games).
2. Comparing the series-win percentage in 4-0 series with that in 7-game series, if a "real" home-ice effect occurred we would expect home teams to win substantially more of the 7-game series than the 4-game series. In fact, the "home" team won 14/19 = 74% of the 4-game series and 8/10 = 80% of the 7-game series. This very slight increase is not significant, substantively or statistically.

We should acknowledge that series making it to seven games will tend to be composed of teams that are relatively evenly matched, so it might be argued that the seventh game at home "restores" to the home (better-record) team a win-percentage that it would otherwise have seen erode through six games' worth of Bayesian updating (observing a 3-3 record in those games). Such an explanation is easily ruled offsides, however, by the fact that the home team's win percentage (see Table 1) in 7-game series (80%) is *identical* to its win percentage in both 5-game series (12/15 = 80%) and in 6-game series (12/15 = 80%), the latter being balanced in home ice (3 home games, 3 away games) for each team.

The 82-game regular season in the National Hockey League does indeed determine which teams get the home-ice advantage...

In summary, although a hot-goalie effect cannot be proved conclusively, it could be readily discarded through various possible patterns of empirical results.

Accordingly there is no noteworthy evidence of a "real" home-ice effect in Stanley Cup final series.

Since the Home-Ice Team Is Demonstrably the Dominant Team, What Happened to the Hot-Goalie Hypothesis?

Recall that the home team won 78% of Stanley Cup finals and, from the series-record frequency table, we estimated that the "better" team's game-victory probability was .73, assuming a homogeneous Bernoulli process. Does the similarity of these two numbers mean that the dominant-team hypothesis—that is, the fact that the home team had a better regular season record—suffices to account for the series records observed and, further, eliminate an apparent hot-goalie effect? Nothing could be further from the truth.

Imagine that the home team in each series were indeed that "better" team in the Bernoulli process, having game-victory probability = .73. Then, in such a process, where repeated games' outcomes are independent (e.g., assuming no hot-goalie effect), the probability that the "better" (home) team wins in four games is $.73^4 =$ .2840; and the probability that the "worse" (away) team wins in four is $(1 - .73)^4 = .0053$. Consequently, among series that go exactly four games the dominant-team hypothesis would expect the percentage of such series won by the home team to be .2840/(.2840 + .0053) = .982.

Table 1 showed that home teams won 73.7% (14 of 19) of the 4-game series played. Even with our modest sample size, this observed percentage is significantly different from the 98.2% expected by the dominant-team hypothesis.

As we noted earlier, such 4-game series are balanced with respect to any real home-ice effect: Each team plays two home games and two away games. The other balanced series are those decided in six games. Absent a hot-goalie effect, we can calculate the probability that the home (better) team wins in six games (with p = .73 as previously) as:

$$\binom{5}{3}(.73)^4\,(.27)^2$$

Similarly the probability that the away team wins in six games is:

$$\binom{5}{3}(.27)^4\,(.73)^2$$

Thus, among six-game series the expected proportion won by the home team is $(.73)^2/(.73^2 + .27^2) = .880$. This is again greater than the actual proportion of six-game series won by the home team (12/15 = .80).

For these six-game series the observed proportion (.8) is only one standard deviation below the theoretical value and so not significantly different from it. Nonetheless, when taken together with the preceding four-game series results it is clear that the dominant-team hypothesis cannot explain the home team's observed inclination to win the Stanley Cup finals. The home team wins too few series, and this decrement is statistically significant. Furthermore, a hot-goalie effect—that is, nonindependence between successive game outcomes within a series—can readily close this gap and account for the data observed. In these relatively short series, either team's goalie is of course eminently capable of getting "hot." Thus, one consequence of a hot-goalie effect would be to equalize (to some degree) the overall series wins between home teams and away teams.

A second observation also supports a hot-goalie interpretation—namely, that the shortage in home-team series victories (relative to the Bernoulli process expectation) is greater both arithmatically and statistically for four-game series than it is for the longer six-game series. Naturally, it is more likely for a goalie to get hot *and stay hot* for a four-game sweep than for the six-game series. Accordingly, our equalization effect stemming from the hot-goalie phenomenon should be greater for series going four games than for those that go six games, and this is indeed what we observed.

In summary, although a hot-goalie effect cannot be proved conclusively, it could be readily discarded through various possible patterns of empirical results. Instead, the empirical patterns observed in this article all point toward, rather than away from, such an effect.

Specifically, we find that the team gaining the series home-ice advantage through possessing a better season record does indeed fare better in the Stanley Cup finals. The benefit seems to stem entirely from the dominant team effect—that is, a team with a dominant regular season record probably is the more talented of the two. There is no compelling evidence that having home ice for a decisive seventh game (i.e., a "real" home-ice effect) is of any consequence.

Finally, the dominant-team effect could suffice to explain the duration of Stanley Cup finals—so many series ending so quickly—but only if we accept that the better team has three-to-one odds of beating the weaker team game in and game out. This seems unlikely, to say the least. Furthermore, the dominant-team effect does not satisfactorily explain the percent of series won by the home team: Home teams win too few, especially in four-game series. Both of these latter observations are predicted and explained by a hot-goalie phenomenon.

Conclusion

The 82-game regular season in the National Hockey League does indeed determine which teams get the home ice advantage if a seventh game is necessary. The duration pattern of the 59 Stanley Cup finals played, however, rejects the notion that home ice is in fact a significant advantage. The hot-goalie theory is at least strongly consistent with the observed data. The Bernoulli p = .73 model would be more plausible if data showed that one of the finalists typically had more key players injured than the other team—but, of course, we have no data on that. The best that we can say on the Stanley Cup finals is the following:

1. The home ice advantage is not a factor overall;
2. The "home" team (i.e., team getting home-ice for game 7) does dominate these series but due solely to its inharant ability, reflected by its having had the better, record during the regular season;
3. The hot goalie hypothesis is alive and well.

Finally, what was Vernon's reward for being the hot goalie who led the Red Wings in from 42 years in the Stanley Cupless wilderness? He was released: The Red Wings could only protect two goalies in the upcoming expansion draft. They kept the much younger (and probably better) Chris Osgood and a young backup goalie. Mike with his championship ring and hot goalie résumé—not to mention his large salary—was sent to San José for goal tending duties with the Sharks. Thus, the Red Wings implicitly bought into the notion of the hot goalie—as a transient phenomenon that only lasts for the time it takes to stage the Stanley Cup playoffs.

Part V
Statistical Methodologies and Multiple Sports

Chapter 27

Introduction to the Methodologies and Multiple Sports Articles

Scott Berry

27.1 Introduction

This section includes a wonderful blend of papers of a general nature that address multiple sports, multiple topics, and nontraditional sports. They address important sports problems, important statistical problems, and all are very interesting and well done. They are sorted into three broad classes: hypothesis tests, prediction, and estimation.

27.1.1 Hypothesis Tests

Over the last 20 years the issue of the existence of a "hot hand" has captured the focus and imagination of the statistics-in-sports community. The hot hand is essentially an example of the age-old cliché that success breeds success. Statistically it can be modeled in many ways, but the classical idea is whether success or failure on one trial changes the probability of success or failure on the next trial. Tversky and Gilovich (1989) (also Chapter 21 in this volume) wrote that there was no evidence for the existence of the hot hand effect in basketball. To the sports community the existence of the hot hand is a tautology. This is part of what made the questioning of the existence of such an effect such a powerful idea.

Addressing the existence of the hot hand is a challenge statistically. The standard method is to assume no hot hand effect exists—this assumption forms the null hypothesis. Data is then collected to see if it agrees or, more conclusively, disagrees with what is expected. This classical hypothesis testing is the standard approach that is used in looking for a hot hand effect. Hooke (Chapter 31 in this volume) was one of two papers in the 1989 Fall *Chance* issue to examine the hot hand (also see Larkey, Smith, and Kadane (Chapter 19 in this volume)). Hooke addresses the approach of others, specifically Tversky and Gilovich (Chapter 21), in which they conclude that the data they collected are consistent with the null hypothesis of no hot hand effect. Hooke discusses a historically very difficult problem of making conclusions when classical tests find the data consistent with the null hypothesis. He states that the existence of the hot hand is still very much up in the air. Its effect, if it exists, is smaller than many may have thought, but it clearly could still exist. Not only is this paper important for sports statisticians, but for researchers in every field.

27.1.2 Prediction

This second class includes three papers on the prediction of sports outcomes. In a horse race the outcome of interest is the order of finish of the horses, a permutation of the entrants. Harville (Chapter 30) models the probabilities for the various permutations using a probability that each horse finishes first. He assumes that the odds on a horse winning represent the true probability for a horse finishing first. To reduce the hopelessly large class of permutations, he uses the assumption that the probability of finishing second for horse B, given that horse A has won the race, is proportional to the probability that horse B would finish first in a race with the remaining horses. Checking his model with results from 335 races, Harville finds a nice match—with some small deviations. This can result in bets that are expected money winners, which is rare in horse racing; in particular, show and place betting reaped a positive expected payout in some races (an amazing thing in the world of 16% take pari-mutuel betting!).

Except for possibly a big win by the underdog, the most dramatic event in sports is the big comeback. Games where a team trails by a large margin and comes back to win are analyzed forever. Was it the winning team believing in themselves or was it a choke by the losing team? How difficult was the comeback, really? Stern (Chapter 34) models the progress of a game using a Brownian motion model. While others have modeled the abilities of teams and looked at the probabilities of teams winning based on a normal distribution, this paper looks at these same quantities continuously throughout the game. The Brownian motion approach models the difference between the scores for each team as a normal distribution. The time remaining in the game determines the mean and standard deviation for the remaining point differential. Using this approach, and by modeling the ability of each team, Stern finds that the probability that each team wins a game is conditional based on the score and the time remaining.

He applies the method to National Basketball Association games and Major League Baseball games. The model works very well in basketball because the Brownian motion model assumes a continuous time and scoring structure. While these assumptions are not exactly true in basketball, they are very close. The model works reasonably well in baseball, but does have shortcomings near the end of the game. Stern compares the results to those developed by Lindsey (1961) (Lindsey appears as Chapter 16 in the baseball section (Part II) of this volume).

Mosteller (Chapter 32) presents a collection of sports-related analyses that he has done over his distinguished career. He describes each of these analyses and the lessons each of them have taught him. First he addresses the seemingly innocuous problem of estimating the relative ability of two teams, conditional on the results of a seven-game series. Because the stopping rule for the series depends on the results (when one team wins four games), the unbiased estimator for the probability that one team beats another has some undesirable properties. This result was very important in the history of statistics. As Mosteller says in Chapter 32: "The existence of such unreasonable results has downgraded somewhat the importance of unbiasedness."

Next, he describes a robust statistical approach for ranking National Football League teams using the 1972 season as an example. Interestingly, he did not rank as first the undefeated Super Bowl champion, the Miami Dolphins. In spirit his rankings are similar to those that are now used to partially determine the NCAA champion football team. He writes the following about a lesson he learned: "The nation is so interested in robust sports statistics that it can hog the newspaper space even at an AAAS annual meeting."

He continues with modeling the number of runs in a half-inning of baseball and the 18-hole score of a professional golfer. Throughout the article, Mosteller brings clarity to the main goals of statistical analyses in sports. He highlights very important statistical ideas, answers critical questions about statistics in sports, and provides a means of bringing statistics to the general public.

27.1.3 Estimation

The evaluation of athletes is the focus of this third class of papers. While articles and studies rating athletes are ubiquitous in sports today, there is a lack of good papers on the experimental design of studies for optimizing performance. Roberts (1993) in Chapter 33 brings the revolutionary ideas of Total Quality Management (TQM) to the optimization of athletic performance. The methods of TQM are a natural choice for this application because of its design and analysis for optimizing the performance of specific processes. Roberts discusses and presents data for two interesting examples. The first is an example in which a golfer alters his putting technique. By using intervention analysis Roberts shows clearly that one of the two grips results in a better putting performance. In the second example, a billiards player designs and carries out a 2×2 experiment on the eye position and bridge used for the shot. Roberts also informally describes his career as a distance runner and the thinking he has used to hone his training techniques. One of the interesting aspects of his approach is that one technique may be better for one athlete and worse for another. Each of these examples addresses the challenging problem of finding the correct technique for a single athlete—not a population of athletes.

Efron and Morris (1975) in Chapter 29 address an example of estimating the true batting average of baseball players based on partial season data. The maximum likelihood estimator (MLE) is the player's current batting average. Stein showed that this estimate is inadmissible—there existed estimators which were uniformly better with respect to a squared-error loss function. It turns out that regressing each player's batting average toward the mean of all the players' averages is a uniformly better estimator, a 350% improvement in the example considered.

This idea is now commonly accepted by statisticians, but at the time was shocking. Efron and Morris's seminal paper was a forerunner to the modern movement of hierarchical models and random effects models, which are now ubiquitous in statistics. It is generally thought of as the paper that brought this type of problem—the empirical Bayes problem—into mainstream statistical use. Consequently, this paper is one of the most important in statistics literature, and certainly the most important to statistics in sports literature. While the paper's significance extends well beyond the arena of sports, as in Mosteller's work, it shows that sports can provide the inspiration for many novel statistical applications.

One of the hot button topics in sports is that of comparing players from different eras. How does Babe Ruth compare to today's sluggers? How does Rocket Richard compare to today's goal scorers? How do Ben Hogan and Jack Nicklaus compare to Tiger Woods? These questions frequently bring about heated arguments which almost always end with "It's impossible to compare players from different eras!" In Chapter 28, Berry, Reese, and Larkey (1999) use statistical tools to compare the players from different eras in Major League Baseball, the National Hockey League, and professional golf. While Babe Ruth never played with Mark McGwire, Ruth played with players, who played with players, who played with players, who did play with McGwire. These overlaps are bridges to estimate the relative abilities for all players in the history of a sport. Using a Bayesian hierarchical model approach, Berry, Reese, and Larkey simultaneously estimate the aging effects, season effects, and abilities of the players.

Comparisons of players are interesting, and their conclusions about the changing nature of the talent pool in each sport are fascinating. They find that many more good home run hitters are playing today than played years ago. Babe Ruth would still be a great home run hitter (third best of all-time players), but there are many more players similar to him in today's game. In each of the three sports, the talent pool gets better and better through time. In particular, the middle-level player is getting much better through time while the best players of the different eras are comparable.

Comments by Jim Albert, Jay Kadane, and Michael Schell on this article provide fuel for discussion (for the comments, see JASA, 94 (1999), pp. 677–686). Their enthusiasm and interest lend further evidence to the powerful nature of the problem of comparing players from different eras.

27.2 Summary

As stimulating, inventive, and revolutionary as the papers presented in this section are, they do not provide the last word on the questions addressed. The reader is encouraged to continue exploring these research topics with the following papers. The hot hand: Albright (1993), Berry (1997), Berry (1999a), Jackson and Mosurski (1997), and Stern (1995). Probabilities of ranked results: Graves, Reese, and Fitzgerald (2001), Stern (1990), and Stern (1998). Prediction of games from intermediate results: Berry (2000), Cooper, DeNeve, and Mosteller (1992), and Zaman (2001). Distance running performance: Martin and Buoncristiani (1999). Regression to mean player performance: Berry (1999b) and Schall and Smith (2000b). Comparison of players from different eras: Schell (1999). Aging effects on player performance: Albert (1992) and Schall and Smith (2000a).

References

Albert, J. (1992), "A Bayesian analysis of a Poisson random effects model for homerun hitters," The American Statistician, 46, 246–253.

Albright, S. C. (1993), "A statistical analysis of hitting streaks in baseball," Journal of the American Statistical Association, 88, 1175–1183 (with discussion).

Berry, S. M. (1997), "Judging who's hot and who's not," Chance, 10 (2), 40–43.

Berry, S. M. (1999a), "Does 'the zone' exist for home-run hitters?" Chance, 12 (1), 51–56.

Berry, S. M. (1999b), "How many will Big Mac and Sammy hit in '99," Chance, 12 (2), 51–55.

Berry, S. M. (2000), "My Triple Crown," Chance, 13 (3), 56–61.

Berry, S. M., Reese, C. S., and Larkey, P. D. (1999), "Bridging different eras in sports," Journal of the American Statistical Association, 94, 661–676 (with discussion).

Cooper, H., DeNeve, K. M., and Mosteller, F. (1992), "Predicting professional game outcomes from intermediate game scores," Chance, 5 (3/4), 18–22.

Efron, B. and Morris, C. (1975), "Data analysis using Stein's estimator and its generalizations," Journal of the American Statistical Association, 70, 311–319.

Graves, T., Reese, C. S., and Fitzgerald, M. (2001), *Hierarchical Models for Permutations: Analysis of Auto Racing Results*, Los Alamos National Laboratory Technical Report, Los Alamos, NM.

Harville, David A. (1973), "Assigning probabilities to the outcomes of multi-entry competitions," Journal of the American Statistical Association, 68, 312–316.

Hooke, R. (1989), "Basketball, baseball, and the null hypothesis," Chance, 2 (4), 35–37.

Jackson, D. and Mosurski, K. (1997), "Heavy defeats in tennis: Psychological momentum or random effect?" Chance, 10 (2), 27–34.

Larkey, P. D., Smith, R. A., and Kadane, J. B. (1989), "It's okay to believe in the 'hot hand,'" Chance, 2 (4), 22–30.

Lindsey, G. R. (1961), "The progress of the score during a baseball game," American Statistical Association Journal, September, 703–728.

Martin, D. E. and Buoncristiani, J. F. (1999), "The effects of temperature on marathon runners' performance," Chance, 12 (4), 20–24.

Mosteller, F. (1997), "Lessons from sports statistics," The American Statistician, 51, 305–310.

Roberts, H. V. (1993), "Can TQM improve athletic performance?" Chance, 6 (3), 25–29; 69.

Schall, T. and Smith, G. (2000a), "Career trajectories in baseball," Chance, 13 (4), 35–38.

Schall, T. and Smith, G. (2000b), "Do baseball players regress to the mean?" The American Statistician, 54, 231–235.

Schell, M. J. (1999), *Baseball's All-Time Best Hitters*, Princeton, NJ: Princeton University Press.

Stern, H. (1990), "Models for distributions on permutations," Journal of the American Statistical Association, 85, 558–564.

Stern, H. (1994), "A Brownian motion model for the progress of sports scores," Journal of the American Statistical Association, 89, 1128–1134.

Stern, H. (1995), "Who's hot and who's not," in *Proceedings of the Section on Statistics in Sports*, Alexandria, VA: American Statistical Association, 26–35.

Stern, H. (1998), "How accurate are the posted odds?" Chance, 11 (4), 17–21.

Tversky, A. and Gilovich, T. (1989), "The cold facts about the 'hot hand' in basketball," Chance, 2 (1), 16–21.

Zaman, Z. (2001), "Coach Markov pulls goalie Poisson," Chance, 14 (2), 31–35.

Chapter 28

Bridging Different Eras in Sports

Scott M. BERRY, C. Shane REESE, and Patrick D. LARKEY

This article addresses the problem of comparing abilities of players from different eras in professional sports. We study National Hockey League players, professional golfers, and Major League Baseball players from the perspectives of home run hitting and hitting for average. Within each sport, the careers of the players overlap to some extent. This network of overlaps, or bridges, is used to compare players whose careers took place in different eras. The goal is not to judge players relative to their contemporaries, but rather to compare all players directly. Hence the model that we use is a statistical time machine. We use additive models to estimate the innate ability of players, the effects of aging on performance, and the relative difficulty of each year within a sport. We measure each of these effects separated from the others. We use hierarchical models to model the distribution of players and specify separate distributions for each decade, thus allowing the "talent pool" within each sport to change. We study the changing talent pool in each sport and address Gould's conjecture about the way in which populations change. Nonparametric aging functions allow us to estimate the league-wide average aging function. Hierarchical random curves allow for individuals to age differently from the average of athletes in that sport. We characterize players by their career profile rather than a one-number summary of their career.

KEY WORDS: Aging function; Bridge model; Hierarchical model; Population dynamics; Random curve.

1. INTRODUCTION

This article compares the performances of athletes from different eras in three sports: baseball, hockey, and golf. A goal is to construct a statistical time machine in which we estimate how an athlete from one era would perform in another era. For examples, we estimate how many home runs Babe Ruth would hit in modern baseball, how many points Wayne Gretzky would have scored in the tight-checking National Hockey League (NHL) of the 1950s, and how well Ben Hogan would do with the titanium drivers and extra-long golf balls of today's game.

Comparing players from different eras has long been pub fodder. The topic has been debated endlessly, generally to the conclusion that such comparisons are impossible. However, the data available in sports are well suited for such comparisons. In every sport there is a great deal of overlap in players' careers. Although a player that played in the early 1900s never played against contemporary players, they did play against players, who played against players, . . . , who played against contemporary players. This process forms a *bridge* from the early years of sport to the present that allows comparisons across eras.

A complication in making this bridge is that the overlapping of players' careers is confounded with the players' aging process; players in all sports tend to improve, peak, and then decline. To bridge the past to the present, the effects of aging on performance must be modeled. We use a nonparametric function to model these effects in each sport.

An additional difficulty in modeling the effects of age on performance is that age does not have the same effect on all players. To handle such heterogeneity, we use random effects for each player's aging function, which allows for modeling players that deviate from the "standard" aging pattern. A desirable effect of using random curves is that each player is characterized by a career profile, rather than by a one-number summary. Player A may be better than player B when they are both 23 years old, and player A may be worse than player B when they are both 33 years old. Section 3.4 discusses the age effect model.

By modeling the effects of age on the performance of each individual, we can simultaneously model the difficulty of each year and the ability of each player. We use hierarchical models (see Draper et al. 1992) to estimate the innate ability of each player. To capture the changing pool of players in each sport, we use separate distributions for each decade. This allows us to study the changing distribution of players in each sport over time. We also model the effect that year (season) has on player performance. We find that, for example, in the last 40 years, improved equipment and course conditions in golf have decreased scoring by approximately 1 shot per 18 holes. This is above and beyond any improvement in the abilities of the players over time. The estimated innate ability of each player, and the changing evolution of each sport is discussed in Section 7.

Gould (1996) has hypothesized that the population of players in sport is continually improving. He claimed there is a limit to human ability—a wall that will never be crossed. There will always be players close to this wall, but as time passes and the population increases, more and more players will be close to this wall. He believes there are great players in all eras, but the mean players and lower end of the tail players in each era are closer to the "wall." By separating out the innate ability of each player, we study the dynamic nature of the population of players. Section

Scott M. Berry is Assistant Professor, Department of Statistics, Texas A&M University, College Station, TX 77843. C. Shane Reese is Technical Staff Member, Statistical Sciences, Los Alamos National Laboratory, Los Alamos, NM, 87545. Patrick D. Larkey is Professor, H. John Heinz III School of Public Policy and Management, Carnegie Mellon University, Pittsburgh, PA 15213. The authors thank Wendy Reese and Tammy Berry for their assistance in collecting and typing in data, Sean Lahman for providing the baseball data, and Marino Parascenzo for his assistance in finding birth years for the golfers. The authors are also grateful for discussions with Jim Calvin, Ed Kambour, Don Berry, Hal Stern, Jay Kadane, Brad Carlin, Jim Albert, and Michael Schell. The authors thank the editor, the associate editor, and three referees for encouraging and helpful comments and suggestions.

Journal of the American Statistical Association
September 1999, Vol. 94, No. 447, Applications and Case Studies

8 describes our results regarding the population dynamics. We provide a discussion of Gould's claims as well.

We have four main goals:

1. To describe the effects of aging on performance in each of the sports, including the degree of heterogeneity among players. Looking at the unadjusted performance of players over their careers is confounded with the changing nature of the players and the changing structure of the sports. We separate out these factors to address the aging effects.

2. To describe the effects of playing in each year in each of the sports. We want to separate out the difficulty of playing in each era from the quality of players in that era. These effects may be due to rule changes, changes in the quality of the opponents, changes in the available (and legal) equipment, and the very nature of the sport.

3. To characterize the talent of each player, independent of the era or age of the player.

4. To characterize the changing structure of the population of players. In a sport involving one player playing against an objective measure with the same equipment that has always been used (e.g. throwing a shot put, lifting weights), it is clear that the quality of players is increasing. We want to know if that is true in these three professional sports.

Addressing any factor that affects performance requires addressing all such factors. If the league-wide performance is used as a measure of the difficulty of a particular year, then this is confounded with the players' ability in that year. In hockey, if an average of 3 goals are scored per game in 1950 and 4 goals scored per game in 1990, it is not clear whether scoring is easier or if the offensive players are more talented. Our aim is to separate out each effect, while accounting for the other effects.

We have found little research in this area. Riccio (1994) examined the aging pattern of golfer Tom Watson in his U.S. Open performances. Berry and Larkey (1998) compared the performance of golfers in major tournaments. Albert (1998) looked at the distribution of home runs by Mike Schmidt over his career. Schell (1998) ranked the greatest baseball players of all time on their ability to hit for average. He used a z-score method to account for the changing distribution of players and estimated the ballpark effects. He ignored the aging effects by requiring a minimum number of at bats to qualify for his method. Both of these effects are estimated separately without accounting for the other changing effects. Our goal is to construct a comprehensive model that makes the necessary adjustments simultaneously, rather than a series of clever adjustments.

The next section examines the measures of performance in each sport and the available data for each. Section 3 describes the models used and the key assumptions of each. Section 4 discusses the Markov chain Monte Carlo (MCMC) algorithms used. The algorithms are standard MCMC successive substitution algorithms. Section 5 looks at the goodness of fit for each of the models. To address the aging effects, Section 6 presents nonparametric aging functions. Random curves are used to allow for variation in aging across individuals. Section 7 discusses the results for each sport, including player aging profiles, top peak performers, and the changes over time for each sport. Section 8 discusses the population dynamics within each sport, and Section 9 discusses the results and possible extensions.

2. SPORTS SPECIFICS AND AVAILABLE DATA

In our study of hockey, we model the ability of NHL players to score points. In hockey, two teams battle continuously to shoot the puck into the other team's goal. Whoever shoots the puck into the goal receives credit for scoring a *goal.* If the puck was passed from teammates to the goal scorer, then up to the last two players to pass the puck on the play receive credit for an *assist.* A player receives credit for a *point* for either a goal or an assist. In hockey there are three categories of players: forwards, defensemen, and goalies. A main task of defensemen and goalies is to prevent the other team from scoring. A main task of forwards is to score goals. Therefore, we consider forwards only. We recorded the number of points in each season for the 1,136 forwards playing at least 100 games between 1948 and 1996. All hockey data are from Hollander (1997). We deleted all seasons played for players age 40 and older. There were very few such seasons, and thus the age function was not well defined greater than 40. Any conclusions in this article about these players is based strictly on their ability to score points, which is not necessarily reflective of their "value" to a hockey team. Some forwards are well known for their defensive abilities; thus their worth is not accurately measured by their point totals.

Considered among the most physically demanding of sports, hockey requires great physical endurance, strength, and coordination. As evidence of this, forwards rotate throughout the game, with three or four lines (sets of three forwards) playing in alternating shifts. In no other major sport do players participate such a small fraction of the time. We do not have data on which players were linemates. The NHL has undergone significant changes over the years. The league has expanded from 6 teams in 1948 to 30 teams in 1996. Recent years have brought a dramatic increase in the numbers of Eastern European and American players, as opposed to almost exclusively Canadians in the early years. Technological developments have made an impact in the NHL. The skates that players use today are vastly superior to those of 25 years ago. The sticks are stronger and curved, helping players control the puck better and shoot more accurately. The style of play has also changed. At different times in NHL history coaches have stressed offense or stressed defense.

In golf, it takes a long time for a player to reach his or her peak. Golf requires a great deal of talent, but it does not take the physical toll that hockey does. It seems reasonable to expect that the skills needed to play golf do not deteriorate as quickly in aging players as do speed and strength in hockey. Therefore, the playing careers of golfers are much longer. Technology is believed to have played an enormous role in golf. Advances in club and ball design have aided modern players. The conditions of courses today are far superior to conditions of 50 years ago: Modern professional golfers experience very few bad lies of the ball on the fairways of today's courses. The speed of the greens has increased over the years, which may increase scores, but this may be offset by a truer roll. The common perception is that technology has made the game easier.

We model the scoring ability of male professional golfers in the four major tournaments, considered the most important events of each golf season. We have individual round scores for every player in the Masters and U.S. Open from 1935–1997 and in the Open Championship (labeled the British Open by Americans) and the PGA of America Championship from 1961–1997. (The Masters and U.S. Open were not played in 1943–1945 because of World War II.) A major tournament comprises four rounds each of 18 holes of play. A "cut" occurs after the second round, and thus playing in a major generally consists of playing either two or four rounds. We found the birth years for 488 players who played at least 10 majors in the tournaments we are considering. We did not find the ages of 38 players who played at least 10 majors. The birth years for current players were found at various web sites (*pgatour.com;www.golfweb.com*). For older players, we consulted golf writer Marino Parascenzo. We had trouble finding the birth years for marginal players from past eras. This bias has consequences in our analysis of the population dynamics in Section 8.

Baseball is rich in data. We have data on every player (nonpitcher) who has batted in Major League Baseball (MLB) in the modern era (1901–1996). We have the year of birth and the home ballpark for each player during each season. The number of at bats, base hits, and home runs are recorded for each season. An official at bat is one in which the player reaches base safely from a base hit or makes an out. An at bat does not include a base on balls, sacrifice, or hit by pitch. (Interestingly, sacrifices were considered at bats before 1950 but not thereafter.) A player's batting average is the proportion of at bats in which he gets a base hit. We also model a player's home run average, which is the proportion of at bats a player hits a home run.

In terms of player aging, baseball is apparently between golf and hockey. Hand-eye coordination is crucial, but the game does not take an onerous physical toll on players. A common perception is that careers in baseball are longer than in hockey, but shorter than in golf. Baseball prides itself on being a traditional game, and there have been relatively few changes in the rules during the twentieth century. Some changes include lowering the mound, reducing the size of the strike zone, and modifications to the ball. The first 20 years of this century were labeled the "deadball era." The most obvious change in the population of players came in the late 1940s, when African-Americans were first allowed to play in the major leagues. MLB has historically been played mainly by U.S. athletes, although Latin Americans have had an increasing influence over the last 40 years.

3. MODELS

In this section we present the bridging model, with details of the model for each sport. To compare players from different eras, we select the most recent season in the dataset as the *benchmark* season. All evaluations of players are relative to the benchmark season. The ability of every player that played during the benchmark season can be estimated by their performance in that season. In the home run example, this includes current sluggers like Mark McGwire (1987–present), Ken Griffey, Jr. (1989–present), and Mike Piazza (1992–present). The ability of players whose careers overlapped with the current players can be estimated by comparing their performances to the current players' performances in common years. In the home run example, this includes comparing players like Reggie Jackson (1967–1987), Mike Schmidt (1972–1989), and Dale Murphy (1976–1992) to McGwire, Griffey, and Piazza. The careers of Jackson, Schmidt, and Murphy overlapped with the careers of players who preceded them, such as Mickey Mantle (1951–1968), Harmon Killebrew (1954–1975), and Hank Aaron (1954–1976). The abilities of Mantle, Killebrew, and Aaron can be estimated from their performances relative to Jackson, Schmidt, and Murphy in their common years. The network of thousands of players with staggered careers extends back to the beginning of baseball. All three sports considered in this article have similar networks.

We estimate a league-wide age effect by comparing each player's performance as they age with their estimated ability. The difficulty of a particular season can be estimated by comparing each player's performance in the season with their estimated ability and estimated age effect during that season. We can estimate other effects, such as ball park and individual rounds in golf, in an analogous fashion. This explanation is an iterative one, but the estimates of these effects are produced simultaneously.

There are two critical assumptions for each model used in this article. The first is that outcomes across events (games, rounds, and at bats) are independent. A success or failure in one trial does not affect the results of other trials. One example of dependence between trials is the "hot-hand" effect: Success breeds success, and failure breeds failure. This topic has received a great deal of attention in the statistics literature. We have found no conclusive evidence of a hot-hand effect. (For interesting studies of the hot-hand, see Albert 1993, Albright 1993, Jackson and Mosurski 1997, Larkey, Smith, and Kadane 1989, Stern 1995, Stern and Morris 1993, and Tversky and Gilovich 1989a,b.) We do not take up this issue here, but we do believe that golf is the most likely sport to have a hot-hand effect (and we have found no analysis of the hot-hand effect in golf).

All of the models used are additive. Therefore, the second critical assumption is that there are no interactions. An interaction in this context would mean that the performances of different players are affected differently by a predictor. For example, if player A is more successful in the modern game than player B, then had they both played 50 years ago player A would still have been better than player B.

We address the question of interactions in the discussion section.

We use the same parameters across sports to represent player and year effects. When necessary, superscripts h, g, a, and r are used to represent hockey, golf, and batting averages and home runs in baseball.

3.1 Hockey

For the hockey data, we have $k = 1{,}136$ players. The number of seasons played by player i is n_i, and the age of player i in his jth season is a_{ij}. The year in which player i played his jth season is y_{ij}, the number of points scored in that season is x_{ij}, and the games played is g_{ij}. Counting the number of points for a player in a game is counting rare events in time, which we model using the Poisson distribution.

Per game scoring for a season is difficult to obtain. To address the appropriateness of the Poisson distribution for one player, we collected data on the number of points scored in each game for Wayne Gretzky in the 1995–1996 season, as shown in Table 1. The Poisson appears to be a reasonable match for the points scored per game (the chi-squared goodness-of-fit test statistic is 5.72, with a p value of .22).

We assume that the points scored in a game are independent of those scored in other games, conditionally on the player and year, and that the points scored by one player are independent of the points scored by other players. The model is

$$x_{ij} \sim \text{Poisson}(\lambda_{ij} g_{ij}), \qquad i = 1, \ldots, k; \qquad j = 1, \ldots, n_i,$$

where the x_{ij} are independent conditional on the λ_{ij}'s. Assume that

$$\log(\lambda_{ij}) = \theta_i + \delta_{y_{ij}} + f_i(a_{ij}).$$

In this log-linear model, θ_i represents the player-specific ability; that is, $\exp(\theta_i)$ is the average number of points per game for player i when he is playing at his peak age ($f_i = 0$) in 1996 ($\delta_{1996} = 0$). There are 49 δ_l's, one for each year in our study. They represent the difficulty of year l relative to 1996. Therefore, we constrain $\delta_{1996} \equiv 0$. We refer to 1996 as the benchmark year. The function f_i represents the aging effects for player i. We use a random curve to model the aging, as discussed in Section 3.4. The function f_i is restricted to be 0 for some age a (player i's peak age).

A conditionally independent hierarchical model is used for the θ's. To allow for the distribution of players to change over time, we model a separate distribution for the θ's for each decade. Let d_i be the decade in which player i was born. In the hockey example, the first decade is 1910–1919, the second decade is 1920–1929, and the last decade, the seventh, is 1970–1979. The model is

$$\theta_i \sim \text{N}(\mu_\theta(d_i), \sigma_\theta^2(d_i)), \qquad i = 1, \ldots, 1{,}136,$$

where $\text{N}(\mu, \sigma^2)$ refers to a normal distribution with a mean of μ and a variance of σ^2. The hyperparameters have the distributions

$$\mu_\theta(d_i) \sim \text{N}(m, s^2), \qquad d_i = 1, \ldots, 7$$

and

$$\sigma_\theta^2(d_i) \sim \text{IG}(a, b), \qquad d_i = 1, \ldots, 7,$$

where $\text{IG}(a, b)$ refers to an inverse gamma distribution with mean $1/b(a-1)$ and variance $1/b^2(a-1)^2(a-2)$. The θ_i are independent conditional on the μ_θ's and σ_θ^2's. The δ_l's are independent with prior distributions

$$\delta_l \sim \text{N}(0, \tau^2), \qquad l = 1948, \ldots, 1995.$$

The average forward scores approximately 40 points in a season, or approximately .5 points per game. Thus we set $m = \log(.5)$, and allow for substantial variability around this number by setting $s = .5$. For the distribution of σ_θ^2, we set $a = 3$ and $b = 3$. This distribution has mean .167 and standard deviation .167. We chose $\tau = 1$. We specified prior distributions that we thought were reasonable and open minded. This prior represents the notion that σ_θ^2 is not huge, but is flexible enough so that the posterior is controlled by the data. We find little difference in the results for the priors that we considered reasonable.

Table 1. The Points Scored in Each of Wayne Gretzky's 83 Games in the 1995–96 Season

Points	Gretzky	Poisson
0	23/83 = .28	.31
1	34/83 = .41	.36
2	18/83 = .22	.21
3	4/83 = .05	.08
4	4/83 = .05	.02
5	0/83 = 0	.006

NOTE: For each point total the probability of that occurrence, assuming a Poisson distribution with a mean of 1.18, is shown in the second column.

3.2 Golf

The golf study involves $k = 488$ players, with player i playing n_i rounds of golf (a round consisting of 18 holes). For the jth round of player i, the year in which the round is played is y_{ij}, the score of the round is x_{ij}, the age of the player is a_{ij}, and the round number in year y_{ij} is r_{ij}. The round number ranges from 1 to 16 in any particular year, corresponding to the chronological order.

We adopt the following model for golf scores:

$$x_{ij} \sim \text{N}(\beta_{ij}, \sigma_x^2),$$

where the x_{ij} are independent given the β_{ij}'s and σ_x^2. Assume that

$$\beta_{ij} = \theta_i + \delta_{y_{ij}} + \gamma_{y_{ij}, r_{ij}} + f_i(a_{ij}).$$

Parameter θ_i represents the mean score for player i when that player is at his peak ($f_i = 0$), playing a round of average difficulty in 1997 ($\delta = 0$ and $\gamma = 0$). The benchmark year is 1997; thus $\delta_{1997} \equiv 0$, and each δ_l represents the difficulty of that year's major tournaments relative to 1997. There is variation in the difficulty of rounds within a year. Some courses are more difficult than others; the course setup can be relatively difficult or relatively easy, and the weather plays a major role in scoring. The γ's represent the difficulty of rounds within a year. Thus $\gamma_{u,v}$ is the mean difference, in strokes, for round v from the average round in year u. To preserve identifiability, and thus interpretability, we restrict

$$\sum_{v=1}^{16} \gamma_{u,v} \equiv 0.$$

The aging function f_i is discussed in Section 3.4. A decade-specific hierarchical model is used for the θ's. Let d_i be the decade in which a golfer was born. There are seven decades: 1900–1909, 1910–1919, ... , 1960+. Only three players in the dataset were born in the 1970s, so they were combined into the 1960s. Let the θ_i's be independent conditional on the μ_θ's and σ_θ^2's and be distributed as

$$\theta_i \sim \mathrm{N}(\mu_\theta(d_i), \sigma_\theta^2(d_i)), \qquad i = 1, \ldots, 488,$$

where

$$\mu_\theta(d_i) \sim \mathrm{N}(m, s^2), \qquad d_i = 1, \ldots, 7$$

and

$$\sigma_\theta^2(d_i) \sim \mathrm{IG}(a, b), \qquad d_i = 1, \ldots, 7.$$

The δ_l's are independent with prior distributions

$$\delta_l \sim \mathrm{N}(0, \tau^2), \qquad l = 1935, \ldots, 1996$$

and the $\gamma_{u,v}$'s are independent with the priors

$$\gamma_{u,v} \sim \mathrm{N}(0, \phi^2).$$

We specify the hyperparameters as $m = 73, s = 3, a = 3, b = 3, \tau = 3$, and $\phi = 3$. As in the hockey study, here the results from priors similar to this one are virtually identical. The distribution of golf scores has been discussed by Mosteller and Youtz (1993). They modeled golf scores as 63 plus a Poisson random variable. Their resulting distribution looked virtually normal, with a slight right skew. They developed their model based on combining the scores of all professional golfers. Scheid (1990) studied the scores of 3,000 amateur golfers and concluded that the normal fits well, except for a slightly heavier right tail. There are some theoretical reasons why normality is attractive. Each round score is the sum of 18 individual hole scores. The distribution of scores on each hole is somewhat right-skewed, because scores are positive and unlimited from above. A score of 2, 3, or 4 over par on one hole is not all that rare, whereas 2, 3, or 4 under par on a hole is extremely rare, if not impossible. A residual normal probability plot shown in Section 5 demonstrates the slight right-skewed nature of golf scores. We checked models with a slight right skew, and found the results to be virtually identical (not shown). The only resulting difference that we noticed was in predicting individual scores (in which we are not directly interested). Because of its computational ease and reasonable fit, we adopt the normality assumption.

3.3 Baseball

The baseball studies involve $k = 7{,}031$ players, with player i playing in n_i seasons. For player i in his jth season, x_{ij} is the number of hits, h_{ij} is the number of home runs, m_{ij} is the number of at bats, a_{ij} is the player's age, y_{ij} is the year of play, and t_{ij} is the player's home ballpark. (Players play half their games in their home ballpark and the other half at various ballparks of the other teams.)

We model at bats as independent Bernoulli trials, with the probability of success for player i in his jth year equal to π_{ij}. We study both hits and home runs as successes; therefore, we label π_{ij}^a and π_{ij}^r as the probability of getting a hit and of hitting a home run. Thus

$$x_{ij} \sim \mathrm{binomial}(m_{ij}, \pi_{ij}^a),$$

where

$$\log\left(\frac{\pi_{ij}^a}{1-\pi_{ij}^a}\right) = \theta_i^a + \delta_{y_{ij}}^a + \xi_{t_{ij}}^a + f_i^a(a_{ij}).$$

For the baseball home run study, we use a similar model,

$$h_{ij} \sim \mathrm{binomial}(m_{ij}, \pi_{ij}^r),$$

where

$$\log\left(\frac{\pi_{ij}^r}{1-\pi_{ij}^r}\right) = \theta_i^r + \delta_{y_{ij}}^r + \xi_{t_{ij}}^r + f_i^r(a_{ij}).$$

The δ parameters are indicator functions for seasons, and the ξ parameters are indicator functions for home ballparks. We include the ξ parameters to account for the possibility that certain stadiums are "hitters'" parks and others are "pitchers'" parks. The aging function f_i is discussed in the following subsection.

Let d_i be the decade in which player i was born. There are 12 decades for the baseball players: 1860–1869, ... , 1970+. A decade-specific conditionally independent hierarchical model is used:

$$\theta_i^a \sim \mathrm{N}(\mu_\theta^a(d_i), (\sigma_\theta^a)^2(d_i)), \qquad i = 1, \ldots, 7{,}031,$$

where the θ_i^a's are independent conditional on the μ_θ^a's and $(\sigma_\theta^a)^2$'s. Assume that

$$\mu_\theta^a(d_i) \sim \mathrm{N}(m^a, (s^a)^2), \qquad d_i = 1, \ldots, 12$$

and

$$(\sigma_\theta^a)^2(d_i) \sim \mathrm{IG}(a^a, b^a), \qquad d_i = 1, \ldots, 12.$$

The δ_l^a's are independent with prior distributions

$$\delta_l^a \sim \mathrm{N}(0, (\tau^a)^2), \qquad l = 1, \ldots, 1995,$$

and the ξ_q^a's are independent with prior distributions

$$\xi_q^a \sim \mathrm{N}(0, (\phi^a)^2).$$

The parameters are selected as $m^a = -1, s^a = 1, a^a = 3, b^a = 3, \tau^a = 1$, and $\phi^a = 1$.

For the home run data, the following decade-specific hierarchical model is used:

$$\theta_i^r \sim \mathrm{N}(\mu^r(d_i), (\sigma_\theta^r)^2(d_i)), \qquad i = 1, \ldots, 7{,}031,$$

where the θ_i^r's are independent conditional on the μ_θ^r's and $(\sigma_\theta^r)^2$'s. Assume that

$$\mu_\theta^r(d_i) \sim \text{N}(m^r, (s^r)^2), \qquad d_i = 1, \ldots, 12$$

$$(\sigma_\theta^r)^2(d_i) \sim \text{IG}(a^r, b^r), \qquad d_i = 1, \ldots, 12.$$

The δ_l^r's are independent with prior distributions

$$\delta_l^r \sim \text{N}(0, (\tau^r)^2), \qquad l = 1901, \ldots, 1995$$

and the ξ_q^r's are independent with prior distributions

$$\xi_q^r \sim \text{N}(0, (\phi^r)^2).$$

We set $m^r = -3.5, s^r = 1, a^r = 3, b^r = 3, \tau^r = 1$, and $\phi^r = 1$. In both the average and the home run studies, the selection of the parameters in the priors have essentially no effect on the final conclusion.

3.4 Aging Functions

A common perception of aging functions is that players improve in ability as they mature, up to a peak level, then slowly decline. It is generally believed that players improve faster while maturing than they decrease in ability while declining. The aging curve is clearly different for different sports with regard to both peak age and the rate of change during maturity and decline. Moreover, some players tend to play at near-peak performance for a long period of time, whereas others have short periods of peak performance. This may be due to conditioning, injuries, or genetics. We assume a *mean aging curve* for each sport. We model the variation in aging for each player using hierarchical models, with the mean aging curve as the standard. In each model, θ_i represents the ability of player i at peak performance in a benchmark year. Thus each player's ability is characterized by a profile rather than one number; it may be that player A is better than player B when they are both 22 years old, but player B is better than player A when they are both 35. For convenience we round off ages, assuming that all players were born on January 1. Lindley and Smith (1972) proposed using random polynomial curves. Shi, Weiss, and Taylor (1996) used random spline curves to model CD4 cell counts in infants over time. Their approach is similar to ours in that it models an effect in longitudinal data with a flexible random curve.

We let $g(a)$ denote the mean aging curve in each sport. We let $\breve{a}$ be the peak age for a player. Without loss, we assume that $g(\breve{a}) = 0$. We use the following model for player i's aging curve:

$$f_i(a) = \begin{cases} g(a)\psi_{1i} & \text{if } a < a_{\text{M}} \\ g(a) & \text{if } a_{\text{M}} \le a \le a_{\text{D}} \\ g(a)\psi_{2i} & \text{if } a > a_{\text{D}}. \end{cases}$$

The parameter $\psi_i = (\psi_{1i}, \psi_{2i})$ represents player i's variation from the mean aging curve. We define the maturing period as any age less than a_{M} and the declining period as any age greater than a_{D}. To preserve the interpretation of ψ_1 and ψ_2 as aging parameters, we select a_{M} and a_{D} where the aging on each side becomes significant. We fit the mean aging function for every player, then select ages (or knots) a_{M} and a_{D}, to represent players after their rise and before their steady decline. For ages a, such that, $a_{\text{M}} \le a \le a_{\text{D}}$, each player ages the same. This range was determined from initial runs of the algorithm. We selected a region in which the players' performance was close to the peak performance. Part of the motivation for a range of values unaffected by individual aging patterns is to ensure stability in the calculations. In each study we use the hierarchical model

$$(\psi_{1i}, \psi_{2i})' \sim \text{N}_2((1,1)^T, \ \text{diag}(\sigma_{\text{M}}^2, \sigma_{\text{D}}^2)),$$

which is a bivariate normal distribution. We use IG(10, 1) priors for σ_{M}^2 and σ_{D}^2 (mean .11 and standard deviation .039). Due to the large number of players in each example, the priors that we considered reasonable had virtually identical results.

In the golf model, $g(a)$ represents the additional number of strokes worse than peak level for the average professional golfer at age a. The maturing and declining parameters for each player have a multiplicative effect on the additional number of strokes. A player with $\psi_1 = 1$ matures the same as the average player. If $\psi_1 > 1$, then the player averages more strokes over his peak value than the average player would at the same age $a < a_{\text{M}}$. If $\psi_1 < 1$, then the player averages fewer strokes over his peak value than the average player would at the same age $a < a_{\text{M}}$. The same interpretation holds for ψ_2, only representing players of age $a > a_{\text{D}}$.

The quantity $\exp(f_i(a))$ has a multiplicative effect on the mean points per game parameter in hockey and on the log-odds of success in baseball. Therefore, $\exp(g(a))$ represents the proportion of peak performance for the average player at age a.

We use a nonparametric form for the mean aging function in each sport:

$$g(a) = \alpha_a, \qquad a = \min(\text{age}), \ldots, \max(\text{age}), \qquad (1)$$

where the α's are parameters. The only restriction is that $\alpha_a \equiv 0$ for some value a. We select $\alpha_{\breve{a}} = 0$ by initial runs of the algorithm to find $\breve{a}$. This preserves the interpretation for the θ's as the peak performance values. This model allows the average aging function to be of arbitrary form on both sides of peak age. In particular, the aging function may not be monotone. Although this may be nonintuitive, it allows for complete flexibility. A restriction is that the age of peak performance is the same across a sport. We believe that this is a reasonable assumption. The model is robust against small deviations in the peak age, because the aging function will reflect the fact that players performed well at those ages. By allowing players to age differently during the maturing and declining stages, each player's aging function can better represent good performance away from the league-wide peak. An alternative would be to model the peak age as varying across the population using a hierarchical model.

We tried alternative aging functions that were parametric. We used a quadratic form and an exponential decay (growth) model. Both of these behaved very similar to the nonparametric form close to the peak value. The parametric forms behaved differently for very young and very old players. The parametric form was too rigid in that it predicted far worse performance for older players. A piecewise parametric form may be more reasonable.

4. ALGORITHMS

In this section we describe the Markov chain Monte Carlo algorithms used to calculate the posterior distributions. The structure of the programs is to successively generate values one at a time from the complete conditional distributions (see Gelfand and Smith 1990; Tierney 1994).

In the golf model, all of the complete conditional distributions are available in closed form. In the hockey, batting average, and home run models, a Metropolis–Hastings step is used for most of the complete conditional distributions (see Chib and Greenberg 1995). In all of the models, generating the decade specific means and standard deviations are available in closed form.

Our results are based on runs with burn-in lengths of 5,000. Every third observation from the joint distribution is selected from one chain until 10,000 observations are collected. We used Fortran programs on a 166 MHz Sun Sparc Ultra 1. The golf programs took about 15 minutes, the hockey programs about 30 minutes, and each baseball program took about 80 minutes. With thousands of parameters, monitoring convergence is difficult. We found that most of the parameters depended on the year effects, and so concentrated our diagnostic efforts on the year effects. The algorithm appeared to converge very quickly to a stable set of year effects. Waiting for convergence for 5,000 observations is probably overkill. Monitoring the mixing of the chain is also difficult. Again the year effects were important. We also monitored those effects generated with a Metropolis step. We varied the candidate distributions to assure that the chain was mixing properly.

To validate our approach, we designed simulations and compared the results with the known values. We set up scenarios where players were constant over time, getting gradually better, and getting gradually worse. We crossed this with differing year effects. Some of the aspects of the models were developed using this technique. For example, we adopted different means and standard deviations for each decade based on their increased performance in the simulations. Our models did very well in the simulations. In particular, we found no systematic bias from these models.

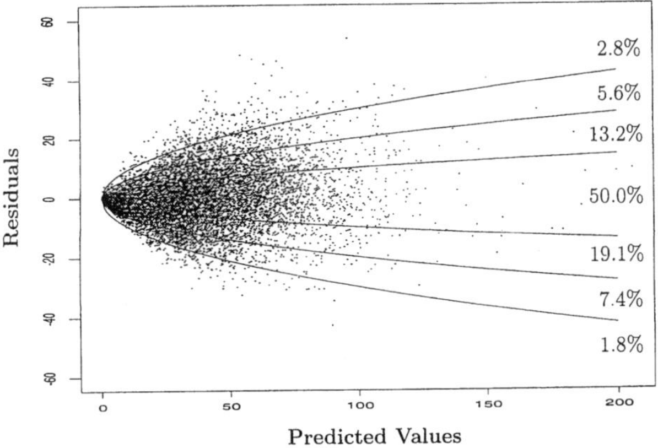

Figure 1. The Residuals in the Hockey Study Plotted Against the Fitted Values. The lines are ±1, 2, and 3 square root of the predicted values. These are the standard deviations assuming the model and parameters are correct, and the data are truly Poisson. The percentage of observations in each of the regions partitioned by the ±1, 2, and 3 standard deviations are reported on the graph.

5. GOODNESS OF FIT

In this section we consider the appropriateness of our models and address their overall fit. For each sport we present an analysis of the residuals. We look at the sum of squared errors for our fitted model (referred to as the *full* model) and several alternative models. The *no individual aging* model is a subset of the full model, with the restriction that $\psi_{1i} = 1$ and $\psi_{2i} = 1$, for all i. The *no aging effects* model assumes that $f_i(a) = 0$, for all i and a. The *null* model is a one-parameter model that assumes all players are identical and there are no other effects. Although this one-parameter model is not taken seriously, it does provide some information about the fit of the other models.

The objective sum of squares is the expected sum of squares if the model and parameters are correct. This is an unattainable goal in practice but gives an overall measure of the combined fit of the model and the parameters. We provide an analog to R^2, which is the proportion of sum of squares explained by each model. For each model M, this is defined as $1 - \text{SS}_M/\text{SS}_N$, where SS_M refers to the sum of squared deviations using model M and subscript N indexes the null model. Myers (1990) discussed R^2 in the context of log-linear models.

In calculating the sum of squares, we estimate the parameters with their posterior means.

5.1 Hockey

Figure 1 plots the residual points per player season are plotted against the predicted points per player season. We include curves for $\pm 1, 2$, and 3 times the square root of the predicted values. These curves represent the $\pm 1, 2$, and 3 standard deviations of the residuals, assuming that the parameters and model are correct. The percent of residuals in each region is also plotted. The residual plot demonstrates a lack of fit of the model.

Table 2 presents the sum of squared deviations for each model. The sum of squares for each model is the sum of the squared difference between the model estimated point total and the actual point total, over every player season.

Table 2. The Sum of Squared Deviations (SS) Between the Predicted Point Totals for Each Model and the Actual Point Totals in the Hockey Example

Model	SS	R^2
Objective	346,000	.91
Full	838,000	.79
No individual aging	980,000	.75
No aging effects	1,171,000	.70
Null	3,928,000	

The objective sum of squares is $\sum_{ij} g_{ij}\hat{\lambda}_{ij}$, where $\hat{\lambda}_{ij}$ is the estimate of the points per game parameter from the full model. This represents the expected sum of squares if the model and parameters are exactly correct.

We feel that the model is reasonable but clearly demonstrates a lack of fit. The full model is a huge improvement over the null model, but it still falls well short of the objective. Of the three examples (golf has no objective), hockey represents the biggest gap between the objective and the full model. We believe that this is because strong interactions are likely in hockey. Of the three sports studied, hockey is the most team oriented, in which the individual statistics of a player are the most affected by the quality of his teammates. For example, Bernie Nicholls scored 78 points in the 1987–88 season without Wayne Gretzky as a teammate, and scored 150 points the next season as Gretzky's teammate.

There is also strong evidence that the aging effects and the individual aging effects are important. The R^2 is increased by substantial amounts by adding the age effects and additionally, the individual aging effects. We think that the aging functions have a large effect because hockey is a physically demanding sport in which a slight loss of physical skill and endurance can have a big impact on scoring ability. Two players who have slight differences in aging patterns can exhibit large differences in point totals (relative to their peaks).

5.2 Golf

Figure 2 is a normal probability plot of the standardized residuals in the golf example. The pattern in the residual q-q plot is interesting, showing a deviation from normality. The left tail is "lighter" than that of a normal distribution, and the right tail is "heavier" than that of a normal distribution. As discussed in Section 3.2, this makes intuitive sense. It is very difficult to score low, and it is reasonably likely to score high. We tried various right-skewed distributions but found little difference in the results. The only difference we can see is in predicting individual scores, which is not a goal of this article.

Table 3 presents the sum of squared deviations between the estimated scores from each model and the actual scores.

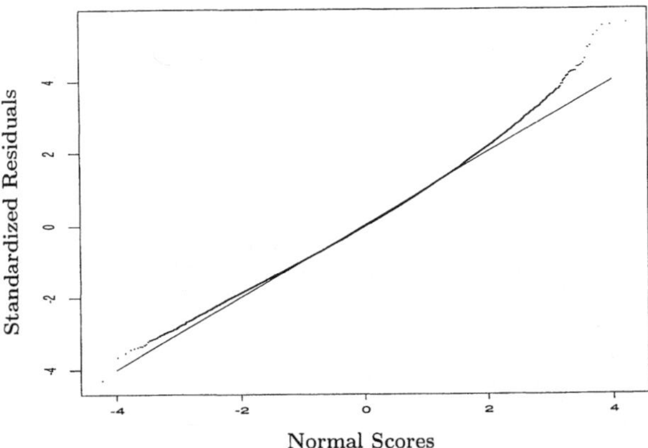

Figure 2. Normal Probability Plot of the Residuals From the Golf Model.

Table 3. The Sum of Squared Deviations (SS) Between the Predicted Score for Each Model and the Actual Score in the Golf Example

Model	SS	R^2
Full	366,300	.30
No individual aging	366,600	.30
No aging effects	372,000	.29
Null	527,000	

Because of the normal model there is no objective sum of squares to compare the fit. The variance in the scores, σ_x^2, is a parameter fitted by the model and thus naturally reflects the fit of the model. Despite the small improvement between the null model and the full model, we feel that this is a very good-fitting model. This conclusion is based on the estimate of σ_x, which is 2.90. The R^2 for this model is only .30, which is small, but we believe there is a large amount of variability in a golf score, which will never be modeled. We were pleased with a standard error of prediction of 2.90. There is little evidence that aging plays an important role in scoring. This is partly due to the fact that most of the scores in the dataset are recorded when players are in their prime. Few players qualified for the majors when they were very old or very young, and for these ages there is an effect. There is also little evidence that individual aging effects are needed, but this suffers from the same problem just mentioned.

5.3 Baseball

The residual plot for each baseball example indicated no serious departures from the model. The normal probability plots showed almost no deviations from normality. Table 4 presents the home run sum of squares; Table 5, the batting average sum of squares. The batting average example presents the sum of squared deviations of the predicted number of base hits from the model from the actual number of base hits. The R^2 is .60 for the full model, very close to the objective sum of squares of .62. We believe the batting average model is a good-fitting model. The home run model does not fit as well as the batting average example. Despite an R^2 of .80, it falls substantially short of the objective sum of squares. The high R^2 is due to the large spread in home run ability across players, for which the null model does not capture.

Aging does not play a substantial role in either measure. This is partly due to the large number of observations close to peak, where aging does not matter, but also can be attributed to the lack of a strong effect due to aging. The contrast between the four examples in the role of aging

Table 4. The Sum of Squared Deviations (SS) Between the Predicted Number of Home Runs for Each Model and the Actual Number of Home Runs in the Home Run Example

Model	SS	R^2
Objective	171,000	.86
Full	238,000	.80
No individual aging	242,500	.80
No aging effects	253,700	.78
Null	1,203,000	

Table 5. The Sum of Squared Deviations (SS) Between the Predicted Number of Hits for Each Model and the Actual Number of Hits in the Batting Average Example

Model	*SS*	R^2
Objective	1,786,000	.62
Full	1,867,000	.60
No individual aging	1,897,000	.60
No aging effects	1,960,000	.58
Null	4,699,000	

and the individual aging effects is interesting. In the most physically demanding of the sports, hockey, aging plays the greatest role. In the least physically demanding sport, golf, the aging effect plays the smallest role.

6. AGE EFFECT RESULTS

Figures 3–6 illustrate the four mean age effect (g) functions. Figure 3 shows the hockey age function. The y-axis represents the proportion of peak performance for a player of age a. Besides keeping track of the mean of the aging function for each age, we also keep track of the standard deviation of the values of the curve. The dashed lines are the ± 2 standard deviation curves. This graph is very steep on both sides of the peak age, 27. The sharp increase during the maturing years is surprising—20- to 23-year-old players are not very close to their peak. Because of the sharp peak, we specified 29 and older as declining and 25 and younger as maturing.

Figure 4 presents the average aging function for golf. In this model g represents the average number of strokes from the peak. The peak age for golfers is 34, but the range 30–35 is essentially a "peak range." The rate of decline for golfers is more gradual than the rate of maturing. An average player is within .25 shots per round (1 shot per tournament) from peak performance when they are in the 25–40 age range. An average 20-year-old and an average 50-year-old are both 2 shots per round off their peak performance. Because of the peak range from 30–35, we specified the declining stage as 36 and older and the maturing phase as 29 and younger.

Figures 5 and 6 present the aging functions for home runs and batting averages. The home run aging function presents the estimated number of home runs for a player who is a 20–home run hitter at his peak. The peak age for home runs is 29. A 20–home run hitter at peak is within 2 home runs of his peak at 25–35 years old. There is a sharp increase for maturers. Apparently, home run hitting is a talent acquired through experience and learning, rather than being based on brute strength and bat speed. The ability to hit home runs does not decline rapidly after the peak level—even a 40-year-old 20–home run-at-peak player is within 80% of peak performance.

The age effects for batting average are presented for a hitter who is a .300 hitter at his peak. Hitting for average does differ from home run hitting—27 is the peak age, and younger players are relatively better at hitting for average than hitting home runs. An average peak .300 hitter is expected to be a .265 hitter at age 40. For batting average and home runs, the maturing phase is 25 and younger and the declining phase is 31 and older.

7. PLAYER AND SPORT RESULTS

This section presents the results for the individual players and the year effects within each sport. To understand the rankings of the players, it is important to see the relative difficulty within each sport over the years. Each player is characterized by θ, his value at peak performance in a benchmark year, and by his aging profile. We present tables that categorize players by their peak performance, but we stress that their career profiles are a better categorization of the players. For example, in golf Jack Nicklaus is the best player when the players are younger than 43, but Ben Hogan is the best for players over 43. The mean of their maturing and declining parameters are presented for comparison.

7.1 Hockey

The season effect in hockey is strong. Figure 7a shows the estimated multiplicative effects, relative to 1996. From 1948–1968 there were only six teams in the NHL, and the game was defensive in nature. In 1969 the league added six teams. The league continued to expand to the present 30 teams. With this expansion, goal scoring increased. The 1970s and early 1980s were the height of scoring in the NHL. As evidence of the scoring effects over the years, many players who played at their peak age in the 1960s with moderate scoring success played when they were "old" in the 1970s and scored better than ever before (e.g., Gordie Howe, Stan Mikita, and Jean Beliveau). In 1980 the wide-open offensive-minded World Hockey Association, a competitor to the NHL, folded and the NHL absorbed some of the teams and many of the players. This added to the offensive nature and style of the NHL. In the 1980s the NHL began to attract players from the Soviet block, and the United States also began to produce higher-caliber players. This influx again changed the talent pool.

Scoring began to wane beginning in 1983. This is attributed in part to a change in the style of play. Teams went from being offensive in nature to defensively oriented. "Clutching and grabbing" has become a common term to describe the style of play in the 1990s. As evidence of this, in 1998 the NHL made rule changes intended to increase scoring. The seasonal effects are substantial. The model predicts that a player scoring 100 points in 1996 is would have scored 140 points in the mid-1970s or early 1980s.

Table 6 presents the top 25 players, rated on their peak level. Figure 8 presents profiles of some of these best players. It demonstrates the importance of a profile over a one-number summary. Mario Lemieux is rated as the best peak-performance player, but Wayne Gretzky is estimated to be better when they are young, whereas Lemieux is estimated to be the better after peak. The fact that Lemieux is ahead of Gretzky at peak may seem surprising. Gretzky played during the most wide-open era in the NHL, whereas Lemieux played more of his career in a relatively defensive-minded era. Lemieux's career and season totals are a bit misleading, because he rarely played a full season. He missed many games throughout his career, and we rate players on their per game totals.

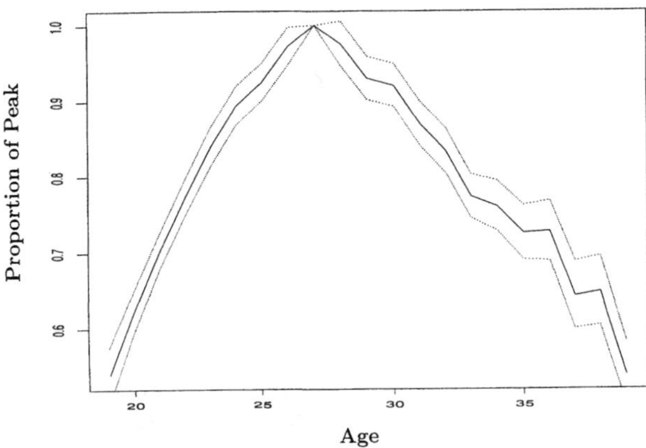

Figure 3. The Estimated Mean Aging Function and Pointwise ±2 Standard Deviation Curves for the Hockey Study. The y-axis is the proportion of peak for a player of age a.

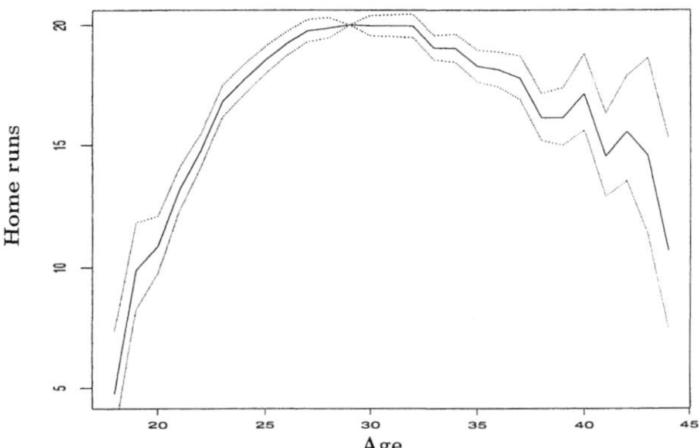

Figure 5. The Estimated Mean Aging Function and Pointwise ±2 Standard Deviation Curves for the Home Run Study. The y-axis is the number of home runs for a player who is a 20–home run hitter at peak performance.

As a cross-validation, we present the model-predicted point totals for the 10 highest-rated peak players who are still active. We estimated the 1997 season effect, $\hat{\delta}_{1997} = -.075$, by the log of the ratio of goals scored in 1997 to goals scored in 1996. Table 7 presents the results. We calculated the variance of each predicted point total using the variance of the Poisson model and the points per game parameter, λ_{ij} (the standard deviation is reported in Table 7). With the exception of Pavel Bure, the predictions are very close.

7.2 Golf

Figure 7b shows the estimate for the difficulty of each round of the Masters tournament. The mean of these years is also plotted. We selected the Masters because it is the one tournament played on the same course (Augusta National) each year and the par has stayed constant at 72. These estimates measure the difficulty of each round, separated from the ability of the players playing those rounds. These estimates may account for weather conditions, course difficulty, and the equipment of the time. Augusta in the 1940s played easier than in the 1950s or 1960s; we are unsure why. There is approximately a 1 shot decrease from the mid-1950s to the present. We attribute the decrease from the 1950s to the present to the improved equipment available to the players. Although it does appear that Augusta National is becoming easier to play, the effects of improved equipment do not appear to be as strong as public perception would have one believe. Augusta is a challenging course in part because of the speed and undulation of the greens. It may be that the greens have become faster, and more difficult, over the years. If this is true, then the golfers are playing a more difficult course and playing it 1 shot better than before. Such a case would imply the equipment has helped more than 1 shot.

Table 8 shows the top 25 players of all time at their peak. Figure 9 shows the profile of six of the more interesting careers. The y-axis is the predicted mean average for each player when they are the respective age. Jack Nicklaus at his peak is nearly .5 shot better than any other player. Nicklaus essentially aged like the average player. Ben Hogan, who is .7 shot worse than Nicklaus at peak, aged very well. He is estimated to be better than Nicklaus at age 43 and older. The beauty of the hierarchical models comes through in the estimation of Tiger Woods' ability. Woods has played very

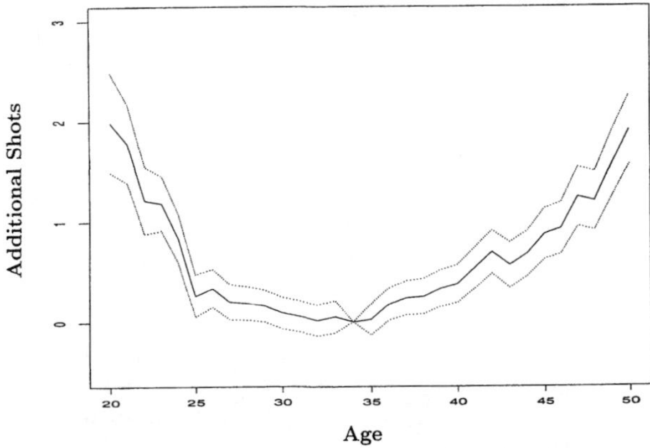

Figure 4. The Estimated Mean Aging Function and Pointwise ±2 Standard Deviation Curves for the Golf Study. The y-axis is the number of shots more than peak value for a player of age a.

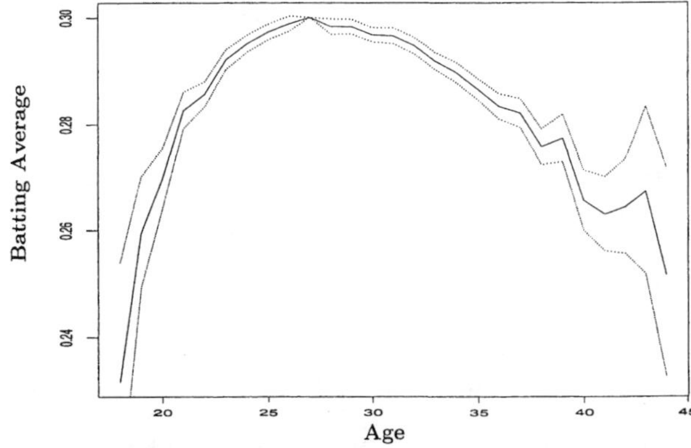

Figure 6. The Estimated Mean Aging Function and Pointwise ±2 Standard Deviation Curves for the Batting Average Study. The y-axis is the batting average for a player who is a .300 hitter at peak performance.

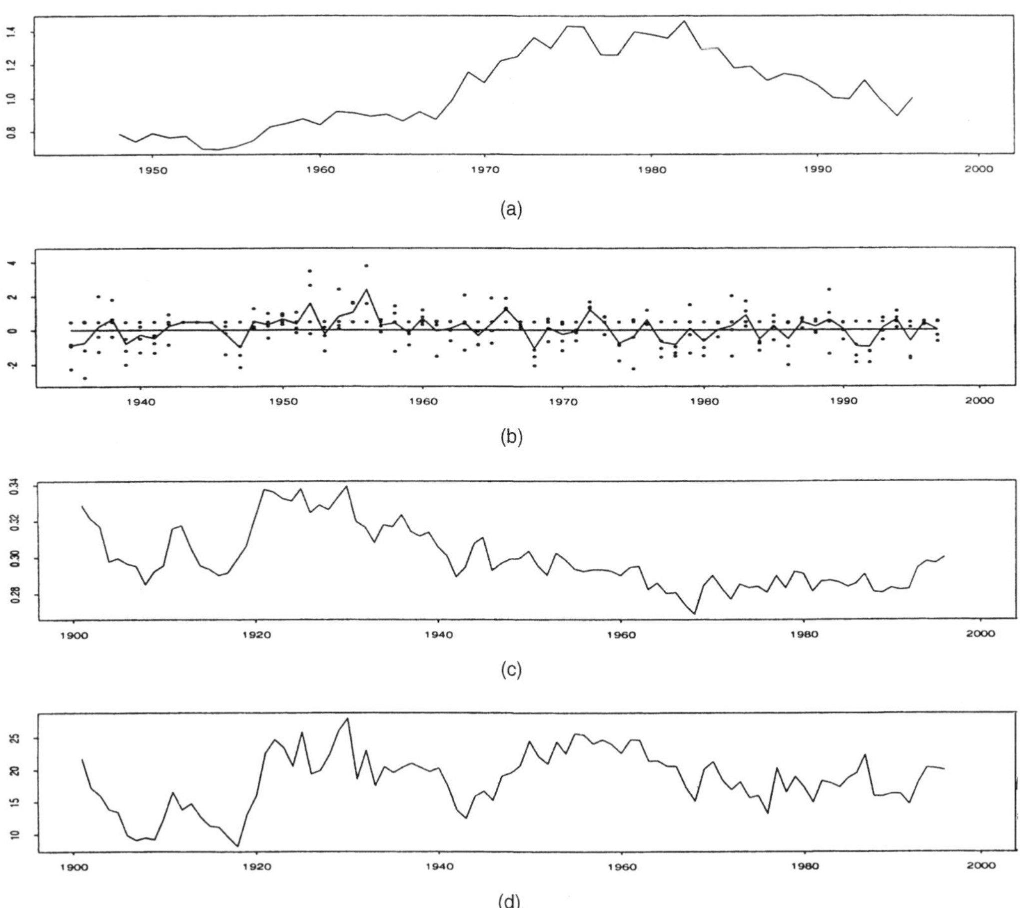

Figure 7. The Yearly Effects for the Hockey (a), Golf (b), Batting Average (c), and Home Run (d) Studies. The hockey plot shows the multiplicative effect on scoring for each year, relative to 1996. The golf plot shows the additional number of strokes for each round in the Masters, relative to the average of 1997. The line is the average for each year. The home run plot shows the estimated number of home runs for a 20–home run hitter in 1996. The batting average plot shows the estimated batting average for a player who is a .300-hitter in 1996.

well as a 21-year-old player (winning the Masters by 12 shots). He averaged 70.4 during the benchmark 1997 year. If he aged like the average player, which means during 1997 he was 1.8 shots per round off his peak performance, then he would have a peak performance of 68.6. If he aged like the average player, then he would be by far the best player of all time. This is considered very unlikely because of the distribution of players. It is more likely that he is a quick maturer and is playing closer to his peak than the average 21 year old. Thus his maturing parameter is estimated to be .52. The same phenomenon is seen for both Ernie Els and Justin Leonard, who are performing very well at young ages.

7.3 Baseball

Figures 7c and 7d illustrate the yearly effects for home runs and batting average. They show that after 1920, when the dead-ball era ended, the difficulty of hitting home runs has not changed a great deal. A 20–home run hitter in 1996 is estimated to have hit about 25 in the mid-1920s. Home run hitting has slowly decreased over the years, perhaps because of the increasing ability of pitchers. The difficulty of getting a base hit for a batter of constant ability has also decreased since the early 1920s. The probability of getting a hit bottomed out in 1968 and has increased slightly since then. The slight increase after 1968 has been attributed to the lowering of the pitcher's mound, thus decreasing the pitchers' ability, and also to expansion in MLB. Most baseball experts believe that umpires are now using a smaller strike zone, which may also play a role. We attribute part of the general decrease over the century in the difficulty of getting a base hit to the increasing depth and ability of pitchers.

Tables 9 and 10 show the top 25 peak batting average and home run hitters. The posterior means for peak performance in the benchmark year of 1996, the maturing parameter (ψ_1), and the declining parameter (ψ_2) are provided. Figures 10 and 11 show the career profiles for the batting average and home run examples. The model selects Mark McGwire as the greatest home run per at bat hitter in history. The model estimates that Babe Ruth's at his prime would hit 5 fewer home runs than McGwire in 1996. Interestingly the all-time career home run king (with 755) Hank Aaron, is only 23rd on the peak performance list. Aaron declined very slowly (the slowest of the top 100). He is higher on the batting average list (13) than on the home run list!

Table 6. The Top 25 Peak Players in the Hockey Study

Rank	Name	Born	Points in 1996	ψ_1	ψ_2
1	M. Lemieux	1965	187 (7)	1.18	.89
2	W. Gretzky	1961	181 (5)	.66	1.66
3	E. Lindros	1973	157 (16)	.93	1
4	J. Jagr	1972	152 (9)	1.37	1
5	P. Kariya	1974	129 (15)	.95	1
6	P. Forsberg	1973	124 (10)	.84	1
7	S. Yzerman	1965	120 (5)	.91	1.43
8	J. Sakic	1969	119 (6)	.95	1
9	G. Howe	1928	119 (7)	1.04	.69
10	T. Selanne	1970	113 (6)	.78	1
11	P. Bure	1971	113 (8)	.81	1
12	J. Beliveau	1931	112 (5)	.67	.90
13	P. Esposito	1942	112 (5)	1.82	1.36
14	A. Mogilny	1969	112 (6)	1.18	1
15	P. Turgeon	1969	110 (6)	.95	1
16	S. Federov	1969	110 (5)	1.05	1
17	M. Messier	1961	110 (4)	1.51	.55
18	P. LaFontaine	1965	109 (5)	1.20	1.32
19	Bo. Hull	1939	108 (4)	.94	1.29
20	M. Bossy	1957	108 (4)	.86	1.02
21	Br. Hull	1964	107 (5)	1.15	1.12
22	M. Sundin	1971	106 (7)	.99	1
23	J. Roenick	1970	106 (6)	.67	1
24	P. Stastny	1956	105 (4)	1.20	1.12
25	J. Kurri	1960	105 (4)	1.11	1.30

NOTE: The means of ψ_1 and ψ_2 are also presented. The Points in 1996 column represents the mean points (with standard deviations given in parentheses) for the player in 1996 if the player was at his peak performance.

Willie Stargell and Darryl Strawberry provide an interesting contrast in profiles. At peak they are both considered 41 home run hitters. Strawberry is estimated to have matured faster than Stargell, whereas Stargell maintained a higher performance during the declining phase.

Ty Cobb, who played in his prime about 80 years ago, is still considered the best batting average hitter of all time. Tony Gwynn is estimated to decline slowly ($\psi_2 = .78$) and is considered a better batting average hitter than Cobb after age 34. Paul Molitor is estimated to be the best decliner of the top 100 peak players. At age 40, in 1996, he recorded a batting average of .341. Alex Rodriguez exhibits the same regression to the mean characteristics as Tiger Woods does in golf. In Rodriguez's second year, the benchmark year of 1996, he led the American League in hitting at .358. The model predicts that at his peak in 1996 he would hit .336. Because of the shrinkage factor, as a result of the hierarchical model, it is more likely that Rodriguez is closer to his peak than the average player (i.e., is a rapid maturer) and that 1996 was a "lucky" year for him.

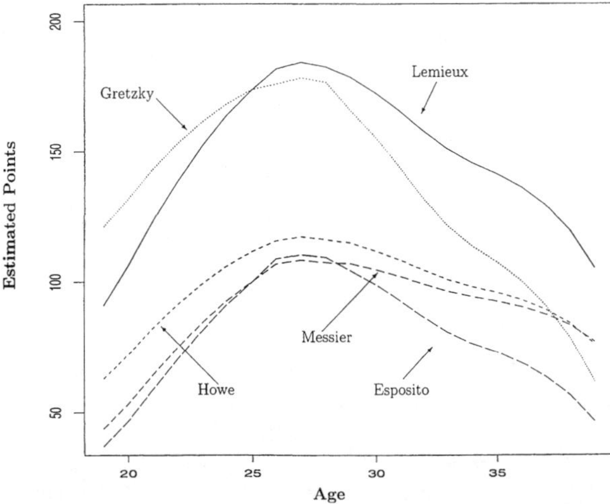

Figure 8. A Profile of Some of the Best Players in the Hockey Study. The estimated mean number of points for each age of the player, if that season were 1996.

We recorded 78 ballparks in use in MLB beginning in 1901. When a ballpark underwent significant alterations, we included the "before" and "after" parks as different. The constraint for the parks is that $\xi_{\{\text{new Fenway}\}} = 0$ (There is an old Fenway, from 1912–1933, and a new Fenway, 1933–. Significant changes were made in 1933, including moving the fences in substantially.) We report the three easiest and three hardest ballparks for home runs and batting average. (We ignore those with less than 5 years of use unless they are current ballparks.) For a 20–home run hitter in new Fenway, the expected number of home runs in the three easiest home run parks are 30.1 in South End Grounds (Boston Braves, 1901–1914), 28.6 in Coors Field (Colorado, 1995–), and 26.3 in new Oakland Coliseum (Oakland, 1996–). The 20–home run hitter would be expected to hit 14.5 at South Side (Chicago White Sox, 1901–1909), 14.8 at old Fenway (Boston, 1912–1933), and 15.9 at Griffith Stadium (Washington, 1911–1961), which are the three most difficult parks. The average of all ballparks for a 20–home run hitter at new Fenway is 20.75 home runs.

For a .300 hitter in new Fenway, the three easiest parks in which to hit for average are .320 at Coors Field, .306 at Connie Mack Stadium (Philadelphia, 1901–1937), and .305 at Jacobs Field (Cleveland, 1994–). The three hardest parks in which to hit for average are .283 at South Side, .287 at old Oakland Coliseum (Oakland, 1968–1995), and .287 at old Fenway. New Fenway is a good (batting average) hitters' park. A .300 hitter at new Fenway would be a .294 hitter in the average of the other parks. Some of the changes to the ballparks have been dramatic. Old Fenway was a very difficult park in which to hit for average or home runs, but after the fences were moved in, the park became close to average. The Oakland Coliseum went from a very difficult park to a very easy park after the fences were moved in and the outfield bleachers were enclosed in 1996.

Table 7. The Predicted and Actual Points for the Top 10 Model-Estimated Peak Players Who Played in 1997

Rank	Name	Age in 1997	Games played	Predicted points	Actual points
1	M. Lemieux	32	76	135 (14.9)	122
2	W. Gretzky	36	82	103 (13.9)	97
3	E. Lindros	24	52	84 (12.0)	79
4	J. Jagr	25	63	99 (13.1)	97
5	P. Kariya	23	69	86 (12.8)	99
6	P. Forsberg	24	65	83 (12.5)	86
7	S. Yzerman	32	81	84 (13.1)	85
8	J. Sakic	28	65	87 (12.6)	74
10	T. Selanne	27	78	99 (13.6)	109
11	P. Bure	27	63	81 (12.3)	55

NOTE: The model used data only from 1996 and prior to predict the point totals.

Table 8. The Top 25 Peak Players in the Golf Study

Rank	Name	Born	θ	ψ_1	ψ_2
1	J. Nicklaus	1940	70.42 (.29)	1.03	.99
2	T. Watson	1949	70.82 (.23)	.92	1.19
3	B. Hogan	1912	71.12 (.29)	1.13	.27
4	N. Faldo	1957	71.19 (.21)	1.19	1.21
5	A. Palmer	1929	71.33 (.28)	1.19	.95
6	G. Norman	1955	71.39 (.19)	1.21	.64
7	J. Leonard	1972	71.40 (.45)	.68	1
8	E. Els	1969	71.45 (.34)	.78	1
9	G. Player	1935	71.45 (.23)	.87	.62
10	F. Couples	1959	71.50 (.21)	1.00	.97
11	H. Irwin	1945	71.56 (.26)	1.02	.68
12	C. Peete	1943	71.56 (.36)	1	.80
13	J. Boros	1920	71.62 (.37)	1	.61
14	R. Floyd	1942	71.63 (.24)	1.22	.38
15	L. Trevino	1939	71.63 (.29)	1.00	.72
16	S. Snead	1912	71.64 (.27)	1.10	.21
17	J. Olazabal	1966	71.69 (.39)	.74	1
18	T. Kite	1949	71.71 (.23)	.98	.70
19	B. Crenshaw	1952	71.74 (.22)	.43	1.22
20	T. Woods	1975	71.77 (.64)	.52	1
21	B. Casper	1931	71.77 (.26)	1.00	1.09
22	B. Nelson	1912	71.78 (.31)	1.00	1.11
23	P. Mickelson	1970	71.79 (.44)	.79	1
24	L. Wadkins	1949	71.79 (.22)	1.13	.78
25	T. Lehman	1959	71.82 (.30)	1.05	.79

NOTE: The standard deviations are in parentheses. The means of ψ_1 and ψ_2 are also presented.

As cross-validation we present the model predictions for 1997 performance. Recall, the baseball study uses data from 1996 and earlier in the estimation. We estimate the season effect of 1997 by the league-wide performance relative to 1996. For batting average, the estimated year effect for 1997 is −.01. Table 11 presents the model-predicted batting average for the 10 highest-rated peak batting average players of all time who are still active. The estimates are good, except for Piazza and Gwynn, both of whom had batting averages approximately two standard deviations above the predicted values.

Table 12 presents the model-predicted and actual number of home runs, conditional on the number of at bats, for

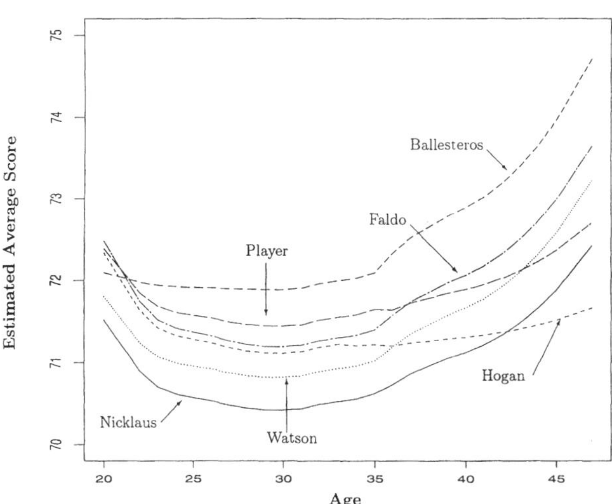

Figure 9. A Profile of Some of the Best Players in the Golf Study. The estimated mean score for each age of the player, if that round were an average 1997 round.

Table 9. The Top 25 Peak Players for the Batting Average Study

Rank	Name	Born	Average	ψ_1	ψ_2
1	T. Cobb	1886	.368 (.005)	1.14	1.31
2	T. Gwynn	1960	.363 (.006)	1.08	.78
3	T. Williams	1918	.353 (.006)	.95	.93
4	W. Boggs	1958	.353 (.005)	1.05	1.17
5	R. Carew	1945	.351 (.005)	1.06	.92
6	J. Jackson	1889	.347 (.007)	.86	1.12
7	N. Lajoie	1874	.345 (.009)	1	1.36
8	S. Musial	1920	.345 (.005)	.98	1.16
9	F. Thomas	1968	.344 (.008)	.99	1
10	E. Delahanty	1867	.340 (.001)	1	1.02
11	T. Speaker	1888	.339 (.006)	.99	1.32
12	R. Hornsby	1896	.338 (.005)	1.02	1.02
13	H. Aaron	1934	.336 (.006)	.89	1.25
14	A. Rodriguez	1975	.336 (.001)	.85	1
15	P. Rose	1941	.335 (.004)	1.25	.89
16	H. Wagner	1874	.333 (.007)	1	1.30
17	R. Clemente	1934	.332 (.005)	1.37	.50
18	G. Brett	1953	.331 (.005)	.92	1.16
19	D. Mattingly	1961	.330 (.006)	.88	1.07
20	K. Puckett	1961	.330 (.006)	1.14	.93
21	M. Piazza	1968	.330 (.009)	1.04	1
22	E. Collins	1887	.329 (.004)	.96	1.01
23	E. Martinez	1963	.328 (.008)	1.22	.79
24	P. Molitor	1956	.328 (.005)	.94	.31
25	W. Mays	1931	.328 (.005)	.99	1.19

NOTE: The standard deviations are in parentheses. The means of ψ_1 and ψ_2 are presented.

the 10 highest-rated peak home run hitters of all time who are still active. The estimated year effect for 1997 is −.06. Palmer, Belle, and Canseco did worse than their projected values. The model provided a nice fit for Griffey and McGwire, each of whom posted historical years that were not so unexpected by the model. Standard errors of prediction were calculated using the error of the binomial model and the error in the estimates of player abilities, age effects, and ballpark effects.

Table 10. The Top 25 Peak Players in the Home Run Study

Rank	Name	Born	θ	ψ_1	ψ_2
1	M. McGwire	1963	.104 (.006)	.97	1.12
2	J. Gonzalez	1969	.098 (.008)	1.05	1
3	B. Ruth	1895	.094 (.004)	.72	.93
4	D. Kingman	1948	.093 (.004)	.96	1.05
5	M. Schmidt	1949	.092 (.005)	.99	1.18
6	H. Killebrew	1936	.090 (.005)	.87	1.13
7	F. Thomas	1968	.089 (.007)	.99	1
8	J. Canseco	1964	.088 (.004)	1.05	1.01
9	R. Kittle	1958	.086 (.006)	1.08	.96
10	W. Stargell	1940	.084 (.003)	1.24	.79
11	W. McCovey	1938	.084 (.004)	1.04	1.22
12	D. Strawberry	1962	.084 (.005)	.70	1.10
13	B. Jackson	1962	.083 (.006)	1.06	1.04
14	T. Williams	1918	.083 (.004)	.88	.97
15	R. Kiner	1922	.083 (.004)	1.01	1.05
16	P. Seerey	1923	.081 (.009)	.91	1
17	R. Jackson	1946	.081 (.004)	.83	1.11
18	K. Griffey	1969	.080 (.006)	1.03	1
19	A. Belle	1966	.080 (.006)	1.12	1
20	R. Allen	1942	.080 (.004)	1.16	1.12
21	B. Bonds	1964	.079 (.004)	1.27	1.05
22	D. Palmer	1968	.079 (.007)	1.07	1
23	H. Aaron	1934	.078 (.003)	1.26	.53
24	J. Foxx	1907	.078 (.003)	1.34	1.16
25	M. Piazza	1968	.078 (.006)	.95	1

NOTE: The standard deviations are in parentheses. The means of ψ_1 and ψ_2 are presented.

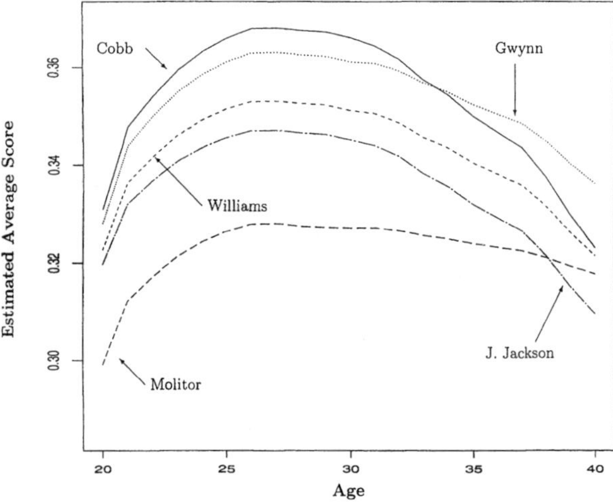

Figure 10. A Profile of Some of the Best Players in the Batting Average Study. The estimated batting average for each age of the player, if that year were 1996.

8. POPULATION DYNAMICS

In this section we address the changing distribution of players within each study. Figures 12–15 present graphs of the peak value estimate for each player, graphed against the year the player was born. These player effects are separated from all the other effects; thus the players can be compared directly.

In hockey there is some slight bias on each end of the population distribution (see Fig. 12). Players born early in the century were fairly old when our data began (1948). They are in the dataset only if they are good players. The restriction that each player plays at least 100 games was harder for a player to reach earlier in this century because a season consisted of 48 games, rather than the current 82 games. Therefore, there is a bias, overestimating the percentiles of the distribution of players for the early years.

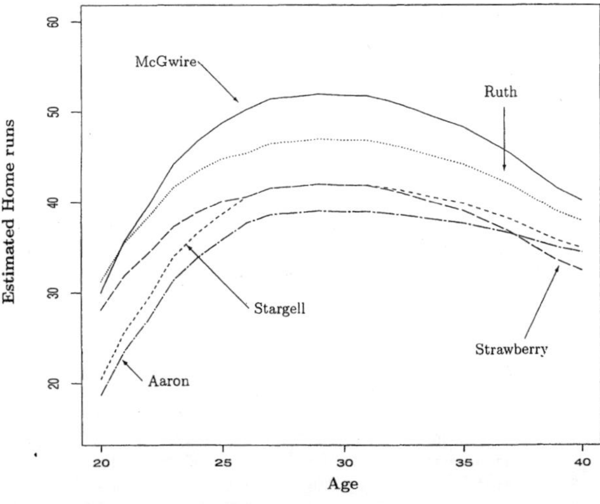

Figure 11. A Profile of Some of the Best Players in the Home Run Study. The estimated number of home runs, conditional on 500 at bats, for each age of the player, if that year were 1996.

Table 11. The Predicted and Actual Batting Averages (BA) for the Top 10 Model-Estimated Peak Players Who Played in 1997

Rank	Name	Age in 1997	At bats	Predicted BA	Actual BA
2	T. Gwynn	37	592	.329 (.021)	.372
4	W. Boggs	39	353	.318 (.027)	.292
9	F. Thomas	29	530	.328 (.023)	.347
14	A. Rodriguez	22	587	.312 (.022)	.300
21	M. Piazza	29	556	.316 (.023)	.362
23	E. Martinez	34	542	.309 (.022)	.330
24	P. Molitor	41	538	.290 (.022)	.305
29	R. Alomar	29	412	.316 (.026)	.333
39	K. Griffey	28	608	.313 (.022)	.304
47	M. Grace	33	555	.308 (.022)	.319

NOTE: The model used 1901–1996 data to predict 1997 totals. Standard deviations are in parentheses.

Of the players born late in this century (after 1970), it is more likely that the good ones are included. Thus the percentiles are probably overestimated slightly.

For hockey players born after 1940, there is a clear increase in ability. Of the top 25 players, 9 are current players who have yet to reach their peak (36%, where only 8% of the players in our data had not reached their peak). It is hard to address Gould's claim with the hockey distribution. This is because not everyone in this dataset is trying to score points. Many hockey players are role players, with the job of to playing defense or even just picking fights with the opposition! The same is true of the distribution of home run hitters in baseball. Many baseball players are not trying to hit home runs; their role may focus more on defense or on hitting for average or otherwise reaching base. The same type of pattern shows up in home run hitting. The top 10% of home run hitters are getting better with time (see Fig. 13). This could be attributed to the increasing size and strength of the population from which players are produced, the inclusion of African-Americans and Latin Americans, or an added emphasis by major league managers on hitting home runs.

It is easier to address Gould's claim with the batting average and golf studies. In baseball, every player is trying to get a base hit. Every player participates in the offense an equal amount, and even the defensive-minded players try to get hits. In golf, every player tries to minimize his

Table 12. The Predicted and Actual Home Runs (HR) for the Top 10 Model-Estimated Peak Players Who Played in 1997

Rank	Name	Age in 1997	At bats	Predicted HR	Actual HR
1	M. McGwire	34	540	55 (7.63)	58
2	J. Gonzalez	28	541	41 (7.08)	42
7	F. Thomas	29	530	42 (7.17)	35
8	J. Canseco	33	388	37 (6.36)	23
12	D. Strawberry	35	29	2 (1.58)	0
18	K. Griffey	28	608	49 (7.74)	56
19	A. Belle	31	634	46 (7.15)	30
21	B. Bonds	33	532	37 (6.48)	40
22	D. Palmer	29	556	34 (6.51)	23
25	M. Piazza	29	542	36 (6.63)	40

NOTE: The model used 1901–1996 data to predict 1997 totals. Standard deviations are in parentheses.

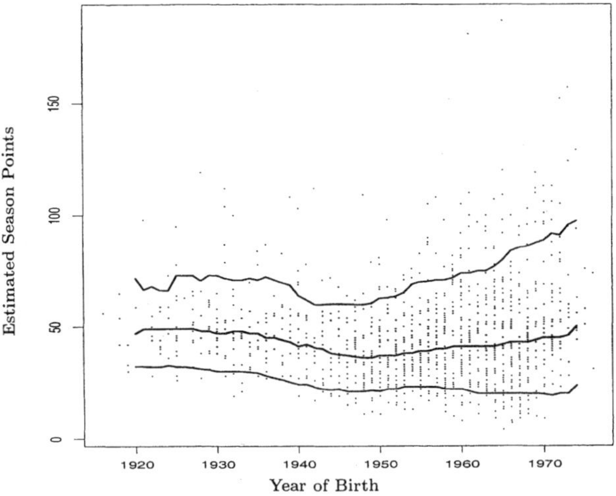

Figure 12. The Estimated Peak Season Scoring Performance of Each Player Plotted Against the Year They Were Born. The y-axis represents the mean number of points scored for each player, at their peak, if the year was 1996. The three curves are the smoothed 10th, 50th, and 90th percentiles.

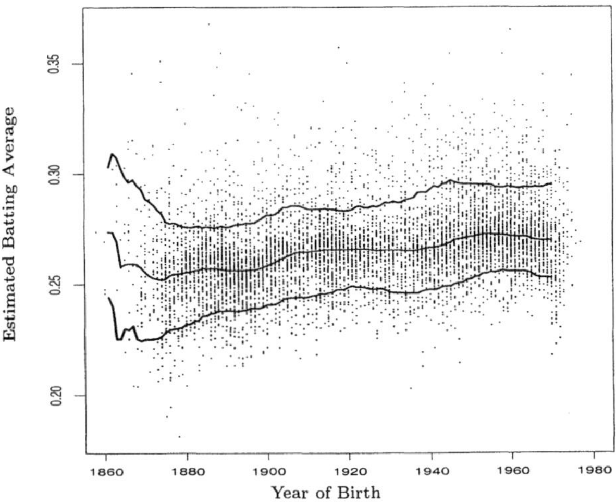

Figure 14. The Estimated Peak Batting Average Performance of Each Player Plotted Against the Year They Were Born. The y-axis represents the mean probability of a hit for each player, at their peak, if the year was 1996. The three curves are the smoothed 10th, 50th, and 90th percentiles.

scores—the only goal for the golfer. In the golf study there is a bias in the players who are in our dataset: Only players with 10 majors are included. It was harder to achieve this in the early years, because we have data on only two majors until 1961. It was also hard to find the birth dates for marginal players from the early years. We believe we have dates for everyone born after 1940, but we are missing dates for about 25% of the players born before then. There is also a slight bias on each end of the batting average graph. Only the great players born in the 1860s were still playing after 1900, and only the best players born in the early 1970s are in the dataset.

Except for the tails of Figures 12 and 13, there is an clear increase in ability. The golf study supports Gould's conjecture. The best players are getting slightly better, but there are great players in every era. The median and 10th percentile are improving rapidly (see Fig. 15). The current 10th percentile player is almost 2 shots better than the 10th percentile fifty years ago. This explains why nobody dominates golf the way Hogan, Snead, and Nelson dominated in the 1940s and 1950s. The median player, and even the marginal player, can have a good tournament and win. Batting average exhibits a similar pattern. The best players are increasing in ability, but the 10th percentile is increasing faster than the 90th percentile (see Fig. 14). It appears as though batting averages have increased steadily, whereas golf is in a period of rapid growth.

These conclusions coincide with the histories of these sports. American sports are experiencing increasing diversity in the regions from which they draw players. The globalization has been less pronounced in MLB, where players

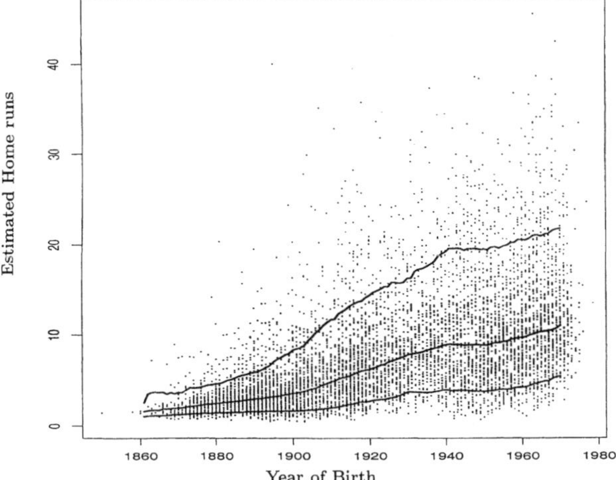

Figure 13. The Estimated Peak Home Run Performance of Each Player Plotted Against the Year They Were Born. The y-axis represents the mean number of home runs for each player, at their peak, if the year was 1996. The three curves are the smoothed 10th, 50th, and 90th percentiles.

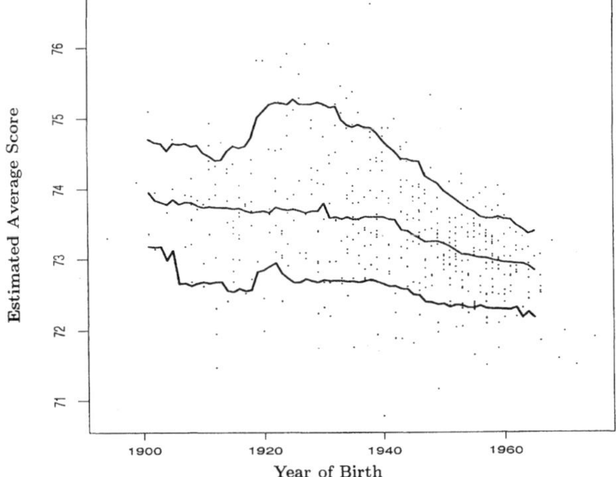

Figure 15. The Estimated Peak Scoring Performance of Each Player Plotted Against the Year They Were Born. The y-axis represents the mean score for each player, at their peak, if the year was 1997. The three curves are the smoothed 10th, 50th, and 90th percentiles.

are drawn mainly from the United States and other countries in the Americas. Baseball has remained fairly stable within the United States, where it has been an important part of the culture for more than a century. On the other hand, golf has experienced a huge recent boom throughout the world.

9. DISCUSSION

In this article we have developed a model for comparing players from different eras in their performance in three different sports. Overlapping careers in each sport provide a network of bridges from the past to the present.

For each sport we constructed additive models to account for various sources of error. The ability of each player, the difficulty of each year, and the effects of aging on performance were modeled. To account for different players aging differently, we used random curves to represent the individual aging effects. The changing population in each sport was modeled with separate hierarchical distributions for each of the decades.

Because of multiple sources of variation not accounted for in scoring, the model for the scoring ability of NHL players did not fit as well as the model in the other three studies. It still provided reasonable estimates, however, and the face validity of the results is very high. The different years in hockey play an important role in scoring. Career totals for individuals are greatly influenced by the era in which they played. Wayne Gretzky holds nearly every scoring record in hockey and yet we estimate him to be the second-best scorer of all time. The optimal age for a hockey player is 27, with a sharp decrease after age 30. A hockey player at age 34, the optimal golf age, is at only 75% of his peak value. Many of the greatest scorers of all time are playing now, NHL hockey has greatly expanded its talent pool in the last 20 years, and the number of great players has increased as well.

The golf model provided a very good fit, with results that are intuitively appealing. Players' abilities have increased substantially over time, and the golf data support Gould's conjecture. The best players in each era are comparable, but the median and below-average players are getting much better over time. The 10th percentile player has gotten about 2 shots better over the last 40 years. The optimal age for a professional golfer is 34, though the range 30–35 is nearly optimal. A golfer at age 20 is approximately equivalent to the same golfer at age 50—both are about 2 shots below their peak level. We found evidence that playing Augusta National now, with the equipment and conditions of today, is about 1 shot easier than playing it with the equipment and conditions of 1950. Evidence was also found that golf scores are not normal. The left tail of scores is slightly shorter than a normal distribution and the right tail slightly heavier than a normal distribution.

The baseball model fit very well. The ability of players to hit home runs has increased dramatically over the century. Many of the greatest home run hitters ever are playing now. Batting average does not have the same increase over the century. There is a gradual increase in the ability of players to hit for average, but the increase is not nearly as dramatic as for home runs. The distribution of batting average players lends good support to Gould's conjecture. The best players are increasing in ability, but the median and 10th percentile players are increasing faster over the century. It has gotten harder for players of a fixed ability to hit for average. This may be due to the increasing ability of pitchers.

Extensions of this work include collecting more complete data in hockey and golf. The aging curve could be extended to allow for different peak ages for the different players. Model selection could be used to address how the populations are changing over time—including continuously indexed hierarchical distributions.

[Received May 1998. Revised January 1999.]

REFERENCES

Albert, J. (1993), Comment on "A Statistical Analysis of Hitting Streaks in Baseball," by S. Albright, *Journal of the American Statistical Association*, 88, 1184–1188.

——— (1998), "The Homerun Hitting of Mike Schmidt," *Chance*, 11, 3–11.

Albright, S. (1993), "A Statistical Analysis of Hitting Streaks in Baseball," *Journal of the American Statistical Association*, 88, 1175–1183.

Berry, S., and Larkey, P. (1998), "The Effects of Age on the Performance of Professional Golfers," *Science and Golf III*, London: E & FN SPON.

Chib, S., and Greenberg, E. (1995), "Understanding the Metropolis–Hastings Algorithm," *The American Statistician*, 49, 327–335.

Draper, D., Gaver, D., Goel, P., Greenhouse, J., Hedges, L., Morris, C., Tucker, J., and Waternaux, C. (1992), *Combining Information: Statistical Issues and Opportunities for Research*, (Vol. 1 of *Contemporary Statistics*), Washington, DC: National Academy Press.

Gelfand, A., and Smith, A. (1990), "Sampling-Based Approaches to Calculating Marginal Densities," *Journal of the American Statistical Association*, 85, 398–409.

Gilks, W., Richardson, S., and Spiegelhalter, D. (Eds.) (1996), *Markov Chain Monte Carlo in Practice*, London: Chapman and Hall.

Gould, S. (1996), *Full House: The Spread of Excellence from Plato to Darwin*, New York: Three Rivers Press.

Hollander, Z. (Ed.) (1997), *Inside Sports Hockey*, Detroit: Visible Ink Press.

Jackson, D., and Mosurski, K. (1997), "Heavy Defeats in Tennis: Psychological Momentum or Random Effect?," *Chance*, 10, 27–34.

Larkey, P., Smith, R., and Kadane, J. (1989), "It's Okay to Believe in the Hot Hand," *Chance*, 2, 22–30.

Lindley, D., and Smith, A. (1972), "Bayes Estimates for the Linear Model," *Journal of the Royal Statistical Society*, Ser. B, 34, 1–41.

Mosteller, F., and Youtz, C. (1993), "Where Eagles Fly," *Chance*, 6, 37–42.

Myers, R. (1990), *Classical and Modern Regression With Applications*, Belmont, CA: Duxbury Press.

Riccio, L. (1994), "The Aging of a Great Player; Tom Watson's Play," *Science and Golf II*, London: E & FN SPON.

Scheid, F. (1990), "On the Normality and Independence of Golf Scores," *Science and Golf*, London: E & FN SPON.

Schell, M. (1999), *Baseball's All-Time Best Hitters*, Princeton, NJ: Princeton University Press.

Shi, M., Weiss, R., and Taylor, J. (1996), "An Analysis of Paediatric CD4 Counts for Acquired Immune Deficiency Syndrome Using Flexible Random Curves," *Applied Statistics*, 45, 151–163.

Stern, H. (1995), "Who's Hot and Who's Not," in *Proceedings of the Section on Statistics in Sports, American Statistical Association.*

Stern, H., and Morris, C. (1993), Comment on "A Statistical Analysis of Hitting Streaks in Baseball," by S. Albright, *Journal of the American Statistical Association*, 88, 1189–1194.

Tierney, L. (1994), "Markov Chains for Exploring Posterior Distributions," *The Annals of Statistics*, 22, 1701–1762.

Tversky, A., and Gilovich, T. (1989a), "The Cold Facts About the "Hot Hand" in Basketball," *Chance*, 2, 16–21.

——— (1989b), "The "Hot Hand": Statistical Reality or Cognitive Illusion," *Chance*, 2, 31–34.

Chapter 29

Data Analysis Using Stein's Estimator and Its Generalizations

BRADLEY EFRON and CARL MORRIS*

In 1961, James and Stein exhibited an estimator of the mean of a multivariate normal distribution having uniformly lower mean squared error than the sample mean. This estimator is reviewed briefly in an empirical Bayes context. Stein's rule and its generalizations are then applied to predict baseball averages, to estimate toxomosis prevalence rates, and to estimate the exact size of Pearson's chi-square test with results from a computer simulation. In each of these examples, the mean square error of these rules is less than half that of the sample mean.

1. INTRODUCTION

Charles Stein [15] showed that it is possible to make a uniform improvement on the maximum likelihood estimator (MLE) in terms of total squared error risk when estimating several parameters from independent normal observations. Later James and Stein [13] presented a particularly simple estimator for which the improvement was quite substantial near the origin, if there are more than two parameters. This achievement leads immediately to a uniform,, nontrivial improvement over the least squares (Gauss-Markov) estimators for the parameters in the usual formulation of the linear model. One might expect a rush of applications of this powerful new statistical weapon, but such has not been the case. Resistance has formed along several lines:

1. Mistrust of the statistical interpretation of the mathematical formulation leading to Stein's result, in particular the sum of squared errors loss function;
2. Difficulties in adapting the James-Stein estimator to the many special cases that invariably arise in practice;
3. Long familiarity with the generally good performance of the MLE in applied problems;
4. A feeling that any gains possible from a "complicated" procedure like Stein's could not be worth the extra trouble. (J.W. Tukey at the 1972 American Statistical Association meetings in Montreal stated that savings would not be more than ten percent in practical situations.)

We have written a series of articles [5, 6, 7, 8, 9, 10, 11] that cover Points 1 and 2. Our purpose here, and in a lengthier version of this report [12], is to illustrate the methods suggested in these articles on three applied problems and in that way deals with Points 3 and 4. Only one of the three problems, the toxoplasmosis data, is "real" in the sense of being generated outside the statistical world. The other two problems are contrived to illustrate in a realistic way the genuine difficulties and rewards of procedures like Stein's. They have the added advantage of having the true parameter values available for comparison of methods. The examples chosen are the first and only ones considered for this report, and the favorable results typify our previous experience.

To review the James-Stein estimator in the simplest setting, suppose that for given θ_i

$$X_i \mid \theta_i \overset{\text{ind}}{\sim} N(\theta_i, 1), \quad i = 1, \cdots, k \geq 3 \;, \qquad (1.1)$$

meaning the $\{X_i\}$ are independent and normally distributed with mean $E_{\theta_i} X_i \equiv \theta_i$ and variance $\mathrm{Var}_{\theta_i}(X_i) = 1$. The example (1.1) typically occurs as a reduction to this canonical form from more complicated situations, as when X_i is a sample mean with known variance that is taken to be unity through an appropriate scale transformation. The unknown vector of means $\boldsymbol{\theta} \equiv (\theta_1, \cdots, \theta_k)$ is to be estimated with loss being the sum of squared component errors

$$L(\boldsymbol{\theta}, \hat{\boldsymbol{\theta}}) \equiv \sum_{i=1}^{k} (\hat{\theta}_i - \theta_i)^2 \;, \qquad (1.2)$$

where $\hat{\boldsymbol{\theta}} \equiv (\hat{\theta}_1, \cdots, \hat{\theta}_k)$ is the estimate of $\boldsymbol{\theta}$. The MLE, which is also the sample mean, $\boldsymbol{\delta}^0(\mathbf{X}) \equiv \mathbf{X} \equiv (X_1, \cdots, X_k)$ has constant risk k,

$$R(\boldsymbol{\theta}, \boldsymbol{\delta}^0) \equiv E_\theta \sum_{i=1}^{k} (X_i - \theta_i)^2 = k \;, \qquad (1.3)$$

E_θ indicating expectation over the distribution (1.1). James and Stein [13] introduced the estimator $\boldsymbol{\delta}^1(\mathbf{X}) = (\delta_1{}^1(\mathbf{X}), \cdots, \delta_k{}^1(\mathbf{X}))$ for $k \geq 3$,

$$\delta_i{}^1(\mathbf{X}) \equiv \mu_i + (1 - (k-2)/S)(X_i - \mu_i) \;, \quad i = 1, \cdots, k \qquad (1.4)$$

with $\boldsymbol{\mu} \equiv (\mu_i, \cdots, \mu_k)'$ any initial guess at $\boldsymbol{\theta}$ and $S \equiv \sum (X_j - \mu_j)^2$. This estimator has risk

$$R(\boldsymbol{\theta}, \boldsymbol{\delta}^1) \equiv E_\theta \sum_{i=0}^{k} (\delta_i{}^1(\mathbf{X}) - \theta_i)^2 \qquad (1.5)$$

$$\leq k - \frac{(k-2)^2}{k - 2 + \sum (\theta_i - \mu_i)^2} < k \;, \qquad (1.6)$$

being less than k for all $\boldsymbol{\theta}$, and if $\theta_i = \mu_i$ for all i the risk is two, comparing very favorably to k for the MLE.

* Bradley Efron is professor, Department of Statistics, Stanford University, Stanford, Calif. 94305. Carl Morris is statistician, Department of Economics, The RAND Corporation, Santa Monica, Calif. 90406.

June 1975, Volume 70, Number 350
Applications Section

The estimator (1.4) arises quite naturally in an empirical Bayes context. If the $\{\theta_i\}$ themselves are a sample from a prior distribution,

$$\theta_i \overset{\text{ind}}{\sim} N(\mu_i, \tau^2), \quad i = 1, \cdots, k , \tag{1.7}$$

then the Bayes estimate of θ_i is the *a posteriori* mean of θ_i given the data

$$\delta_i^*(X_i) = E\theta_i | X_i = \mu_i + (1 - (1 + \tau^2)^{-1})(X_i - \mu_i) . \tag{1.8}$$

In the empirical Bayes situation, τ^2 is unknown, but it can be estimated because marginally the $\{X_i\}$ are independently normal with means $\{\mu_i\}$ and

$$S = \sum (X_j - \mu_j)^2 \sim (1 + \tau^2)\chi_k^2 , \tag{1.9}$$

where χ_k^2 is the chi-square distribution with k degrees of freedom. Since $k \geq 3$, the unbiased estimate

$$E(k - 2)/S = 1/(1 + \tau^2) \tag{1.10}$$

is available, and substitution of $(k - 2)/S$ for the unknown $1/(1 + \tau^2)$ in the Bayes estimate δ_i^* of (1.8) results in the James-Stein rule (1.4). The risk of δ_i^1 averaged over both $\mathbf{X}$ and $\boldsymbol{\theta}$ is, from [6] or [8],

$$E_\tau E_\theta(\delta_i^1(\mathbf{X}) - \theta_i)^2 = 1 - (k - 2)/k(1 + \tau^2) , \tag{1.11}$$

E_τ denoting expectation over the distribution (1.7). The risk (1.11) is to be compared to the corresponding risks of 1 for the MLE and $1 - 1/(1 + \tau^2)$ for the Bayes estimator. Thus, if k is moderate or large δ_i^1 is nearly as good as the Bayes estimator, but it avoids the possible gross errors of the Bayes estimator if τ^2 is misspecified.

It is clearly preferable to use min $\{1, (k - 2)/S\}$ as an estimate of $1/(1 + \tau^2)$ instead of (1.10). This results in the simple improvement

$$\delta_i^{1+}(\mathbf{X}) = \mu_i + (1 - (k - 2)/S)^+(X_i - \mu_i) \tag{1.12}$$

with $a^+ \equiv \max(0, a)$. That $R(\boldsymbol{\theta}, \boldsymbol{\delta}^{1+}) < R(\boldsymbol{\theta}, \boldsymbol{\delta}^1)$ for all $\boldsymbol{\theta}$ is proved in [2, 8, 10, 17]. The risks $R(\boldsymbol{\theta}, \boldsymbol{\delta}^1)$ and $R(\theta, \delta^{1+})$ are tabled in [11].

2. USING STEIN'S ESTIMATOR TO PREDICT BATTING AVERAGES

The batting averages of 18 major league players through their first 45 official at bats of the 1970 season appear in Table 1. The problem is to predict each player's batting average over the remainder of the season using only the data of Column (1) of Table 1. This sample was chosen because we wanted between 30 and 50 at bats to assure a satisfactory approximation of the binomial by the normal distribution while leaving the bulk of at bats to be estimated. We also wanted to include an unusually good hitter (Clemente) to test the method with at least one extreme parameter, a situation expected to be less favorable to Stein's estimator. Batting averages are published weekly in the *New York Times,* and by April 26, 1970 Clemente had batted 45 times. Stein's estimator requires equal variances,[1] or in this situation, equal at bats, so the remaining 17 players are all whom either the April 26 or May 3 *New York Times* reported with 45 at bats.

Let Y_i be the batting average of Player i, $i = 1, \cdots, 18$ $(k = 18)$ after $n = 45$ at bats. Assuming base hits occur according to a binomial distribution with independence between players, $nY_i \overset{\text{ind}}{\sim} \text{Bin}(n, p_i)$ $i = 1, 2, \cdots, 18$ with p_i the true season batting average, so $EY_i = p_i$. Because the variance of Y_i depends on the mean, the arc-sin transformation for stabilizing the variance of a binomial distribution is used: $X_i \equiv f_{45}(Y_i)$, $i = 1, \cdots, 18$ with

$$f_n(y) \equiv (n)^{\frac{1}{2}} \arcsin(2y - 1) . \tag{2.1}$$

Then X_i has nearly unit variance[2] independent of p_i. The mean[3] θ_i of X_i is given approximately by $\theta_i = f_n(p_i)$. Values of X_i, θ_i appear in Table 1. From the central limit theorem for the binomial distribution and continuity of f_n we have approximately

$$X_i | \theta_i \overset{\text{ind}}{\sim} N(\theta_i, 1), \quad i = 1, 2, \cdots, k , \tag{2.2}$$

the situation described in Section 1.

We use Stein's estimator (1.4), but we estimate the common unknown value $\mu = \sum \mu_i/k$ by $\bar{X} = \sum X_i/k$, shrinking all X_i toward $\bar{X}$, an idea suggested by Lindley [6, p. 285–7]. The resulting estimate of the ith component θ_i of $\boldsymbol{\theta}$ is therefore

$$\tilde{\delta}_i^1(\mathbf{X}) = \bar{X} + (1 - (k - 3)/V)(X_i - \bar{X}) \tag{2.3}$$

with $V \equiv \sum (X_i - \bar{X})^2$ and with $k - 3 = (k - 1) - 2$ as the appropriate constant since one parameter is estimated. In the empirical Bayes case, the appropriateness of (2.3) follows from estimating the Bayes rule (1.8) by using the unbiased estimates $\bar{X}$ for μ and $(k - 3)/V$ for $1/(1 + \tau)^2$ from the marginal distribution of $\mathbf{X}$, analogous to Section 1 (see also [6, Sec. 7]). We may use the Bayesian model for these data because (1.7) seems at least roughly appropriate, although (2.3) also can be justified by the non-Bayesian from the suspicion that $\sum (\theta_i - \bar{\theta})^2$ is small, since the risk of (2.3), analogous to (1.6), is bounded by

$$R(\boldsymbol{\theta}, \tilde{\boldsymbol{\delta}}^1) \leq k - \frac{(k - 3)^2}{k - 3 + \sum (\theta_i - \bar{\theta})^2}, \quad \bar{\theta} \equiv \sum \theta_i/k . \tag{2.4}$$

For our data, the estimate of $1/(1 + \tau^2)$ is $(k - 3)/V = .791$ or $\hat{\tau} = 0.514$, representing considerable *a priori* information. The value of $\bar{X}$ is -3.275 so

$$\tilde{\delta}_i^1(\mathbf{X}) = \hat{\theta}_i = .791\bar{X} + .209X_i = .209X_i - 2.59 . \tag{2.5}$$

[1] The unequal variances case is discussed in Section 3.

[2] An exact computer computation showed that the standard deviation of X_i is within .036 of unity for $n = 45$ for all p_i between 0.15 and 0.85.

[3] For most of this discussion we will regard the values of p_i of Column 2, Table 1 and θ_i as the quantities to be estimated, although we actually have a prediction problem because these quantities are estimates of the mean of Y_i. Accounting for this fact would cause Stein's method to compare even more favorably to the sample mean because the random error in p_i increases the losses for all estimators equally. This increases the errors of good estimators by a higher percentage than poorer ones.

1. 1970 Batting Averages for 18 Major League Players and Transformed Values X_i, θ_i

i	Player	Y_i = batting average for first 45 at bats (1)	p_i = batting average for remainder of season (2)	At bats for remainder of season (3)	X_i (4)	θ_i (5)
1	Clemente (Pitts, NL)	.400	.346	367	−1.35	−2.10
2	F. Robinson (Balt, AL)	.378	.298	426	−1.66	−2.79
3	F. Howard (Wash, AL)	.356	.276	521	−1.97	−3.11
4	Johnstone (Cal, AL)	.333	.222	275	−2.28	−3.96
5	Berry (Chi, AL)	.311	.273	418	−2.60	−3.17
6	Spencer (Cal, AL)	.311	.270	466	−2.60	−3.20
7	Kessinger (Chi, NL)	.289	.263	586	−2.92	−3.32
8	L. Alvarado (Bos, AL)	.267	.210	138	−3.26	−4.15
9	Santo (Chi, NL)	.244	.269	510	−3.60	−3.23
10	Swoboda (NY, NL)	.244	.230	200	−3.60	−3.83
11	Unser (Wash, AL)	.222	.264	277	−3.95	−3.30
12	Williams (Chi, AL)	.222	.256	270	−3.95	−3.43
13	Scott (Bos, AL)	.222	.303	435	−3.95	−2.71
14	Petrocelli (Bos, AL)	.222	.264	538	−3.95	−3.30
15	E. Rodriguez (KC, AL)	.222	.226	186	−3.95	−3.89
16	Campaneris (Oak, AL)	.200	.285	558	−4.32	−2.98
17	Munson (NY, AL)	.178	.316	408	−4.70	−2.53
18	Alvis (Mil, NL)	.156	.200	70	−5.10	−4.32

The results are striking. The sample mean **X** has total squared prediction error $\sum (X_i - \theta_i)^2$ of 17.56, but $\tilde{\delta}^1(\mathbf{X}) \equiv (\tilde{\delta}_1^1(\mathbf{X}), \cdots, \tilde{\delta}_k^1(\mathbf{X}))$ has total squared prediction error of only 5.01. The efficiency of Stein's rule relative to the MLE for these data is defined as $\sum (X_i - \theta_i)^2 / \sum (\tilde{\delta}_i^1(\mathbf{X}) - \theta_i)^2$, the ratio of squared error losses. The efficiency of Stein's rule is 3.50 (=17.56/5.01) in this example. Moreover, $\tilde{\delta}_i^1$ is closer than X_i to θ_i for 15 batters, being worse only for Batters 1, 10, 15. The estimates (2.5) are retransformed in Table 2 to provide estimates $\hat{p}_i^1 = f_n^{-1}(\hat{\theta}_i)$ of p_i.

Stein's estimators achieve uniformly lower aggregate risk than the MLE but permit considerably increased risk to individual components of the vector $\boldsymbol{\theta}$. As a function of $\boldsymbol{\theta}$, the risk for estimating θ_1 by $\tilde{\delta}_1^1$, for example, can be as large as $k/4$ times as great as the risk of the MLE X_1. This phenomenon is discussed at length in [5, 6], where "limited translation estimators" $\tilde{\delta}^s(\mathbf{X})$ $0 \leq s \leq 1$ are introduced to reduce this effect. The MLE corresponds to $s = 0$, Stein's estimator to $s = 1$. The estimate $\tilde{\delta}_i^s(\mathbf{X})$ of θ_i is defined to be as close as possible to $\tilde{\delta}_i^1(\mathbf{X})$ subject to the condition that it not differ from X_i by more than $[(k-1)(k-3)/kV]^{\frac{1}{2}} D_{k-1}(s)$ standard deviations of X_i, $D_{k-1}(s)$ being a constant taken from [6, Table 1]. If $s = 0.8$, then $D_{17}(s) = 0.786$, so $\tilde{\delta}_i^{0.8}(\mathbf{X})$ may differ from X_i by no more than

$$0.786\ (17 \times 0.791/18)^{\frac{1}{2}} = .68$$

This modification reduces the maximum component risk of 4.60 for $\tilde{\delta}_i^1$ to 1.52 for $\tilde{\delta}_i^{0.8}$ while retaining 80 percent of the savings of Stein's rule over the MLE. The retransformed values $\hat{p}_i^{0.8}$ of the limited translation estimates $f_n^{-1}(\tilde{\delta}_i^{0.8}(\mathbf{X}))$ are given in the last column of Table 2, the estimates for the top three and bottom two batters being affected. Values for $s = 0.9$ are also given in Table 2.

2. Batting Averages and Their Estimates

i	Batting average for season remainder p_i	Maximum likelihood estimate Y_i	Retransform of Stein's estimator $\hat{p}_i^1$	Retransform of $\tilde{\delta}^{0.9}$ $\hat{p}_i^{0.9}$	Retransform of $\tilde{\delta}^{0.8}$ $\hat{p}_i^{0.8}$
1	.346	.400	.290	.334	.351
2	.298	.378	.286	.313	.329
3	.276	.356	.281	.292	.308
4	.222	.333	.277	.277	.287
5	.273	.311	.273	.273	.273
6	.270	.311	.273	.273	.273
7	.263	.289	.268	.268	.268
8	.210	.267	.264	.264	.264
9	.269	.244	.259	.259	.259
10	.230	.244	.259	.259	.259
11	.264	.222	.254	.254	.254
12	.256	.222	.254	.254	.254
13	.303	.222	.254	.254	.254
14	.264	.222	.254	.254	.254
15	.226	.222	.254	.254	.254
16	.285	.200	.249	.249	.242
17	.316	.178	.244	.233	.218
18	.200	.156	.239	.208	.194

Clemente ($i = 1$) was known to be an exceptionally good hitter from his performance in other years. Limiting translation results in a much better estimate for him, as we anticipated, since $\tilde{\delta}_1^1(\mathbf{X})$ differs from X_1 by an excessive 1.56 standard deviations of X_1. The limited translation estimators are closer than the MLE for 16 of the 18 batters, and the case $s = 0.9$ has better efficiency (3.91) for these data relative to the MLE than Stein's rule (3.50), but the rule with $s = 0.8$ has lower efficiency (3.01). The maximum component error occurs for Munson ($i = 17$) with all four estimators. The Bayesian effect is so strong that this maximum error $|\hat{\theta}_{17} - \theta_{17}|$ decreased from 2.17 for $s = 0$, to 1.49 for $s = 0.8$, to 1.25 for $s = 0.9$ to 1.08 for $s = 1$. Limiting translation therefore increases the worst error in this example, just opposite to the maximum risks.

3. A GENERALIZATION OF STEIN'S ESTIMATOR TO UNEQUAL VARIANCES FOR ESTIMATING THE PREVALENCE OF TOXOPLASMOSIS

One of the authors participated in a study of toxoplasmosis in El Salvador [14]. Sera obtained from a total sample of 5,171 individuals of varying ages from 36 El Salvador cities were analyzed by a Sabin-Feldman dye test. From the data given in [14, Table 1], toxoplasmosis prevalence rates X_i for City i, $i = 1, \cdots, 36$ were calculated. The prevalence rate X_i has the form (observed minus expected)/expected, with "observed" being the number of positives for City i and "expected" the number of positives for the same city based on an indirect standardization of prevalence rates to the age distribution of City i. The variances $D_i = \text{Var}(X_i)$ are known from binomial considerations and differ because of unequal sample sizes.

These data X_i together with the standard deviations $D_i^{\frac{1}{2}}$ are given in Columns 2 and 3 of Table 3. The prevalence rates satisfy a linear constraint $\sum d_i X_i = 0$ with known coefficients $d_i > 0$. The means $\theta_i = EX_i$, which also satisfy $\sum d_i \theta_i = 0$, are to be estimated from the $\{X_i\}$. Since the $\{X_i\}$ were constructed as sums of independent random variables, they are approximately normal; and except for the one linear constraint on the $k = 36$ values of X_i, they are independent. For simplicity, we will ignore the slight improvement in the independence approximation that would result from applying our methods to an appropriate 35-dimensional subspace and assume that the $\{X_i\}$ have the distribution of the following paragraph.

3. Estimates and Empirical Bayes Estimates of Toxoplasmosis Prevalence Rates

i	X_i	$\sqrt{D_i}$	$\delta_i(\mathbf{X})$	$\hat{A}_i$	k_i	$\hat{B}_i$
1	.293	.304	.035	.0120	1334.1	.882
2	.214	.039	.192	.0108	21.9	.102
3	.185	.047	.159	.0109	24.4	.143
4	.152	.115	.075	.0115	80.2	.509
5	.139	.081	.092	.0112	43.0	.336
6	.128	.061	.100	.0110	30.4	.221
7	.113	.061	.088	.0110	30.4	.221
8	.098	.087	.062	.0113	48.0	.370
9	.093	.049	.079	.0109	25.1	.154
10	.079	.041	.070	.0109	22.5	.112
11	.063	.071	.045	.0111	36.0	.279
12	.052	.048	.044	.0109	24.8	.148
13	.035	.056	.028	.0110	28.0	.192
14	.027	.040	.024	.0108	22.2	.107
15	.024	.049	.020	.0109	25.1	.154
16	.024	.039	.022	.0108	21.9	.102
17	.014	.043	.012	.0109	23.1	.122
18	.004	.085	.003	.0112	46.2	.359
19	−.016	.128	−.007	.0116	101.5	.564
20	−.028	.091	−.017	.0113	51.6	.392
21	−.034	.073	−.024	.0111	37.3	.291
22	−.040	.049	−.034	.0109	25.1	.154
23	−.055	.058	−.044	.0110	28.9	.204
24	−.083	.070	−.060	.0111	35.4	.273
25	−.098	.068	−.072	.0111	34.2	.262
26	−.100	.049	−.085	.0109	25.1	.154
27	−.112	.059	−.089	.0110	29.4	.210
28	−.138	.063	−.106	.0110	31.4	.233
29	−.156	.077	−.107	.0112	40.0	.314
30	−.169	.073	−.120	.0111	37.3	.291
31	−.241	.106	−.128	.0114	68.0	.468
32	−.294	.179	−.083	.0118	242.4	.719
33	−.296	.064	−.225	.0111	31.9	.238
34	−.324	.152	−.114	.0117	154.8	.647
35	−.397	.158	−.133	.0117	171.5	.665
36	−.665	.216	−.140	.0119	426.8	.789

To obtain an appropriate empirical Bayes estimation rule for these data we assume that

$$X_i \mid \theta_i \overset{\text{ind}}{\sim} N(\theta_i, D_i), \quad i = 1, \cdots, k \tag{3.1}$$

and

$$\theta_i \overset{\text{ind}}{\sim} N(0, A), \quad i = 1, \cdots, k\ , \tag{3.2}$$

A being an unknown constant. These assumptions are the same as (1.1), (1.7), which lead to the James-Stein estimator if $D_i = D_j$ for all i, j. Notice that the choice of *a priori* mean zero for the θ_i is particularly appropriate here because the constant $\sum d_i \theta_i = 0$ forces the parameters to be centered near the origin.

We require $k \geq 3$ in the following derivations. Define

$$B_i \equiv D_i/(A + D_i)\ . \tag{3.3}$$

Then (3.1) and (3.2) are equivalent to

$$\theta_i \mid X_i \overset{\text{ind}}{\sim} N((1 - B_i)X_i, D_i(1 - B_i)), \quad i = 1, \cdots, k\ . \tag{3.4}$$

For squared error loss[4] the Bayes estimator is the *a posteriori* mean

$$\delta_i^*(X_i) = E\theta_i \mid X_i = (1 - B_i)X_i\ , \tag{3.5}$$

with Bayes risk $\text{Var}(\theta_i \mid X_i) = (1 - B_i)D_i$ being less than the risk D_i of $\hat{\theta}_i = X_i$.

Here, A is unknown, but the MLE $\hat{A}$ of A on the basis of the data $S_j \equiv X_j^2 \sim (A + D_j)\chi_1^2$, $j = 1, 2, \cdots, k$ is the solution to

$$\hat{A} = \sum_{j=1}^{k} (S_j - D_j) I_j(\hat{A}) / \sum_{j=1}^{k} I_j(\hat{A}) \tag{3.6}$$

with

$$I_j(A) \equiv 1/\text{Var}(S_j) = 1/[2(A + D_j)^2] \tag{3.7}$$

being the Fisher information for A in S_j. We could use $\hat{A}$ from (3.6) to define the empirical Bayes estimator of θ_i as $(1 - D_i/(\hat{A} + D_i))X_i$. However, this rule does not reduce to Stein's when all D_j are equal, and we instead use a minor variant of this estimator derived in [8] which does reduce to Stein's. The variant rule estimates a different value $\hat{A}_i$ for each city (see Table 3). The difference between the rules is minor in this case, but it might be important if k were smaller.

Our estimates $\delta_i(\mathbf{X})$ of the θ_i are given in the fourth column of Table 3 and are compared with the unbiased

[4] Or for any other increasing function of $|\theta_i - \hat{\theta}_i|$.

estimate X_i in Figure A. Figure A illustrates the "pull in" effect of $\delta_i(\mathbf{X})$, which is most pronounced for Cities 1, 32, 34, 35, and 36. Under the empirical Bayes model, the major explanation for the large $|X_i|$ for these cities is large D_i rather than large $|\theta_i|$. This figure also shows that the rankings of the cities on the basis of $\delta_i(\mathbf{X})$ differs from that based on the X_i, an interesting feature that does not arise when the X_i have equal variances.

A. Estimates of Toxoplasmosis Prevalence Rates

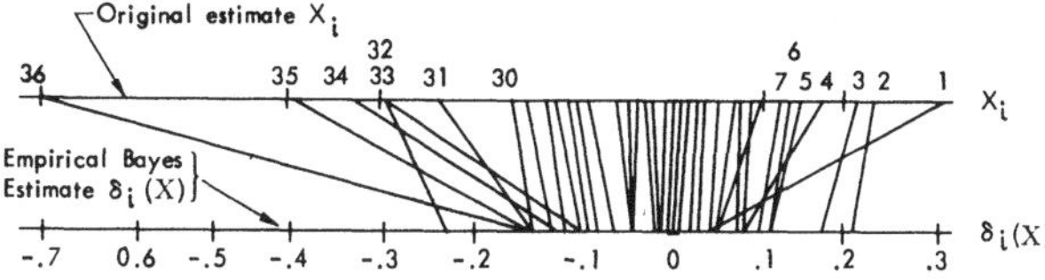

The values $\hat{A}_i$, $\hat{k}_i$, and $\hat{B}_i(S)$ defined in [8] are given in the last three columns of Table 3. The value $\hat{A}$ of (3.6) is $\hat{A} = 0.0122$ with standard deviation $\sigma(\hat{A})$ estimated as 0.0041 (if $A = 0.0122$) by the Cramér-Rao lower bound on $\sigma(\hat{A})$. The preferred estimates $\hat{A}_i$ are all close to but slightly smaller than $\hat{A}$, and their estimated standard deviations vary from 0.00358 for the cities with the smallest D_i to 0.00404 for the city with the largest D_i.

The likelihood function of the data plotted as a function of A (on a log scale) is given in Figures B and C as LIKELIHOOD. The curves are normalized to have unit area as a function of $\alpha = \log A$. The maximum value of this function of α is at $\hat{\alpha} = \log(\hat{A}) = \log(.0122) = -4.40 \equiv \mu_\alpha$. The curves are almost perfectly normal with mean $\hat{\alpha} = -4.40$ and standard deviation $\sigma_\alpha \equiv .371$. The likely values of A therefore correspond to a α differing from μ_α by no more than three standard deviations, $|\alpha - \mu_\alpha| \leq 3\sigma_\alpha$, or equivalently, $.0040 \leq A \leq .0372$.

In the region of likely values of A, Figure B also graphs two risks: BAYES RISK and EB RISK (for empirical Bayes risk), each conditional on the data $\mathbf{X}$. EB RISK[5] is the conditional risk of the empirical Bayes rule defined (with $D_0 \equiv (1/k) \sum_{i=1}^{k} D_i$) as

$$E_A \frac{1}{kD_0} \sum_{i=0}^{k} (\delta_i(\mathbf{X}) - \theta_i)^2 | \mathbf{X} , \tag{3.8}$$

and BAYES RISK is

$$E_A \frac{1}{kD_0} \sum_{i=1}^{k} \left(\frac{A}{A + D_i} X_i - \theta_i \right)^2 \Bigg| \mathbf{X} . \tag{3.9}$$

Since A is not known, BAYES RISK yields only a lower envelope for empirical Bayes estimators, agreeing with EB RISK at $A = .0122$. Table 4 gives values to supplement Figure B. Not graphed because it is too large to fit in Figure B is MLE RISK, the conditional risk of the MLE, defined as

$$E_A \frac{1}{kD_0} \sum_{i=1}^{k} (X_i - \theta_i)^2 | \mathbf{X} . \tag{3.10}$$

MLE RISK exceeds EB RISK by factors varying from 7 to 2 in the region of likely values of A, as shown in Table 4. EB RISK tends to increase and MLE RISK to decrease as A increases, these values crossing at $A = .0650$, about $4\frac{1}{2}$ standard deviations above the mean of the distribution of $\hat{A}$.

4. Conditional Risks for Different Values of A

Risk	A				
	.0040	.0122	.0372	.0650	∞
EB RISK	.35	.39	.76	1.08	2.50
MLE RISK	2.51	1.87	1.27	1.08	1.00
P(EB CLOSER)	1.00	1.00	.82	.50	.04

The remaining curve in Figure B graphs the probability that the empirical Bayes estimator is closer to $\boldsymbol{\theta}$ than the MLE $\mathbf{X}$, conditional on the data $\mathbf{X}$. It is defined as

$$P_A[\sum (\delta_i(\mathbf{X}) - \theta_i)^2 < \sum (X_i - \theta_i)^2 | \mathbf{X}] . \tag{3.11}$$

This curve, denoted P(EB CLOSER), decreases as A increases but is always very close to unity in the region of likely values of A. It reaches one-half at about $4\frac{1}{2}$ standard deviations from the mean of the likelihood function and then decreases as $A \to \infty$ to its asymptotic value .04 (see Table 4).

The data suggest that almost certainly A is in the interval $.004 \leq A \leq .037$, and for all such values of A, Figure B and Table 4 indicate that the numbers $\delta_i(\mathbf{X})$ are much better estimators of the θ_i than are the X_i. Non-Bayesian versions of these statements may be based on a confidence interval for $\sum \theta_i^2/k$.

Figure A illustrates that the MLE and the empirical Bayes estimators order the $\{\theta_i\}$ differently. Define the

B. Likelihood Function of A and Aggregate Operating Characteristics of Estimates as a Function of A, Conditional on Observed Toxoplasmosis Data

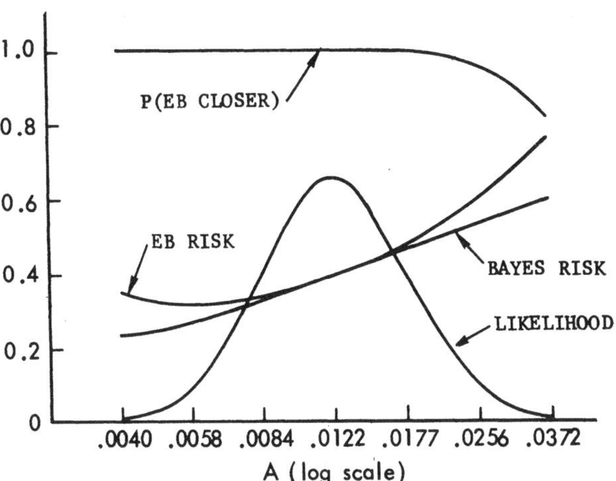

[5] In (3.8) the $\delta_i(\mathbf{X})$ are fixed numbers—those given in Table 3. The expectation is over the *a posteriori* distribution (3.4) of the θ_i.

correlation of an estimator $\hat{\boldsymbol{\theta}}$ of $\boldsymbol{\theta}$ by

$$r(\hat{\boldsymbol{\theta}}, \boldsymbol{\theta}) = \sum \hat{\theta}_i\theta_i/(\sum \hat{\theta}_i^2 \sum \theta_i^2)^{\frac{1}{2}} \quad (3.12)$$

as a measure of how well $\hat{\boldsymbol{\theta}}$ orders $\boldsymbol{\theta}$. We denote $P(r^{\text{EB}} > r^{\text{MLE}})$ as the probability that the empirical Bayes estimate $\boldsymbol{\delta}$ orders $\boldsymbol{\theta}$ better than $\mathbf{X}$, i.e., as

$$P_A\{r(\boldsymbol{\delta}, \boldsymbol{\theta}) > r(\mathbf{X}, \boldsymbol{\theta}) | \mathbf{X}\} \ . \quad (3.13)$$

The graph of (3.13) given in Figure C shows that $P(r^{\text{EB}} > r^{\text{MLE}}) > .5$ for $A \leq .0372$. The value at $A \doteq \infty$ drops to .046.

C. Likelihood Function of A and Individual and Ordering Characteristics of Estimates as a Function of A, Conditional on Observed Toxoplasmosis Data

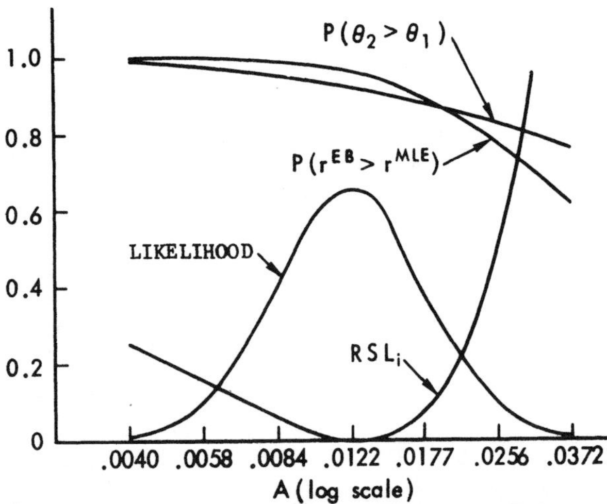

Although $X_1 > X_2$, the empirical Bayes estimator for City 2 is larger, $\delta_2(\mathbf{X}) > \delta_1(\mathbf{X})$. This is because $D_1 \gg D_2$, indicating that X_1 is large under the empirical Bayes model because of randomness while X_2 is large because θ_2 is large. The other curve in Figure C is

$$P_A(\theta_2 > \theta_1 | \mathbf{X}) \quad (3.14)$$

and shows that $\theta_2 > \theta_1$ is quite probable for likely values of A. This probability declines as $A \to \infty$, being .50 at $A = .24$ (eight standard deviations above the mean) and .40 at $A = \infty$.

4. USING STEIN'S ESTIMATOR TO IMPROVE THE RESULTS OF A COMPUTER SIMULATION

A Monte Carlo experiment is given here in which several forms of Stein's method all double the experimental precision of the classical estimator. The example is realistic in that the normality and variance assumptions are approximations to the true situation.

We chose to investigate Pearson's chi-square statistic for its independent interest and selected the particular parameters ($m \leq 24$) from our prior belief that empirical Bayes methods would be effective for these situations. Although our beliefs were substantiated, the outcomes in this instance did not always favor our pet methods.

The simulation was conducted to estimate the exact size of Pearson's chi-square test. Let Y_1 and Y_2 be independent binomial random variables, $Y_1 \sim \text{bin}(m, p')$, $Y_2 \sim \text{bin}(m, p'')$ so $EY_1 = mp'$, $EY_2 = mp''$. Pearson advocated the statistic and critical region

$$T = \frac{2m(Y_1 - Y_2)^2}{(Y_1 + Y_2)(2m - Y_1 - Y_2)} > 3.84 \quad (4.1)$$

to test the composite null hypothesis $H_0: p' = p''$ against all alternatives for the nominal size $\alpha = 0.05$. The value 3.84 is the 95th percentile of the chi-square distribution with one degree of freedom, which approximates that of T when m is large.

The true size of the test under H_0 is defined as

$$\alpha(p, m) \equiv P(T > 3.84 | p, m) \ , \quad (4.2)$$

which depends on both m and the unknown value $p \equiv p' = p''$. The simulation was conducted for $p = 0.5$ and the $k = 17$ values of m with $m_j = 7 + j$, $j = 1, \cdots, k$. The k values of $\alpha_j \equiv \alpha(0.5, m_j)$ were to be estimated. For each j we simulated (4.1) $n = 500$ times on a computer and recorded Z_j as the proportion of times H_0 was rejected. The data appear in Table 5. Since $nZ_j \sim \text{bin}(n, \alpha_j)$ independently, Z_j is the unbiased and maximum likelihood estimator usually chosen[6] to estimate α_j.

5. Maximum Likelihood Estimates and True Values for p = 0.5

	MLE		True values
j	m_j	Z_j	α_j
1	8	.082	.07681
2	9	.042	.05011
3	10	.046	.04219
4	11	.040	.05279
5	12	.054	.06403
6	13	.084	.07556
7	14	.036	.04102
8	15	.036	.04559
9	16	.040	.05151
10	17	.050	.05766
11	18	.078	.06527
12	19	.030	.05306
13	20	.036	.04253
14	21	.060	.04588
15	22	.052	.04896
16	23	.046	.05417
17	24	.054	.05950

Under H_0 the standard deviation of Z_j is approximately $\sigma = \{(.05)(.95)/500\}^{\frac{1}{2}} = .009747$. The variables $X_j \equiv (Z_j - .05)/\sigma$ have expectations

$$\theta_j \equiv EX_j = (\alpha_j - .05)/\sigma$$

[6] We ignore an extensive bibliography of other methods for improving computer simulations. Empirical Bayes methods can be applied simultaneously with other methods, and if better estimates of α_j than Z_j were available then the empirical Bayes methods could instead be applied to them. But for simplicity we take Z_j itself as the quantity to be improved.

and approximately the distribution

$$X_j | \theta_j \overset{\text{ind}}{\sim} N(\theta_j, 1), \quad j = 1, 2, \cdots, 17 = k \ , \qquad (4.3)$$

described in earlier sections.

The average value $\bar{Z} = .051$ of the 17 points supports the choice of the "natural origin" $\bar{\alpha} = .05$. Stein's rule (1.4) applied to the transformed data (4.3) and then retransformed according to $\hat{\alpha}_j = .05 + \sigma\hat{\theta}_j$ yields

$$\hat{\alpha}_j = (1 - \hat{B})Z_j + .05\hat{B} \ , \quad \hat{B} = .325 \ , \qquad (4.4)$$

where $\hat{B} \equiv (k - 2)/S$ and

$$S \equiv \sum_{j=1}^{17} (Z_j - .05)^2/\sigma^2 = 46.15 \ .$$

All 17 true values α_j were obtained exactly through a separate computer program and appear in Figure D and Table 5, so the loss function, taken to be the normalized sum of squared errors $\sum (\hat{\alpha}_j - \alpha_j)^2/\sigma^2$, can be evaluated.[7] The MLE has loss 18.9, Stein's estimate (4.4) has loss 10.2, and the constant estimator, which always estimates α_j as .05, has loss 23.4. Stein's rule therefore dominates both extremes between which it compromises.

Figure D displays the maximum likelihood estimates, Stein estimates, and true values. The true values show a surprising periodicity, which would frustrate attempts at improving the MLE by smoothing.

D. MLE, Stein Estimates, and True Values for p = 0.5

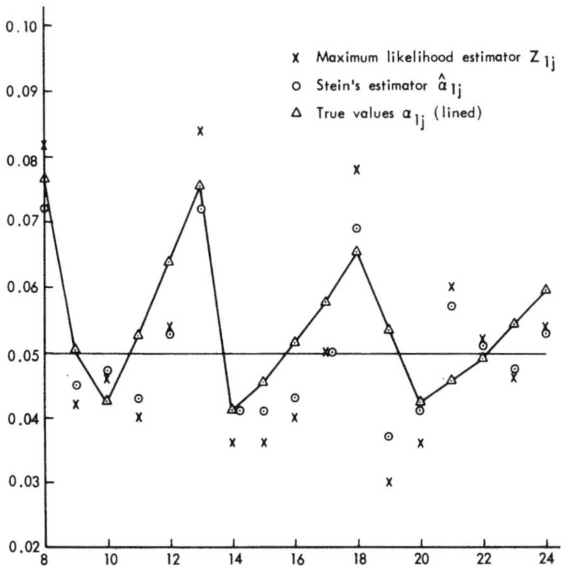

On theoretical grounds we know that the approximation $\alpha(p, m) = .05$ improves as m increases, which suggests dividing the data into two groups, say $8 \leq m \leq 16$ and $17 \leq m \leq 24$. In the Bayesian framework [9] this disaggregation reflects the concern that A_1, the expectation of $A_1^* \equiv \sum_{j=1}^{9} (\alpha_j - .05)^2/9\sigma^2$ may be much larger than A_2, the expectation of $A_2^* = \sum_{j=10}^{17} (\alpha_j - .05)^2/8\sigma^2$, or equivalently that the pull-in factor $B_1 = 1/(1 + A_1)$ for Group 1 really should be smaller than $B_2 = 1/(1 + A_2)$ for Group 2.

[7] Exact rejection probabilities for other values of p are given in [12].

The combined estimator (4.4), having $\hat{B}_1 = \hat{B}_2$, is repeated in the second row of Table 6 with loss components for each group. The simplest way to utilize separate estimates of B_1 and B_2 is to apply two separate Stein rules, as shown in the third row of the table.

6. Values of $\hat{B}$ and Losses for Data Separated into Two Groups, Various Estimation Rules

Rule	$8 \leq m \leq 16$ $\hat{B}_1$	Group 1 loss	$17 \leq m \leq 24$ $\hat{B}_2$	Group 2 loss	Total loss
Maximum Likelihood Estimator	.000	7.3	.000	11.6	18.9
Stein's rule, combined data	.325	4.2	.325	6.0	10.2
Separate Stein rules	.232	4.5	.376	5.4	9.9
Separate Stein rules, bigger constant	.276	4.3	.460	4.6	8.9
All estimates at .05	1.000	18.3	1.000	5.1	23.4

In [8, Sec. 5] we suggest using the bolder estimate

$$\hat{B}_i = (k_i - .66)/S_i \ , \quad S_1 \equiv \sum_{j=1}^{9} (Z_j - .05)^2/\sigma^2 \ ,$$

$$S_2 \equiv S - S_1 \ , \quad k_1 = 9 \ , \quad k_2 = 8 \ .$$

The constant $k_i - .66$ is preferred because it accounts for the fact that the positive part (1.12) will be used, whereas the usual choice $k_i - 2$ does not. The fourth row of Table 6 shows the effectiveness of this choice.

The estimate of .05, which is nearly the mean of the 17 values, is included in the last row of the table to show that the Stein rules substantially improve the two extremes between which they compromise.

The actual values are

$$A_1^* = \sum_{j=1}^{9} (\alpha_j - .05)^2/9\sigma^2 = 2.036$$

for Group 1 and

$$A_2^* = \sum_{j=10}^{17} (\alpha_j - .05)^2/8\sigma^2 = .635 \ ,$$

so $B_1^* = 1/(1 + A_1^*) = .329$ and $B_2^* = 1/(1 + A_2^*) = .612$. The true values of B_1^* and B_2^* are somewhat different, as estimates for separate Stein rules suggest. Rules with $\hat{B}_1$ and $\hat{B}_2$ near these true values will ordinarily perform better for data simulated from these parameters $p = 0.5$, $m = 8, \cdots, 24$.

5. CONCLUSIONS

In the baseball, toxoplasmosis, and computer simulation examples, Stein's estimator and its generalizations increased efficiencies relative to the MLE by about 350 percent, 200 percent, and 100 percent. These examples were chosen because we expected empirical Bayes methods to work well for them and because their efficiencies could be determined. But we are aware of other successful applications to real data[8] and have suppressed no negative results. Although blind application of these methods would gain little in most instances, the statistician who uses them sensibly and selectively can expect major improvements.

Even when they do not significantly increase efficiency, there is little penalty for using the rules discussed here because they cannot give larger total mean squared error than the MLE and because the limited translation modification protects individual components. As several authors have noted, these rules are also robust to the assumption of the normal distribution, because their operating characteristics depend primarily on the means and variances of the sampling distributions and of the unknown parameters. Nor is the sum of squared error criterion especially important. This robustness is borne out by the experience in this article since the sampling distributions were actually binomial rather than normal. The rules not only worked well in the aggregate here, but for most components the empirical Bayes estimators ranged from slightly to substantially better than the MLE, with no substantial errors in the other direction.

Tukey's comment, that empirical Bayes benefits are unappreciable (Section 1), actually was directed at a method of D.V. Lindley. Lindley's rules, though more formally Bayesian, are similar to ours in that they are designed to pick up the same intercomponent information in possibly related estimation problems. We have not done justice here to the many other contributors to multiparameter estimation, but refer the reader to the lengthy bibliography in [12]. We have instead concentrated on Stein's rule and its generalizations to illustrate the power of the empirical Bayes theory, because the main gains are derived by recognizing the applicability of the theory, with lesser benefit attributable to the particular method used. Nevertheless, we hope other authors will compare their methods with ours on these or other data.

The rules of this article are neither Bayes nor admissible, so they can be uniformly beaten (but not by much; see [8, Sec. 6]). There are several published, admissible, minimax rules which also would do well on the baseball data, although probably not much better than the rule used there, for none yet given is known to dominate Stein's rule with the positive part modification. For applications, we recommend the combination of simplicity, generalizability, efficiency, and robustness found in the estimators presented here.

The most favorable situation for these estimators occurs when the statistician wants to estimate the parameters of a linear model that are known to lie in a high dimensional parameter space H_1, but he suspects that they may lie close to a specified lower dimensional parameter space $H_0 \subset H_1$.[9] Then estimates unbiased for every parameter vector in H_1 may have large variance, while estimates restricted to H_0 have smaller variance but possibly large bias. The statistician need not choose between these extremes but can instead view them as endpoints on a continuum and use the data to determine the compromise (usually a smooth function of the likelihood ratio statistic for testing H_0 versus H_1) between bias and variance through an appropriate empirical Bayes rule, perhaps Stein's or one of the generalizations presented here.

We believe many applications embody these features and that most data analysts will have good experiences with the sensible use of the rules discussed here. In view of their potential, we believe empirical Bayes methods are among the most under utilized in applied data analysis.

[*Received October 1973. Revised February 1975.*]

REFERENCES

[1] Anscombe, F., "The Transformation of Poisson, Binomial and Negative-Binomial Data," *Biometrika*, 35 (December 1948), 246–54.

[2] Baranchik, A.J., "Multiple Regression and Estimation of the Mean of a Multivariate Normal Distribution," Technical Report No. 51, Stanford University, Department of Statistics, 1964.

[3] Carter, G.M. and Rolph, J.E., "Empirical Bayes Methods Applied to Estimating Fire Alarm Probabilities," *Journal of the American Statistical Association*, 69, No. 348 (December 1974), 880–5.

[4] Efron, B., "Biased Versus Unbiased Estimation," *Advances in Mathematics*, New York: Academic Press (to appear 1975).

[5] ——— and Morris, C., "Limiting the Risk of Bayes and Empirical Bayes Estimators—Part I: The Bayes Case," *Journal of the American Statistical Association*, 66, No. 336 (December 1971), 807–15.

[6] ——— and Morris, C., "Limiting the Risk of Bayes and Empirical Bayes Estimators—Part II: The Empirical Bayes Case," *Journal of the American Statistical Association*, 67, No. 337 (March 1972), 130–9.

[7] ——— and Morris, C., "Empirical Bayes on Vector Observations—An Extension of Stein's Method," *Biometrika*, 59, No. 2 (August 1972), 335–47.

[8] ——— and Morris, C., "Stein's Estimation Rule and Its Competitors—An Empirical Bayes Approach," *Journal of the American Statistical Association*, 68, No. 341 (March 1973), 117–30.

[9] ——— and Morris, C., "Combining Possibly Related Estimation Problems," *Journal of the Royal Statistical Society*, Ser. B, 35, No. 3 (November 1973; with discussion), 379–421.

[10] ——— and Morris, C., "Families of Minimax Estimators of the Mean of a Multivariate Normal Distribution," P-5170, The RAND Corporation, March 1974, submitted to *Annals of Mathematical Statistics* (1974).

[11] ——— and Morris, C., "Estimating Several Parameters Simultaneously," to be published in *Statistica Neerlandica*.

[12] ——— and Morris, C., "Data Analysis Using Stein's Estimator and Its Generalizations," R-1394-OEO, The RAND Corporation, March 1974.

[13] James, W. and Stein, C., "Estimation with Quadratic Loss,"

[8] See, e.g., [3] for estimating fire alarm probabilities and [4] for estimating reaction times and sunspot data.

[9] One excellent example [17] takes H_0 as the main effects in a two-way analysis of variance and $H_1 - H_0$ as the interactions.

Proceedings of the Fourth Berkeley Symposium on Mathematical Statistics and Probability, Vol. 1, Berkeley: University of California Press, 1961, 361–79.

[14] Remington, J.S., *et al.*, "Studies on Toxoplamosis in El Salvador: Prevalence and Incidence of Toxoplasmosis as Measured by the Sabin-Feldman Dye Test," *Transactions of the Royal Society of Tropical Medicine and Hygiene*, 64, No. 2 (1970), 252–67.

[15] Stein, C., "Inadmissibility of the Usual Estimator for the Mean of a Multivariate Normal Distribution," *Proceedings of the Third Berkeley Symposium on Mathematical Statistics and Probability*, Vol. 1, Berkeley: University of California Press, 1955, 197–206.

[16] ———, "Confidence Sets for the Mean of a Multivariate Normal Distribution," *Journal of the Royal Statistical Society*, Ser. B, 24, No. 2 (1962), 265–96.

[17] ———, "An Approach to the Recovery of Inter-Block Information in Balanced Incomplete Block Designs," in F.N. David, ed., *Festschrift for J. Neyman*, New York: John Wiley & Sons, Inc., 1966, 351–66.

Chapter 30

Assigning Probabilities to the Outcomes of Multi-Entry Competitions

DAVID A. HARVILLE*

The problem discussed is one of assessing the probabilities of the various possible orders of finish of a horse race or, more generally, of assigning probabilities to the various possible outcomes of any multi-entry competition. An assumption is introduced that makes it possible to obtain the probability associated with any complete outcome in terms of only the 'win' probabilities. The results were applied to data from 335 thoroughbred horse races, where the win probabilities were taken to be those determined by the public through pari-mutuel betting.

1. INTRODUCTION

A horse player wishes to make a bet on a given horse race at a track having pari-mutuel betting. He has determined each horse's 'probability' of winning. He can bet any one of the entires to win, place (first or second), or show (first, second, or third). His payoff on a successful place or show bet depends on which of the other horses also place or show. Our horse player wishes to make a single bet that maximizes his expected return. He finds that not only does he need to know each horse's probability of winning, but that, for every pair of horses, he must also know the probability that both will place, and, for every three, he must know the probability that all three will show. Our better is unhappy. He feels that he has done a good job of determining the horses' probabilities of winning; however he must now assign probabilities to a much larger number of events. Moreover, he finds that the place and show probabilities are more difficult to assess. Our better looks for an escape from his dilemma. He feels that the probability of two given horses both placing or of three given horses all showing should be related to their probabilities of winning. He asks his friend, the statistician, to produce a formula giving the place and show probabilities in terms of the win probabilities.

The problem posed by the better is typical of a class of problems that share the following characteristics:

1. The members of some group are to be ranked in order from first possibly to last, according to the outcome of some random phenomena, or the ranking of the members has already been effected, but is unobservable.
2. The 'probability' of each member's ranking first is known or can be assessed.
3. From these probabilities alone, we wish to determine the probability that a more complete ranking of the members will equal a given ranking or the probability that it will fall in a given collection of such rankings.

Dead heats or ties will be assumed to have zero probability. For situations where this assumption is unrealistic, the probabilities of the various possible ties must be assessed separately.

We assign no particular interpretation to the 'probability' of a given ranking or collection of rankings. We assume only that the probabilities of these events satisfy the usual axioms. Their interpretation will differ with the setting.

Ordinarily, knowledge of the probabilities associated with the various rankings will be of most interest in situations like the horse player's where only the ranking itself, and not the closeness of the ranking, is important. The horse player's return on any bet is completely determined by the horses' order of finish. The closeness of the result may affect his nerves but not his pocketbook.

2. RESULTS

We will identify the n horses in the race or members in the group by the labels 1, 2, $\cdots$, n. Denote by $p_k[i_1, i_2, \cdots, i_k]$ the probability that horses or members i_1, i_2, $\cdots$, i_k finish or rank first, second, $\cdots$, kth, respectively, where $k \leq n$. For convenience, we use $p[i]$ interchangeably with $p_1[i]$ to represent the probability that horse or member i finishes or ranks first. We wish to obtain $p_k[i_1, i_2, \cdots, i_k]$ in terms of $p[1]$, $p[2]$, $\cdots$, $p[n]$, for all $i_1, i_2, \cdots, i_k$ and for $k = 2, 3, \cdots, n$. In a sense, our task is one of expressing the probabilities of elementary events in terms of the probabilities of more complex events.

Obviously, we must make additional assumptions to obtain the desired formula. Our choice is to assume that, for all $i_1, i_2, \cdots, i_k$ and for $k = 2, 3, \cdots, n$, the conditional probability that member i_k ranks ahead of members $i_{k+1}, i_{k+2}, \cdots, i_n$ given that members $i_1, i_2, \cdots, i_{k-1}$ rank first, second, $\cdots$, $(k-1)$th, respectively, equals the conditional probability that i_k ranks ahead of i_{k+1}, $i_{k+2}, \cdots, i_n$ given that $i_1, i_2, \cdots, i_{k-1}$ do not rank first. That is,

$$\frac{p_k[i_1, i_2, \cdots, i_k]}{p_{k-1}[i_1, i_2, \cdots, i_{k-1}]} = \frac{p[i_k]}{q_{k-1}[i_1, i_2, \cdots, i_{k-1}]}, \quad (2.1)$$

* David A. Harville is research mathematical statistician, Aerospace Research Laboratories, Wright-Patterson Air Force Base, Ohio 45433. The author wishes to thank the Theory and Methods editor, an associate editor and a referee for their useful suggestions.

© Journal of the American Statistical Association
June 1973, Volume 68, Number 342
Applications Section

where

$$q_k[i_1, i_2, \cdots, i_k] \equiv 1 - p[i_1] - p[i_2] - \cdots - p[i_k],$$

so that, for the sought-after formula, we obtain

$$p_k[i_1, i_2, \cdots, i_k] \equiv \frac{p[i_1]p[i_2]\cdots p[i_k]}{q_1[i_1]q_2[i_1, i_2]\cdots q_{k-1}[i_1, i_2, \cdots, i_{k-1}]}. \quad (2.2)$$

In the particular case $k = 2$, the assumption (2.1) is equivalent to assuming that the event that member i_2 ranks ahead of all other members, save possibly i_1, is stochastically independent of the event that member i_1 ranks first.

The intuitive meaning and the reasonableness of the assumption (2.1) will depend on the setting. In particular, our horse player would probably not consider the assumption appropriate for every race he encounters. For example, in harness racing, if a horse breaks stride, the driver must take him to the outside portion of the track and keep him there until the horse regains the proper gait. Much ground can be lost in this maneuver. In evaluating a harness race in which there is a horse that is an 'almost certain' winner unless he breaks, the bettor would not want to base his calculations on assumption (2.1). For such a horse, there may be no such thing as an intermediate finish. He wins when he doesn't break, but finishes 'way back' when he does.

In many, though not all, cases, there is a variate (other than rank) associated with each member of the group such that the ranking is strictly determined by ordering their values. For example, associated with each horse is its running time for the race. Denote by X_i the variate corresponding to member i, $i = 1, 2, \cdots, n$. Clearly, the assumption (2.1) can be phrased in terms of the joint probability distribution of $X_1, X_2, \cdots, X_n$. It seems natural to ask whether there exist other conditions on the distribution of the X_i's which imply (2.1) or which follow from it, and which thus would aid our intuition in grasping the implications of that assumption. The answer in general seems to be no. In particular, it can easily be demonstrated by constructing a counterexample that stochastic independence of the X_i's does not in itself imply (2.1). Nor is the converse necessarily true. In fact, in many situations where assumption (2.1) might seem appropriate, it is known that the X_i's are not independent. For example, we would expect the running times of the horses to be correlated in most any horse race. An even better example is the ordering of n baseball teams according to their winning percentages over a season of play. These percentages are obviously not independent, yet assumption (2.1) might still seem reasonable.

The probability that the ranking belongs to any given collection of rankings can be readily obtained in terms of $p[1], p[2], \cdots, p[n]$ by using (2.2) to express the probability of each ranking in the collection in terms of the $p[i]$'s, and by then adding. For example, the horse player can compute the probability that both entry i and entry j place from

$$p_2[i, j] + p_2[j, i] = \frac{p[i]p[j]}{1 - p[i]} + \frac{p[j]p[i]}{1 - p[j]}.$$

A probability of particular interest in many situations is the probability that entry or member r finishes or ranks kth or better, for which we write

$$p_k^*[r] = \sum p_k[i_1, i_2, \cdots, i_k], \quad (2.3)$$

where the summation is over all rankings $i_1, i_2, \cdots, i_k$ for which $i_u = r$ for some u. If assumption (2.1) holds, then $p_k^*[r] > p_k^*[s]$ if and only if $p[r] > p[s]$. This statement can be proved easily by comparing the terms of the right side of (2.3) with the terms of the corresponding expression for $p_k^*[s]$. Each term of (2.3), whose indices are such that $i_u = r$ and $i_v = s$ for some u, v, appears also in the second expression. Thus, it suffices to show that any term $p_k[i_1, i_2, \cdots, i_k]$, for which $i_j \neq s$, $j = 1, 2, \cdots, k$, but $i_u = r$ for some u, is made smaller by putting $i_u = s$ if and only if $p[r] > p[s]$. That the latter assertion is true follows immediately from (2.2).

3. APPLICATION

In pari-mutuel betting, the payoffs on win bets are determined by subtracting from the win pool (the total amount bet to win by all bettors on all horses) the combined state and track take (a fixed percentage of the pool—generally about 16 percent, but varying from state to state), and by then distributing the remainder among the successful bettors in proportion to the amounts of their bets. (Actually, the payoffs are slightly smaller because of 'breakage,' a gimmick whereby the return on each dollar is reduced to a point where it can be expressed in terms of dimes.) In this section, we take the 'win probability' on each of the n horses to be in inverse proportion to what a successful win bet would pay per dollar, so that every win bet has the same 'expected return.' Note that these 'probabilities' are established by the bettors themselves and, in some sense, represent a consensus opinion as to each horse's chances of winning the race. We shall suppose that, in any sequence of races in which the number of entries and the consensus probabilities are the same from race to race, the horses going off at a given consensus probability win with a long-run frequency equal to that probability. The basis for this supposition is that, once the betting on a race has begun, the amounts bet to win on the horses are flashed on the 'tote' board for all to see and this information is updated periodically, so that, if at some point during the course of the betting the current consensus probabilities do not coincide with the bettors' experience as to the long-run win frequencies for 'similar' races, these discrepancies will be noticed and certain of the bettors will place win bets that have the effect of reducing or eliminating them.

By adopting assumption (2.1) and applying the results of the previous section, we can compute the long-run frequencies with which any given order of finish is encountered over any sequence of races having the same number

1. APPLICATION OF THEORETICAL RESULTS TO THIRD RACE OF SEPTEMBER 6, 1971, AT RIVER DOWNS RACE TRACK

	Amounts bet to win, place, and show as percentages of totals			Theoretical probability			Expected payoff per dollar	
Name	Win	Place	Show	Win	Place	Show	Place bet	Show bet
Moonlander	27.6	20.0	22.3	.275	.504	.688	1.11	1.01
E'Thon	16.5	14.2	11.1	.165	.332	.499	.94	1.06
Golden Secret	3.5	4.7	6.3	.035	.076	.126	.58	.42
Antidote	17.3	18.8	20.0	.175	.350	.521	.80	.80
Beviambo	4.0	6.2	7.8	.040	.087	.144	.51	.41
Cedar Wing	11.9	10.4	10.4	.118	.245	.382	.90	.86
Little Flitter	8.5	11.2	9.9	.085	.180	.288	.62	.68
Hot and Humid	10.7	14.4	12.2	.107	.224	.353	.62	.72

of entries and the same consensus win probabilities. In particular, we can compute the 'probability' that any three given horses in a race finish first, second, and third, respectively. As we shall now see, these probabilities are of something more than academic interest, since they are the ones needed to compute the 'expected payoff' for each place bet (a bet that a particular horse will finish either first or second) and each show bet (a bet that the horse will finish no worse than third).

Like the amounts bet to win, the amounts bet on each horse to place and to show are made available on the 'tote' board as the betting proceeds. The payoff per dollar on a successful place (show) bet consists of the original dollar plus an amount determined by subtracting from the final place (show) pool the combined state and track take and the total amounts bet to place (show) on the first two (three) finishers, and by then dividing a half (third) of the remainder by the total amount bet to place (show) on the horse in question. (Here again, the actual payoffs are reduced by breakage.) By using the probabilities computed on the basis of assumption (2.1) and the assumption that consensus win probabilities equal appropriate long-run frequencies, we can compute the expected payoff per dollar for a given place or show bet on any particular race, where the expectation is taken over a sequence of races exhibiting the same number of entries and the same pattern of win, place, and show betting. If, as the termination of betting on a given race approaches, any of the place or show bets are found to have potential expected payoffs greater than one, there is a possibility that a bettor, by making such place and show bets, can 'beat the races'. Of course, if either assumption (2.1) or the assumption that the consensus win probabilities equal long-run win frequencies for races with similar betting patterns is inappropriate, then this system will not work. It will also fail if there tend to be large last-minute adverse changes in the betting pattern, either because of the system player's own bets or because of the bets of others. However, at a track with considerable betting volume, it is not likely that such changes would be so frequent as to constitute a major stumbling block.

In Table 1, we exemplify our results by applying them to a particular race, the third race of the September 6, 1971, program at River Downs Race Track. The final win, place, and show pools were \$45,071, \$16,037, and \$9,740, respectively. The percentage of each betting pool bet on each horse can be obtained from the table. The table also gives, for each horse, the consensus win probability, the overall probabilities of placing and showing, and the expected payoffs per dollar of place and show bets. The race was won by E'Thon who, on a per-dollar basis, paid \$5.00, \$3.00, and \$2.50 to win, place, and show, respectively; Cedar Wing was second, paying \$3.80 and \$2.70 per dollar to place and show; and Beviambo finished third, returning \$3.20 for each dollar bet to show.

In order to check assumption (2.1) and the assumption that the consensus win probabilities coincide with the long-run win frequencies over any sequence of races having the same number of entries and a similar betting pattern, data was gathered on 335 thoroughbred races from several Ohio and Kentucky race tracks. Data from races with finishes that involved dead heats for one or more of the first three positions were not used. Also, in the pari-mutuel system, two or more horses are sometimes lumped together and treated as a single entity for betting purposes. Probabilities and expectations for the remaining horses were computed as though these 'field' entries consisted of single horses and were included in the data, though these figures are only approximations to the 'true' figures. However, the field entires themselves were not included in the tabulations.

As one check on the correspondence between consensus win probabilities and the long-run win frequencies over races with similar patterns of win betting, the horses were divided into eleven classes according to their consensus win probabilities. Table 2 gives, for each class, the associated interval of consensus win probabilities, the average consensus win probability, the actual frequency

2. FREQUENCY OF WINNING—ACTUAL VS. THEORETICAL

Theoretical probability of winning	Number of horses	Average theoretical probability	Actual frequency of winning	Estimated standard error
.00 - .05	946	.028	.020	.005
.05 - .10	763	.074	.064	.009
.10 - .15	463	.124	.127	.016
.15 - .20	313	.175	.169	.021
.20 - .25	192	.225	.240	.031
.25 - .30	114	.272	.289	.042
.30 - .35	71	.324	.394	.058
.35 - .40	49	.373	.306	.066
.40 - .45	25	.423	.640	.096
.45 - .50	12	.464	.583	.142
.50 +	10	.554	.700	.145

3. FREQUENCY OF FINISHING SECOND—ACTUAL VS. THEORETICAL

Theoretical probability of finishing second	Number of horses	Average theoretical probability	Actual frequency of finishing second	Estimated standard error
.00 - .05	776	.030	.046	.008
.05 - .10	750	.074	.095	.011
.10 - .15	548	.124	.128	.014
.15 - .20	426	.175	.155	.018
.20 - .25	283	.223	.170	.022
.25 - .30	164	.269	.226	.033
.30 +	11	.311	.364	.145

of winners, and an estimate of the standard error associated with the actual frequency. The actual frequencies seem to agree remarkably well with the theoretical probabilities, though there seems to be a slight tendency on the part of the betters to overrate the chances of long-shots and to underestimate the chances of the favorites and near-favorites. Similar results, based on an extensive amount of data from an earlier time period and from different tracks, were obtained by Fabricand [1].

Several checks were also run on the appropriateness of assumption (2.1). These consisted of first partitioning the horses according to some criterion involving the theoretical probabilities of second and third place finishes and then comparing the actual frequency with the average theoretical long-run frequency for each class. Tables 3–6 give the results when the criterion is the probability of finishing second, finishing third, placing, or showing, respectively. In general, the observed frequencies of second and third place finishes are in reasonable accord with the theoretical long-run frequencies, though there seems to be something of a tendency to overestimate the chances of a second or third place finish for horses with high theoretical probabilities of such finishes and to underestimate the chances of those with low theoretical probabilities, with the tendency being more pronounced for third place finishes than for second place finishes. A logical explanation for the conformity of the actual place results to those predicted by the theory which is evident in Table 5 is that those horses with high (low) theoretical probabilities of finishing second generally also have high (low) theoretical

4. FREQUENCY OF FINISHING THIRD—ACTUAL VS. THEORETICAL

Theoretical probability of finishing third	Number of horses	Average theoretical probability	Actual frequency of finishing third	Estimated standard error
.00 - .05	587	.032	.049	.009
.05 - .10	713	.074	.105	.011
.10 - .15	691	.124	.126	.013
.15 - .20	838	.175	.147	.012
.20 - .25	115	.212	.130	.031
.25 +	14	.273	.214	.110

5. FREQUENCY OF PLACING—ACTUAL VS. THEORETICAL

Theoretical probability of placing	Number of horses	Average theoretical probability	Actual frequency of placing	Estimated standard error
.00 - .05	330	.034	.036	.010
.05 - .10	526	.074	.091	.013
.10 - .15	404	.125	.121	.016
.15 - .20	358	.174	.179	.020
.20 - .25	268	.224	.257	.027
.25 - .30	240	.274	.271	.029
.30 - .35	193	.326	.306	.033
.35 - .40	175	.375	.354	.036
.40 - .45	117	.425	.359	.044
.45 - .50	109	.472	.440	.048
.50 - .55	73	.525	.425	.058
.55 - .60	51	.578	.667	.066
.60 - .65	48	.623	.625	.070
.65 - .70	29	.673	.621	.090
.70 - .75	22	.724	.909	.095
.75 +	15	.808	.867	.088

6. FREQUENCY OF SHOWING—ACTUAL VS. THEORETICAL

Theoretical probability of showing	Number of horses	Average theoretical probability	Actual frequency of showing	Estimated standard error
.00 - .05	111	.038	.045	.020
.05 - .10	316	.075	.092	.016
.10 - .15	328	.124	.180	.021
.15 - .20	266	.174	.222	.025
.20 - .25	253	.227	.257	.027
.25 - .30	243	.274	.284	.029
.30 - .35	201	.326	.303	.032
.35 - .40	196	.374	.439	.035
.40 - .45	169	.425	.426	.038
.45 - .50	150	.477	.460	.041
.50 - .55	158	.525	.468	.040
.55 - .60	137	.574	.474	.043
.60 - .65	97	.625	.577	.050
.65 - .70	100	.672	.500	.050
.70 - .75	67	.722	.627	.059
.75 - .80	67	.777	.731	.054
.80 - .85	49	.823	.816	.055
.85 - .90	30	.874	.867	.062
.90 +	20	.930	1.000	.056

7. PAYOFFS ON PLACE AND SHOW BETS—ACTUAL VS. THEORETICAL

Expected payoff per dollar	Number of different place and show bets	Average expected payoff per dollar	Average actual payoff per dollar	Estimated standard error
.00 - .25	80	.216	.088	.062
.25 - .35	214	.303	.286	.068
.35 - .45	386	.404	.609	.091
.45 - .55	628	.504	.570	.071
.55 - .65	904	.601	.730	.072
.65 - .75	980	.700	.660	.047
.75 - .85	958	.800	.947	.066
.85 - .95	819	.898	.938	.050
.95 - 1.05	546	.995	.983	.090
1.05 - 1.15	286	1.090	.989	.060
1.15 - 1.25	90	1.186	.974	.108
1.25 +	25	1.320	1.300	.258

probabilities of finishing first, so that the effects of the overestimation (underestimation) of their chances of finishing second are cancelled out by the underestimation (overestimation) of their chances of finishing first. While a similar phenomenon is operative in the show results, the cancellation is less complete and there seems to be a slight tendency to overestimate the show chances of those horses with high theoretical probabilities and to underestimate the chances of those with low theoretical probabilities.

Finally, the possible place and show bets were divided into classes according to the theoretical expected payoffs of the bets as determined from the final betting figures. The average actual payoff per dollar for each class can then be compared with the corresponding average expected payoff per dollar. The necessary figures are given in Table 7. The results seem to indicate that those place and show bets with high theoretical expected payoffs per dollar actually have expectations that are somewhat lower, giving further evidence that our assumptions are not entirely realistic, at least not for some races.

The existence of widely different expected payoffs for the various possible place and show bets implies that either the bettors 'do not feel that assumption (2.1) is entirely appropriate' or they 'believe in assumption (2.1)' but are unable to perceive its implications. Our results indicate that to some small extent the bettors are successful in recognizing situations where assumption (2.1) may not hold and in acting accordingly, but that big differences in the expected place and show payoffs result primarily from 'incorrect assessments' as to when assumption (2.1) is not appropriate or from 'ignorance as to the assumption's implications.'

A further implication of the results presented in Table 7 is that a bettor could not expect to do much better than break even by simply making place and show bets with expected payoffs greater than one.

[*Received January 1972. Revised September 1972.*]

REFERENCE

[1] Fabricand, Burton P., *Horse Sense*, New York: David McKay Company, Inc., 1965.

". . . and thereby return our game to the pure world of numbers, where it belongs."—Roger Angell

Chapter 31

Basketball, Baseball, and the Null Hypothesis

Robert Hooke

Tversky and Gilovich (*Chance*, Winter 1989) and Gould (*Chance*, Spring 1989) write persuasively on the nonexistence of hot and cold streaks in basketball and baseball. As a statistician, I find no fault with their methods, but as a sometime competitor (at very low levels) in various sports and games I feel uncomfortable with their conclusions. Gould speaks of "a little homunculus in my head [who] continues to jump up and down shouting at me" that his intuitive feeling is right regardless of the mathematics. I, too, have such a homunculus, who has goaded me into raising questions about the conclusions of these articles and the use of the null hypothesis in general.

Every statistician knows that people (even statisticians) tend to see patterns in data that are actually only random fluctuations. However, in almost every competitive activity in which I've ever engaged (baseball, basketball, golf, tennis, even duplicate bridge), a little success generates in me a feeling of confidence which, as long as it lasts, makes me do better than usual. Even more obviously, a few failures can destroy this confidence, after which for a while I can't do anything right. If any solid evidence of such experiences can be found, it seemingly must be found outside of the statistical arguments of the aforementioned papers, because there are no apparent holes in these arguments. If the mathematics is all right and the conclusions still seem questionable, the place to look is at the model, which is the connection between the mathematics and reality.

If the model "explains" the data, then the model is correct and unique.

(True or False?)

Everybody knows this is false. (Well, almost everybody.) Tversky and Gilovich seem to know this, because they show that their data do not confirm the existence of the so-called hot hand, and the casual reader might conclude that they have shown its nonexistence, but they don't actually say so. Gould, though, does say about baseball: "Everybody knows about hot hands. The only problem is that no such phenomenon exists."

Statisticians are trained to speak precisely, and usually they remember to do so, being careful to say, perhaps, "The normal distribution is not contradicted by the data at the 5% level, so we may safely proceed as if the normal distribution actually holds." Careful speech may become tiresome, though, and some people are even offended by it. Years of trying to placate such customers sometimes drives statisticians to make statements such as, "The data show that the distribution is normal," hoping that this rash conclusion will not reach the ears of any colleague.

The two *Chance* articles use the standard approach to the problem. First, they look at what would happen if only chance were involved, see what observed data would look like under this assumption, and then compare the result with real data to see if there are major differences. If only chance is involved, the mathematical model for the real situation is the usual coin tossing or "Bernoulli trials" model, in which each event has a probability of success that is constant and independent of previous events. Using this model and observing no significant differences, they conclude that there is no evidence to dispute the null hypothesis that only chance is operating. While we do know very well what happens if only chance is involved, we do not have a good idea of how data should turn out if there really is a psychological basis for "streaks."

In statistical language, we don't have a well-formulated alternative hypothesis to test against the null hypothesis. Thus we invent various measures (such as the serial correlations of Tversky and Gilovich), and we state how we *intuitively* think these measures will behave if the null hypothesis is not true. If they don't appear to behave this way, then we can conclude fairly safely that the null hypothesis is at least approximately true, or that the opposing effect, if true, is at least not very large.

My intuition tells me that the alternative hypothesis is not that there is a "hot hand" effect that is the same for everyone, but that the real situation is much more complex. Some people are slaves to their recent past, some can ignore it altogether, and others lie somewhere in between. The slaves, who become discouraged after a few failures, probably don't make it to the professional level in competition unless they have an unusual excess of talent. If they are in the minority among professional athletes, it would take a very large amount of data to show how their behavior affects the overall statistics. Also, if a player only has a hot hand sometimes, how do we know how many successes are required for the hot hand to take over? With one player this number may be one, with another it may be three or four. A measure that is appropriate in detecting the effect for one of these types may not be very powerful for another.

Why does a statistician continue to look with skepticism on these negative results? For one thing, if there are no hot hands there are also no slumps. Thus no matter how many hits or walks a good pitcher has allowed in a game, the manager should not take him out unless he has some physical problem. Of all the slumps that I've observed, the one of most majestic proportions was endured by Steve Blass, a pitcher for the Pittsburgh Pirates from 1964 to 1974. From a fair start he gradually became a star in 1971 and 1972, but in 1973 he became a disaster. An anecdote such as this is in itself no argument for the existence of slumps, since his numbers, bad as they were, might possibly have occurred by chance. Additional data, though, were available to observers without getting into the box scores: Blass's pitches were often very wild, missing the plate by feet, not inches. In 1974 he tried again, pitched in one game and then retired from baseball. So far as I know, no physical reason for all this was ever found.

I was once asked: "At the end of a baseball season, is there real statistical evidence that the best team won?" The statistician's first attack on this (not the final one,

by any means) is to suppose that for each game the assumptions of Bernoulli trials hold. In some years this null hypothesis is rejected, but often not. Even when the null hypothesis is not rejected, the statistics on such additional measures as runs scored and runs allowed may show conclusively that the top teams (if not the top team) were consistently performing better than the others. Thus, we have statistics that seem to show that the game was all luck, while more detailed statistics may be available to contradict this conclusion.

Then there is the issue of defense. In basketball, some teams are alleged by the experts to play much better defense than others. In a given game, a player takes a series of shots against the same team, whose defensive capabilities may be considerably greater or less than the league's average. Can this be true without some effect showing up in the serial correlations? Do the same statistics that fail to show the existence of the hot hand also show that defense is not important?

Baseball has a similar feature. Gould quotes statistical results from a colleague to the effect that "Nothing ever happened in baseball above and beyond the frequency predicted by coin-tossing models," but he gives no details. One assumes that the effect of various opposing pitchers was not part of the model used, since this would introduce enormous complications. Yet a batter usually faces the same pitcher several times in a row. If the statistics do not show some sort of dependence on the opposition, then the statistical procedures are simply not powerful enough to detect effects of interest such as streakiness.

In short, my conclusion is that the data examined and analyzed to date show that the hot hand effect is probably smaller than we think. No statistician would deny that people, even statisticians, tend to see patterns that are not there. I would not say, however, that the hot hand doesn't exist. Were I a Bayesian, I would assign a very high prior probability to the existence of hot hands and challenge others to produce data that would contradict it.

Additional Reading

Angell, R. (1988), *Season Ticket*, Boston: Houghton Mifflin.

Gould, S.J. (1989), "The Streak of Streaks," *Chance*, 2(2) 10–16.

Hooke, R. (1983), *How to Tell the Liars from the Statisticians*, New York: Marcel Dekker.

Tversky, A. and Gilovich, T. (1989), "The Cold Facts About the 'Hot Hand' in Basketball," *Chance*, 2(1) 16–21.

Chapter 32

General

Lessons from Sports Statistics

Frederick MOSTELLER

The author reviews and comments on his work in sports statistics, illustrating with problems of estimation in baseball's World Series and with a model for the distribution of the number of runs in a baseball half inning. Data on collegiate football scores have instructive distributions that indicate more about the strengths of the teams playing than their absolute values would suggest. A robust analysis of professional football scores led to widespread publicity with the help of professional newswriters. Professional golf players on the regular tour are so close in skill that a few rounds do little to distinguish their abilities. A simple model for golf scoring is "base $+X$" where the base is a small score for a round rarely achieved, such as 64, and X is a Poisson distribution with mean about 8. In basketball, football, and hockey the leader at the beginning of the final period wins about 80% of the time, and in baseball the leader at the end of seven full innings wins 95% of the time. Empirical experience with runs of even and odd numbers in tossing a die millions of times fits closely the theoretical distributions.

KEY WORDS: Baseball; Football; Golf; Dice.

1. WORLD SERIES

My first paper on statistics in sports dealt with the World Series of major-league baseball (Mosteller 1952). At a cocktail party at Walter and Judith Rosenblith's home someone asked: What is the chance that the better team in the series wins? Some people did not understand the concept that there might be a "best" or "better" team, possibly different from the winner. It occurred to me that this question provided an excellent application of work on unbiased estimation for quality control that Jimmie Savage and I completed during World War II. I drafted the World Series paper with the considerable assistance of Doris Entwisle, now a professor at The Johns Hopkins University, and I submitted it to the *Journal of the American Statistical Association*. W. Allen Wallis, then the editor of *JASA*, and who had prompted our work on estimation when he directed the Statistical Research Group of Columbia, sent it out to a number of referees who were intensely interested in baseball. The referees had a variety of good ideas that led to extensions of the paper.

That led me to my first lesson from mixing science and statistics in sports.

Lesson 1. If many reviewers are both knowledgeable about the materials and interested in the findings, they will drive the author crazy with the volume, perceptiveness, and relevance of their suggestions.

The paper doubled in size in response to the first round of suggestions, and the second round lengthened it further.

The second lesson came from the work on unbiased estimates that applied to some models used in the World Series paper.

Lesson 2. If you develop some inferential statistical methods, you are likely to have a use for them in a paper on sports statistics.

In the World Series analysis we are making inferences from the statistics of a truncated series (once a winner is determined, the series stops without carrying out the seven games). Jimmie and I had a theorem about how to get the unique unbiased estimate from binomial processes (Girshick, Mosteller, and Savage 1946). We showed that there were unreasonable results in unbiased estimation. The existence of such unreasonable results has downgraded somewhat the importance of unbiasedness. To see an example of this unreasonableness, I turn to a very short binomial game in which p is the probability of a success on any trial, trials are independent, and you play until either you get one success or two failures, and then stop. The material in Exhibit

Exhibit 1. Play Until You Get a Success or Two Failures

The sequence begins at (0 failures, 0 successes) with stopping points (0, 1), (1, 1), and (2, 0). Let the value of the estimate of p at these three points be x, y, and z, respectively.

Unbiasedness implies

$$xp + yqp + zq^2 \equiv p.$$

Rewriting in terms of q gives

$$x(1-q) + yq(1-q) + zq^2 = 1-q$$

$$1 \cdot x + q(-x+y) + q^2(-y+z) = 1-q.$$

Equating coefficients of the powers of q on the two sides yields $x = 1$, $y = 0$, $z = 0$. It is annoying to many that the estimate of p for (1, 1) is 0 although the observed proportion of successes is 1/2.

Frederick Mosteller is Roger I. Lee Professor of Mathematical Statistics, Emeritus, Department of Statistics, Harvard University, Cambridge, MA 02138. This is a revised version of a talk presented at the Chicago Meeting of the American Statistical Association, August 5, 1996, on the occasion of the award of Sports Statistician of the Year to the author by the Section on Statistics in Sports. A referee and the associate editor have made many suggestions for improving the final paper (see Lesson 1).

Exhibit 2. Grid Showing Boundary Points for Best-of-Seven-Games Series

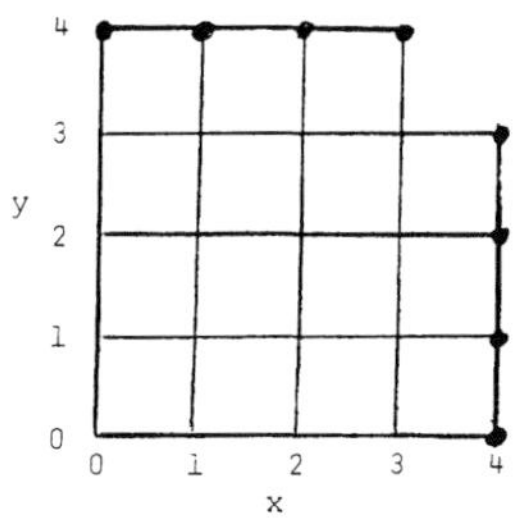

NOTE: The number of paths from (0, 0 to (2, 4) is $\binom{5}{2}$ = 10 because the only way to get to (2, 4) is first to reach (2, 3) and then go to (2, 4) with an American League win.

1 shows that the unique unbiased estimate of p is 1 if the first trial is a success, 0 if the second trial is a success, and 0 if the first two trials are failures. Most people find it unreasonable that when half the trials are successes, the value of the estimate of probability of success is 0.

For a best-of-seven series we can represent the sequences of wins and losses of games in a series by paths in a rectangular grid (see Exhibit 2) consisting of points $(x, y), 0 \leq x, y \leq 4$, excluding (4, 4). The point (x, y) represents x wins by the American League and y wins by the National League. A path starts at (0, 0). The binomial probability p is the chance that the next game is won by the American League, which would add the step from (x, y) to $(x, y+1)$ to the path, or $1-p$ that the National League wins, which would add the step from (x, y) to $(x+1, y)$ to the path. The *boundary points* $(x, 4)$ $x = 0, 1, 2, 3$ correspond to series wins by the American League, and $(4, y)$ $y = 0, 1, 2, 3$ to wins by the National League. Paths stop when they reach a boundary point.

The unique value for the unbiased estimate for p at a given boundary point is given by a ratio:

$$\frac{\text{number of paths from } (0,1) \text{ to the boundary point}}{\text{number of paths from } (0,0) \text{ to the boundary point}}.$$

Table 2 shows the unbiased estimate associated with the boundary points for the best-of-seven series.

Table 1. Number of World Series Won by the American League in 12-Year Intervals

Years	No. won by AL
1903–1915[a]	7 of 12
1916–1927[b]	7 of 12
1928–1939	9 of 12
1940–1951	8 of 12
1952–1963	6 of 12
1964–1975	6 of 12
1976–1987	6 of 12
1988–1995[c]	4 of 7
Totals	53 of 91

[a] No series in 1904.
[b] Includes NL victory in 1919, year of "Black Sox Scandal."
[c] No series in 1994.

Table 2. Outcomes of the 87 Best-of-Seven Games in a World Series

Games won in series NL x	AL y	Unbiased estimate of P (win by AL)	Frequency
4	0	0	7
4	1	1/4	6
4	2	2/5	5
4	3	3/6	18
3	4	3/6	14
2	4	3/5	15
1	4	3/4	14
0	4	1	8
			Total 87

NOTE: Average of unbiased estimates = .540.

Example. Consider the boundary point (2, 4). The number of paths from (0, 1) to (2, 4) is 6, the number from (0, 0) to (2, 4) is 10, and the value at the estimate is 6/10 or 3/5. It is amusing that (3, 4) has the value 1/2, as does (4, 3).

The formula applies to much more general patterns of boundary points than just best-of-n series. We do require that the sum of probabilities of paths hitting boundary points is 1. The uniqueness requires that there be no interior points that can end the series. For example, if we added the rule of the best-of-seven series that if the state is ever (2, 2) we stop the series and declare a tie, then the estimate described above would not be unique.

Getting back to statistics in sports, there have been about 43 World Series since my original paper was written. The American League had been winning more than half the series. Table 1 shows that in the first 48 series (1903–1951) they won 31 or 65%, and in the most recent 43 series (1952–1995) they won 22 or 51% and dropped back to nearly even. Of the 91 series, 87 were best-of-seven-game series, based on games played until one team won four. Four series were based on best-of-nine games played until one team won five.

For the 87 best-of-seven-game series the average value of the unbiased estimates of the probability p of the American League winning a given game is .540 (Table 2). In computing this average value we weighted the estimate associated with each outcome by the number of series of that type that occurred. The model used is independent Bernoulli trials with p fixed for the series.

We should try to examine the model more carefully. I have the examined impression that the independent binomial trials model was reasonable during the first 48 series, but an unexamined impression that it may not be appropriate during the last 43.

In the first 48 series there seemed to be no home-field advantage, and I wonder whether that may have changed.

To close our World Series discussion as of 1952, in answer to the motivating question: we estimated the probability that the better team won the World Series according to Model A (fixed p across years as well as within series) at .80, and for Model B (normally distributed p, but fixed within a series) at .76. How that may have changed, I do not know.

Lesson 3. There is always more to do.

2. RUNS IN AN INNING OF MAJOR LEAGUE BASEBALL

Bernard Rosner has allowed my associate, Cleo Youtz, and me to participate with him as he developed a theory of the distribution of the number of runs in a major league baseball inning (runs per three outs). We started by assigning a negative binomial distribution to the number of persons at bat in a three-out inning, but found that the results underestimated the number of innings with three players at bat and overestimated the number of innings with four players at bat, but otherwise the number of batters faced fitted well. By making a brute-force adjustment that added and subtracted a parameter to correct for these deviations Rosner was able to develop a theory for the number of runs. The expected values in Table 3 show the theoretical distribution of runs in innings (really half innings) as compared with the observed 1990 American League results for 77 principal pitchers (Rosner, Mosteller, and Youtz 1996).

Along the way Rosner developed parameters for the pitchers' properties; but the description of these is too long to include here. However, the parameters did show that in 1990 Roger Clemens had the best pitching record, with an expected earned run average of 1.91. Thus, however disappointing the Red Sox may be, we owe them something for Roger Clemens.

Lesson 4. Wait until next year.

3. COLLEGIATE FOOTBALL

I stumbled across a whole season's collegiate football scores somewhere, and was most impressed with the highest tie score. In 1967 Alabama and Florida State tied at 37–37; in 1968 Harvard and Yale tied 29–29. I wondered what the probability of winning was with a given score. Figure 1 lays that out as the estimate of $P(\text{winning}|\text{score})$ for 1967. Roughly 16 is the median winning score (see Fig. 1) (Mosteller 1970).

I thought that the most interesting regression was of the losing score on the winning score; see Figure 2. It would be clearer if the vertical axis were horizontal, so please look at it sideways. The point seems to be that when one team scores many points, there is no time (and perhaps little ability) for the other team to score. The winning score associated with the highest average losing score is about 32, and then the loser averages around 17.

Table 3. Observed and Expected Distribution of Number of Runs Scored in an Individual Half Inning (x) in Baseball (Based on 77 Starting Pitchers in the 1990 American League Season)

Runs scored (x)	Observed number of innings	(%)	Expected number of innings	(%)
0	4,110	(73)	4,139.9	(73)
1	903	(16)	920.1	(16)
2	352	(6)	319.9	(6)
3	172	(3)	137.5	(2)
4	65	(1)	66.9	(1)
5	29	(1)	33.3	(1)
6	5	(0)	15.6	(0)
7	3	(0)	5.9	(0)
Total	5,639		5,639	

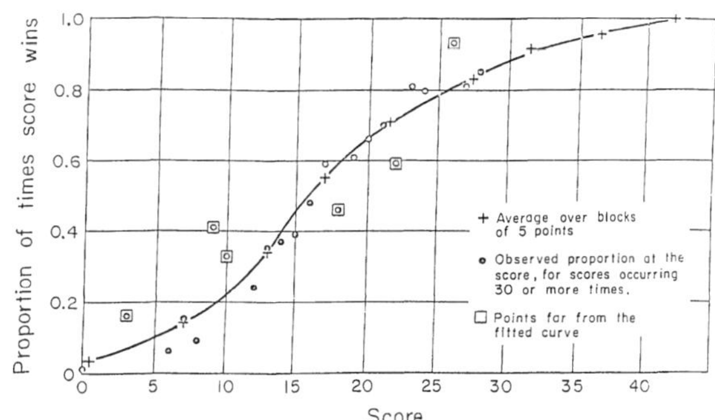

Figure 1. Proportion of Times a Given Score Wins, 1967 Collegiate Football Scores.

These results all seem to fall into the realm of descriptive statistics. Ties occur about half as often as one-point differences. This can be argued from trivial combinatorics.

From Table 4 we can see that some scores are especially lucky and others unlucky. For example, a score of 17 is more than twice as likely to win as to lose, whereas the higher score of 18 won less often than it lost. Again, 21 is lucky, winning 69% of its games, but 22 is not, winning only 59%. Table 4 shows the irregularity of these results rather than a monotonic rise in probability of winning given the absolute score. (Ties gave half a win to each team.)

Lesson 5. Collegiate football scores contain extra information about the comparative performances of a pair of teams beyond the absolute size of the scores.

4. PROFESSIONAL FOOTBALL

I wanted to carry out a robust analysis of professional football scores adjusted for the strength of the opposition

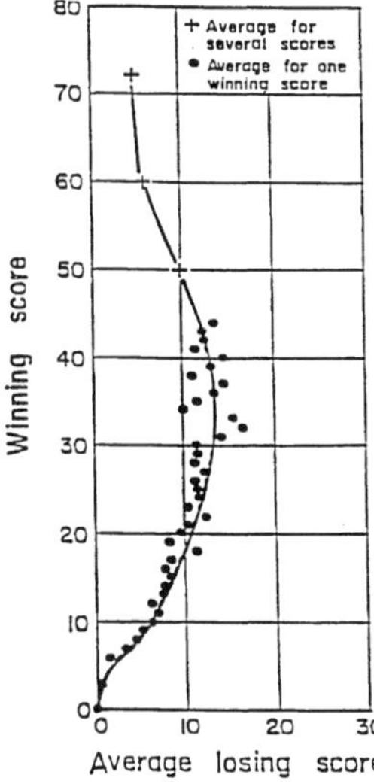

Figure 2. Graph of Average Losing Score for Each Winning Score, Collegiate Football, 1967.

Table 4. Distributions of Team Scores Up to Scores of 29

Score	Winning %	Total	Score	Winning %	Total
0	1.4	222	18	46.2	26
2	.0	6	19	61.1	36
3	16.3	43	20	65.5	84
4	.0	1	21	69.5	118
6	5.8	121	22	58.8	34
7	15.5	220	23	81.4	43
8	8.6	35	24	80.3	66
9	40.6	32	25	61.5	13
10	33.3	72	26	93.1	29
11	20.0	5	27	80.8	52
12	23.6	55	28	84.7	72
13	35.1	111	29	81.8	22
14	36.6	202	—	—	—
15	39.5	38	—	—	—
16	48.2	56	—	—	—
17	69.3	75	Total		2,316

because blowout scores occur when a team is desperately trying to win, often after having played a close game. In 1972 Miami was undefeated, and no other team was, but at least on the surface it looked as if Miami had played weaker teams than had other high-ranking teams. I planned to present this paper at the post-Christmas meetings of the American Association for the Advancement of Science (AAAS). AAAS makes very extensive efforts for the press. Speakers are asked to prepare and deposit materials for newswriters in advance. The head of the Harvard News Office and I were acquainted, and I described these practices to him. He asked if I had ever considered having my paper rewritten for the press by a newswriter, and, of course, I had not. He then suggested that after I had prepared the paper, I should give it to him for someone to do a rewrite, which I did. After a couple of rounds of rewriting by a brilliant writer whose name I wish I could recall, we completed it and sent it to the AAAS News Office. When I arrived at my hotel room in Washington, the phone was already ringing off the hook as various newswriters wanted interviews.

The basic concept of the robust analysis was to create an index for a team that would be the difference between a robust measure of its offensive scoring strength against its opponents and the corresponding index for its opponents' offensive score (or equivalently the team's weakness in defense). The higher the difference, the better the ranking.

I essentially used Tukey's trimeans (Tukey 1977)

$$(1(\text{lower quartile}) + 2(\text{median}) + 1(\text{upper quartile}))/4$$

on both a specific team's scores for the season and on its opponents' scores against the team. And their difference was an index of the performance of the team for the season. We adjusted each team's score for all of the teams it played against. We also made a second-order adjustment for the opponents' scores based on the quality of their opponents.

In the end we used the robust index to obtain a ranking for each team based on its adjusted scores against its opponents and on its opponents' adjusted scores. Our estimates for the ranking adjusted for quality of scheduled opponents ranked Miami second. (Miami did win the Superbowl that season.)

The newswriters spread this information all across the country. I got hate letters from many Miami fans, including a number who claimed to be elderly women.

Lesson 6. Some people are bonded to their local teams.

Lesson 7. The nation is so interested in robust sports statistics that it can hog the newspaper space even at an AAAS annual meeting.

Lesson 8. Maybe it would pay statisticians to have more of their papers rewritten by newswriters.

5. GOLF

Youtz and I were surprised to find that the top professional men golf players were so close in skill that a small number of rounds could do little to distinguish them. Indeed, the standard deviation among players of mean true scores (long-run) for one round at par 72 was estimated to be about .1 of a stroke (Mosteller and Youtz 1992, 1993).

By equating courses using adjusted scores (adding 2 at a par 70 course, adding 1 at a par 71, and using par 72 scores as they stand) we were able to pool data from the last two rounds of all four-round men's professional golf tournaments in 1990 in the U.S.P.G.A tour. We modeled scores as

$$y = \text{base} + X$$

where the base was a low score rarely achieved, such as 62, and X was a Poisson variable. We might think of the base as a score for a practically perfect round. We used for fitting all 33 tournaments:

$$\text{base} = 63 \qquad X: \text{Poisson with mean } 9.3$$

(1 score in 2,500 is less than 63).

For fine-weather days we used

$$\text{base} = 64 \qquad X: \text{Poisson with mean } 8.1.$$

For windy weather we used

$$\text{base} = 62 \qquad X: \text{Poisson with mean } 10.4.$$

The smaller base for windy weather is contrary to intuition, but it may flow partly from the unreliability of the estimate of the base. We found it surprising that the average scores for fine-weather days (72.1) and windy days (72.4) were so close.

Although the negative binomial is attractive as a mixture of Poissons, the variance among professional players' long-run true scores is so small that it offers little advantage.

Figure 3 shows the fit of the distribution of scores to the Poisson model for the season.

Lesson 9. Stay off the course in thunderstorms; Lee Trevino was once struck by lightning while waiting for play to resume.

6. DO WE NEED TO WATCH WHOLE GAMES?

In basketball, football, and hockey the leader at the beginning of the final period (quarter or period) wins the game about 80% of the time (Cooper, DeNeve, and Mosteller 1992). In baseball the leader at the end of seven full innings wins 95% of the time.

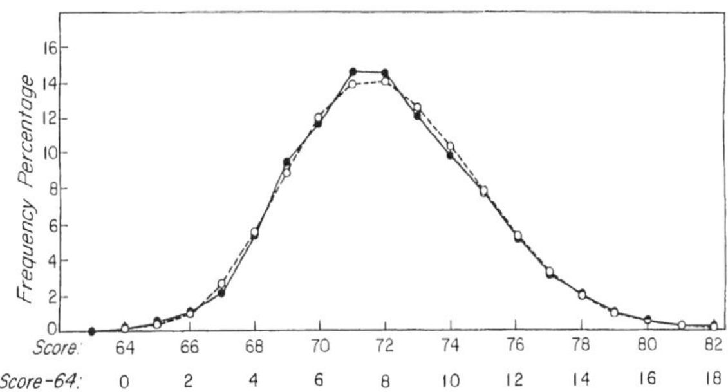

Figure 3. Frequency Distribution of Adjusted Golf Scores in Rounds 3 and 4 in Ten Tournaments in 1990 Having Fine Weather Compared to the Poisson Distribution with Mean 8.1. —— golf scores, - - - Poisson. Source: Mosteller and Youtz (1993, Fig. 2). Reprinted with permission of Springer-Verlag New York Inc.

Lesson 10. We can afford to turn off the TV at the beginning of the final period unless the game is very close.

"Home" teams win about 60% of the time. "Home" teams in basketball make more last-quarter comebacks from behind than "away" teams by a factor of 3 to 1.

The News Office of the National Academy of Sciences wrote a news release about the results given in that paper, and the story appeared in some form in dozens of papers throughout the country.

Lesson 11. Those newswriters know both how to shorten a paper and what will grab readers' attention.

7. RUNS IN TOSSES OF DICE

Statisticians have a strong need to know how accurately their mathematical models of probabilistic events imitate their real-life counterparts. Even for theories of coin tossing and of dice and the distribution of shuffled cards, the gap between theory and practice may be worth knowing. Consequently, I have always been eager to know of demonstrations where these outcomes are simulated.

About 1965 Mr. Willard Longcor came to my office and explained that he had a hobby, and wondered whether this hobby might be adapted to some scientific use because he was going to practice it anyway. He explained that in his retiring years he had, as one intense hobby, recording the results of coin tossing and die tossing. He had discussed applying his hobby with several probabilists and statisticians, and his visit to my office was part of his tour to explore whether his coin and dice throwing might be made more useful than for his personal enjoyment.

I had often thought that, although means and variances might work out all right in tosses of coins and dice, perhaps there were problems with the actual distribution of runs of like outcomes. I was well acquainted with Alexander Mood's paper (1940) on the theory of runs, and so Mr. Longcor's proposal awakened me at once to an opportunity to see some practical results on runs with a human carrying out the tossing and the recording. Mr. Longcor's experience was in tossing a coin or in tossing a die, but recording only whether the outcome was odd or even.

I explained my interest in the distribution of runs. We worked out a plan to record the outcomes of runs of evens in a systematic way. I do not know whether anyone else proposed a project to him, but he told me that he returned to at least one probabilist to get independent assurance that our program might be a useful task.

We planned to use dice, both the ordinary ones with drilled pips that we called Brand X and some of the highest quality to be obtained from professional gambling supply houses. Inexpensive dice with holes for the pips with a drop of paint in each hole might show bias. Precision-made dice have sharp edges meeting at 90 angles, and pips are either lightly painted or back-filled with extremely thin disks. We used three different high-class brands A, B, and C. Each die was to be used for 20,000 tosses and then set aside together with its data. Work began with occasional contact by phone and letter.

Many months later we received a large crate containing the results for millions of throws very neatly recorded, each dataset in the envelope with its own die. We did a lot of checking, and found the results to be in excellent order. Analyzing that crate of data kept a team of four of us busy for a long time.

Some of the findings are shown in Table 5.

Brand X did turn out to be biased, more biased than Weldon's dice in his historical demonstration (as given in Fry 1965). Weldon's dice were more biased than the high-quality dice Mr. Longcor purchased.

Lesson 12. Getting additional people involved in statistical work is a beneficial activity, and they should not have to recruit themselves. Can we do more of this?

As for the number of runs, perhaps a good thing to look at is the mean number of runs of 10 or more per 20,000 tosses. The theoretical value is 9.76. Brands A and B are very close at 10.08 and 9.67, Brand C is a little higher at 10.52, and Brand X is 3.5 standard deviations off at 11.36. However, if we use its observed probability of "even" as .5072, then Brand X has a standard score of only .6, and so its "long-runs" deviation is small, given the bias. This gives us a little reassurance about the model.

8. CONCLUDING REMARKS

My experience with newswriters in relation to sports statistics strongly suggests to me that statisticians should do more about getting statistical material intended for the public written in a more digestable fashion. We tend to give seminars for newswriters so that they will know more about our field. Maybe we should be taking seminars ourselves from newswriters to improve our communication with the consumers of our work. Thus self-improvement might be one direction to go. An alternative would be to have more statistical work that is for public consumption rewritten by newswriters. Such efforts do take a couple of iterations because first translations often run afoul of technical misunderstandings, but with good will on both sides, newswriters can clean out the misunderstandings between statistician and writer. What can we do to take more advantage of the newswriters' skills?

Because the ASA Section on Statistics in Sports has many members, and because many young people have an interest

Table 5. Percentage Distributions for Blocks of 20,000 Throws According to the Number of Even Throws in the Blocks for Theoretical Distributions and for Six Sets of Data, Together with Observed Means, Variances, and Standard Deviations, and Standard Scores for Mean

	Percentage distribution of blocks of 20,000 throws						
		Brands			RAND random numbers	Pseudorandom numbers	Brand
Number of even throws	Theoretical	A	B	C	R	P	X
10,281–10,320							3
10,241–10,280		1					7
10,201–10,240		1				1	12
10,161–10,200	1	1		3	4	2	19
10,121–10,160	3	5	10	6	3	2	16
10,081–10,120	8	6	17	10	7	5	22
10,041–10,080	16	19	7	10	11	17	17
10,001–10,040	22	21	7	19	17	28	3
9,961–10,000	22	20	20	35	21	21	
9,921–9,960	16	15	17	13	19	14	
9,881–9,920	8	6	20	3	12	8	
9,841–9,880	3	4	3		5	1	
9,001–9,840	1	1			1	1	
Total[a]	100%	100%	101%	99%	100%	100%	99%
Number of blocks of 20,000 throws		100	30	31	100	100	58
Mean—10,000	0	9	−1	14	−7	6	145
Variance of block totals	5,000	6,124	6,651	4,776	6,348	4,618	4,933
Standard deviation	71	78	82	69	80	68	70
Standard deviation of mean		7.8	14.9	12.4	8.0	6.8	9.2
Standard score for mean based on observed S.D.		1.15	−.06	1.16	−.82	.86	15.70

[a] Totals may not add to 100 because of rounding.

Source: Iversen et al. (1971, p. 7). Reprinted with permission from Psychometrika and the authors.

in sports, perhaps additional interest in statistics could be promoted by having more statisticians speak about sports statistics to young people, for example, in student mathematics clubs in high schools. Some high-school mathematics teachers do already use sports for illustrating some points in mathematics.

When I think of the number of sports enthusiasts in the United States, I feel that more of these people should be involved in statistical matters. But so far, we do not seem to have organized a practical way to relate their interests to the more general problems of statistics. Perhaps mentioning this will encourage others to think of some ways of improving our connections.

Turning to activities of the Section on Statistics in Sports, I believe that we could profit from a lesson from the mathematicians. They have written out many important problems in lists for their researchers to solve. If we had a list of sports questions, whether oriented to strategies or to questions that may not be answerable, or to problems that might be solved by new methods of data gathering, these questions might attract more focused attention by researchers, and lead to new findings of general interest. And so I encourage the production of some articles oriented to lists of problems.

[Received September 1996. Revised March 1997.]

REFERENCES

Cooper, H., DeNeve, K. M., and Mosteller, F. (1992), "Predicting Professional Sports Game Outcomes from Intermediate Game Scores," *Chance*, 5(3–4), 18–22.

Fry, T. C. (1965), *Probability and Its Engineering Uses* (2nd ed.) Princeton, NJ: Van Nostrand, pp. 312–316.

Girshick, M. A., Mosteller, F., and Savage, L. J. (1946), "Unbiased Estimates for Certain Binomial Sampling Problems with Applications," *Annals of Mathematical Statistics*, 17, 13–23.

Iversen, G. R., Longcor, W. H., Mosteller, F., Gilbert, J. P., and Youtz, C. (1971), "Bias and Runs in Dice Throwing and Recording: A Few Million Throws," *Psychometrika*, 36, 1–19.

Mood, A. M. (1940), "The Distribution Theory of Runs," *Annals of Mathematical Statistics*, 11, 367–392.

Mosteller, F. (1952), "The World Series Competition," *Journal of the American Statistical Association*, 47, 355–380.

——— (1970), "Collegiate Football Scores, U.S.A.," *Journal of the American Statistical Association*, 65, 35–48.

——— (1979), "A Resistant Analysis of 1971 and 1972 Professional Football," in *Sports, Games, and Play: Social and Psychological Viewpoints*, ed. J. H. Goldstein, Hillsdale, NJ: Lawrence Erlbaum Associates, pp. 371–399.

Mosteller, F., and Youtz, C. (1992), "Professional Golf Scores are Poisson on the Final Tournament Days," in *1992 Proceedings of the Section on Statistics in Sports*, American Statistical Association, pp. 39–51.

——— (1993), "Where Eagles Fly," *Chance*, 6(2), 37–42.

Rosner, B., Mosteller, F., and Youtz, C. (1996), "Modeling Pitcher Performance and the Distribution of Runs per Inning in Major League Baseball," *The American Statistician*, 50, 352–360.

Tukey, J. W. (1977), *Exploratory Data Analysis*, Reading, MA: Addison-Wesley.

A statistician and athlete finds a wide range of applications of Total Quality Management.

Chapter 33

Can TQM Improve Athletic Performance?

Harry V. Roberts

Introduction

As a statistician and athlete—much of the athletics coming in age-group distance running and triathlon competition after the age of 50—I have kept careful records of training and competition over the last 20 years in hope that statistical analysis would benefit my performance. Among the many questions I sought to answer were: Is even-pacing during a marathon a good strategy? What type of training regimen is appropriate? Is it really necessary to drink fluids during long distance races in hot weather? Standard statistical methods, such as randomized controlled experiments, have been of limited use in answering such questions. Over the years, I have drawn on techniques from Total Quality Management (TQM)—ideas that have been effectively used in manufacturing and now increasingly in service industries—to improve athletic performance and found the methods not only helped improve my performance but also had a wide range of application.

Group Versus Individual Studies

The title asks "Can *TQM* Improve Athletic Performance?" rather than "Can *Statistics* Improve Athletic Performance?" Statistics can, in principle, improve anything, but the statistician's orientation is often toward *research to obtain knowledge about a population.* TQM, on the other hand, *relies on statistical methods to focus on the improvement of specific processes.* The TQM focus is valuable in helping one take the direct road to process improvement rather than the more leisurely path to new knowledge about a population, which may or may not apply to a specific individual.

A typical research approach to athletic performance is to experiment on subject groups in hope of finding general relationships. For example, after a base period, a group of runners could be randomized into two subgroups, one of which trains twice a day and the other once a day. Subsequent performance relative to the base period could be compared by standard statistical techniques. The lessons from such studies can then be applied to improve individual athletic performance. Studies of this kind, however, have not yet been very helpful.

By contrast, the individual athlete in the TQM approach is viewed as an ongoing process to be improved. Historical data suggest hypotheses on how improvement should be sought, that is, what type of intervention might improve the process. A study is then designed and performance measured; this is essentially the Deming Plan-Do-Check-Act cycle (PDCA) that is central to statistical process control in industry. The basic statistical methodology is often Box and Tiao's "Intervention Analysis," a term they introduced in a classic article in the *Journal of the American Statistical Association* in 1975 entitled "Intervention Analysis with Applications to Economic and Environmental Problems."

When time-ordered performance measurements on an underlying process are in a state of statistical control—that is, the data behave like independent random drawings from a fixed distribution—intervention analysis calls for a comparison of the mean performance level before and after a particular improvement is attempted using techniques for two independent samples. When an underlying process is not in control—for example, when there are autocorrelated variation, trend, or day-of-week effects—intervention analysis essentially uses regression analysis to disentangle the effect of the intervention from the effects of the other sources of variation, random or systematic.

TQM for Improving Technique

The quickest opportunities for TQM to improve athletic performance come in matters of technique; students of mine have often done studies aimed at improvement in this area. Popular subjects have been free-throw shooting and three-point shooting in basketball, tennis service, place kicking, archery, target shooting, skeet shooting, swimming, cycling, and golf. Students often come up with ideas for improvement when recording and

analyzing the results of regular practice.

In most studies, students' performance has been in a state of statistical control during the base period and in control at a higher level after the alteration of technique. A student who was initially in statistical control with a free-throw success rate of 75%, for example, changed his aim point, to the back of the rim, and increased the arch. Subsequent to this change, he found he was in control with a success rate of 82%. This is like starting with a biased coin that has probability of heads of 0.75, independently from toss to toss, and modifying the coin so that the probability of heads rises to 0.82. The success of intervention analysis can be further illustrated in three detailed examples.

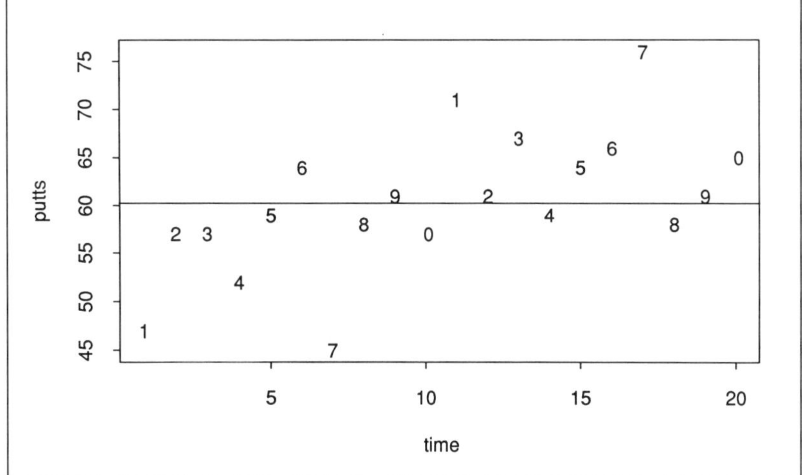

Figure 1. Scatterplot of scores for 20 groups of 100 putts each shows scores improved when stance was modified for the final 10 observations.

Example 1. Improvement of Putting

A student, an excellent golfer who had never been satisfied with his putting, set up an indoor putting green for careful practice and experimentation. Results of the first 2,000 putts of his study, summarized by 20 groups of 100 each, are presented: Putts sunk per 100 trials from fixed distance:

47, 57, 57, 52, 59, 64, 45, 58, 61, 57,

71, 61, 67, 59, 64, 66, 76, 58, 61, 65

At the end of the first 10 groups of 100, he noticed that 136 of 443 misses were left misses and 307 were right misses. He reasoned that the position of the ball relative to the putting stance was a problem. To correct this, he proposed "moving the ball several inches forward in my stance, keeping it just inside the left toe." The final 10 observations were made with the modified stance, and all groups were displayed in a simple time-series plot (see Fig. 1) to help visualize the change.

Examination of the plot suggests that he is improving: On average, the last 10 points are higher than the first. Simple regression analysis with an indicator variable for change of stance suggests that the improvement is genuine; diagnostic checks of adequacy of the regression model are satisfactory. The process appears to have been in statistical control before the intervention and to have continued in statistical control at a higher level subsequent to the intervention. A more subtle question is whether we are seeing the steady improvement that comes with practice, a sharp improvement due to the change in stance, or both combined. Comparison of two regression models, one that incorporates a time trend and one that incorporates a single-step improvement, suggests that the sharp improvement model fits the data better than a trend one.

The estimated improvement is substantial; it translates into several strokes per round. (The student confirmed the results with further work and also discovered that the "baseball grip," recommended by Lee Trevino, was at least as good as the standard "reverse overlap grip.")

This student's example is an application of intervention analysis. There are no randomized controls, as in "true" statistical experimentation, but if one takes care in interpretation, conclusions from intervention analysis rest on nearly as firm ground. When randomized controls are possible, statistical analysis can be even more conclusive.

Example 2. A Pool Experiment

In a study of technique for the game of pool, it was possible to do a randomized experiment on alternative techniques. The techniques being considered were randomized in blocks instead of just making a single transition from one to the other, as was done in the putting example. The student was interested in alterations of technique affecting two aspects of pool and had been using an unconventional approach: an unorthodox upside-down V bridge with eye focused on the object ball. The standard approach was closed bridge and eye focused on the cue ball. Which of these four combinations is best? The experiment that was carried out was a two-level, two-factor design with blocking and with randomization of techniques within blocks. There were five sessions consisting of eight games each. The data set and details of the design are presented in Table 1.

Table 1—Study to Improve Game of Pool

Session 1			Session 2			Session 3			Session 4			Session 5		
Shots	**Bridge**	**Eye**	**Shots**	**Bridge**	**Eye**	**Shots**	**Bridge**	**Eye**	**Shots**	**Bridge**	**Eye**	**Shots**	**Bridge**	**Eye**
50	1	−1	32	1	1	39	1	1	40	1	−1	35	−1	−1
39	1	1	36	−1	1	58	1	−1	41	−1	1	38	−1	1
43	−1	−1	48	−1	−1	44	−1	−1	27	1	1	33	1	−1
78	−1	1	45	1	−1	56	−1	1	46	−1	−1	25	1	1
62	1	−1	54	−1	1	62	1	−1	48	−1	1	40	1	−1
40	−1	−1	46	1	−1	57	−1	1	32	1	1	52	−1	−1
62	−1	1	55	1	1	46	1	1	52	1	−1	45	−1	1
62	1	1	50	−1	−1	52	−1	−1	50	−1	−1	36	1	1

Shots: The number of shots from the break to get all the balls in.
Session: Session, eight games per session on a given day, five sessions.
Bridge: = −1 for unorthodox upside-down V bridge (starting method); = 1 for standard closed bridge (standard method).
Eye: = −1 eye focused on object ball (starting method); = 1 eye focused on cue ball (standard method).

The eight games of each session are blocked into two blocks of four, and the four treatment combinations are randomized within each block of four. Hence, the listing in Table 1 is in the sequence in which the games were played.

Figure 2 displays SHOTS as a time series. A careful examination of the plot in Fig. 2 suggests that several systematic effects are happening in the data. The experimental variations of technique are clearly superimposed on a process that is not in control. In particular:

1. *Variability of SHOTS appears to be higher when level of SHOTS is higher.* A logarithmic or other transformation?
2. *Overall downtrend is evident.* Can this be the result of improvement with practice?
3. *There is an uptrend within each of the five days.* Is this fatigue effect?

It is important to emphasize that this process was not in a state of statistical control, yet the experiment was valid because the systematic factors—the trends within and between days—causing the out-of-control condition were modeled in the statistical analysis, as we now show.

In the final regression by ordinary least squares, the sources of variation are sorted out as follows. LSHOTS is the log of the number of shots to clear the table. STANDARD is an indicator for the standard bridge and eye focus; it takes the value 1 for games played using these techniques and 0 for other games. This variable was defined when preliminary tabular examination of a two-way table of mean LSHOTS suggested that any departure from the standard bridge and focus led to a similar degradation. TIME is the linear trend variable across all 40 observations. ORDER is sequence of games within each day. (Block effects turned out to be insignificant.)

The estimated regression equation is

Fitted LSHOTS =
3.84 − 0.248 (STANDARD)
− 0.00998 (TIME)
+ 0.0518 (ORDER)

Diagnostic checks of model adequacy were satisfactory.

The standard error of STANDARD was 0.06045 and the *t*-ratio was −4.10. The standard technique, not the student's unorthodox technique, worked best. Using the standard technique led to an estimated 22% improvement in estimated SHOTS [exp{−0.248} = 0.78].

Note that there were two trends: an upward trend within each day, presumably reflecting fatigue, and a downward overall trend, presumably reflecting the effects of practice. For given ORDER within a day, the fitted scores drop about 8(0.01) = 0.08 from day to day; because we are working in log units, this translates to a trend improvement due to practice of about 8% per day.

The pool example shows that in trying to improve technique, one can go beyond simple intervention analysis to designed experiments, even when the underlying process is not in a state of statistical control. Moreover, two aspects of technique—bridge and eye focus—were studied simultaneously.

The full advantages of multifactor experimentation, however, are difficult to realize in athletics. This is because change of athletic technique entails considerable effort. In the pool experiment, it was relatively easy to switch around the bridge and eye focus from trial to trial, but this is atypical. A swimmer, for example, might have a much harder time attempting to change technique of each lap with the four combinations of stroke technique defined by (1) moderate versus extreme body roll and (2) ordinary versus bilateral breathing.

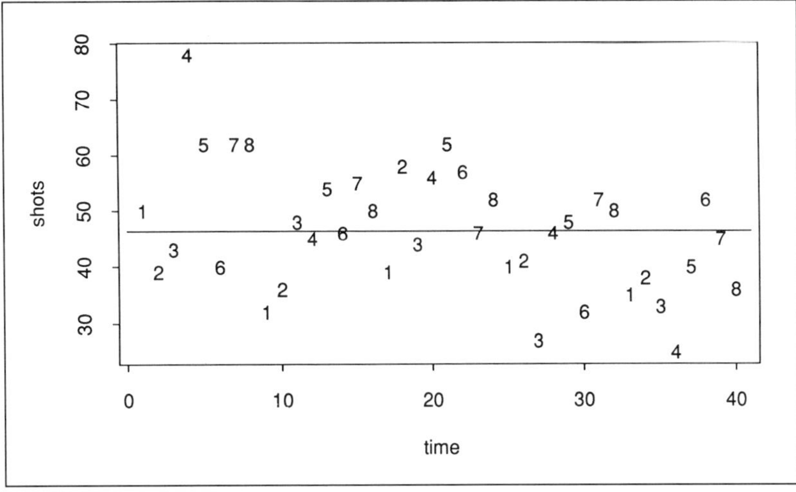

Figure 2. Plot of shots over five sessions of eight games of pool reveals overall downward trend and upward trends in each of the five sessions.

Example 3. Even Pacing in Running

In cross-country and track races in the 1940s, I learned from painful experience that a fast start always led to an agonizing slowing down for the balance of a race. However, the coaching wisdom of the time always called for starting very fast and hoping that eventually one could learn to maintain the pace all the way. (A more refined version was that one should go out fast, slow down in the middle of the race, and then sprint at the end.)

Eventually, a distaste for suffering led me to experiment with a more reasonable starting pace; that was the intervention. In my next cross-country race, a 3-mile run, I was dead last at the end of the first quarter mile in a time of 75 seconds; the leaders ran it in about 60 seconds. (Had they been able to maintain that pace, they would have run a 4-minute mile on the first of the 3 miles.) For the rest of the race, I passed runners steadily and finished in the upper half of the pack, not because I speeded up but because the others slowed down. In the remaining races that season, my performance relative to other runners was much improved.

Decades afterward, as an age-group runner, I was able to validate the youthful experiment. At age 59, in my best marathon, I averaged 7:06 per mile throughout the race with almost no variation. In other marathons, I discovered that an even *slightly* too-fast pace (as little as 5–10 seconds per mile) for the first 20 miles was punished by agony and drastic slowing in the final 10 kilometers, with consequent inflation of the overall marathon time. In my best marathons, all run at an even pace, I was only pleasantly tired in the final 10 kilometers and was able to pass large numbers of runners who had run the first 20 miles too fast.

What worked for me apparently works for others. In recent decades, even-pacing has become the generally accepted practice. Most world distance records are nearly evenly paced.

In my later experiences in age-group competition, I did two other intervention studies that ran against the conventional wisdom of the time. I found that long training runs for the marathon were not necessary so long as my total training mileage was adequate (roughly 42 miles per week). I also discovered that it was unnecessary to drink fluids during long races in hot weather provided I was supersaturated with fluids at the start of the race and then, instead of drinking, doused myself with water at every opportunity. In this way, I kept cool and conserved body fluids but did not have the discomfort and mild nausea that came with drinking while running.

These personal examples illustrate the point that the individual athlete can improve performance without having to locate and draw on group data in the scientific literature. In each instance, the idea for improvement came from a study of my actual experience. The applications also illustrate the usefulness of simple before–after comparisons between what happened before the intervention and what happened subsequently.

In the example of a change in training regimen—no long-distance training runs—the abrupt intervention is the only practicable approach because the effects of training are cumulative and lagged. One cannot switch training regimens on and off for short periods of time and hope to trace the lagged effects; the same observation applies to attempts to improve general fitness.

The same ideas, of course, apply in industrial settings and have been applied there for decades in conjunction with use of Shewhart control charts by workers. The control chart makes it easier to see quickly, even without formal statistical testing, whether an intervention to improve the process has or has not succeeded.

The Limitations of Happenstance Data

All my examples have involved some conscious intervention in an ongoing process. "Happenstance data" also may be helpful in suggesting hypotheses for improvement. For this reason, it seems useful to maintain athletic diaries, recording training methods, workouts, competitive performances, injuries, weight, resting pulse rates, and other pertinent information. But there are dangers in relying on happenstance data if one lacks elementary statistical skills. First, it is often difficult to infer causation from regression analysis of happenstance data. Second, it is tempting to overreact to apparently extreme individual values and jump to a hasty process change; this is what Deming calls "tampering," and it usually makes processes worse, not better. Data plotting and statistical analysis are needed to ensure a proper conclusion based on sound observations and studies.

What Works Best

In improving athletic technique, the appropriate experimental strategy is likely to be intervention analysis because randomized experimentation often is not feasible. Moreover, for practical reasons, we are usually limited to changing one thing at a time. Nevertheless, careful application of the methods of TQM may enhance our ability to make causal judgments about the effects of our interventions. As illustrated in the pool example, it is not necessary for the athletic processes of interest to be in a state of statistical control before we can profitably intervene to improve them.

The examples presented above show that individual athletes can enhance their abilities without having to locate and draw on experimental group data in the scientific literature that may suggest how to improve performance. The ideas for improvement may come from the study of actual experience or even the advice of experts. Individual athletes can then collect and analyze personal data to see if the ideas work for them, and by so doing, determine ways to better their scores, finishing times, or other performance measures.

Chapter 34

A Brownian Motion Model for the Progress of Sports Scores

Hal S. STERN*

The difference between the home and visiting teams' scores in a sports contest is modeled as a Brownian motion process defined on $t \in (0, 1)$, with drift μ points in favor of the home team and variance σ^2. The model obtains a simple relationship between the home team's lead (or deficit) ℓ at time t and the probability of victory for the home team. The model provides a good fit to the results of 493 professional basketball games from the 1991–1992 National Basketball Association (NBA) season. The model is applied to the progress of baseball scores, a process that would appear to be too discrete to be adequately modeled by the Brownian motion process. Surprisingly, the Brownian motion model matches previous calculations for baseball reasonably well.

KEY WORDS: Baseball; Basketball; Probit regression

1. INTRODUCTION

Sports fans are accustomed to hearing that "team A rarely loses if ahead at halftime" or that "team B had just accomplished a miracle comeback." These statements are rarely supported with quantitative data. In fact the first of the two statements is not terribly surprising; it is easy to argue that approximately 75% of games are won by the team that leads at halftime. Suppose that the outcome of a half-game is symmetrically distributed around 0 so that each team is equally likely to "win" the half-game (i.e., assume that two evenly matched teams are playing). In addition, suppose that the outcomes of the two halves of a game are independent and identically distributed. With probability .5 the same team will win both half-games, and in that case the team ahead at halftime certainly wins the game. Of the remaining probability, it seems plausible that the first half winner will defeat the second half winner roughly half the time. This elementary argument suggests that in contests among fairly even teams, the team ahead at halftime should win roughly 75% of the time. Evaluating claims of "miraculous" comebacks is more difficult. Cooper, DeNeve, and Mosteller (1992) estimated the probability that the team ahead after three quarters of the game eventually wins the contest for each of the four major sports (basketball, baseball, football, hockey). They found that the leading team won more than 90% of the time in baseball and about 80% of the time in the other sports. They also found that the probability of holding a lead is different for home and visiting teams. Neither the Cooper, et al. result nor the halftime result described here considers the size of the lead, an important factor in determining the probability of a win.

The goal here is to estimate the probability that the home team in a sports contest wins the game given that they lead by ℓ points after a fraction $t \in (0, 1)$ of the contest has been completed. Of course, the probability for the visiting team is just the complement. The main focus is the game of basketball.

Among the major sports, basketball has scores that can most reasonably be approximated by a continuous distribution. A formula relating ℓ and t to the probability of winning allows for more accurate assessment of the propriety of certain strategies or substitutions. For example, should a star player rest at the start of the fourth quarter when his team trails by 8 points or is the probability of victory from this position too low to risk such a move? In Section 2 a Brownian motion model for the progress of a basketball score is proposed, thereby obtaining a formula for the probability of winning conditional on ℓ and t. The model is applied to the results of 493 professional basketball games in Section 3. In Section 4 the result is extended to situations in which it is known only that $\ell > 0$. Finally, in Section 5 the Brownian motion model is applied to a data set consisting of the results of 962 baseball games. Despite the discrete nature of baseball scores and baseball "time" (measured in innings), the Brownian motion model produces results quite similar to those of Lindsey (1977).

2. THE BROWNIAN MOTION MODEL

To begin, we transform the time scale of all sports contests to the unit interval. A time $t \in (0, 1)$ refers to the point in a sports contest at which a fraction t of the contest has been completed. Let $X(t)$ represent the lead of the home team at time t. The process $X(t)$ measures the difference between the home team's score and the visiting team's score at time t; this may be positive, negative, or 0. Westfall (1990) proposed a graphical display of $X(t)$ as a means of representing the results of a basketball game. Naturally, in most sports (including the sport of most interest here, basketball), $X(t)$ is integer valued. To develop the model, we ignore this fact, although we return to it shortly. We assume that $X(t)$ can be modeled as a Brownian motion process with drift μ per unit time ($\mu > 0$ indicates a μ point per game advantage for the home team) and variance σ^2 per unit time. Under the Brownian motion model,

$$X(t) \sim N(\mu t, \sigma^2 t)$$

and $X(s) - X(t)$, $s > t$, is independent of $X(t)$ with

$$X(s) - X(t) \sim N(\mu(s-t), \sigma^2(s-t)).$$

* Hal S. Stern is Associate Professor, Department of Statistics, Harvard University, Cambridge, MA 02138. Partial support for this work was provided by Don Rubin's National Science Foundation Grant SES-8805433. The author thanks Tom Cover for suggesting the problem and the halftime argument several years ago and Elisabeth Burdick for the baseball data. Helpful comments were received from Tom Belin, Andrew Gelman, and Carl Morris.

Journal of the American Statistical Association
September 1994, Vol. 89, No. 427, Statistics in Sports

The probability that the home team wins a game is $\Pr(X(1) > 0) = \Phi(\mu/\sigma)$, and thus the ratio μ/σ indicates the magnitude of the home field advantage. In most sports, the home team wins approximately 55–65% of the games, corresponding to values of μ/σ in the range .12–.39. The drift parameter μ measures the home field advantage in points (typically thought to be 3 points in football and 5–6 points in basketball).

Under the random walk model, the probability that the home team wins [i.e., $X(1) > 0$] given that they have an ℓ point advantage (or deficit) at time t [i.e., $X(t) = \ell$] is

$$\begin{aligned} P_{\mu,\sigma}(\ell, t) &= \Pr(X(1) > 0 \mid X(t) = \ell) \\ &= \Pr(X(1) - X(t) > -\ell) \\ &= \Phi\left(\frac{\ell + (1-t)\mu}{\sqrt{(1-t)\sigma^2}}\right), \end{aligned}$$

where Φ is the cdf of the standard normal distribution. Of course, as $t \to 1$ for fixed $\ell \neq 0$, the probability tends to either 0 or 1, indicating that any lead is critically important very late in a game. For fixed t, the lead ℓ must be relatively large compared to the remaining variability in the contest for the probability of winning to be substantial.

The preceding calculation treats $X(t)$ as a continuous random variable, although it is in fact discrete. A continuity correction is obtained by assuming that the observed score difference is the value of $X(t)$ rounded to the nearest integer. If we further assume that contests tied at $t = 1$ [i.e., $X(1) = 0$] are decided in favor of the home team with probability .5, then it turns out that

$$P^{cc}_{\mu,\sigma}(\ell, t) = 0.5\,\Phi\left(\frac{\ell - .5 + (1-t)\mu}{\sqrt{(1-t)\sigma^2}}\right) + .5\,\Phi\left(\frac{\ell + .5 + (1-t)\mu}{\sqrt{(1-t)\sigma^2}}\right).$$

In practice, the continuity correction seems to offer little improvement in the fit of the model and causes only minor changes in the estimates of μ and σ. It is possible to obtain a more accurate continuity correction that accounts for the drift in favor of the home team in deciding tied contests. In this case .5 is replaced by a function of μ, σ, and the length of the overtime used to decide the contest.

The Brownian motion model motivates a relatively simple formula for $P_{\mu,\sigma}(\ell, t)$, the probability of winning given the lead ℓ and elapsed time t. A limitation of this formula is that it does not take into account several potentially important factors. First, the probability that a home team wins, conditional on an ℓ point lead at time t, is assumed to be the same for any basketball team against any opponent. Of course, this is not true; Chicago (the best professional basketball team during the period for which data has been collected here) has a fairly good chance of making up a 5-point halftime deficit ($\ell = -5$, $t = .50$) against Sacramento (one of the worst teams), whereas Sacramento would have much less chance of coming from behind against Chicago. One method for taking account of team identities would be to replace μ with an estimate of the difference in ability between the two teams in a game, perhaps the Las Vegas point spread. A second factor not accounted for is whether the home team is in possession of the ball at time t and thus has the next opportunity to score. This is crucial information in the last few minutes of a game ($t > .96$ in a 48-minute basketball game). Despite the omission of these factors, the formula appears to be quite useful in general, as demonstrated in the remainder of the article.

3. APPLICATION TO PROFESSIONAL BASKETBALL

Data from professional basketball games in the United States are used to estimate the model parameters and to assess the fit of the formula for $P_{\mu,\sigma}(\ell, t)$. The results of 493 National Basketball Association (NBA) games from January to April 1992 were obtained from the newspaper. This sample size represents the total number of games available during the period of data collection and represents roughly 45% of the complete schedule. We assume that these games are representative of modern NBA basketball games (the mean score and variance of the scores were lower years ago). The differences between the home team's score and the visiting team's score at the end of each quarter are recorded as $X(.25)$, $X(.50)$, $X(.75)$, and $X(1.00)$ for each game. For the ith game in the sample, we also represent these values as $X_{i,j}$, $j = 1, \ldots, 4$. The fourth and final measurement, $X(1.00) = X_{i,4}$, is the eventual outcome of the game, possibly after one or more overtime periods have been played to resolve a tie score at the end of four quarters. The overtime periods are counted as part of the fourth quarter for purposes of defining X. This should not be a problem, because $X(1.00)$ is not used in obtaining estimates of the model parameters. In a typical game, on January 24, 1992, Portland, playing Atlanta at home, led by 6 points after one quarter and by 9 points after two quarters, trailed by 1 point after three quarters, and won the game by 8 points. Thus $X_{i,1} = 6$, $X_{i,2} = 9$, $X_{i,3} = -1$, and $X_{i,4} = 8$.

Are the data consistent with the Brownian motion model? Table 1 gives the mean and standard deviation for the results of each quarter and for the final outcome. In Table 1 the outcome of quarter j refers to the difference $X_{i,j} - X_{i,j-1}$ and the final outcome refers to $X_{i,4} = X(1.00)$. The first three quarters are remarkably similar; the home team outscores the visiting team by approximately 1.5 points per quarter, and the standard deviation is approximately 7.5 points. The fourth quarter seems to be different; there is only a slight advantage to the home team. This may be explained by the fact that if a team has a comfortable lead, then it is apt to ease up or use less skillful players. The data suggests that the home team is much more likely to have a large lead after three quarters; this may explain the fourth quarter results in Table 1. The normal distribution appears to be a satisfactory approximation to the distribution of score differences in each quarter, as indicated by the QQ plots in Figure 1. The cor-

Table 1. Results by Quarter of 493 NBA Games

Quarter	Variable	Mean	Standard deviation
1	$X(.25)$	1.41	7.58
2	$X(.50) - X(.25)$	1.57	7.40
3	$X(.75) - X(.50)$	1.51	7.30
4	$X(1.00) - X(.75)$	.22	6.99
Total	$X(1.00)$	4.63	13.18

relations between the results of different quarters are negative and reasonably small ($r_{12} = -.13$, $r_{13} = -.04$, $r_{14} = -.01$, $r_{23} = -.06$, $r_{24} = -.05$, and $r_{34} = -.11$). The standard error for each correlation is approximately .045, suggesting that only the correlation between the two quarters in each half of the game, r_{12} and r_{34}, are significantly different from 0. The fact that teams with large leads tend to ease up may explain these negative correlations, a single successful quarter may be sufficient to create a large lead. The correlation of each individual quarter's result with the final outcome is approximately .45. Although the fourth quarter results provide some reason to doubt the Brownian motion model, it seems that the model may be adequate for the present purposes. We proceed to examine the fit of the formula $P_{\mu,\sigma}(\ell, t)$ derived under the model.

The formula $P_{\mu,\sigma}(\ell, t)$ can be interpreted as a probit regression model relating the game outcome to the transformed variables $\ell/\sqrt{1-t}$ and $\sqrt{1-t})$ with coefficients $1/\sigma$ and μ/σ. Let $Y_i = 1$ if the home team wins the ith game [i.e., $X(1) > 0$)] and 0 otherwise. For now, we assume that the three observations generated for each game, corresponding to the first, second, and third quarters, are independent. Next we investigate the effect of this independence assumption. The probit regression likelihood L can be expressed as

$$L = \prod_{i=1}^{493}\prod_{j=1}^{3} \Phi\left(\frac{X_{ij} + \left(1-\frac{j}{4}\right)\mu}{\sqrt{\left(1-\frac{j}{4}\right)\sigma^2}}\right)^{Y_i} \times \left(1 - \Phi\left(\frac{X_{ij} + \left(1-\frac{j}{4}\right)\mu}{\sqrt{\left(1-\frac{j}{4}\right)\sigma^2}}\right)\right)^{(1-Y_i)}$$

$$= \prod_{i=1}^{493}\prod_{j=1}^{3} \Phi\left(\alpha\frac{X_{ij}}{\sqrt{1-\frac{j}{4}}} + \beta\sqrt{1-\frac{j}{4}}\right)^{Y_i} \times \left(1 - \Phi\left(\alpha\frac{X_{ij}}{\sqrt{1-\frac{j}{4}}} + \beta\sqrt{1-\frac{j}{4}}\right)\right)^{(1-Y_i)}$$

where $\alpha = 1/\sigma$ and $\beta = \mu/\sigma$. Maximum likelihood estimates of α and β (and hence μ and σ) are obtained using a Fortran program to carry out a Newton–Raphson procedure. Convergence is quite fast (six iterations), with $\hat{\alpha} = .0632$ and $\hat{\beta} = .3077$ implying

$$\hat{\mu} = 4.87 \quad \text{and} \quad \hat{\sigma} = 15.82.$$

An alternative method for estimating the model parameters directly from the Brownian motion model, rather than through the implied probit regression, is discussed later in this section. Approximate standard errors of $\hat{\mu}$ and $\hat{\sigma}$ are obtained via the delta method from the asymptotic variance and covariance of $\hat{\alpha}$ and $\hat{\beta}$:

$$\text{s.e.}(\hat{\mu}) = .90 \quad \text{and} \quad \text{s.e.}(\hat{\sigma}) = .89.$$

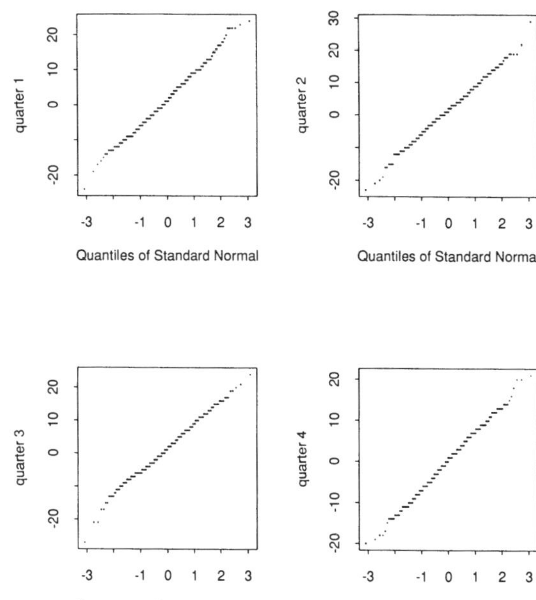

Figure 1. Q-Q Plots of Professional Basketball Score Differences by Quarter. These are consistent with the normality assumption of the Brownian motion model.

These standard errors are probably somewhat optimistic, because they are obtained under the assumption that individual quarters contribute independently to the likelihood, ignoring the fact that groups of three quarters come from the same game and have the same outcome Y_i. We investigate the effect of the independence assumption by simulation using two different types of data. "Nonindependent" data, which resemble the NBA data, are obtained by simulating 500 Brownian motion basketball games with fixed μ, σ and then using the three observations from each game (the first, second, and third quarter results) to produce data sets consisting of 1,500 observations. Independent data sets consisting of 1,500 independent observations are obtained by simulating 1,500 Brownian motion basketball games with fixed μ, σ and using only one randomly chosen quarter from each game. Simulation results using "nonindependent" data suggest that parameter estimates are approximately unbiased but the standard errors are 30–50% higher than under the independence condition. The standard errors above are computed under the assumption of independence and are therefore too low. Repeated simulations, using "nonindependent" data with parameters equal to the maximum likelihood estimates, yield improved standard error estimates, s.e.$(\hat{\mu}) = 1.3$ and s.e.$(\hat{\sigma}) = 1.2$.

The adequacy of the probit regression fit can be measured relative to the saturated model that fits each of the 158 different (ℓ, t) pairs occurring in the sample with its empirical probability. Twice the difference between the log-likelihoods is 134.07, which indicates an adequate fit when compared to the asymptotic chi-squared reference distribution with 156 degrees of freedom. As is usually the case, there is little difference between the probit regression results and logistic regression results using the same predictor variables. We use probit regression to retain the easy interpretation of the regression coefficients in terms of μ, σ. The principal contribution of the Brownian motion model is that regressions based on the transformations of (ℓ, t) suggested by the Brownian motion model, $(\ell/\sqrt{1-t}, \sqrt{1-t})$, appear to provide a better fit than models based on the untransformed variables. As mentioned in Section 2, it is possible to fit the Brownian motion model with a continuity correction. In this case the estimates for μ and σ are 4.87 and 15.80, almost identical to the previous estimates. For simplicity, we do not use the continuity correction in the remainder of the article.

Under the Brownian motion model, it is possible to obtain estimates of μ, σ without performing the probit regression. The game statistics in Table 1 provide direct estimates of the mean and standard deviation of the assumed Brownian process. The mean estimate, 4.63, and the standard deviation estimate, 13.18, obtained from Table 1 are somewhat smaller than the estimates obtained by the probit model. The differences can be attributed in part to the failure of the Brownian motion model to account for the results of the fourth quarter. The probit model appears to produce estimates that are more appropriate for explaining the feature of the games in which we are most interested—the probability of winning.

Table 2 gives the probability of winning for several values of ℓ, t. Due to the home court advantage, the home team has a better than 50% chance of winning even if it is behind by two points at halftime ($t = .50$). Under the Brownian motion model, it is not possible to obtain a tie at $t = 1$ so this cell is blank; we might think of the value there as being approximately .50. In professional basketball $t = .9$ corresponds roughly to 5 minutes remaining in the game. Notice that home team comebacks from 5 points in the final 5 minutes are not terribly unusual. Figure 2 shows the probability of winning given a particular lead; three curves are plotted corresponding to $t = .25, .50, .75$. In each case the empirical probabilities are displayed as circles with error bars ($\pm$ two binomial standard errors). To obtain reasonably large sample sizes for the empirical estimates, the data were divided into bins containing approximately the same number of games (the number varies from 34 to 59). Each circle is plotted at the median lead of the observations in the bin. The model appears consistent with the pattern in the observed data.

Figure 3 shows the probability of winning as a function of time for a fixed lead ℓ. The shape of the curves is as expected. Leads become more indicative of the final outcome as time passes and, of course, larger leads appear above smaller leads. The $\ell = 0$ line is above .5, due to the drift in favor of the home team. A symmetric graph about the horizontal line at .5 is obtained if we fix $\mu = 0$. Although the probit regression finds μ is significantly different than 0, the no drift model $P_{0,\sigma}(\ell, t) = \Phi(\ell/\sqrt{(1-t)\sigma^2})$ also provides a reasonable fit to the data with estimated standard deviation 15.18.

Figure 4 is a contour plot of the function $P_{\hat{\mu},\hat{\sigma}}(\ell, t)$ with time on the horizontal axis and lead on the vertical axis. Lines on the contour plot indicate game situations with equal probability of the home team winning. As long as the game is close, the home team has a 50–75% chance of winning.

4. CONDITIONING ONLY ON THE SIGN OF THE LEAD

Informal discussion of this subject, including the introduction to this article, often concerns the probability of winning given only that a team is ahead at time t ($\ell > 0$) with the exact value of the lead unspecified. This type of partial information may be all that is available in some circumstances. Integrating $P_{\mu,\sigma}(\ell, t)$ over the distribution of the lead ℓ at time t yields (after some transformation)

$$P_{\mu/\sigma}(t) = \Pr(X(1) > 0 \mid X(t) > 0)$$

$$= \int_0^\infty \Phi\left(y\sqrt{\frac{t}{1-t}} + \sqrt{1-t}\,\frac{\mu}{\sigma}\right)\frac{1}{\sqrt{2\pi}} \times \exp\left(-\left(y - \sqrt{t}\,\frac{\mu}{\sigma}\right)^2 \Big/ 2\right) dy,$$

which depends only on the parameters μ and σ through the ratio μ/σ. The integral is evaluated at the maximum likelihood estimates of μ and σ using a Fortran program to implement Simpson's rule. The probability that the home team wins given that it is ahead at $t = .25$ is .762, the probability at $t = .50$ is .823, and the probability at $t = .75$ is .881. The corresponding empirical values, obtained by considering only those games in which the home team led at the appropriate time point, (263 games for $t = .25$, 296 games for $t = .50$, 301 games for $t = .75$) are .783, .811, and .874, each within a single standard error of the model predictions.

If it is assumed that $\mu = 0$, then we obtain the simplification

$$P_0(t) = \frac{1}{2} + \frac{1}{\pi}\tan^{-1}\left(\sqrt{\frac{t}{1-t}}\right),$$

with $P_0(.25) = 2/3$, $P_0(.50) = 3/4$, and $P_0(.75) = 5/6$. Because there is no home advantage when $\mu = 0$ is assumed, we combine home and visiting teams together to obtain empirical results. We find that the empirical probabilities (based on 471, 473, and 476 games) respectively are .667, .748, and .821. Once again, the empirical results are in close agreement with the results from the probit model.

Table 2. $P_{\hat{\mu},\hat{\sigma}}(\ell, t)$ for Basketball Data

	Lead						
Time t	$\ell = -10$	$\ell = -5$	$\ell = -2$	$\ell = 0$	$\ell = 2$	$\ell = 5$	$\ell = 10$
.00				.62			
.25	.32	.46	.55	.61	.66	.74	.84
.50	.25	.41	.52	.59	.65	.75	.87
.75	.13	.32	.46	.56	.66	.78	.92
.90	.03	.18	.38	.54	.69	.86	.98
1.00	.00	.00	.00		1.00	1.00	1.00

5. OTHER SPORTS

Of the major sports, basketball is best suited to the Brownian motion model because of the nearly continuous nature of the game and the score. In this section we report the results of applying the Brownian motion model to the results of the 1986 National League baseball season. In baseball, the teams play nine innings; each inning consists of two half-innings, with each team on offense in one of the half-innings. The half-inning thus represents one team's opportunity to score. The average score for one team in a single half-inning is approximately .5. More than 70% of the half-innings produce 0 runs. The data consist of 962 games (some of the National League games were removed due to data entry errors or because fewer than nine innings were played).

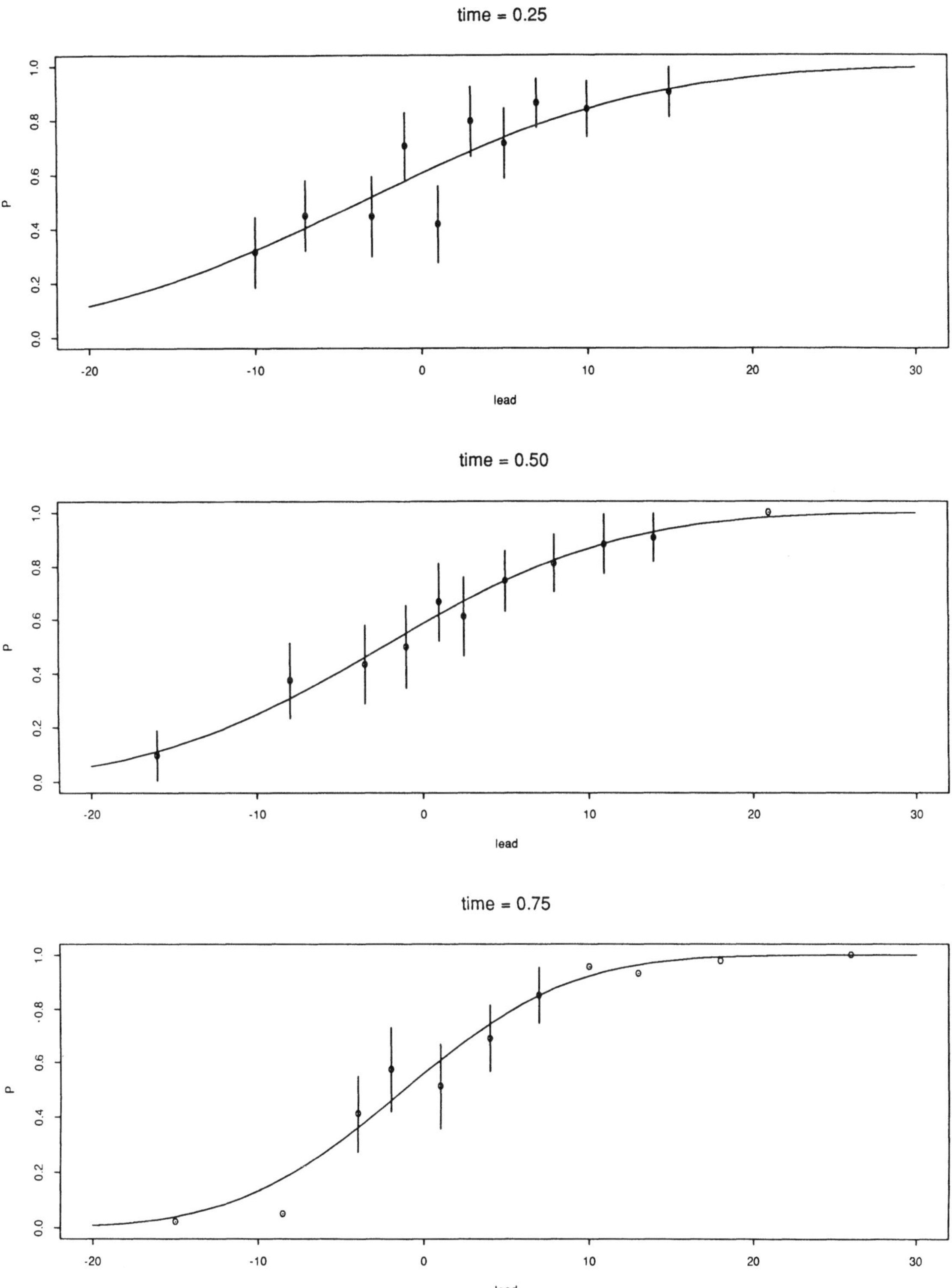

Figure 2. Smooth Curves Showing Estimates of the Probability of Winning a Professional Basketball Game, $P_{\hat{\mu},\hat{\sigma}}(\ell, t)$, as a Function of the Lead ℓ under the Brownian Motion Model. The top plot is $t = .25$, the middle plot is $t = .50$; and the bottom plot is $t = .75$. Circles $\pm$ two binomial standard errors are plotted indicating the empirical probability. The horizontal coordinate of each circle is the median of the leads for the games included in the calculations for the circle.

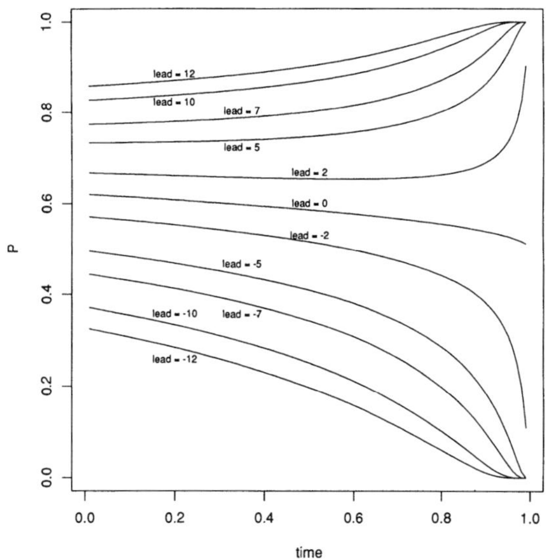

Figure 3. Estimated Probability of Winning a Professional Basketball Game, $P_{\hat{\mu},\hat{\sigma}}(\ell, t)$, as a Function of Time t for Leads of Different Sizes.

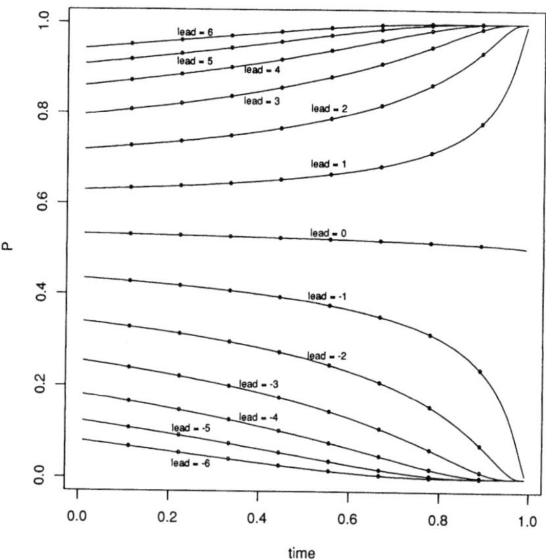

Figure 5. Estimated Probability of Winning a Baseball Game, $P_{\hat{\mu},\hat{\sigma}}(\ell, t)$, as a Function of Time t for Leads of Different Sizes.

Clearly, the Brownian motion model is not tailored to baseball as an application, although one might still consider whether it yields realistic predictions of the probability of winning given the lead and the inning. Lindsey (1961, 1963, 1977) reported a number of summary statistics, not repeated here, concerning the distribution of runs in each inning. The innings do not appear to be identically distributed due to the variation in the ability of the players who tend to bat in a particular inning. Nevertheless, we fit the Brownian motion model to estimate the probability that the home team wins given a lead ℓ at time t (here $t \in \{1/9, \ldots, 8/9\}$). The probit regression obtains the point estimates $\hat{\mu} = .34$ and $\hat{\sigma} = 4.04$. This mean and standard deviation are in good agreement with the mean and standard deviation of the margin of victory for the home team in the data. The asymptotic standard errors for $\hat{\mu}$ and $\hat{\sigma}$ obtained via the delta method are .09 and .10. As in the basketball example, these standard errors are optimistic, because each game is assumed to contribute eight independent observations to the probit regression likelihood, when the eight observations from a single game share the same outcome. Simulations suggest that the standard error of $\hat{\mu}$ is approximately .21 and the standard error of $\hat{\sigma}$ is approximately .18. The likelihood ratio test statistic, comparing the probit model likelihood to the saturated model, is 123.7 with 170 degrees of freedom. The continuity correction again has only a small effect.

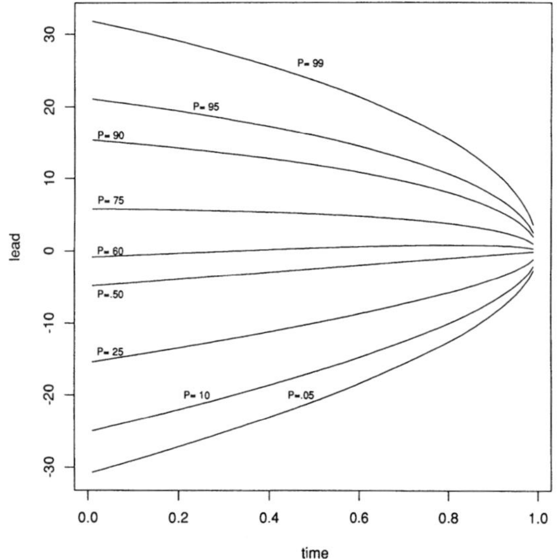

Figure 4. Contour Plot Showing Combinations of Home Team Lead and Fraction of the Game Completed for Which the Probability of the Home Team Winning is Constant for Professional Basketball Data.

Figure 5 shows the probability of winning in baseball as a function of time for leads of different sizes; circles are plotted at the time points corresponding to the end of each inning,

Table 3. $P_{\hat{\mu},\hat{\sigma}}(\ell, t)$ for Baseball Compared to Lindsey's Results

t	ℓ	$\hat{\mu} = .34$, $\hat{\sigma} = 4.04$	$\hat{\mu} = .0$, $\hat{\sigma} = 4.02$	Lindsey
3/9	0	.53	.50	.50
3/9	1	.65	.62	.63
3/9	2	.75	.73	.74
3/9	3	.84	.82	.83
3/9	4	.90	.89	.89
5/9	0	.52	.50	.50
5/9	1	.67	.65	.67
5/9	2	.79	.77	.79
5/9	3	.88	.87	.88
5/9	4	.94	.93	.93
7/9	0	.52	.50	.50
7/9	1	.71	.70	.76
7/9	2	.86	.85	.88
7/9	3	.95	.94	.94
7/9	4	.98	.98	.97

$t \in \{1/9, \ldots, 8/9\}$. Despite the continuous curves in Figure 5, it is not possible to speak of the probability that the home team wins at times other than those indicated by the circles, because of the discrete nature of baseball time. We can compare the Brownian motion model results with those of Lindsey (1977). Lindsey's calculations were based on a Markov model of baseball with transition probabilities estimated from a large pool of data collected during the late 1950s. He essentially assumed that $\mu = 0$. Table 3 gives a sample of Lindsey's results along with the probabilities obtained under the Brownian motion model with $\mu = 0$ ($\hat{\sigma} = 4.02$ in this case) and the probabilities obtained under the Brownian motion model with μ unconstrained. The agreement is fairly good. The inadequacy of the Brownian motion model is most apparent in late game situations with small leads. The Brownian motion model does not address the difficulty of scoring runs in baseball, because it assumes that scores are continuous. Surprisingly, the continuity correction does not help. We should note that any possible model failure is confounded with changes in the nature of baseball scores between the late 1950s (when Lindsey's data were collected) and today. The results in Table 3 are somewhat encouraging for more widespread use of the Brownian motion model.

[Received May 1993. Revised July 1993.]

REFERENCES

Cooper, H., DeNeve, K. M., and Mosteller, F. (1992), "Predicting Professional Game Outcomes From Intermediate Game Scores," *CHANCE*, 5, 3-4, 18–22.

Lindsey, G. R. (1961), "The Progress of the Score During a Baseball Game," *Journal of the American Statistical Association*, 56, 703–728.

——— (1963), "An Investigation of Strategies in Baseball," *Operations Research*, 11, 477–501.

——— (1977), "A Scientific Approach to Strategy in Baseball," in *Optimal Strategies in Sports*, eds. R. E. Machol and S. P. Ladany, Amsterdam: North-Holland, pp. 1–30.

Westfall, P. H. (1990), "Graphical Representation of a Basketball Game," *The American Statistician*, 44, 305–307.

Part VI
Statistics in Miscellaneous Sports

Chapter 35

Introduction to the Miscellaneous Sports Articles

Donald Guthrie

35.1 Introduction

Statistical studies of sports seem to fall into three categories: analysis of outcomes, analysis of rules and strategies, and analysis of extent of participation. The nine chapters of this section fall into the first two categories; analyses of participation are rare in statistical literature.

Americans, and not just the statistically inclined, seem fascinated with the collection of data about our sports. Every day the sports pages of newspapers are full of raw and summary data about various sports. Consider baseball as a model: elaborate data collection methods have been developed from which we have derived summary statistics, and these statistics have been used to understand and interpret the games. The rules of baseball have remained substantially unchanged for over a century, and data collection on the standard scorecard has left a historical archive of the game. Football, basketball, and hockey have been more subject to rule changes and less amenable to longitudinal measurement, but fans' seemingly relentless need for detailed analysis has led to instrumentation of data as well. Each sport has its repertoire of summary statistics familiar to its fans.

But the collection and interpretation of data is not necessarily limited to the major sports. In the following, we reprint several contributions from sports that occupy less newspaper space but are every bit as important to the participants. Temptation is great to organize them according to the particular game, but that would not reflect the true clustering along categorical lines.

With the possible exception of golf, the sports analyzed in these chapters do not have strong followings in the United States. Worldwide, however, football (soccer to Americans) is certainly at the forefront. Numerical analysis of football has perhaps suffered from the absence of American zeal for detail. Indeed, with the increasing popularity of soccer in the United States and increasing worldwide television coverage, we have seen more and more ad hoc statistics associated with soccer.

Instead of the popularity of the games discussed, the papers in the following chapters have been chosen to reflect their application of statistical reasoning and for their innovative use in understanding the sport discussed. Participation in these games ranges from extensive (golf, tennis) to limited (figure skating, darts). All of the chapters, however, offer insight into the games and into the corresponding statistical methodology. In all chapters, the authors are proposing a model for the conduct of the competition, applying that model, and interpreting it in the context of the sport.

The chapters are organized into two groups, the first dealing with rules and scoring and the second concerned with results.

35.2 Rules and Strategy of Sports

Chapters 36, 38, 39, 41, and 44 of this section focus on rules and strategy. In Chapter 39, "Rating Skating," Bassett and Persky give an elegant description of the scoring rules for high-level ice skating competition. These authors formalize two requirements for a function determining placing, and they show that the median rank rule provides the proper placing given their two requirements. Finally, they argue convincingly that the current system for rating ice skating correctly captures the spirit of majority rule, and that the method effectively controls for measurement error. They conclude by commending skating officials for having arrived at a reasonable system, even though there is no evidence that it was determined with these or similar criteria in mind.

In Chapter 36, Stern and Wilcox, in the *Chance* article "Shooting Darts," explore alternative strategies for skillful darts competitors. Using some empirical evidence (presumably collected in an unnamed British pub), they model the error distribution of the dart locations using Weibull-based methods and then use simulation to identify optimal target areas. The results of this chapter suggest aiming strategies based on the accuracy of the shooter.

In Chapter 41, "Down to Ten: Estimating the Effect of a Red Card in Soccer," Ridder, Cramer, and Hopstaken estimate the effect of player disqualification (receiving a "red card") in soccer/football. They gather data from 140 games in a Dutch professional league, develop a model for the progress of the score in a match that adjusts for possible differences in team strength, and estimate the likelihood that a team losing a player eventually loses the game. The interesting result is that a red card, especially early but even later in a match, is devastating. They then use their scoring model to estimate the effect of a penalty period rather than disqualification. They conclude that a 15-minute penalty would have an effect similar to that of disqualification with 15 minutes remaining in the match, but the effect on the game outcome would be the same regardless of when the penalty occurred.

Scheid and Calvin, both consultants to the United States Golf Association (USGA), contribute Chapter 38, "Adjusting Golf Handicaps for the Difficulty of the Course." Handicaps are designed for matching players of unequal skills in order to provide level competition. Handicaps are widely used by golfers below the highest level, and are thus a vital element of the popularity of the game. Ordinarily, a player will have a rating (his/her handicap) determined at the course he/she plays most frequently; this rating may or may not be appropriate on another course. This chapter performs the practical service of explaining clearly the history of the USGA method of rating courses and suggesting corrections in handicaps. It describes the history of the Slope System and its application to golf courses. Every golfing statistician—and there are many—should benefit from this information.

In Chapter 44, Wainer and De Veaux, both competitive swimmers, argue in "Resizing Triathlons for Fairness" that the classical triathlon is unfair. These authors give statistical evidence that the swimming portion of the competition is unduly less challenging than the running and cycling segments. They suggest a method of determining segment lengths that would provide fairer competition. Although one may not accept all of their arguments (e.g., basing marathon standards on a record set at Rotterdam, the flattest course known), they make a good case for adjustment of distances, but they only casually mention energy expenditure as a standard. Like darts and golf, racing is a popular form of competition, and a triathlon combines several types of racing (running, biking, swimming) requiring several talents. The paper has stimulated and will continue to stimulate vigorous discussions among competitors.

35.3 Analysis of Sports Results

Chapters 37, 40, 42, and 43 of this section evaluate the results of competition between players and teams. Although head-to-head competition is a standard method, many observers are not convinced that the winner of this type of competition is actually the best team or player. For example, one major league baseball team may win the largest number of games and be declared the "best team," even though this team may lose a large fraction of the face-to-face encounters with particular opponents. Chapters 40 and 43 in this section address the issue of team comparison in the presence of direct competition, but with consideration given to play against common opponents or in different settings. While these chapters look at soccer/football and sprint foot racing, it seems likely that similar analyses could be applied to other sports. Occasional upsets, thankfully, add interest to competition; both individual sports (e.g., tennis, golf, running) and team sports bring upsets. The purpose of these statistical analyses is to add credibility to declaring the overall leader.

In Chapter 40, "Modeling Scores in the Premier League: Is Manchester United *Really* the Best?," Lee uses Poisson regression to model team scores in the 1995/96 Premier League soccer league. As a general opinion, Manchester United has been regarded as the soccer equivalent of the New York Yankees—they do not win all of their games, but are consistently at the top of the standings. One thousand replications of that season are simulated using the fitted Poisson model, and 38% show United to be the league winner, although their number of standings points in the actual season (82) is somewhat greater than the average simulated points (75.5). The simulations, however, justify the disappointment of Liverpool fans since their team actually scored only 71 points, but the simulations suggested that the team should have scored 74.9, nearly equal to United; Liverpool "won" 33% of the simulated seasons. Football as played in the Premier League has lower scores than are typical in Major League Baseball, yet the methods used in this analysis could possibly be applied to baseball or ice hockey. Perhaps the Boston Red Sox could be vindicated for their difficulties with the Yankees!

In Chapter 43 Tibshirani addresses the question of "Who Is the Fastest Man in the World?" He tries to compare the brilliant sprinters Michael Johnson (USA) and Donovan Bailey (Canada) based on the 1996 Olympic outcomes at 100 and 200 meters. After building and applying several models to account for the fact that the two did not compete against one another, he concludes by presenting arguments favoring each runner, and thus leaves the question unanswered. This discussion, however, reflects one good aspect of statistical practice. That is, it is useful to reformulate questions in a more accessible way and in the context of the problem at hand. There is little doubt that Bailey attained the faster maximum speed (he needed to run only 100 meters), but Johnson maintained his speed over a longer time (he ran only the 200 meter race).

Berry's analysis of data provided by the Professional Golf Association (PGA) in Chapter 37, "Drive for Show and Putt for Dough," opens some interesting issues. He addresses the relative value of longer and shorter shots in the accumulation of the scores of high-level professional golfers. His analysis is very thoughtfully conducted given that data collected by the PGA are rather tangential to his objective, and indeed to an understanding of the game. Nevertheless, he manages to give support to the slogan in his title and to quantify these components. His graphical summary vividly illustrates the compromise between driving distance and accuracy and the effects of each of these qualities on the final score. Finally, he gently encourages the collection of more specific data on the golf tour. Let's imagine the fun we could have in data analysis if each golf shot were monitored by a global positioning system (GPS) device to set its exact position.

Jackson and Mosurski continue the "hot hand" debate in Chapter 42, "Heavy Defeats in Tennis: Psychological Momentum or Random Effect?" These authors look at data from a large number of sets played at the U.S. Open and Wimbledon tennis tournaments. They question whether the heavy defeats observed in these matches are due to a dependence structure in the data, represented by a psychological momentum (PM) model, or due to random variation in players' abilities from match to match, represented by a random effects model. In their analysis, the PM model provides a better fit to these data than the random effects model. However, the PM is not as successful in explaining the variation in the results in some of the titanic rivalries in professional tennis.

Chapter 36

A STATISTICIAN READS THE SPORTS PAGES

Hal S. Stern,
Column Editor

Shooting Darts

Hal S. Stern and Wade Wilcox

The major team sports (baseball, basketball, football, hockey) receive most of the attention from the popular media and have been the subject of much statistical research. By contrast, the game of darts is rarely covered by television or written about on the sports pages of the local newspaper. A project carried out while one of the authors (Wilcox) was an undergraduate demonstrates that even the friendly, neighborhood game of darts can be used to demonstrate the benefits of measuring and studying variability. Of course, darts is more than just a friendly, neighborhood game; there are also professional players, corporate sponsors, leagues, and tournaments. An effort to model dart throwing was motivated by a series of observations of computerized dart games—when the game was running on "automatic" its performance was too good. The computer occasionally missed its target but did not produce the occasional sequences of mistakes that haunt even top players. We wondered how one would build a realistic model for dart playing and how such a model could be used.

A Brief Introduction to Darts

We begin by describing the dart board in some detail and then briefly describing how dart games are played. A picture of a dart board is presented in Fig. 1. Darts are thrown at the board and points are awarded for each dart based on where the dart lands. The small circle in the center corresponds to the double bullseye (denoted DB), which is worth 50 points, and the ring directly surrounding this is the single bullseye (SB) worth 25 points. Beyond those areas, the circular board is divided into 20 equal-sized arcs with basic point values given by the numbers printed on the arcs (which run from 1 to 20). There are two rings that pass through each of the 20 arcs, one ring in the middle of each arc, and one ring at the outside edge of each arc. A dart landing in the middle ring receives triple value—that is, three times the number of points indicated on the relevant arc. A dart landing in the outer ring receives double value—that is, twice the number of points indicated on the relevant arc. We refer to the point values associated with different regions by referring to the arc and the multiple of points that are awarded so that T18 refers to triple value in the arc labeled 18 (54 points), D18 refers to double value in the arc labeled 18 (36 points), and S18 refers to single value in the arc labeled 18 (18 points). It follows that the most points that can be earned with a single dart is 60 by hitting the triple value ring of the arc valued at 20 points—that is, T20.

A fundamental aspect of the most popular dart games (called 301 and 501) is that players try to accumulate points as fast as possible (some other games place a greater premium on accuracy). The fastest way to accumulate points is to continually hit the triple 20 (T20), which is worth 60 points. Note that this is considerably more valuable than the double bullseye (50 points). In fact, the next most valuable target is T19. A closer examination of the board shows that the arcs next to the arc labeled 20 are worth only 1 and 5 points, but the arcs that surround the arc labeled 19 are worth 3 and 7 points. Is it possible that a player would accumulate more points in the long run by aiming for T19 rather than T20? Of course, the answer may well depend on the skill level of the player. An additional aspect of these games puts a premium on accuracy; for example, the game 301 requires a player to accumulate exactly 301 points but the player must begin and end by placing a dart in the double-value ring. Are some strategies for playing this game more effective than others? A model that takes into account the skill level of a player and produces realistic throws would allow us to answer questions like those posed here.

Column Editor: Hal S. Stern, Department of Statistics, Iowa State University of Science and Technology, Snedecor Hall, Ames, Iowa 50011-1210, USA; hstern@iastate.edu.

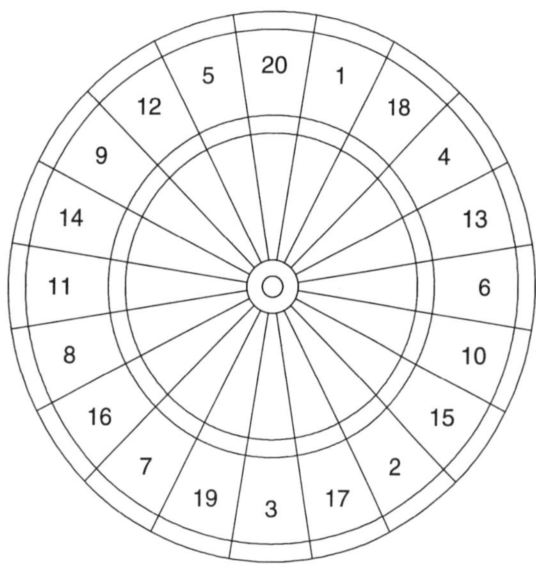

Figure 1. Picture of the point-scoring area of a dart board. The innermost circle is the double bullseye worth 50 points; this is surrounded by the single-bullseye area worth 25 points. The remainder of board is split into 20 arcs with basic point values indicated on each arc. The small area on the outermost ring of each arc has double point value and the area in the intermediate ring of each arc has triple point value.

How to Model Dart Throws

A Sherlock Holmes quote from *The Adventure of the Copper Beeches* seems relevant here: "'Data! data! data!' he cried impatiently. 'I can't make bricks without clay.'" We will have a hard time developing or testing a probability model for dart throws without some data. Four different points on the dart board were chosen (the center of the regions corresponding to T20, T19, D16, and DB) and approximately 150 throws made at each of the targets. Naturally not every throw hits the target point or even the target region. The darts distribute themselves in a scatter around the target point. The distance from the target was measured for each dart (to the nearest millimeter) and the angle at which the dart landed in relation to the target was also recorded. There were no obvious differences among the four sets of throws (corresponding to the four different targets), so the data have been combined—in all, 590 throws were made. The error distribution for the 590 throws is provided in Fig. 2. Notice that 57% of the darts are 25 mm or less from the target (this is about one inch) and 76% of the darts are 33 mm or less from the target. The angles appeared consistent with a uniform distribution; no direction appeared to occur more frequently than any other. The only formal statistical test that was performed found that the proportion of throws in each of the four quadrants surrounding the target point did not differ significantly from what would be expected under a uniform distribution for the angle of error.

The error distribution of Fig. 2 clearly does not follow a normal distribution. This is to be expected because the errors as measured are restricted to be positive and are not likely to follow a symmetric distribution. An alternative model immediately suggests itself. Suppose that horizontal and vertical throwing errors are independent of each other with both having normal distributions with mean 0 and variance σ^2. The radial distance R that we have measured is the square root of the sum of the squared horizontal and vertical errors. Under the assumptions that we have made, the radial error would follow a Weibull distribution (with shape parameter two) with probability density function

$$\frac{r}{\sigma^2} e^{-r/2\sigma^2}$$

for $r > 0$. The Weibull distribution is a flexible two-parameter distribution (the shape parameter and a scale parameter) that is used often in research related to the reliability of products or machine parts. It should be pointed out that, as often occurs, the continuous Weibull model is used even though the data are only recorded to the nearest millimeter.

The scale parameter of the Weibull distribution, σ is a measure of player accuracy. (Technically the scale parameter for the Weibull distribution that we have set up is $\sqrt{2}\sigma$, but we find it more convenient to focus our attention on σ the vertical/horizontal standard deviation.) The value of σ will differ from player to player with smaller values of σ indicating better (more accurate) throwers. There are several statistical techniques that can be used to estimate the parameter σ for a given dataset. The most natural estimate is based on the fact that the radial error squared, R^2, should be near on average. Therefore we can estimate σ by taking the square root of one-half of the average squared radial error. The esti-

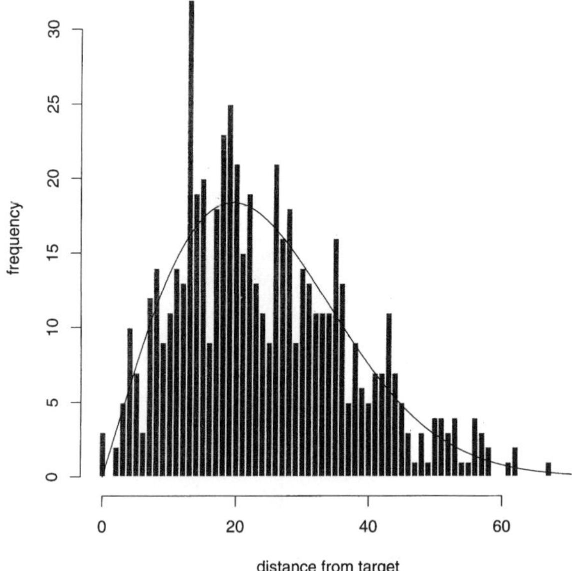

Figure 2. Histogram indicating the distribution of the distance from the target point. The total number of tosses is 590 spread over four different targets. The superimposed curve is the Weibull distribution with $\sigma = 19.5$ (the formula for the probability density function is provided in the text).

Table 1—Actual and Simulated Distribution of Outcomes From Dart Tosses Aimed at Triple-Value Twenty (T20)

Outcome	Actual %	Simulated %
T20	12.0	10.9
S20	52.0	46.7
T5	2.7	3.6
S5	8.0	15.4
T1	6.7	4.1
S1	17.3	16.1
Other	1.3	3.2

Note: The actual distribution is based on 150 throws the simulated distribution is based on 10,000 simulated throws.

mate of σ for the data in Fig. 2 is 19.5 with approximate error bounds ±.8. The Weibull distribution with s set equal to this value is superimposed on the histogram in Fig. 2. We have intentionally used a display with very narrow bins to preserve the information about the data values.

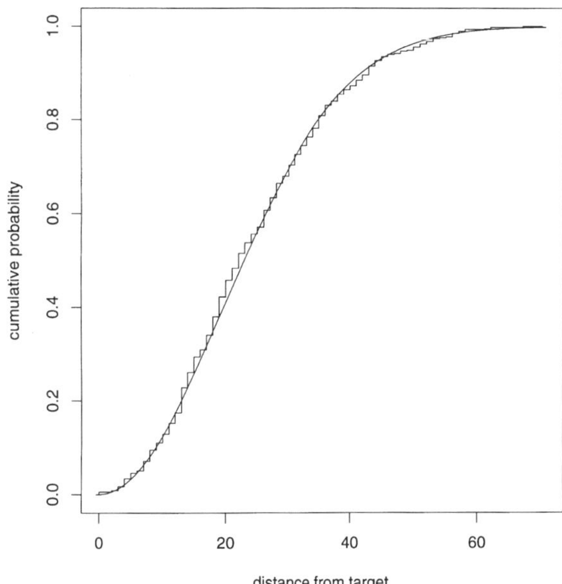

Figure 3. Cumulative distribution function for the observed error distribution (step function) and the Weibull distribution with σ = 19.5 (smooth curve).

Does the Model Fit?

The fit of the Weibull model to the errors observed in throwing darts can be assessed in several ways. The histogram of Fig. 2 provides some evidence that the model fits. A histogram with wider bins (not shown) provides a smoother picture of the error distribution and the Weibull curve provides a closer match in that case. Figure 3 shows the information from the sample in terms of the cumulative distribution function that measures the proportion of throws with errors less than or equal to a given value. The smooth curve is the Weibull model and the step function that accompanies it is based on the data. As an additional check on the fit of the model, three other dart players agreed to throw 600 darts at the same targets. We have created versions of Figs. 2 and 3 for these players (not shown), and the resulting data appear to be consistent with the Weibull model. Two professional players (these are players who are basically supported by sponsors and prize money) had standard deviations of 13.4 and 13.2. Errors bigger than 40 mm (about 1.5 inches) occur in only 1 of 100 tosses for such players.

As a final check on the fit of the model, we wrote a computer program to simulate dart throwing. The simulation program can be used to simulate throws at any target point (e.g., the target point T20 represents the central point in the small arc associated with triple value and 20 points). A radial error is generated randomly from the Weibull distribution and then a random angle is generated uniform over the circle. A simulated outcome is obtained by computing the number of points that correspond to the computer-generated position. Table 1 gives the frequency of the most relevant outcomes for 150 actual tosses and 10,000 simulated tosses at T20. It should be pointed out that the exact probabilities of each outcome under the Weibull model could be evaluated exactly using numerical integration. Simulation provides a relatively fast and accurate approximation in this case. In general the agreement between the actual and simulated outcomes is quite good, an overall chi-squared goodness-of-fit test indicates that the actual tosses are not significantly different than that expected under the Weibull model (treating the simulated values as if they were exact). The largest discrepancy that we observe is that the number of S5 outcomes observed is smaller than expected under the Weibull model. There is no obvious explanation for this discrepancy based on reviewing the location of the actual tosses.

Table 2—Expected Number of Points per Dart When Aiming at T19 and T20 Based on 10,000 Simulated Dart Throws (Standard Errors are in the Range .16–.19)

σ	Pts per dart if aimed at T19	Pts per dart if aimed at T20
20.0	18.0	17.6
19.5	18.2	17.9
19.0	18.5	18.2
18.5	18.6	18.3
18.0	18.9	18.7
17.5	19.4	19.3
17.0	20.2	20.2
16.5	20.3	20.4
16.0	20.9	21.0
15.5	21.4	21.5
15.0	22.0	22.2

Applying the Model

If we accept the model then it is possible to begin addressing strategy issues. What is the fastest way to accumulate points? The highest point value on the board is T20 which is worth 60 points. This target point is surrounded by the arc worth five points and the arc worth one point, however, but T19 (worth 57 points) is surrounded by arcs worth seven and three points. An extremely accurate thrower will accumulate points fastest by aiming at T20. Is there an accuracy level (as measured by s) for which it is better to aim for the T19? Table 2 indicates that there is such a point! Players with high standard deviations ($s \geq 17$) appear to earn more points per dart by aiming at T19, but players with lower standard deviations (more accurate throwers) are better off aiming for T20. The difference between the two strategies is relatively small over the range given in Table 2, the largest difference in means is .34 per dart, which translates into only about 8 points total over the approximately 25 throws that would be required to accumulate 500 points. Even this small amount might be important in a game between evenly matched players.

Of course, to apply Table 2 one needs to know his or her own standard deviation. This can be obtained directly by measuring the error distance for a number of throws at a single target as we have done here. A somewhat easier approach to estimating one's standard deviation would be to throw a number of darts at T20, compute the average score per dart, and use Table 2 to find the corresponding value of σ. For example, an average score of 19.5 (e.g., 1,950 points in 100 darts) would suggest that $\sigma \approx 17.5$ or perhaps a bit lower.

It is possible to apply our approach to more detailed questions of strategy, but we have not yet pursued that area. For example, suppose that a player in the game 301 has 80 points remaining (recall the last throw must be a double). There are several possible strategies; for example, T16, D16 (i.e., triple 16 followed by double 16) requires only two darts, whereas S20, S20, D20 requires three darts. Which is a better choice? The answer may well depend on the accuracy of a player because aiming for S20 has a much greater probability of success then any double or triple.

As always, refinements of the model are possible: We have ignored the possibility of systematic bias by assuming a uniform distribution for the angle of error, we have assumed independence of the horizontal and vertical errors, and we have not addressed the issue of darts bouncing off the metal rims that separate areas. Our relatively simple, one-parameter model appears to perform quite well in describing the variability observed in dart throws and provides useful information for determining strategy. In particular, it appears that aiming for the obvious target, T20, is not the best point-accumulating strategy for some players.

Reference and Further Reading

Townsend, M. S. (1984), *Mathematics in Sport*, Chichester, UK: Wiley.

A STATISTICIAN READS THE SPORTS PAGES

Scott M. Berry, Column Editor

Chapter 37

Drive for Show and Putt for Dough

I hit a beautiful tee shot, 250 yards right down the middle of the fairway. I follow that with a well-struck 5-iron to the green of the difficult 420-yard par 4. I have 20 feet for a birdie. I hit a firm putt that rolls by the hole, and I am left with a four-footer coming back. I push the par putt and tap in for my bogey. It took two shots to travel 420 yards and then three putts to go the remaining 20 feet. All the while my playing partner sliced his drive, mishit a 3-wood to the right rough, then chipped it 12 feet from the hole and made the par-saving putt. If he is kind he says nothing, but usually he utters the well-known cliché, "You drive for show and putt for dough!"

How accurate is this well-known cliché? In this article I investigate the different attributes in golf and their importance to scoring. Many different types of shots are hit by golfers. Is it the awesome length of players or the deft touch around the greens that differentiate golfers in their ability to score well? I study the very best golfers--those on the United States Professional Golfer's Association Tour (PGA Tour). Clearly there are many differences between beginners, 10-handicappers, and professionals. I want to study what differentiates the very best golfers in the world.

Terminology and Available Data

There are par 3, par 4, and par 5 holes on a golf course. Par for a hole is the number of shots it should take, if the hole is played well, to get the ball in the hole. Par is determined by the number of shots it should take to get on the green and then two putts to get the ball in the hole. Therefore, on a par 4, the most common par for a hole, it should reasonably take two shots to get on to the green and then two

Column Editor: Scott M. Berry, Department of Statistics, Texas A&M University, 410B Blocker Building, College Station, TX 77843–3143, USA; E-mail *berry@stat.tamu.edu.*

Photo by Dave Thompkins, © 1994

putts for a par. If a golfer hits his ball on the green in two shots less than par, this is referred to as hitting the green in regulation. The first shot on a hole is called the tee shot. For a par 4 or par 5, the tee shot is called a drive—the goal of which is to hit the ball far and straight. For each par 4 and par 5 there is a fairway from the tee-off position to the green. The fairway has short grass that provides an optimal position for the next shot. On either side of the fairway can be rough, water, sand traps, and various other hazards. These hazards make it more difficult for the next shot or add penalty strokes to your score. For a par 3, the green is close enough that the golfer attempts to hit his tee shot on the green. This is usually done with an iron, which provides precision in accuracy and distance.

I decompose a golfer's skill into six different attributes. These are driving distance, driving accuracy, iron-shot skill, chipping ability, sand-trap skill, and putting ability. There are other skills for a golfer, but these six are the most important. A goal of this article is to find the importance of each

Is it the awesome length of players or the deft touch around the greens that differentiate golfers in their ability to score well?

of these attributes for professional golfers. The PGA Tour keeps various statistics for each golfer, and these will be used to measure each of the preceding attributes.

For each round played (this consists of an 18-hole score), two holes are selected that are flat and are situated in opposite directions. Holes are selected where players generally hit drivers. The hope is that by measuring the distance traveled for tee shots on these holes the wind will be balanced out and the slope of the terrain will not affect the results. The distance of the tee shots for each of the golfers on these holes is measured, and the average of these drives is reported (I label it distance). For each of the par 4 and par 5 holes the PGA Tour records whether the golfer hit his tee shot in the fairway or not. This variable, the percentage of times the golfer hits the fairway with his tee shot, is a good measure of driving accuracy (labeled accuracy). The PGA Tour has recently changed the way it rates putting ability. It used to keep track of how many putts each golfer had per round. This is a poor measure of putting because there may be very different attempts by each player. One player may hit every green in regulation and therefore his first putt is usually a long one. Another player may miss many greens in regulation and then chip the ball close to the hole. This leaves many short putts. Therefore, the PGA Tour records the average number of putts per hole when the green is hit in regulation. There can still be variation in the length of the putts, but it is a much better measure of putting ability (labeled putts).

Measuring chipping ability is not as straightforward. The PGA Tour records the percentage of times that a player makes par or better when the green is not hit in regulation. The majority of the time this means that the ball is around the green and the player must attempt to chip it close to the hole. If the player sinks the next putt, then a par is made. This statistic, *scramble, is* a measure of a player's chipping ability but also of the player's putting ability (I delete all attempts from a sand trap; these will be used to measure sand ability). The player may be a bad chipper but a great putter; thus he is successful at scrambling. To measure the chipping, separated from the putting ability, I create a new statistic, labeled chipping. I ran a simple linear regression, estimating scramble from putts (technically, scramble is a percent and a logistic regression would be more reasonable, but the values for *scramble* are between 40%–70% and the linear regression does just as well):

$$\text{scramble} = 249.3 - 106.0 \text{ putts}$$

The goal of this method is to estimate how much of the scramble statistic is explained by their putting ability. For each player, the aspect of scramble that cannot be explained by putting is his chipping ability, which is represented by the residuals from this regression. These residuals are used to measure chipping (and labeled chipping). In Fig. 1, scramble is plotted against putts. The regression line is also plotted. As this figure shows, some of the variability in scramble is because of putting ability. The residuals from this regression, the distance between the points and the line, is the part of scramble that is not explained by putting. This residual is what is used as the measure of chipping ability. As can be seen in Fig. 1, Brad Faxon is a mediocre player for scramble, but because of his great putting ability he should be a much better scrambler. Therefore, his residual is negative and large in magnitude, showing poor chipping ability. Robert Allenby has a pretty good scramble value, which is that much more amazing because he is not a great putter. Therefore, his residual is quite large, and positive, showing great chipping skill.

The same problem exists for the sand saving ability of a player. The PGA Tour records the percentage of times that a player saves par when he hits his ball in a green side sand trap. Again putting ability helps to save par. The residuals from the following regression are used as a measure of sand play (and labeled sand):

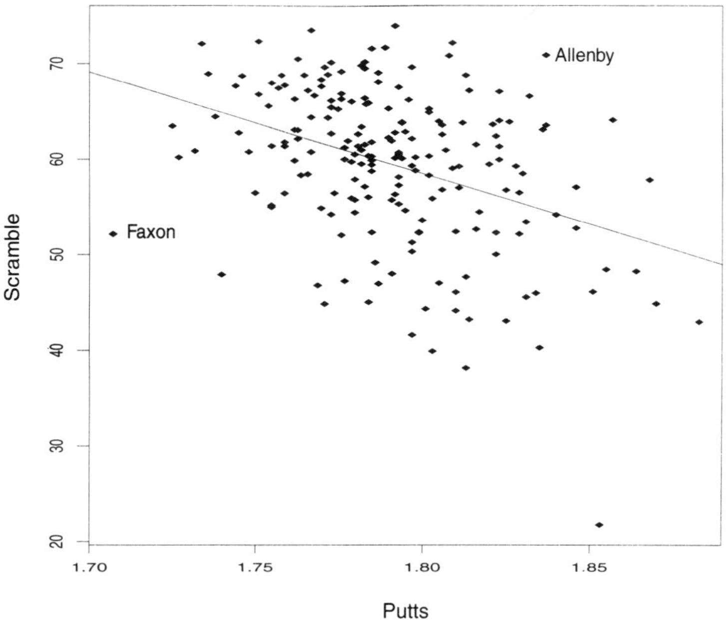

Figure 1. A scatterplot of scramble against putts. The regression line is shown. The residuals from this regression line are used to measure chipping ability.

sand save = 190.3-78.4 putts

The PGA Tour does not keep any good statistics to measure iron-play ability. What would be ideal is to know how close a player hits the ball to the hole when he has a reasonable chance. For example, in the U.S. Open Championship, the sponsoring organization, the United States Golf Association, kept track of the percentage of greens hit in regulation when the fairway was hit. The PGA Tour does not keep track of the same statistic. The PGA Tour does keep track of the performance for each player on par 3 holes. The relation to par for each player on all the par 3's is reported. Virtually every course has four par 3 holes. Therefore, the average score on each par 3 can be calculated. Each player has an ideal starting position, similar to a second shot on a par 4 when the player has hit a fairway. This score is also dependent on the putting, chipping, and sand play of the players. A player may be a bad iron player yet score well because he chips and putts well. Therefore, as with the chipping and sand variables, I regress the average score on the par 3 holes on the putts, chipping, and sand variables:

par 3 = 1.89 + 0.67 putts −.0010 sand −.0020 chipping

The residuals from this regression are used as the measure of iron play (labeled irons). A positive value of chipping is good, a positive value of sand is good, and a negative value of irons is good.

I downloaded each of the preceding statistics for the 1999 season from the PGA Tour Web site, *pgatour.com*. This was done after the first 28 tournaments of the 1999 season—the Mercedes Championship through the Greater Milwaukee Open. Each of the 195 players that played at least 25 rounds during this time are included in the dataset. The scoring averages for each of these 195 players are also recorded.

The main goal of this article is to show the importance to scoring of each of the attributes—not necessarily a judgment of the players. This is important because the players do not all play the same tournaments, and thus the courses have varying difficulty. Although this may affect comparisons across players it does not create as many difficulties in evaluating the effect of the attributes. Courses are easier or harder because the fairways are smaller, the greens more difficult, or the course longer. Therefore, if a player plays an "easy" course, he will hit more fairways, make more putts, and so forth, but his scores will be lower. That golfer may not be as good as another who has a higher scoring average but plays harder courses, but the effect of the attributes will still hold. The PGA has a scoring average measure adjusted by the difficulty of each round, but I do not use this measure for the reasons just described. These adjusted scores do not vary much from the unadjusted scoring average—thus the fact that different courses are played should have minimal effect on this analysis.

Descriptive Analysis and the Model

Tables 1 and 2 present some summary numbers for the population of players for each of the variables. Figure 2 presents the scatterplot for each of the attributes and scoring. Seemingly there is not a large variation in putting across players. The standard deviation is .03 putts per green in regulation. If a player hits 10 greens per round, this is a difference of .3 shots per round. Although this does not seem that large, for this group of top-notch players putting is the most highly correlated attribute with scoring. This strong pattern is also demonstrated in the scatterplot. Sand is the least correlated with scoring. This is most likely because hitting sand-trap shots is not very common. Interestingly, distance has the second smallest correlation with scoring, though the direction of the correlation is intuitive--the longer the drive the smaller the average scores. Driving distance and accuracy have the highest absolute correlation among the attributes. This correlation is negative, as would be expected; the farther the ball travels, the less accuracy there will be. Irons is uncorrelated with sand, chipping, and putts because of its construction as the residuals from the regression of par 3 using these three variables. Likewise sand and chipping are uncorrelated with putts.

The linear model with no interactions for modeling scoring from the six attributes has a residual standard deviation of .346 with an R^2 of .829. Each of the attributes is highly significant. I tested each of the two-way interactions. The only significant interaction is between distance and accura-

Table 1—Descriptive Statistics for the Seven Variables

Statistic	Mean	Std. Dev.	Min	Max
Scoring	71.84	.82	69.76	74.69
Distance	272.2	8.12	250	306
Accuracy	67.4%	5.23%	50.7%	80.90%
Putts	1.79	.030	1.71	1.88
Sand	0	6.01	-16.7	14.34
Chipping	0	7.58	-31.17	16.30
Iron	0	.051	-.11	.18

Table 2—Correlation Matrix for the Seven Variables

	Score	Distance	Accuracy	Putts	Sand	Chipping	Iron
Score	1.000	-.194	-.378	.657	-.163	-.366	.298
Distance	-.194	1.000	-.413	-.047	-.091	-.068	.108
Accuracy	-.378	-.413	1.000	-.045	-.012	.265	-.258
Putts	.657	-.047	-.045	1.000	.000	.000	.000
Sand	-.366	-.068	.265	.000	.057	1.000	.000
Chipping	-.366	-.068	.265	.000	.057	1.000	.000
Iron	.298	.108	-.258	.000	.000	.000	1.000

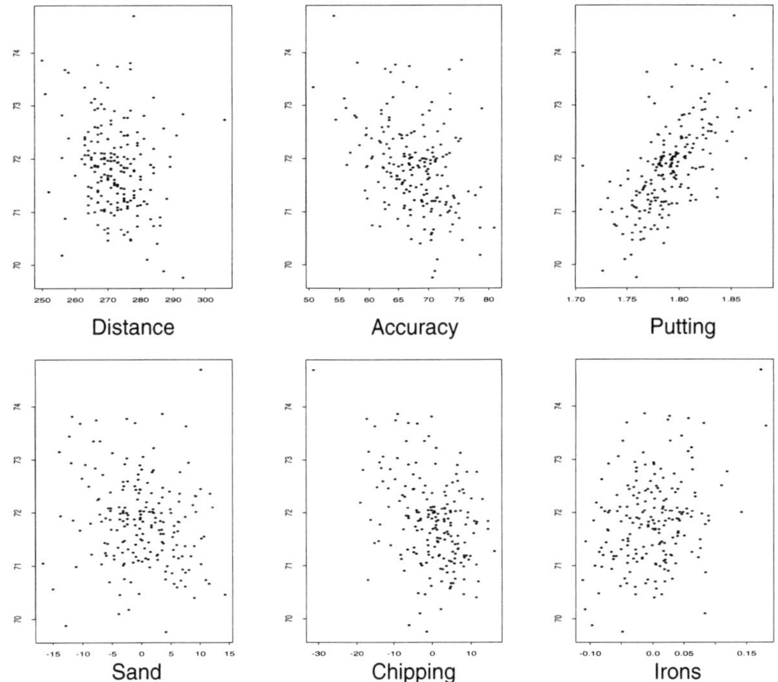

Figure 2. Scatterplots for each of the six attributes plotted against scoring average.

Table 3—Summary of Multiple Regression Model for the Scoring Average Based on the Six Attributes

	Coefficient	Std. error	T value	P value
Intercept	27.5	8.46	3.24	.0014
Distance	.067	.032	2.13	.0347
Accuracy	.373	.128	2.91	.0040
Putts	16.87	.83	20.32	.0000
Sand	-.0257	.0041	-6.35	.0000
Chipping	-.0303	.0033	-9.11	.0000
Iron	3.75	.49	7.65	.0000
Dist*Accu	-.0016	.0005	-3.37	.0009

cy. The resulting model, including this interaction, is summarized in Table 3. Although there is not a huge increase in the model fit over the "no interaction" model, it makes sense that there would be an interaction between distance and accuracy, and thus I use the model with the interaction. The implications of this interaction are discussed in the next section.

David Seawell, one golfer, is an outlier in both scoring and scrambling. He has by far the worst chipping statistic, –31.17 (the second lowest is –19.0), and by far the worst scoring average, 74.69 (the second lowest is 73.86). There was virtually no change in the model when Seawell was removed; therefore, I left him in to fit the model.

To give some notion of importance to these attributes, I present seven different golfers. "Median Mike" has the population median for each of the attributes. Each of the following players are at the median in every category except one—in which they are at the 90th percentile (or the 10th percentile if low values are good). These players are Distance Don, Accurate Al, Putting Pete, Sandy Sandy, Chipping Chip, and Iron Ike. The category in which they are better than the median should be clear!

The estimated scoring average for each of these players is Putting Pete 71.14, Distance Dan 71.29, Accurate Al 71.32, Iron Ike 71.43, Chipping Chip 71.46, Sandy Sandy 71.49, and Median Mike 71.70. As the title cliche claims, you putt for dough! Putting Pete is the best of these players. Thus, if a player is mediocre at everything—except one thing at which he is pretty good—he is best off if that one thing is putting. Distance is not just for show and neither is accuracy. Distance Dan is the second best player, and Accurate Al is the third best. I was surprised that chipping was not more important. This may be because the quality of these players is such that they do not need to chip much. Their ability to save par when they miss a green is also partially explained by their putting ability, which is important.

Total Driving

This analysis clearly shows that putting is very important for PGA Tour players—if not the most important attribute. This does not mean that what separates them from the 20-handicapper is their putting. I think you would find putting relatively unimportant when comparing golfers of a wider class of overall ability. The long game-driving ability and iron play would be very important. There is another interesting question that can be addressed from this model: How can you characterize driving ability? The PGA Tour ranks players in *total driving*. It ranks each of the golfers in distance and accuracy separately and then sums the ranks. I think combining ranks is a poor method of combining categories, but aside from this, it weighs each of them equally. From the model presented here, I can characterize the importance of each. I fix each attribute at the population median value, except distance and accuracy. Figure 3 shows the contours for distance and accuracy in which the estimated scoring average is the same. Two players on the same contour have the same estimated mean score and thus have the same driving ability.

This graph demonstrates the importance of the interaction between distance and accuracy. If there were no interaction in the model, these contours would be linear, with a slope of .66. The linear contours would imply that, regardless of your current driving status, you will improve your score an identical amount if you gain one yard in average driving distance or a .66% increase in fairways hit percentage. Figure 3 shows that this is not true with the interaction present. Compare John Daly (distance of 306 yards) to Fred Funk (accuracy of 80.9%). Daly would need a huge increase in distance to improve, whereas a smaller increase in accuracy would result in a substantial improvement. The opposite is true for Funk; a large increase in accuracy would be needed to improve, but adding distance would result in an appreciable increase. Funk hits the fairway so often that adding yardage would improve virtually every iron shot (making it shorter). When Daly misses the fairway, it limits the second shot, and, with him missing so often, it doesn't matter how much closer he is when he hits the fairways. He does well on those holes. It is surely the holes on which he misses the fairway that are causing troubles.

I think golf, like baseball, is an ideal game for statistics because each event is an isolated discrete event that can easily be characterized.

Most likely, if John Daly tries to improve his accuracy, it will decrease distance, or if Fred Funk tries to increase distance, it will decrease his accuracy. This function will enable them to decide whether the trade-off is beneficial or not. For example, Daly's current estimated scoring average is 72.34 (actual is 72.73). If he could sacrifice 10 yards distance for a 5% increase in accuracy, his estimated scoring average would improve by .30 shots per round. Fred Funk's estimated scoring average is 70.75 (actual is 70.70). If he could sacrifice 5% in accuracy for an extra 10 yards in distance (the opposite of Daly) his estimated scoring average would improve by .25 shots per round. In this example, 10 yards to Daly is worth less than 5% in accuracy, but 5% in accuracy is worth less than 10 yards to Funk.

Tiger Woods (distance 293 and accuracy 70.9%) is the highest-rated driver according to this model. He is rated second in the PGA Tour total driving statistic, behind Hal Sutton (distance 277 and accuracy 75.5%).

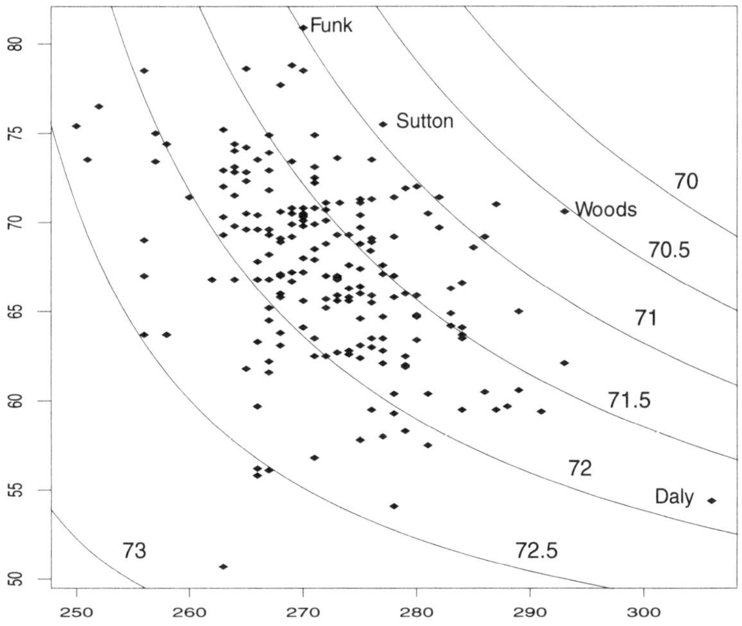

Figure 3. The scatterplot of distance versus accuracy. The lines represent contours of the same estimated scoring average based on these two variables and the player being at the median for putts, sand, chipping and irons.

Discussion

From this model I have concluded that you drive for dough and putt for more dough! A crucial assumption for this conclusion is that you believe that the variables I have used adequately measure the attribute they are designed to measure. I believe they are currently the best available measures. Larkey and Smith (1998) discussed the adequacy of these measures and made suggestions about ways to improve them. A great reference for any statistics in sports topic is *Statistics in Sports* (Bennett 1998). Within this volume, Larkey (1998) gave an overview of statistics research within golf.

I believe the PGA Tour could keep much better records of the performance of players. For example, the National Football League keeps track of every play. It reports the down and distance and the yard line for every play. It reports whether it was a run or a pass, the yards gained and lost, and so forth. Any statistic I want can easily be constructed from this data. The PGA Tour could do the same. A record could be kept on each shot taken—whether the shot was in the fairway, the rough, a trap, and so forth. The resulting position of the ball could be reported—this includes the distance to the hole and whether it is on the green, the fringe, in water, in a sand trap, and so forth. From this dataset it would be easy to construct any statistic of interest. Ultimately I would like to construct a dynamic programming model that could give value to each position of the ball. This would be an amazing tool for golfers and the PGA Tour. It would enable a decision model to be constructed as to whether a player should go for the green on a long par 5: Is the risk worth it to hit over the water or should I lay the ball up short of the water? It would also help in setting up courses for play. I think golf, like baseball, is an ideal game for statistics because each event is an isolated discrete event that can easily be characterized. While I wait for such data to be kept, I will continue playing golf and telling my wife that it is all in the name of research!

References and Further Reading

Bennett, J. (1998), Statistics in Sports, New York: Oxford University Press.

Larkey, P. D. (1998), "Statistics in Golf," in Statistics in Sports, ed. J. Bennett, New York: Oxford University Press.

Larkey, P. D., and Smith, A. A. (1998), "Improving the PGA Tour's Measures of Player Skill," in Science and Golf III, London: E & FN SPON.

Where is the Data?

There are numerous golf sites on the Web. Many will sell you clubs or give information on thousands of different courses. Here are the two best Web sites for information and data. The data used in the article are available for easy download on my web site, *stat.tamu.edu/~berry.*

- *pgatour.com*: This is the official site of the PGA Tour. It has the official statistics for the PGA Tour, the Nike Tour, and the PGA Senior Tour. The data are not in a good format to download, but they are available. This site also provides live scoring updates during tournaments.

- *www.golfweb.com*: This site, affiliated with CBS Sports, provides information for the PGA, LPGA, Senior, European, and Asian Tours, as well as amateur events. The statistics are not as extensive, but there is information on more than just the U.S. PGA. This site also has live scoring.

Chapter 38

ADJUSTING GOLF HANDICAPS FOR THE DIFFICULTY OF THE COURSE

Francis Scheid, Professor Emeritus, Boston University
135 Elm Street, Kingston, MA. 02364
Lyle Calvin, Professor Emeritus, Oregon State University

KEY WORDS: Golf, slope, ratings, handicaps.

1. The problem.

In many sports differences in playing ability are wider under difficult circumstances and more narrow when the going is easier. In golf, ability is measured by a player's handicap which estimates the difference between his or her ability and that of a standard called scratch, in this country the level of the field in the Amateur Championships. Handicaps are generally larger at difficult courses, meaning that differences from scratch are larger. This magnification of ability differences introduces inequity when players move from one course to another.

Early in this century British golfers developed a procedure to ease this inequity but it was abandoned as ineffective. In the early 1970s the problem was revived[1, 2] and in 1973 GOLF DIGEST published an article[3] in which handicap adjustments were described. In the late 1970s the United States Golf Association (USGA) organized a handicap research team (HRT) to study problems of equity in play. In one of its recent studies[4] the magnification of ability differences was consistently detected for courses differing in length by only 400 yards, about the difference between back tee and front tee length at many courses. The present report describes the process developed by the HRT and used by the USGA to deal with the magnification problem. It is called the Slope System.

2. Slope.

Plots of expected score against home course (local) handicap for play at dozens of courses strongly suggest a straight line relationship and regressions usually lead to correlations in the .90s. Although the data for extremely high and low handicaps is thin it is assumed that linearity prevails across the entire handicap spectrum. No evidence to the contrary has been found.

The problem of achieving handicap portability has led to the idea of measuring golfers by their performance on a standardized golf course, affectionately called Perfect Valley. It is assumed that the resulting relationship between expected scores and standardized handicaps, to be called indexes, will still be linear. Figure 1 exhibits this assumption in a typical plot of expected scores against index at golf course C.

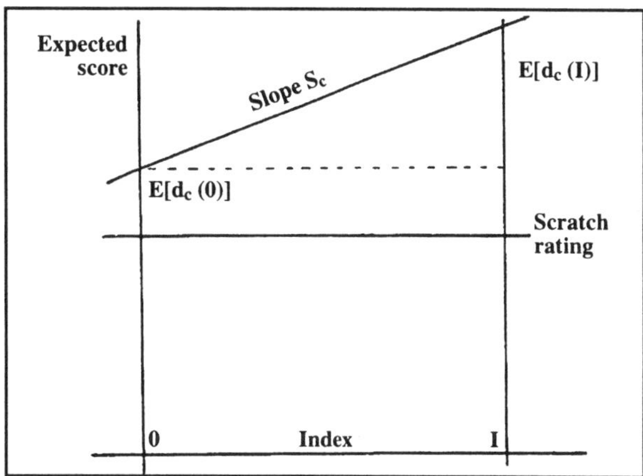

Figure 1.

The line for course C has slope S_c and the product 100 S_c is the slope of the course. Expected differentials, which are scores minus scratch rating (a measure of course difficulty), for players with indexes I and 0 are shown starting at the level of the rating and running upward to the line and we see that

$$(1) \quad E[d_c(I)] - E[d_c(0)] = S_c * I$$

where d is a differential. It holds for any course and any I, assuming only the linearity. It follows easily from (1) that

$$(1a) \quad E[d_c(I_1)] - E[d_c(I_2)] = S_c(I_1 - I_2)$$

for any indexes I_1 and I_2. This is the magnification effect. If course A has higher slope than course B the performance measure on the left side of (1a) will be larger at A, for the same I_1 and I_2.

3. The mapping.

Since (1) holds for any course we can take C to be our reference course Perfect Valley, for which S_{pv} has been defined as 1.13, the average figure found when scores were first regressed against local handicaps (now closer to 1.18).

(2) $E[d_{pv}(I)] - E[d_{pv}(0)] = 1.13 * I$

In this special case the index I also serves as the local handicap. From (1) and (2) it follows that

$$E[d_{pv}(I)] - E[d_{pv}(0)] = (1.13/S_c)\{E[d_c(I)] - E[d_c(0)]\}$$

or

(3) $E[d_{pv}(I)] = (1.13/S_c)\, E[d_c(I)] + T$

where $T = E[d_{pv}(0)] - (1.13/S_c)\, E[d_c(0)]$

depends only on the 0 index player, called scratch. Equation (3) holds for any course C and any index I.

Now assume that (3) can be applied not only to the expected differentials but to any pair $d_c(I)$ and $d_{pv}(I)$. This is a broad assumption with interesting consequences. It is based largely on intuition and amounts to saying that a good day at course C would have been a good day at PV, and likewise for bad days. For each value of I it provides a mapping of the distribution of differentials at C to one at the reference course PV. (See Figure 2.)

The mapping is thus defined by

(4) $d_{pv}(I) = (1.13/S_c)\, d_c(I) + T$

The fact that expected values map to one another is the content of (3).

A bit of background will be useful in simplifying this mapping. Golf handicaps are not found by averaging all of a player's scores relative to scratch ratings, that is, not all the differentials. For a number of reasons, to emphasize a player's potential rather than average ability and to preserve some bonus for excellence, only the better half are used. And for consistency, golf courses are not rated using all scores at the Amateur Championships, only the better half. Scratch players are those whose better half performances match the scratch ratings. It follows that these players will have handicap 0 on all courses. This "scratch on one course, scratch on all" is the scratch principle. (It must also be mentioned that better half averages are multiplied by .96 to produce handicaps, adding slightly to the emphasis on potential ability.)

Now, since (4) is assumed for all differentials it also holds for the expected better halves at each I level, again taken over all players with index I.

(5) $E[bh_{pv}(I)] = (1.13/S_c)\, E[bh_c(I)] + T$

Choose I = 0, the scratch level. Both better half terms in (5) are then 0 by the scratch principle, which means that T = 0 and (4) simplifies.

(6) $d_{pv}(I) = (1.13/S_c)\, d_c(I)$

4. The Slope System.

Equation (6) is the input procedure for the Slope System. It maps differentials shot at course C to images at PV for use in obtaining or updating a player's index I. Inverting it brings

(7) $d_c(I) = (S_c/1.13)\, d_{pv}(I)$

Since this applies to all differentials it can be applied to a player's expected better half

$$E[bh_c(I)] = (S_c/1.13)\, E[bh_{pv}(I)]$$

which, multiplied by .96, becomes

(8) $Hcap_c(I) = (S_c/1.13)\, I$

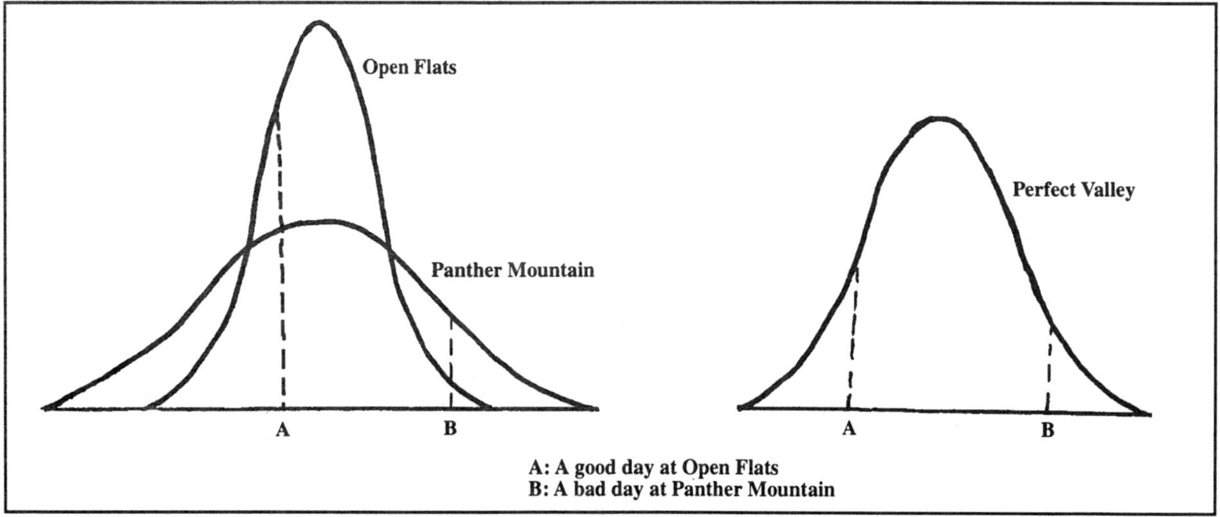

Figure 2.

since a Perfect Valley handicap is an index. Equation (8) is the system's output procedure. It converts the player's index I to a suitable handicap for play at course C. Together (6) and (8) convert scores shot at any set of courses into a standardized handicap, and the standardized handicap into a local handicap for local purposes.

In summary, the assumptions made in this development of the Slope System are:

(a) linearity of expected scores vs. index

(b) the mapping

(c) the scratch principle.

The first of these is supported by all the evidence in hand and implies that the system treats all handicap levels in the same way. The second is a broad leap, supported largely by intuition. The last is a pillar of handicapping. (An earlier approach[5,6] made somewhat different assumptions which led to the same input and output procedures.)

5. Implementation.

Designing a system and making it operational over thousands of golf courses are two different problems. The measurement of course difficulty has traditionally been done by teams of raters who note bunkers and water holes, measure the widths of fairways and the depth of rough, judge the difficulty of greens and other obstacles. In the past this was done for scratch players only and the scratch rating was the standard from which all others were measured. The Slope System also requires a bogey rating, an estimate of course difficulty for the more or less average player, which doubles the task of the rating teams.

Efforts using multiple regression to assess the importance of the various obstacles that golfers encounter were a first step, the main difficulty being that length alone explains well over ninety percent of scores, leaving very little to distribute among bunkers and such. The dominance of length also has a positive side, since it makes it harder to be too far wrong. The work of refining ratings based on yardage alone has fallen to the traditional rating teams but a detailed (inch thick) field manual[7] has been prepared to assist them and training seminars are regularly offered. As a result ratings made by the various regional associations have been found[8] to be very consistent. Tests conducted by regional associations, in particular the Colorado Golf Association, have provided feedback and led to revisions, a process which is ongoing. Regressions have also been used[9] to detect the more serious outliers and to provide smoothing. These are described in the following section.

6. Outliers and smoothing.

Scratch and slope ratings are subject to errors from a number of sources. In addition to the errors which may result from inaccurate values assigned by raters or from improper weighting of these values in calculating the scratch and bogey ratings, other sources of variation include measurement error, both intra- and inter-team errors and model error caused by the unique placement of obstacles on a course. With ratings taken at only two handicap levels, scratch and bogey, this latter error is not immediately apparent. If ratings were taken at a number of handicap levels, rather than at only two, it would easily be recognized that the best fitting line would not necessarily pass through the two points established for the scratch and bogey golfers.

Although such a line is a reasonable basis for the estimation of slope, other estimates might also be used. In particular, one might assume that all courses are drawn from a population with a common relationship between scratch and bogey ratings. An estimate of scratch rating for a course might then be obtained from the regression of the scratch rating on the bogey rating. A still better estimate might be obtained by combining the two, the original from the rating team and the one from the regression, as obtained from the population of courses in the association (or state, region or country).

The same procedure can also be used to obtain a weighted estimate of the bogey rating, combining the original from the rating team and an estimate from the regression of bogey rating on scratch. These two estimates can then be used to obtain another estimate of the slope rating by substituting the weighted estimates of scratch and bogey ratings into the slope formula. The slope rating so obtained will be called the weighted slope rating.

This procedure has been tried at the golf association level and at the national level. As an example, the plot of bogey rating against scratch rating is shown in Figure 3 for all courses in the Colorado Golf Association. The correlation coefficient for the two ratings is .959 and has comparable values in other associations. The ratings were those assigned for the mens' primary tees, usually designated as the white tees. The regression estimate of the scratch rating for any course is

$$SR_r = a + b\,BR_o$$

where the subscripts o and r refer to the original and regression estimates respectively. For the Colorado Golf Association this equation was

$$SR_r = 15.14 + .588\,BR_o$$

The regression estimate of the bogey rating for any course is

$$BR_r = c + d\,SR_o$$

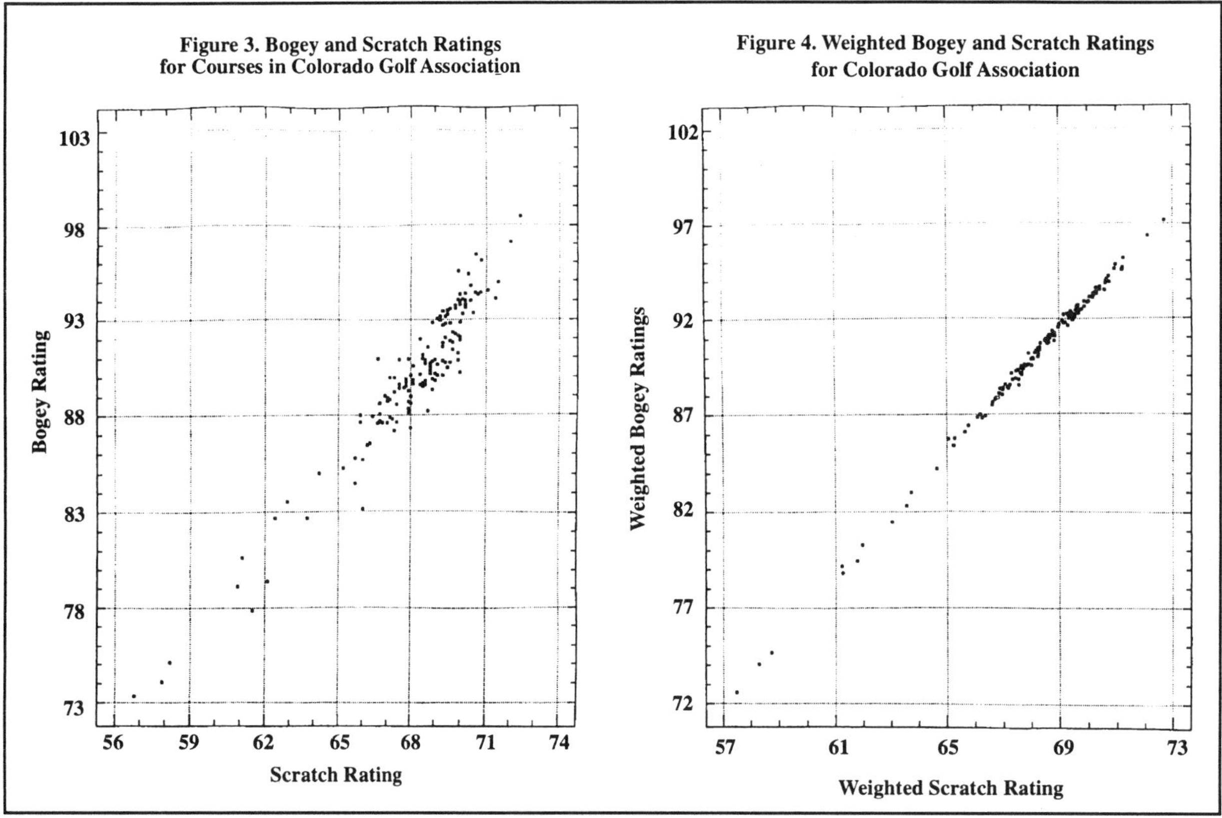

which, for Colorado, is

$$BR_r = -16.43 + 1.563\ SR_o$$

and the combined estimates are

$$SR_w = w_o\ SR_o + w_r SR_r$$

and $$BR_w = w_o BR_o + w_r BR_r$$

where $w_o + w_r = 1$.

Weights were assigned proportional to the inverse of the variances of the original and regression estimates of the ratings. Direct estimates of variances for the original ratings were not available; indirect estimates were made from the deviations from regression of the scratch and bogey ratings against yardage. Yardage is responsible for about 85% of the variation so this estimate should include most of the error variation. Variances for the regression estimates were taken from the residual mean squares for the scratch and bogey ratings regressed against each other. We do not claim that these variance estimates are the only ones, or perhaps even the best, but they do appear to be reasonable proxies. For Colorado, the variance estimates for scratch ratings are .4366 and .5203 for the regressions on yardage and bogey ratings respectively and the corresponding variance estimates for bogey ratings are 3.8534 and 1.3850. From these values, the weights for scratch and bogey combined estimates were calculated to be

	Scratch	Bogey
w_o	.544	.264
w_r	.456	.736

Using these, weighted estimates of scratch and bogey ratings, and from them the weighted slope ratings, were obtained. Figure 4 shows the plot of the weighted scratch and bogey ratings. The correlation coefficient for these ratings has increased to .998. We have some concern that the variance estimates as given by the regression of the scratch and bogey ratings on yardage may be too large and, therefore, that the weights for w_o may be too low. If so, the increase in the correlation, as a result of the weighting, would be less than is shown.

It is not suggested that these weighted estimates of course and slope ratings should replace the original estimates. Where the revised estimates have been tried, less than ten percent of the slope ratings are different by more than six points and only occasionally is there any appreciable change in the Course Rating. Since the executive directors of the golf associations already have the prerogative of changing the course and slope ratings when they believe an error has been made, it would be preferable for them to use the revised estimates as alternatives to support a change. This has been tried in a few cases and seems to work well. This procedure gives them a quantitative basis for modifying the ratings when they believe there is a need.

Since the two parameters used in the Slope System are the Course Rating (equal to the scratch rating) and slope rating, one might consider what effect this revision has on the relationship between the two. Figure 5 shows the plot of the original slope and scratch ratings while Figure 6 shows the plot of the weighted slope and scratch ratings. The correlation coefficient has increased from .773 to .966. This means that the weighted scratch rating could be used as a very good predictor of the slope rating.

This raises an interesting question. If Course Rating can be used to predict the Slope Rating, why shouldn't the Course Rating and the Slope Rating for a course be simply calculated from the scratch rating, without bothering to take a bogey rating? The estimate can be made with rather high precision with about half as much work on the part of the rating team. One concern, however, might be that one cannot be sure that the relationship between course and slope ratings would necessarily remain the same without having the bogey ratings to continually test and estimate the relationship. This idea will be examined further and tested on a number of golf associations.

7. References.

(1) Soley and Bogevold; Proposed Refinement to the USGA Golf Handicap System; report to the USGA, 1971. (2) Scheid; Does your handicap hold up on tougher courses?; GOLF DIGEST, 1973 (3) Riccio; How Course Ratings may Affect the Handicap Player, report to the HRT, 1978. (4) Scheid; On the Detectability of M.A.D; report to the HRT, 1995. (5) Stroud; The Slope Method — A Technique for Accurately Handicapping Golfers of all Abilities Playing on Various Golf Courses; report to the HRT, 1982. (6) Stroud and Riccio; Mathematical Underpinnings of the Slope Handicap System; in Science and Golf, edited by A. J. Cochran, 1990. (7) Knuth and Simmons (principally); USGA Course Rating System Manual. (8) Calvin; The Consistency of Ratings Among the Golf Associations; report to the HRT, 1994. (9) Calvin; A Modified Model for Slope and Course Ratings; report to the HRT, 1992.

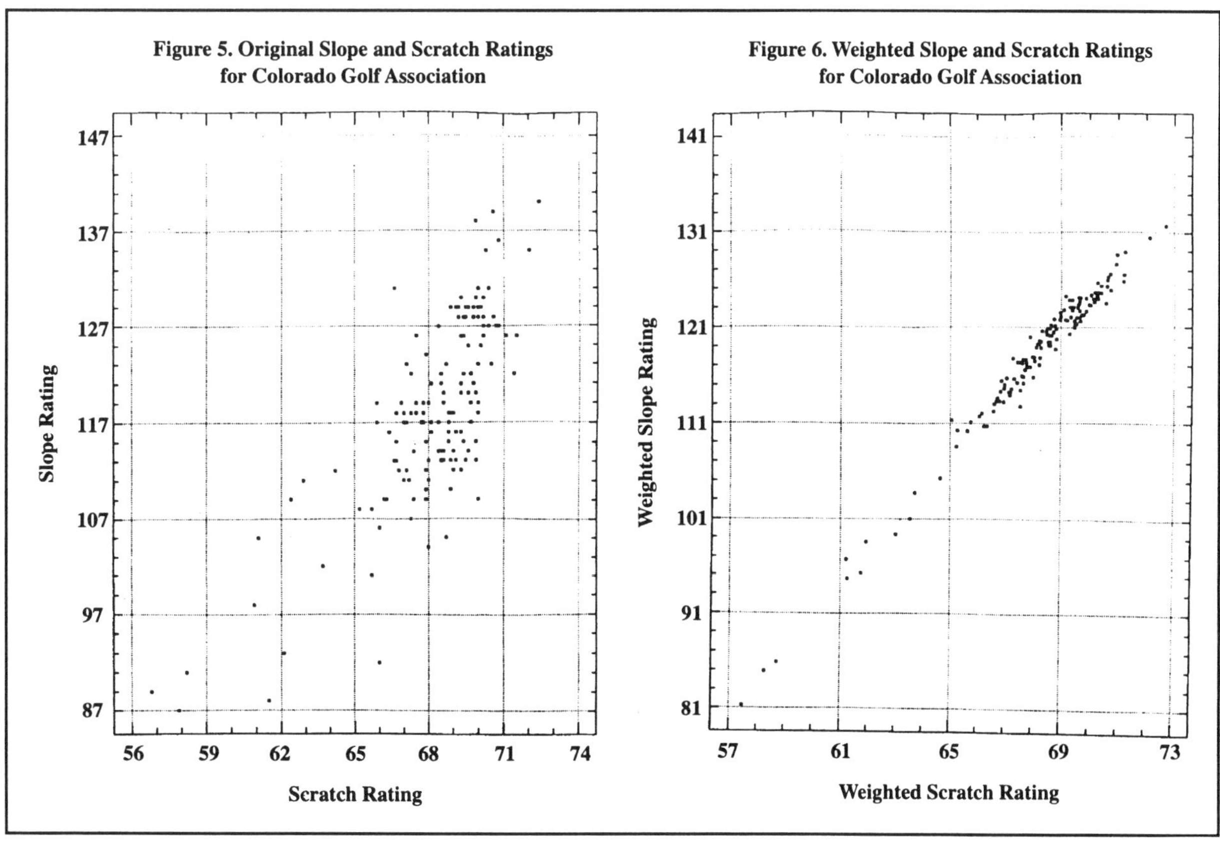

Chapter 39

Rating Skating

Gilbert W. BASSETT,* Jr. and JOSEPH PERSKY*

Among judged sports, figure skating uses a unique method of median ranks for determining placement. This system responds positively to increased marks by each judge and follows majority rule when a majority of judges agree on a skater's rank. It is demonstrated that this is the only aggregation system possessing these two properties. Median ranks provide strong safeguards against manipulation by a minority of judges. These positive features do not require the sacrifice of efficiency in controlling measurement error. In a Monte Carlo study, the median rank system consistently outperforms alternatives when judges' marks are significantly skewed toward an upper limit.

KEY WORDS: Breakdown; Majority rule; Median; Ranks.

1. INTRODUCTION

Early during the 1992 Winter Olympics, Scott Hamilton, the former Olympic champion now working as an announcer for CBS, made a valiant but unsuccessful try to explain how judges' marks are aggregated to determine the placements of figure skaters. Once burned, he gingerly avoided the subject for the duration. Hamilton's difficulties reflected the complex and at first glance arcane procedures dictated by the rulebooks. Yet officials of the United States Figure Skating Association (USFSA) and the International Skating Union (ISU) have long claimed that their approach to judging avoids arbitrary mechanisms in favor of the logic of majority rule. The purpose of this article is to explore this claim.

Figure skating is one of the most graceful and aesthetic of sports. It also involves quick movements and subtle variations. These characteristics of skating, similar in many respects to diving and gymnastics, make the appropriate ranking of competitors quite difficult. Not surprisingly, all three of these sports rely on expert judging to determine placements in competitions. Unwilling to trust such responsibility to a single individual, the rules call for a panel of judges. Yet using such a panel creates a difficult and interesting question: how best to combine the judges' marks?

In considering this question, it seems reasonable to first ask why judges' rankings should ever differ. Why don't all the judges in a competition agree? What we want is a "model" of the generating process that gives rise to such differences. Unfortunately, even a cursory acquaintance with figure skating suggests a whole spectrum of possibilities.

At one extreme, we might imagine that the quality of each skater's performance is an objective entity subject only to errors in measurement. In such a view, judges' marks differ in the same way as clocks timing a swimming event might differ. This judge might have blinked at an important moment, or that judge might have concentrated on one member of a pair of skaters just when the partner wobbled. In this model, judges are imperfect measuring devices. The aggregation problem here is essentially one of measurement error.

At the other extreme, the differences among judges might represent not errors of measurement but rather genuine differences in tastes. Judges' preferences would presumably reflect real aesthetic differences, although they might also be influenced by national pride and other less relevant motivations. In a world of complex aesthetics, we face all the aggregation problems of collective decision making (see Arrow 1963 or Sen 1970). Moreover, skating officials worry continually about the strategic behavior of their judges. Where judges hold strongly to their preferences or have a personal stake in the outcome, we must also worry about the distortions that can be produced by strategic voting behavior.

Both of these models are obviously oversimplifications. Yet even at this level, it is easy to appreciate that a system of aggregation that showed highly desirable properties with respect to one of these might fail to perform well with respect to the other. But both models have some plausibility; at any given competition, one or the other might dominate. At prestigious competitions such as the Olympics or the World Championships, judges' marks are very likely influenced by both tastes and problems of measurement. At local or regional meets measurement problems probably dominate. Any system of aggregating judges' rankings must be considered in light of both models. Thus the rating skating problem requires us to search for a set of aggregation criteria relevant to both measurement error and preference aggregation.

Skating officials have long maintained that their placement system was desirable because it embodied the principle of majority rule. Although the concept of majority rule is open to a number of interpretations, we show that the system that has been adopted is essentially a median rank method with tie-breaking rules. The identification of the method with median ranks does not seem to have been noticed in the skating literature. Further, although the system has evolved only informally, we show that it is actually the only method that satisfies a majority rule and incentive compatibility (or monotonicity) requirement in which a skater's final rank cannot be decreased by a judge who gives the skater a better mark. Hence the skating associations have settled on the unique aggregation method of median ranks, which is resistent to manipulation by a minority subset of judges and also satisfies a reasonable incentive compatibility requirement.

These results relate to the problem of aggregating tastes. One might expect that a ranking system well tuned to handling such aggregation might perform poorly when evaluated

* Gilbert Bassett and Joseph Persky are Professors of Economics, University of Illinois at Chicago, Chicago, IL 60607. The authors would like to thank two anonymous reviewers, Benjamin Wright of the International Skating Union, Dale Mitch of the United States Figure Skating Association, Tammy Findlay, Shiela Lonergan, Nicole Persky, and all the mothers of the McFetridge Ice Skating Rink in Chicago.

Journal of the American Statistical Association
September 1994, Vol. 89, No. 427, Statistics in Sports

Skaters	Judges 1	2	3	4	5	6	7	8	9	Place of Lowest Majority	Size of Lowest Majority	Final Place
A	5	2	2	2	4	4	1	2	4	2	5	3
B	2	1	3	1	2	2	3	1	3	2	6	2
C	3	3	5	5	3	3	2	5	2	3	6	4
D	1	6	1	3	1	1	5	3	1	1	5	1
E	6	5	4	4	6	6	4	4	6	5	5	5
F	4	4	6	6	5	5	6	6	5	5	5	6

Figure 1. Hypothetical Ordinals for a Component Event.

from the perspective of measurement error. Yet we find that the official aggregation procedures deal effectively with the persistent measurement problem of "score inflation" that skews marks toward the upper limit of the scoring range. Thus we conclude that even in competitions where no manipulation is likely but judge's marks are subject to random error, the system performs well.

In Section 2 we provide a brief overview of the official scoring system used by the ISU and the USFSA. We explain that the system is essentially median ranks, but with somewhat arbitrary rules for breaking ties. In Section 3 we show that the median rank method can be justified in terms of a majority rule incentive requirement. In Section 4 we go on to consider the relative performance of the system compared to alternatives in the context of a simple model of error generation. Finally, we summarize results in Section 5.

2. THE RULES

At all major USFSA events, as well as the World Championships and the Olympics, a skater's place is determined by a weighted average of two component events. The short, or original, program is weighted one-third, and the long free-skating program is weighted two-thirds. Each component event is scored by a panel consisting of nine judges. At lesser competitions, there are still three component events: compulsory figures (20%), original program (30%), and free skating (50%). Moreover, there often are fewer judges (but always an odd number) at such events. The compulsory figures component has been dropped from more prestigious competitions, because the slow and tedious etching of school figures makes a less than dramatic video scene.

For each component, a judge gives two cardinal marks on a scale of 1 to 6. For the original program the marks are for required elements and presentation; for free-skating the marks are for technical merit and composition style. These marks are the ones displayed prominently at competitions. But there is a long trail from marks to placement.

Take, for example, the placements in the original program. (Placements in the free-skating program are determined in exactly the same way.) First, for each judge an ordinal ranking of skaters is determined from total marks, the sum of the two subcomponent cardinal scores. These ordinal ranks and not the raw scores become the basis for determining placements.

As presented by the USFSA rulebook, the procedure continues as follows: "The competitor(s) placed first by the absolute majority (M) of judges is first; the competitor(s) placed second or better by an absolute majority of judges is second, and so on" (USFSA CR 26:32). Note here the expression "second or better." In calculating majorities for second place, both first and second ranks are included. In calculating a majority for third place, first, seconds, and thirds are included, and so on for lower places. If for any place there is no majority, then that place goes to the skater with a majority for the nearest following place.

Now of course below first place, there can be numerous ties in this system. (If a judge has given more than one first because of a tied mark, then there can also be a tie for first.) The basic rule for breaking ties is that the place goes to the competitor with the greater majority. If after the application of the greater majority rule there is still a tie, then the place goes to the skater with the "lowest total of ordinals from those judges forming the majority." And if this does not work, then the place goes to the skater with the lowest total of ordinals from all judges. In all cases of ties, the skaters involved in the ties must be placed before other skaters are considered.

To demonstrate how this all works, consider Figure 1, which contains hypothetical ordinal rankings for a component event. Notice that the tie for second between A and B goes to skater B because of the greater size of B's majority for second. Skater A then gets third place, because A must be placed before anyone else is considered. Because no one has a majority of fourths, we go on to consider E and F, each of whom has a majority of fifths. Because each has five judges in their majority, breaking the tie depends on the sum of ordinals in each majority. E then wins with the lower sum, 21, as compared to F's 23.

After just a bit of reflection, it is clear that the placement system used in figure skating starts from a ranking of the median ordinals received by skaters. As defined by the rules, a skater's initial placement depends on the "lowest majority." But this "lowest majority" is just equal to the median ordinal. A majority of judges ranked the skater at the skater's median ordinal or better. It is true of course that a number of tie-breaking devices are applied. These rules involve several other concepts. But under the current procedures, a skater with a lower (better) median will never be ranked worse than one with a higher (worse) median. Such a result is explicitly ruled out, because all tied skaters must be placed before any remaining skaters are considered. In particular, all skaters tied at a given "lowest majority" or median rank must be placed before any other skaters are considered. Notice that in the absence of this rule, a reversal vis-a-vis the median rank rule could easily occur. For example, referring to Figure 1, if after failing in a tie-breaking situation for second place, skater A had to compete with skater C for third place, then the winner would be skater C (despite A's median of 2) because of a greater majority of "3s or better"; C has six "3s or better," whereas A has only five.

Although over the years there have been a number of changes in the various tie-breaking mechanisms, since 1895 the ISU has used its concept of majority rule to determine placements. The only exception we have discovered was an experiment in 1950 that used a trimmed mean. The system has now evolved to a point where it is clearly one of median ranks.

3. MEDIAN RANKS AND MAJORITY RULE

Why use the median rather than the average ordinal, the sum of the raw scores, or a trimmed mean? As in other sports involving subjective judging, ice skating has been plagued by charges of strategic manipulation. This problem is a common one in the theory of constitution building. (There is of course a large literature addressing this issue; see Arrow 1963.) The most obvious reason for using medians is to limit the effect of one or two outliers on the final rankings. But there are any number of ways to begin to guard against such manipulation. In defense of their system, skating officials from the ISU and USFSA have often claimed that it embodies the essence of majority rule. The heart of their argument is that a skater ranked best by a majority of judges should be ranked best overall.

In addition to its relation to majority rule, a system of median ranks has at least one other attractive property: If an individual judge raises a skater's mark, then that action will never decrease that skater's placement. Thus if a judge raises the mark of one skater, that skater will either move up or stay the same in overall placement.

These two properties are attractive characteristics of median ranks that suggest it for serious consideration. But in fact we can make a stronger statement. If these two simple conditions are considered to be necessary, median ranking is the only system that will satisfy both. The result follows from the median as a high-breakdown estimator and the fact that such estimates satisfy an exact-fit property that is equivalent to a majority requirement in the aggregation context (see Bassett 1991).

To formally demonstrate the result, let $m_j(s)$ denote the raw mark and let $r_j(s)$ denote the rank of the sth skater by the jth judge, where $s = 1, \ldots, S$ and $i = 1, \ldots, J$. We suppose that higher-valued cardinal marks are assigned to better performances, and skaters are then ranked with "1" as best. (We assume that there are no tied marks, so that the marks yield a complete ordering for each judge.) The final rank of the sth skater is denoted by RANK(s).

An initial ranking is determined by a place function, denoted by P. The P function takes the matrix of marks and produces a vector, $\mathbf{p}$ with elements $p(s)$, which provides a partial order of skaters. The ranking RANK(s) is obtained by breaking the ties of $\mathbf{p}$.

The total mark is a particularly simple example of a P function. Here $\mathbf{p}(s) = (m_1(s) + m_2(s) + \cdots + m_J(s))$. Observe that this rule can be "manipulated" by a single judge. The skater from the "good" country who is clearly best in the eyes of all but the judge from the "bad" country can lose a competition if the "bad" judge gives that skater a very low mark. A trimmed mean is also a placement function, and of course trimming can eliminate the influence of a single "bad" country judge. But despite this, trimming can still violate our conception of majority rule.

We now formalize the requirements of a place function:

1. Incentive compatibility. A skater's final rank cannot be made worse by a judge who improves the skater's mark. In terms of P functions, this says that if $d > 0$ and $m_j(s) + d$ is substituted for $m_j(s)$, then $p(s)$ cannot fall.
2. Rank majority. If the rank matrix is such that skater s has rank r_0 for at least half the judges and skaters s' has rank q_0, for at least half the judges, where $r_0 < q_0$, then $p(s) < p(s')$.

Note that the rank majority requirement considers only situations in which more than half of the judges agree on the precise rank of skater s and more than half (not necessarily the same "more than half") agree on the precise rank of skater s'. The rank majority sets no explicit conditions on any other situation.

Many placement functions meet requirement 2; for example, the shortest half or least median of squares (LMS (see Rousseeuw 1984). The LMS identifies for each skater the half subset with the most similar or closest ranks and assigns as an initial placement function the midpoint of that interval. Clearly this satisfies the rank majority rule; but it does not satisfy Requirement 1. To illustrate this fact, consider a skater with the following ranks given by five judges: 1, 1, 3, 4, and 7. With LMS, this skater's placement function value is 2. But if the last judge improves the seventh place rank to a fourth place finish, then the skater's placement function actually falls to 3.5.

Theorem. Any place function that satisfies Requirements 1 and 2 is equivalent to the median rank place function.

Proof. It is easy to see that the median satisfies Requirements 1 and 2. To see that only the median rank and no other placement function satisfies these two requirements, we proceed by contradiction. Let M and R be marks and ranks evaluated by a P function satisfying Requirements 1 and 2, where

$$p(1) \leq p(2) \tag{1}$$

but

$$\text{med}_j\{r_1(1), \ldots, r_J(1)\} = x_0 > y_0 = \text{med}_j\{r_1(2), \ldots, r_J(2)\}. \tag{2}$$

We are going to change these marks without affecting either the relative placement of skaters 1 and 2 or their median ranks; however, after the change, a majority of judges will have given an identical rank score to skater 1 that is greater than an identical rank score given by a majority of judges to skater 2. But this will violate the majority requirement of a place function.

Consider the set of judges whose rank for skater 1 is $\geq x_0$; notice that this set includes a majority of judges. For each such judge, adjust marks so that (a) if $r_j(1) = x_0$, then do nothing; leave the mark and rank at their original values, or (b) if $r_j(1) > x_0$, then increase skater 1's mark so that the rank is decreased to x_0. It can be verified that this remarking and reranking leaves the median relation (2) unchanged, and, because the rank value for skater 1 goes down—the relation (1) also still holds (by the incentive requirement). Further there are now a majority of judges for whom the rank of 1 is x_0.

We now perform a similar operation for skater 2. Consider the set of judges whose rank for skater 2 is $\leq y_0$; notice that this set includes a majority of judges. For each judge in this majority set, (a) if $r_j(2) = y_0$, then do nothing, or (b) if $r_j(2) < y_0$, then decrease skater 2's mark so that her rank is decreased to y_0. It can again be verified that this does not change either (1) or (2). Further, there now is a majority of judges for whom the rank of skater 2 is y_0. Hence, by majority rule, $p(1) > p(2)$, which contradicts (1) and completes the proof.

We conclude that the median rank is the only placement function that possesses these two desirable properties. Of course, median ranks cannot perform miracles. Like all social welfare functions, this choice rule will, under specific circumstances, violate Arrow's list of properties. In particular, the winner of a competition as judged by USFSA rules can easily depend on "irrelevant alternatives." A new entrant into a competition can change the outcome, just as a spoiler entering a three-way election can upset a favored candidate.

At the same time, we should also note that our choice of Requirement 2 to represent majority rule is subject to dispute. This is only one of the possible interpretations of majority rule. Indeed the more familiar representation of this concept performs pairwise comparisons between alternatives. If the majority prefers **x** to **y**, then society prefers **x** to **y**. This is a different idea of majority rule than that contained in median ranks, and it is easy to construct examples (see Fig. 2) in which a majority of judges prefer **x** to **y** but **x** obtains a worse median rank than **y**. The well-known problem here is that such a ranking generally will not be transitive.

4. RANKING AS A MEASUREMENT ERROR PROBLEM

The median ranks used in placing figure skaters capture an interesting meaning of majority rule and offer obvious advantages in limiting strategic manipulation. Yet in the vast majority of competitions where there is little concern with such issues of preference, one can reasonably ask whether the present system is unnecessarily cumbersome or worse. For most competitions, the problem is not one of preference aggregation but rather one of statistical estimation, where concern is measurement error. Our first thought was that in these settings, the USFSA system would be less attractive than simpler aggregates, because its emphasis on median ranks ignores considerable information in determining placement. To look at this question, we conducted a series of Monte Carlo experiments comparing the official system to one of simple addition of cardinal marks. For completeness, we also included a trimmed mean similar to that used in diving competitions.

Skaters were assigned a "true" point score, which in turn defined a "true" ranking. The scores measured by individual judges were set equal to the true score plus a random error term. A normal error distributions was used, but as in actual meets, all scores were truncated at 6.0. As a simple measure of how well a system did, we calculated both the proportion of times that it picked the true winner and the average absolute error of placements.

In our first set of meet simulations, we treated the competition as consisting of only one component event judged on a simple six-point scale. Each meet consisted of five judges and six skaters (one through six), with true scores ranging in .2-point intervals from 5.8 to 4.8. The random error was taken to be normal with mean 0 and variance 1. (But as noted earlier, judges' scores were truncated at 6.0, thus skewing the distribution of scores). We ran 20,000 "meets" of this type. The simple addition of judges' scores correctly identified the true first place in 46% of the meets. But the USFSA system picked the correct first place finisher in 54% of meets. The trimmed mean did about the same as the sum, picking 45% of the correct first place finishers. The straightforward sum of ranks had an average absolute error in the estimated rank of a skater of 1.10, a figure identical to that for the USFSA system. The trimmed mean did only a tad worse, with an average absolute error of 1.12.

The result surprised us initially. But in hindsight, we realized that the success of the median ranking system was largely due to the mark ceiling imposed on the judges. In this situation, downward measurement errors for a good skater cannot easily be offset by upward measurement errors. Hence the average or total judges' score of a very good skater is systematically biased downward.

To demonstrate, we redid the simulation, but this time the highest skater had a true score of only 3.6 and the other skaters had scores again at .2-point intervals. The result, as we now expected, was that the USFSA system found a lower

	Judge 1	Judge 2	Judge 3	Judge 4	Judge 5
.					
.					
.					
fourth place finisher	4	4	4	7	6
fifth place finisher	3	3	5	5	5

Figure 2. Conflicting Conceptions of Majority Rule. Notice that every judge but Judge 3 prefers the fifth place finisher, who has a majority of fives, to the fourth place finisher, who has a majority of fours. Also, in this case the fifth place finisher has a better (lower) total score and a better (lower) trimmed mean.

percentage of appropriate winners than the total system (46% vs. 48%). The trimmed mean also came in at 46%. The average absolute error in placement was now a good deal higher for USFSA, 1.15, as compared to 1.07 for the sum of marks and 1.11 for the trimmed mean.

Although hardly conclusive, these simulations suggest that the USFSA system may actually help in distinguishing among skaters of different performance levels when questions of preference are not seriously at issue. This result depends critically on the mark ceiling of six points, which strongly skews judges' marks. The median rank method works well with skewed scores.

5. SUMMARY

Like gymnastics and diving, figure skating requires a method to aggregate judges' marks. Unlike other judged sports, however, figure skating has adopted a system based on median ranks. Skating officials have often bragged that their system represents majority rule. We have shown that median ranks uniquely captures an important meaning of majority rule and provides strong protection against manipulation by a minority of judges.

One might have expected that these positive features would have required the scoring system to sacrifice efficiency in the more mundane world of measurement error. Yet, somewhat accidentally as the result of persistent mark inflation, we find that median ranks do a better job in controlling measurement error than two alternatives, total marks and the trimmed mean.

Although we can find no historical evidence that skating officials ever had this end in mind, they have picked a system particularly well suited to serve as both a method of statistical estimation and a means of preference aggregation as the situation warrants.

[Received April 1993. Revised November 1993.]

REFERENCES

Arrow, K. (1963). *Social Choice and Individual Values* (2nd ed.), New York: John Wiley.

Bassett. G. W. (1991). "Equivariant, Monotone, 50% Breakdown Estimators," *The American Statistician,* May, 135–137.

Rousseeuw, P. J. (1984), "Least Median of Squares Regression," *Journal of the American Statistical Association,* 79,

Sen, A. (1970), *Collective Choice and Social Welfare,* Oakland: Holden-Day.

United States Figure Skating Association (1992), *USFSA Rulebook,* Colorado Springs, CO: Author.

Chapter 40

A game of luck or a game of skill?

Modeling Scores in the Premier League: Is Manchester United *Really* the Best?

Alan J. Lee

In the United Kingdom, Association football (soccer) is the major winter professional sport, and the Football Association is the equivalent of the National Football League in the United States. The competition is organized into divisions, with the Premier League comprising the best clubs. There are 20 teams in the league. In the course of the season, every team plays every other team exactly twice. Simple arithmetic shows that there are 380 = 20 × 19 games in the season. A win gets a team three points and a draw one point. In the 1995/1996 season, Manchester United won the competition with a total of 82 points. Did they deserve to win?

On one level, clearly Manchester United deserved to win because it played every team twice and got the most points. But some of the teams are very evenly matched, and some games are very close, with the outcome being essentially due to chance. A lucky goal or an unfortunate error may decide the game.

The situation is similar to a game of roulette. Suppose a player wins a bet on odds/evens. This event alone does not convince us that the player is more likely to win (is a better team) than the house. Rather, it is the long-run advantage expressed as a probability that is important, and this favors the house, not the player. In a similar way, the team that deserves to win the Premier League could be thought of as the team that has the highest probability of winning. This is not necessarily the same as the team that actually won.

How can we calculate the probability that a given team will win the Premier League? One way of doing this is to consider the likely outcome when two teams compete. For example, when Manchester United plays, what is the probability that it will win? That there will be a draw? Clearly these probabilities will depend on which team Manchester United is playing and also on whether the game is at home or away. (There are no doubt many other pertinent factors, but we shall ignore them.)

If we knew these probabilities for every possible pair of teams in the league, we could in principle calculate the probability that a given team will "top the table." This is an enormous calculation, however, if we want an exact result. A much simpler alternative is to use simulation to estimate this probability to any desired degree of accuracy. In essence, we can simulate as many seasons as we wish and estimate the "top the table" probability by the proportion of the simulated seasons that Manchester United wins. We can then rate the teams by ranking their estimated probabilities of winning the competition.

The Data

The first step in this program is to gather some data. The Internet is a good source of sports data in machine-readable form. The Web site http://dspace.dial.pipex.com/r-johnson/home.html has complete scores of all 380 games played in the 95/96 season, along with home and away information.

Modeling the Scores

Let's start by modeling the distribution of scores for two teams, say Manchester United playing Arsenal at home. We will assume that the number of goals scored by the home team (Manchester United) has a Poisson distribution with a mean λ_{HOME}. Similarly, we will assume that the number of goals scored by the away team (Arsenal) also has a Poisson distribution, but with a different mean λ_{AWAY}. Finally, we will assume that the two scores are independent so that the number of goals scored by the home team doesn't affect the distribution of the away team's score.

This last assumption might seem a bit far-fetched. If we cross-tabulate the home and away scores for all 380 games (not just games between Manchester U and Arsenal), however, we get the following table:

		Home team score				
		0	1	2	3	4+
	0	27	29	10	8	2
Away	1	59	53	14	12	4
team	2	28	32	14	12	4
score	3	19	14	7	4	1
	4+	7	8	10	2	0

A standard statistical test, the χ^2 test, shows that there is no evidence against the assumption of independence (χ^2 = 8.6993 on 16 df, p = .28). Accordingly, we will assume independence in our model.

The next step is to model the distribution of the home team's score. This should depend on the following factors:

Using Poisson Regression to Model Team Scores

We will assume that the score X of a particular team in a particular game has a Poisson distribution so that

$$Pr[X = x] = \frac{e^{-\lambda}\lambda^x}{x!}$$

We want the mean λ of this distribution to reflect the strength of the team, the quality of the opposition, and the home advantage, if it applies. One way of doing this is to express the logarithm of each mean to be a linear combination of the factors. This neatly builds in the requirement that the mean of the Poisson has to be positive. Our equation for the logarithm of the mean of the home team is (say, when Manchester U plays Arsenal at home)

$$\log(\lambda_{HOME}) = \beta + \beta_{HOME} + \beta_{OFFENSE}(\text{Manchester U}) + \beta_{DEFENSE}(\text{Arsenal})$$

Similarly, to model the score of the away team, Arsenal, we assume the log of the mean is

$$\log(\lambda_{AWAY}) = \beta + \beta_{OFFENSE}(\text{Arsenal}) + \beta_{DEFENSE}(\text{Manchester U})$$

We have expressed these mean scores λ_{HOME} and λ_{AWAY} in terms of "parameters," which can be interpreted as follows. First, there is an overall constant β, which expresses the average score in a game, then a parameter β_{HOME}, which measures the home-team advantage. Next comes a series of parameters $\beta_{OFFENSE}$, one for each team, that measure the offensive power of the team. Finally, there is a set of parameters $\beta_{DEFENSE}$, again one for each team, that measures the strength of the defense.

The model just described is called a generalized linear model in the theory of statistics. Such models have been intensively studied in the statistical literature. We can estimate the values of these parameters, assuming independent Poisson distributions, by using the method of maximum likelihood. The actual calculations can be done using a standard statistical computer package. We used S-Plus for our calculations.

The parameters calculated by S-Plus are shown in Table 2, and they allow us to compute the distribution of the joint score for any combination of teams home and away. For example, if Manchester U plays Arsenal at home, the probability that Manchester scores h goals and Arsenal scores a goals is

$$\frac{e^{-\lambda_{HOME}}\lambda_{HOME}^{h}}{h!} \times \frac{e^{-\lambda_{AWAY}}\lambda_{AWAY}^{a}}{a!}$$

where λ_{HOME} and λ_{AWAY} are given by

$$\begin{aligned}\lambda_{HOME} &= \exp(\beta + \beta_{HOME} + \beta_{OFFENSE}\ (\text{Manchester U}) + \beta_{DEFENSE}(\text{Arsenal}))\\ &= \exp(.0165 + .3518 + .4041 - .4075)\\ &= \exp(.0165) \times \exp(.3518) \times \exp(.4041) \times \exp(-.4075)\\ &= 1.4405\end{aligned}$$

and

$$\begin{aligned}\lambda_{AWAY} &= \exp(\beta + \beta_{OFFENSE}\ (\text{Arsenal}) + \beta_{DEFENSE}\ (\text{Manchester U}))\\ &= \exp(.0165 + .0014 - .2921)\\ &= \exp(.0165) \times \exp(.0014) \times \exp(-.2921)\\ &= .7602\end{aligned}$$

Thus, if Manchester U played Arsenal at Manchester many times, on average Manchester U would score 1.44 goals and Arsenal .76 goals. To calculate the probability of a home-side win, we simply total the probabilities of all combination of scores (h,a) with $h > a$. Similarly, to calculate the probability of a draw, we just total all the probabilities of scores where $h = a$ and, for a loss, where $h < a$. A selection of these probabilities are shown in Table 3.

- How potent is the offense of the home team? We expect Manchester U to get more goals than Bolton Wanderers, at the bottom of the table.
- How good is the away team's defense? A good opponent will not allow the home team to score so many goals.
- How important is the home-ground advantage?

We can study how these factors contribute to a team's score against a particular opponent by fitting a statistical regression model, which includes an intercept to measure the average score across all teams, both home and away, a term to measure the offensive capability of the team, a term to measure the defensive capability of the opposition, and finally an indicator for home or away. A similar model is used for the mean score of the away team.

These models are Poisson regression models, which are special cases of *generalized linear models*. The Poisson regression model is described in more detail in the sidebar.

Data Analysis

Before we fit the Poisson regression model, let us calculate some averages that shed light on the home-ground advantage, the strength of the team, and the strength of the opposition. First, if we average the "home" scores in each of the 380 games, we get a mean of 1.53 goals per game. The corresponding figure for the "away" scores is 1.07, so the home-team advantage is about .46 goals per game—a significant advantage.

What about the offensive strength of each team? We can measure this in a crude way by calculating the average number of goals scored per game by each team. Admittedly, this takes no account of who played whom. Similarly, we can evaluate the defensive strength of each team by calculating the number of goals scored against each team. These values are given in Table 1. We see that Manchester United has the best offense, but Arsenal has the best defense.

Table 1—Average Goals for and Against

Team	Average goals for	Average goals against	Team record (W	L	D)	Competition points
Arsenal	1.29	.84	17	9	12	63
Aston Villa	1.37	.92	18	11	9	63
Blackburn R.	1.61	1.24	18	13	7	61
Bolton Wan.	1.03	1.87	18	25	5	29
Chelsea	1.21	1.16	12	12	14	50
Coventry C.	1.11	1.58	8	16	14	38
Everton	1.68	1.16	17	11	10	61
Leeds U.	1.05	1.50	12	19	7	43
Liverpool	1.84	.89	20	7	11	71
Man. City	.87	1.53	9	18	11	38
Man. U.	1.92	.92	25	6	7	82
Middlesbro	.92	1.32	11	17	10	43
Newcastle U.	1.74	.97	24	8	6	78
Nottm. Forest	1.32	1.42	15	10	13	58
QPR	1.00	1.50	9	23	6	33
Sheff. Wed.	1.26	1.61	10	18	10	40
Southampton	.89	1.37	9	18	11	38
Tottenham H.	1.32	1.00	16	9	13	61
West Ham. U.	1.13	1.37	14	15	9	51
Wimbledon	1.45	1.84	10	17	11	41

Table 2—Team and Opposition Parameters From Fitting the Generalized Linear Model

Team	Offensive parameter	Offensive multiplier	Defensive parameter	Defensive multiplier
Arsenal	.00	1.00	-.41	.67
Aston Villa	.06	1.07	-.31	.73
Blackburn R.	.24	1.27	-.01	.99
Bolton Wan.	-.19	.83	.38	1.46
Chelsea	-.05	.95	-.09	.91
Coventry C.	-.12	.88	.22	1.24
Everton	.28	1.33	-.07	.93
Leeds U.	-.18	.84	.16	1.18
Liverpool	.36	1.43	-.32	.72
Man. City	-.37	.69	.17	1.19
Man. U.	.40	1.50	-.29	.75
Middlesbro	-.32	.73	.02	1.03
Newcastle U.	.31	1.36	-.24	.78
Nottm. Forest	.05	1.05	.12	1.13
QPR	-.23	.80	.16	1.17
Sheff. Wed.	.01	1.01	.24	1.27
Southampton	-.34	.71	.06	1.07
Tottenham H.	.03	1.03	-.23	.79
West Ham. U.	-.11	.90	.07	1.08
Wimbledon	.16	1.17	.38	1.47

Now we "fit the model" and estimate the parameters. The intercept is .0165, and the home-team advantage parameter is .3518. The first value means that a "typical" away team will score 1.0166 (= $e^{.0165}$) goals, and the second means that, on average, the home team can expect to score 100 × $e^{.3518}$ = 142% of the goals scored by their opposition. This agrees with the preceding crude estimate; 1.5263 is 142% of 1.0737.

Next we come to the offensive and defensive parameters. The estimates of these are contained in Table 2. We see that Manchester United has the largest offensive parameter (.4041) and Arsenal the smallest defensive parameter (− .4075), which is consistent with the preceding preliminary analysis. To get the expected score for a team, we multiply the "typical away team" score (1.0166) by the offensive multiplier and by the defensive multiplier. In addition, if the team is playing at home, we multiply by 1.4216 (= $e^{.3518}$). Note that these parameters are relative rather than absolute: The average of the offensive and defensive parameters has been arbitrarily set to 0 and the "typical team" parameter adjusted accordingly.

What do we get from this more complicated analysis that we didn't get from the simple calculation of means? First, the model neatly accounts for the offensive and defensive strengths of both the home team and the opposition. In addition, using the model, we can calculate the chance of getting any particular score for any pair of teams. In particular, the model gives us the probability of a win, a loss, or a draw.

The results in Tables 1 and 2 are in agreement, giving the same orderings for offense and defense. This is a consequence of every team playing every other team the same number of times.

If we perform the calculations described in the sidebar on page 18, we can calculate the probability of win, lose, and draw for any pair of teams, home and away. For example, Table 3 gives these probabilities for the top few teams. To continue our example, we see from these tables that when Manchester United plays Arsenal at Manchester, they will win with probability .53, draw with probability .27, and lose with probability .20.

Table 3—Probabilities of a Win, Draw, or Loss for Selected Match-ups

Home team	Away team	Prob. of win	Prob. of draw	Prob. of loss
Man. U.	Liverpool	.48	.25	.26
Liverpool	Man. U.	.47	.25	.27
Man. U.	Newcastle U.	.53	.24	.23
Newcastle U.	Man. U.	.44	.26	.30
Newcastle U.	Liverpool	.43	.26	.31
Liverpool	Newcastle	.52	.25	.23
Man. U.	Arsenal	.53	.27	.20
Arsenal	Man. U.	.37	.30	.33
Arsenal	Liverpool	.37	.30	.33
Liverpool	Arsenal	.52	.28	.20
Arsenal	Newcastle U.	.41	.30	.29
Newcastle U.	Arsenal	.49	.29	.22

Table 4—Results From Simulating the Season

Team	Actual points 95/96	Poisson model expected points	Simulated mean points	Simulated std. dev. points	Proportion at top of table
Man. U.	82	75.7	75.5	7.1	.38
Newcastle U.	78	70.7	70.5	7.8	.16
Liverpool	71	74.9	74.9	7.5	.33
Arsenal	63	63.8	63.6	7.7	.03
Aston Villa	63	63.7	63.6	7.4	.03
Blackburn R.	61	61.2	61.4	7.4	.03
Everton	61	64.9	65.0	7.5	.04
Tottenham H.	61	60.2	60.8	7.5	.01
Nottm. Forest	58	50.0	49.5	7.4	.00
West Ham. U.	51	46.3	46.1	7.7	.00
Chelsea	50	53.4	53.5	7.4	.00
Leeds U.	43	41.4	41.4	7.4	.00
Middlesbro	43	41.5	41.8	7.4	.00
Wimbledon	41	44.7	44.7	7.6	.00
Sheff. Wed.	40	44.8	44.9	7.2	.00
Coventry C.	38	41.2	41.4	7.6	.00
Man. City	38	35.7	35.4	6.9	.00
Southampton	38	39.6	39.5	7.0	.00
QPR	33	39.9	40.1	7.3	.00
Bolton Wan.	29	33.9	34.0	7.2	.00

Simulating the Season

Now we can approach the problem of whether or not Manchester United *was* lucky to top the table in the 95/96 season. As we noted previously, the Poisson regression approach allows us to calculate the chance of a win, loss, or draw for a game between any pair of teams. In principle, this allows us to calculate exactly the chance a given team will top the table. The calculation is too large to be practical, however, so we resort instead to simulation.

For each of the 380 games played, we can simulate the outcome of each game. Essentially, for each game, we throw a three-sided die (conceptually only) whose faces are win, lose, and draw. The probabilities of these three outcomes are similar to those given in the preceding tables. From these 380 simulated games, we can calculate the points table for the season, awarding three points for a win and one for a draw, and see which team topped the table.

In fact we used a computer program to simulate the 95/96 season 1,000 times. We can calculate the mean and standard deviation of 1,000 simulated points totals for each team and also the expected number of points under the Poisson model described previously. We can also count the proportion of times each team topped the table in the 1,000 simulated seasons, which gives an estimate of the probability of topping the table. Table 4 gives this information.

Manchester seems to have been a little lucky, but it still has the highest average score. Liverpool was definitely unlucky and according to our model is really a better team than Newcastle United, who actually came second.

Of course, our approach to modeling the scores is a little simplistic. We have taken no account of the fact that teams differ from game to game due to injuries, trades, and suspensions. In addition, we are assuming that our model leads to reasonable probabilities for winning/losing/drawing games. Teams that tend to "run up the score" against weak opponents may be overrated by a model that looks only at scores, and teams that settle into a "defensive shell" once they have got the lead may be underrated. Still, our results do seem to correspond fairly well to the historical result of the 95/96 season.

References and Further Reading

Groeneveld, R. A. (1990), "Ranking Teams in a League With Two Divisions of t Teams," *The American Statistician*, 44, 277–281.

Hill, I. D. (1974), "Association Football and Statistical Inference," *Applied Statistics*, 23, 203–208.

Keller, J. B. (1994), "A Characterization of the Poisson Distribution and the Probability of Winning a Game," *The American Statistician*, 48, 294–298.

McCullagh, P., and Nelder, J. A. (1989), *Generalised Linear Models*, London: Chapman and Hall.

Schwertman N. C., McCready, T. A., and Howard, L. (1991), "Probability Models for the NCAA Basketball Tournaments," *The American Statistician*, 45,179–183.

Stern, H. S. (1995), "Who's Number 1 in College Football? . . . And How Might We Decide?," *Chance*, 8(3), 7–14.

Chapter 41

Down to Ten: Estimating the Effect of a Red Card in Soccer

G. RIDDER, J. S. CRAMER, and P. HOPSTAKEN*

We investigate the effect of the expulsion of a player on the outcome of a soccer match by means of a probability model for the score. We propose estimators of the expulsion effect that are independent of the relative strength of the teams. We use the estimates to illustrate the expulsion effect on the outcome of a match.

KEY WORDS: Conditional likelihood; Poisson process; Soccer; Unobserved heterogeneity.

1. INTRODUCTION

Professional soccer (known outside the United States as football) is popular all over the world; in Europe and South America it is the dominant spectator sport. Because soccer is a low scoring game, the rules have been often revised so as to raise the number of goals scored by either side and thus increase the play's appeal. Since 1990, players can be expelled for the rest of a match for illegal defensive actions, such as repeated flagrant fouls and preventing an adverse goal by illegal means. The referee expels the player by showing him a red card.

In this article we investigate the effect of such an expulsion on the outcome of a match. Popular opinion holds widely different views on the effectiveness of the red card, but as far as we know the question has not been submitted to empirical research. We propose a model for the effect of the red card that allows for initial differences in the strengths of the teams and for variation in the scoring intensity during the match. More specifically, we propose a time-inhomogeneous Poisson model with a match-specific effect for the score of either side. We estimate the differential effect of the red card by a conditional maximum likelihood (CML) estimator that is independent of the match-specific effects. This estimator was introduced in econometrics by Hausman, Hall, and Griliches (1984), building on ideas of Andersen (1973).

In Section 2 we specify the model, in Section 3 we discuss estimation, and in Section 4 we give the results. We consider some implications of the estimates in Section 5.

2. A MODEL FOR THE SCORE IN A SOCCER MATCH

First, we introduce some notation. The subscript i denotes a match, and $j = 1, 2$ denotes the two sides in that match; a team in a match is thus identified by two subscripts ij. We restrict attention to matches with a red card, and we always take it that the red card is given against the second side, $j = 2$. Time is measured in minutes from 0 to 90, which is the official duration of a match. In soccer the clock is not stopped when play is interrupted, but the referee can allow for lost time at the end of the first and second halfs, after 45 and 90 minutes. Recorded time is measured from the beginning of the match and from its resumption after the interval, however. As a result, there may be some minutes when there is no play at all, whereas the 45th and 90th minutes may last longer than a full minute; but this is a minor distortion.

Let

τ_i = minute in which a player is expelled from team 2,
N_{ij} = total number of goals scored in match i by team j,
K_{ij} = number of goals scored before τ_i,
M_{ij} = number of goals scored after τ_i,
$\lambda_{ij}(t)$ = scoring rate or intensity of team j in match i at the tth minute of play,
θ_j = multiplicative effect on $\lambda_{ij}(t)$ of expulsion of player from team 2, and
γ_{ij} = relative strength of team j in match i as compared with the overall average scoring rate, $\lambda(t)$.

We make the following three assumptions:

1. The two teams score according to two independent Poisson processes. As a consequence, the number of goals scored by team 1 is stochastically independent of the number of goals scored by team 2. Moreover, the time intervals between subsequent goals are stochastically independent. The scoring intensities are not constant during the match; thus the Poisson processes are nonhomogeneous.
2. The ratio of the scoring intensities of the two full teams is a constant for each game; that is, $\lambda_{ij}(t) = \gamma_{ij}\lambda(t)$ for matches of 11 against 11 players, with $\lambda(t)$ the average scoring intensity at the tth minute of play of full sides of 11 against 11.
3. After the red card, for $t > \tau_i$, team 2 has 10 players, and the scoring intensities are $\theta_j\gamma_{ij}\lambda(t), j = 1, 2$.

In Assumption 1 we describe the score in a match as a random phenomenon that is only partly predictable. It depends on the playing time, on the relative strength of the teams, and on the effect of the red card. As we show, the scoring intensity increases with the time played. If we do not allow for this, then the effect of the red card will be overstated, because we confound it with the time effect. Of course the score is strongly affected by the relative strength of the teams. The incidence of red cards may be related to the relative strength, so that a comparison of red card games to uninterrupted games gives a biased estimate of the effect. In ad-

* G. Ridder is Professor, Department of Econometrics, Free University, Amsterdam, The Netherlands. J. S. Cramer is Professor, Department of Economics, and P. Hopstaken is Senior Research Fellow, Foundation for Economic Research, University of Amsterdam, The Netherlands. The authors thank Gusta Renes for helpful comments, Tony Lancaster for spotting an embarrassing error in a previous version, and the editor and two referees for comments that have improved the article considerably.

Journal of the American Statistical Association
September 1994, Vol. 89, No. 427, Statistics in Sports

dition, the timing of red cards may also be related to the relative strength of the teams, and again this biases the effect. The third factor is the effect of the red card, which by Assumption 3 is measured by θ_1 and θ_2.

It is not our aim to predict the outcome of soccer matches, which requires an estimate of γ_{ij}. Our estimate of the effect of the red card is independent of γ_{ij}, which is of great help, because finding a good estimate of γ_{ij} is difficult as experience shows.

The Poisson assumption and its implications form Assumption 1. It is not difficult to relax the Poisson assumption at the cost of a more complex statistical model, but our limited number of observations will not support this.

3. STATISTICAL ANALYSIS

3.1 Estimation of the Average Scoring Intensity

In Table 1 goals scored in 340 full matches in the two professional soccer divisions in the Netherlands in the 1991–1992 season are classified by 15-minute intervals of play. This shows that the rate of scoring increases monotonically over the match, as has also been observed in England by Morris (1981).

If we assume that the average scoring intensity increases linearly during the match, then the expected number of goals scored by a team j in match i in interval s is

$$E(N_{ijs}) = \gamma_{ij}(15\alpha + 112.5\beta(2s-1)) \qquad s = 1, \ldots, 6, \quad (1)$$

so that the average number of goals scored by one team in interval s is

$$E(\bar{N}_s) = 15\alpha + 112.5\beta(2s-1) \qquad s = 1, \ldots, 6, \quad (2)$$

where we take $\bar{\gamma} = 1$. This implicitly defines a scale for γ_{ij}; for example, if $\gamma_{ij} = 2$, then team j has a scoring intensity in match i that is two times the average.

The average number of goals per minute in time interval s equals the entry in Table 1 divided by 680, twice the number of contests. Estimates of α and β are then easily obtained by ordinary least squares (OLS) regression. With $\lambda(t)$ as the scoring intensity for a 90-minute game, we find ($R^2 = .95$; standard errors in parentheses)

$$\hat{\alpha} = 1.050(.024) \quad \text{and} \quad \hat{\beta} = .00776(.00072).$$

Note that the reported standard errors are consistent in the presence of heteroscedasticity. Inclusion of a quadratic term did not improve the fit. In the sequel we ignore the sampling variance of these estimates. This simplifies the computation of variances and is an acceptable approximation, as they are small. The estimates imply that the scoring intensity increases during a 90-minute game from 1.05 in the first minute to 1.75 in the final minute.

3.2 A Conditional Maximum Likelihood Estimator

Because the incidence of red cards is probably related to the relative strength γ_{ij}, a comparison of red card matches with other matches may give a biased estimate of the red card effect. For that reason, we propose an estimator that does not depend on the γ_{ij}'s or on their distribution. This estimator is based on a comparison of the number of goals scored by the same team before and after the red card. More precisely, we consider the fraction of the goals scored after the red card, which we denote by y_{ij}. It is intuitively clear that this fraction is independent of the time-constant match-specific effect.

Table 1. Goals Scored in the 1991–1992 Season by 15-Minute Intervals

Time interval (min)	Number of goals
0–15	128
16–30	140
31–45	147
46–60	169
61–75	170
76–90	198

Under Assumptions 1 to 3 (with P denoting the Poisson distribution),

$$K_{ij} \sim P\left(\gamma_{ij}\int_0^{T_i}\lambda(t)\,dt\right) \quad \text{and} \quad M_{ij} \sim P\left(\theta_j\gamma_{ij}\int_{T_i}^{90}\lambda(t)\,dt\right). \quad (3)$$

In the sequel we denote

$$A_i = \int_0^{T_i}\lambda(t)\,dt \quad \text{and} \quad B_i = \int_{T_i}^{90}\lambda(t)\,dt. \quad (4)$$

The conditional distribution of M_{ij}, given N_{ij}, is

$$M_{ij}|N_{ij} \sim B(N_{ij}, g_{ij}(\theta)), \quad (5)$$

where B denotes the binomial distribution and

$$g_{ij}(\theta) = \frac{\theta_j B_i}{A_i + \theta_j B_i}. \quad (6)$$

The conditional distribution is degenerate if $N_{ij} = 0$, and y_{ij} is defined only if $N_{ij} \geq 1$. In the CML procedure we omit observations with $N_{ij} = 0$. The estimator of the red card effect is not biased by this restriction, as we shall see presently.

In the conditional distribution (5) and in the conditional likelihood, the match-specific effects γ_{ij} cancel. Up to an additive constant that does not depend on θ_j, the log-likelihood is

$$\log L_j = \sum_{i=1}^{n_j} M_{ij}\log(g_{ij}(\theta_j)) + (N_{ij} - M_{ij})\log(1 - g_{ij}(\theta_j)), \quad (7)$$

with n_1, n_2 denoting the number of observations on teams that do not and do receive a red card. Because we condition on the total scores, N_{ij}, we can treat them as nonstochastic constants. Hence omitting observations with a given total score—in particular, observations with $N_{ij} = 0$—does not affect the CML estimator.

The likelihood equation is

$$\sum_{i=1}^{n_j} N_{ij}g_{ij}(\hat{\theta}_{\text{CML}_j}) = \sum_{i=1}^{n_j} N_{ij}y_{ij}. \quad (8)$$

This is a moment equation, equating a weighted average of the y_{ij}'s to a weighted average of their expectations. The weights are the total scores N_{ij}, which by conditioning can be treated as known constants.

In deriving the properties of the CML estimator, we note that the binomial parameter $g_{ij}(\theta_j)$ can be written in the logit form. Hence the log-likelihood is globally concave in $\log(\theta_j)$, so that the CML estimator for θ_j is uniquely defined. The asymptotic variance of the CML estimator can be obtained in the usual way.

3.3 OLS Estimation

With an additional assumption, we can estimate the effect of the red card by linear regression. From (3),

$$K_{ij} = \bar{\gamma}_j A_i + (\gamma_{ij} - \bar{\gamma}_j)A_i + (K_{ij} - E(K_{ij}|\gamma_{ij})) = \bar{\gamma}_j A_i + v_{1ij}$$

and

$$M_{ij} = \bar{\gamma}_j \theta_j B_i + (\gamma_{ij} - \bar{\gamma}_j)\theta_j B_i + (M_{ij} - E(M_{ij}|\gamma_{ij})) = \bar{\gamma}_j \theta_j B_i + v_{2ij}. \quad (9)$$

In (9) we allow the average relative strength in red card games to differ from the overall average 1. $\bar{\gamma}_1$ and $\bar{\gamma}_2$ indicate the average strengths of the teams before a player of team 2 is expelled. The disturbances v_{1ij} and v_{2ij} are independent, and an additional assumption is required for consistent estimates, viz. $\text{cov}(\gamma_{ij}, A_i^2) = \text{cov}(\gamma_{ij}, B_i^2) = 0$. A sufficient condition for this is that τ_i and γ_{ij} are stochastically independent. Under this assumption, we can estimate θ_j by the ratio of the regression coefficients in (9).

4. ESTIMATION RESULTS

We apply CML estimation to data on 140 red card games in the seasons 1989–1990, 1990–1991, and 1991–1992 in both divisions of the Dutch professional football league. In 13 of these matches, two or more red cards were given. Because we estimate the effect of being one player up or down, the part after the second expulsion is omitted. In only two matches were a red card and a penalty kick given jointly. Because for the CML estimator we must omit observations where a team has not scored at all, the effective number of observations is 112 for teams with 11 players and 93 for teams with 10 players. We obtain the following results (standard errors in parentheses):

$$\hat{\theta}_{\text{CML1}} = 1.88\ (.29) \quad \text{and} \quad \hat{\theta}_{\text{CML2}} = .95\ (.20).$$

According to the CML estimates, the scoring intensity increases by 88% for the team with 11 players; this effect is statistically significant. The scoring intensity for the team with 10 players (team 2) hardly changes; the effect is not significantly different from 1.

The OLS estimator gives rather different results (with the standard errors consistent in the presence of heteroscedasticity):

$$\hat{\theta}_{\text{OLS1}} = 1.43\ (.03) \qquad \hat{\theta}_{\text{OLS2}} = 1.14\ (.03).$$

The estimated increase in the scoring intensity for team 1 is much smaller than for the CML estimator (but highly significant). More surprisingly, the OLS estimator shows a statistically significant increase in the scoring intensity for team 2. Hence using between-game information gives rather different estimates that moreover are hard to interpret. The first-stage regressions of the OLS estimator show that teams that receive the red card have the same scoring intensity as the average ($\bar{\gamma}_2 = 1.03\ (.09)$), but the opposing team is much stronger ($\bar{\gamma}_1 = 1.33\ (.09)$). Hence the red card usually is given to the already weaker team.

By stratifying our sample, we can investigate whether the estimates are robust against changes in the specification. First, we test whether the red card effect depends on the venue of play. This captures, among other things, the home advantage, and the estimate should be invariant to this distinction. The LR statistic is .53 for the team with 11 players and .66 for the team with 10 players; hence we can not reject invariance. The estimates are $\theta_{1,\text{home}} = 2.00\ (.36)$, $\theta_{1,\text{away}} = 1.56\ (.46)$, and $\theta_{2,\text{home}} = .73\ (.28)$, $\theta_{2,\text{away}} = 1.07\ (.27)$. The estimates are also invariant to stratification on the total score in a match. In the sequel we use the CML estimates to illustrate the effect of the red card on the outcome of a match.

5. IMPLICATIONS OF THE ESTIMATES

We can use the results to illustrate the effect of the red card on a soccer match. In Table 2 we give the probabilities of the three possible outcomes of the match between equally strong teams as a function of τ. The last row of the table shows that the probability of a draw between two teams of average strength is .25. This is an indication of the role of chance in the outcome of a soccer match. The role of chance was also stressed by Osmond (1993). A red card early in the match increases team 1's probability of victory substantially.

Table 2. Probabilities of the Outcome of the Match by Minute of the Red Card

Minute of red card τ	Pr(team of 11 wins)	Pr(draw)	Pr(team of 10 wins)
0	.65	.17	.18
15	.62	.18	.20
30	.58	.20	.22
45	.54	.21	.25
60	.49	.23	.28
75	.44	.24	.32
90	.375	.25	.375

Table 3. Probabilities of the Outcome of a Match with a 15-Minute Exclusion Starting at τ

Minute of start of penalty τ	Pr(team of 11 wins)	Pr(draw)	Pr(team of 10 wins)
0	.42	.24	.34
15	.42	.24	.34
30	.43	.24	.33
45	.43	.24	.33
60	.43	.24	.33
75	.44	.24	.32

Table 4. Expected Number of Goals in Match by Minute of Red Card

Minute of red card τ	*Expected number of goals*
0	3.95
15	3.80
30	3.63
45	3.45
60	3.25
75	3.03
90	2.80

Team 2's probability of victory decreases even more, whereas the change in the probability of a draw is relatively small.

With a red card, a player is expelled for the remainder of the match. In indoor soccer and ice hockey, a player can be excluded for a certain period. In Table 3 we show the effect of a 15-minute time penalty for equally strong teams. Although the effect depends on the time at which the penalty is imposed, this dependence is rather weak.

As noted in Section 1, a motivation for the more frequent use of the red card is to increase the number of goals scored in a match. Table 4 shows that it has the desired effect.

We also consider the dilemma of a defender who faces a player who threatens to break through the defense. If the opposing player has a clear way to the goal, tripping up the player results in a red card for the defender. If the player goes past the last defender, he will score with a high probability. In our calculation we assume that the objective of the defender is to minimize the probability of losing the match. There is a unique moment in a contest at which the optimal action of the defender changes. After that moment, it is optimal to trip up the opposing player. These times, which depend on the probability that the attacker will score and on the relative strength of the defender's team (with the attacker's team of average strength, $\gamma = 1$), are reported in Table 5. The weaker side has a stronger incentive to resort to illegal defense. This is consistent with our observation that the red card is usually given to the weaker side. It may also induce a correlation between τ and γ, and such a correlation biases the OLS estimates.

Table 5. Time (Minute of Game) After Which a Defender Should Stop a Breaking-Away Player by Probability of Score and Relative Strength of the Defender's Team

Relative strength of teams, γ	*Probability of score* .3	.6	1
.5	70	42	0
1	71	48	16
2	72	52	30

APPENDIX: DATA USED IN ANALYSIS

The symbols are introduced in Section 2.

K1	*K2*	*M1*	*M2*	τ	*K1*	*K2*	*M1*	*M2*	τ	*K1*	*K2*	*M1*	*M2*	τ
1	0	5	1	10	1	1	0	1	58	0	0	1	0	82
0	0	0	0	11	1	2	0	0	58	6	0	1	0	83
1	1	3	3	15	2	0	0	1	58	1	1	0	0	83
0	0	1	0	17	1	0	3	0	60	0	1	0	1	83
0	0	1	2	20	0	2	1	1	60	1	1	0	0	84
0	0	0	0	25	1	1	2	1	61	1	0	0	0	84
0	0	0	0	25	1	1	2	0	61	1	3	0	0	84
0	0	3	0	26	1	0	0	0	62	1	1	0	0	85
0	0	2	0	30	2	1	2	0	62	2	2	0	0	85
1	0	5	1	32	0	0	1	0	65	0	2	0	0	85
0	0	3	0	33	2	0	1	0	65	4	0	0	0	85
0	0	1	2	33	0	1	0	0	65	1	1	0	0	85
0	1	1	0	33	2	0	1	1	65	1	1	0	0	85
0	0	2	0	33	0	0	0	0	66	2	1	1	0	86
0	0	1	1	35	0	2	0	0	67	1	2	0	0	86
1	0	2	0	36	0	1	2	2	68	2	0	0	1	86
1	0	1	2	36	1	0	1	0	68	2	1	0	1	86
0	0	3	0	37	0	1	1	0	68	2	1	1	0	87
0	1	1	0	39	0	1	2	0	70	2	2	0	1	87
0	1	0	0	39	0	1	2	0	70	2	2	0	0	87
1	0	3	0	40	1	0	1	0	70	0	0	0	0	88
3	0	2	2	44	0	1	0	0	70	3	1	1	0	88
0	0	2	0	44	1	2	0	0	70	2	1	1	0	88
1	0	1	0	44	3	0	0	0	70	2	1	1	0	88
0	0	1	1	44	0	0	0	0	71	2	0	0	0	88
0	1	0	0	44	1	1	0	0	71	4	0	0	0	88
0	0	0	1	44	1	1	0	0	73	2	1	0	0	88
0	1	1	0	45	1	0	1	0	73	2	1	0	0	88
1	1	1	0	45	0	0	0	0	74	2	0	0	0	88
0	0	0	1	46	1	0	1	0	74	1	2	0	0	88
0	0	4	1	47	1	1	0	0	74	2	1	1	0	89
1	0	2	0	47	0	3	0	0	75	1	0	0	0	89
0	0	0	0	48	1	1	1	0	77	0	1	0	0	89
2	0	2	0	50	1	1	0	0	78	3	0	0	0	28*
1	0	2	2	50	3	1	0	0	78	0	1	1	1	30*
0	0	0	0	52	2	0	0	1	79	2	0	0	2	36*
0	0	2	0	52	2	1	1	0	80	0	1	1	0	36*
2	0	1	0	52	1	1	1	0	80	0	1	2	0	40*
3	2	4	0	52	1	2	1	1	80	1	1	2	2	43*
0	2	0	0	53	1	0	0	0	80	2	3	1	0	45*
2	1	2	0	55	0	3	0	0	80	0	0	0	0	60*
0	1	0	1	55	0	1	0	0	80	1	1	0	0	63*
0	0	0	1	55	3	2	0	0	81	1	2	1	0	64*
1	0	3	1	55	0	0	0	0	82	1	1	0	0	69*
0	0	0	0	56	1	1	0	0	82	0	0	1	0	71*
2	0	1	0	56	1	0	0	1	82	2	1	0	0	78*
1	1	1	2	56	1	3	0	0	82					

* Matches with two or more red cards. Second red card in 77, 85, 78, 73, 60, 89, 68, 67, 65, 88, 76, 77, 82.

[Received January 1993. Revised March 1994.]

REFERENCES

Andersen, E. B. (1973), *Conditional Inference and Models for Measuring,* Copenhagen: Mental Hygiejnisk Forlag.

Hausman, J. A., Hall, B. H., and Grilliches, Z. (1984), "Econometric Models for Count Data With an Application to the Patents–R&D Relationship," *Econometrica,* 52, 909–938.

Morris, D. (1981), *The Soccer Tribe,* London: Jonathan Cabe.

Osmond, C. (1993), "Random Premiership?," *RSS News,* November, 5.

Chapter 42

Getting slammed during your first set might affect your next!

Heavy Defeats in Tennis: Psychological Momentum or Random Effect?

David Jackson and Krzysztof Mosurski

Sports statistics is a very diverse area. This article is concerned with (1) contests between individuals that are decided not by a single trial but by a series of trials and (2) the dependency structure that may exist between trials in such contests.

Psychological Momentum

There is a widespread belief in many walks of life, not just sports, that "success breeds success and failure breeds failure." If winning a trial increases the probability of winning the next trial, then that kind of dependency structure is quite properly called psychological momentum (PM). Unfortunately "momentum" is, at present, a much abused word that has found its way into the vocabulary of practically every sports commentator and fan alike to account for even the most mundane sequences of successes or failures. If one can demonstrate that PM is truly a factor in a given sport, however, then heavy defeats are a consequence of that dependency structure. There is clearly a strong positive relationship between PM and sequences of successes or failures.

A best-of-five-sets tennis match is a good example of the type of contest that is of interest. And the interest is in the possibly changing probability of winning a set as the match progresses. If PM is a factor, we are talking about a true dependency structure for the probability of winning a set, not merely an updated estimate of an unchanging probability based on additional data. The reason, of course, that one doubts that the sets of a best-of-five-sets tennis match are independent is that the memoryless property, from which assumptions of independence usually gain their strength, is missing in such a series of trials. No matter how much either participant might wish otherwise, the outcomes of the previous sets that have led to the present score in the match are known to both contestants. Perhaps we are made of such stuff that knowledge of what has happened earlier does not affect our probability of winning the next set. But perhaps it does. It is regrettable, but nonetheless true, that in analyzing data, and not just sports data, assumptions of independence are often very casually made in the literature.

The Search for Psychological Momentum in Sport

It is generally accepted in contests that are decided by a series of trials that PM can play a major role in the outcome. It is a long road, however, from being "generally accepted" to being "well known," and the search by authors for evidence of the existence of PM in sport has generated a fair amount of sometimes heated debate in recent years. It is an area that has seen a considerable research effort with numerous works, mainly on basketball, baseball or tennis, since the seminal article on the subject in the statistics literature by Tversky and Gilovich (1989). Their analysis of consecutive shots in basketball shows that contrary to popular belief the chances of a player hitting a shot are as good after a miss as after a hit. In baseball, analysis of hitting streaks (Albert 1993; Albright 1993; Stern 1993) also failed to detect any significant effect on the probability of making a hit, due to a player's recent history of success or failures. According to Stern (1995), the most credible evidence, so far, for the existence of psychological momentum in sport has been provided by tennis (Jackson 1993, 1995). Those works show that, when the odds in the first set of a match are estimated from explanatory variables, then a "success-breeds-success" model provides a much better fit to data from the 1987 Wimbledon and the U.S. Open tennis tournaments than an independent-sets model. These data exhibit far more heavy defeats than can be accommodated by the independence model, which assumes that the probability of winning a set remains constant in a given match. The success-breeds-success model—that is, PM—explains the tennis data extremely well.

Random Variation in Player Ability and Heavy Defeats

There is a possible alternative explanation, however, for the apparent overabundance of heavy defeats that we observe in tennis, and that is random variation in player ability from day to day. A random-effects model for player ability provides a good explanation of a common occurrence in sport in which a player inflicts a heavy defeat on his opponent on one day but himself suffers a heavy defeat from the same opponent on the next day. If a player's ability varies randomly from day to day (but remains relatively constant on any given day), then such apparent reversals of form are to be expected because, for the same two players, the probability of winning a set may vary substantially from day to day. Of course, PM explains such reversals of form equally well. The question we are posing in the title of this article is: Should attributed overabundance of heavy defeats that we observe in tennis(3/0 to either player) be put down to PM or could it equally well be attributed solely to a random day-to-day fluctuation in the ability of the contestants? We answer this question by comparing these alternatives on the basis of two years of data from the Wimbledon and U.S. Open tennis tournaments. In addition the models are fitted to a dataset containing the career "head-to-head" records of Ivan Lendl versus Jimmy Connors and John McEnroe versus Björn Borg.

Table 1—Model Comparisons

Wimbledon and U.S. Open tennis tournaments 1987–1988			
Data for 1847 sets from 501 matches, which includes current rankings of players in each match.			
"Simple" independence	Odds model	Independence with a normal random effect	Odds model with a normal random effect
Model (A)	Model (B)	Model (AR)	Model (BR)
$\hat{\alpha} = .510$	$\hat{\alpha} = .441$	$\hat{\alpha} = .532$	$\hat{\alpha} = .459$
s.e. = .03	s.e. =.035	s.e. = .04	s.e. = .036
	$\log(\hat{k}) = .391$	$\hat{\delta}^2 = .625$	$\log(\hat{k}) = .332$
	s.e. = .05	s.e. = .12	s.e. = .05
	$\Rightarrow \hat{k} = 1.48$		$\Rightarrow \hat{k} = 1.32$
			$\hat{\delta}^2 = .142$
			s.e. = .09
Degrees of freedom = 1,846	Degrees of freedom = 1,845	Degrees of freedom = 1,845	Degrees of freedom = 1,844
Deviance = 2,329	Deviance = 2,264	Deviance = 2,291	Deviance = 2,261

NOTE: Some parameter estimates with standard errors and goodness-of-fit statistics.

The Evidence for Psychological Momentum in Tennis

We were fortunate that, at the same time the basketball and baseball work was taking place, we were trying to detect these psychological effects in tennis (Jackson 1993, 1995)—fortunate because in the other two sports the magnitude of any psychological effect was likely to be small and hence difficult to detect. Even if successive attempts at a shot in basketball or at making a hit in baseball were independent, there was never any possibility that they were identically distributed; that is, the probability of success in both those sports depends to a large extent on the situational variables. In tennis we were not faced with this latter problem because the sets are supposedly identical in the sense that the format is for practical purposes identical and designed not to convey an advantage to either player. And as it turned out, for the models we fit to our tennis data, the magnitude of the psychological effect is considerable.

The Wimbledon and U.S. Open Tennis Data

The main dataset of Jackson (1993) consists of the 251 completed best-of-five-sets matches from the men's singles at Wimbledon and the U.S. Open in 1987. Matches lasted of necessity 3, 4, or 5 sets and in total 918 sets were played. For each match the order in which the sets were won is available, which allows the score in sets at the commencement of a set to be included in any model. Moreover, the official rankings of the players as given by the Association of Tournament Professionals (ATP) are available. These ranks are treated as explanatory variables from which information on the relative abilities of the players in each match is extracted. In particular this prior information is used to obtain an estimate of the odds in the first set of each match. This allows a more thorough investigation of the possible dependency structure between the outcome of sets within a match.

The Relationship Between Ranks and Odds for Professional Tennis Players

The professional players in our dataset are the elite players from a large population of tennis players. What this implies is that, if we treat tennis ability as an attribute that has some standard but unknown distribution, then the expected relationships between the varying amounts of this attribute for the elite players are just those that apply in the tail of this unknown distribution. How one estimates odds in the first set of a contest between two such players when the ranks of the players are known is not central to the issues we are addressing here. Suffice it to say that there are standard procedures available in ranking and rating theory, which depend only on what form one assumes for the tail of the distribution. If we define $O(r,s)$ to be the odds that a player ranked r beats a player ranked s in the first set, then the particular estimator that we use is

$$O(r,s) = \left(\frac{s}{r}\right)^{\alpha} = (\text{ratio of ranks})^{\alpha} \qquad (1)$$

or equivalently

$$\log(\text{odds of success}) = \alpha \cdot \log(\text{ratio}) \qquad (1a)$$

where α is a parameter to be determined from the data.

Odds Model

$$O_{ij} = k^{i-j}\, O_{00} \qquad (2)$$

where (i,j) is the score in sets and O_{00} are the odds in the first set.

Or taking logs

$$\log(O_{ij}) = \log(O_{00}) + \log(k) * (i-j) \qquad (3)$$

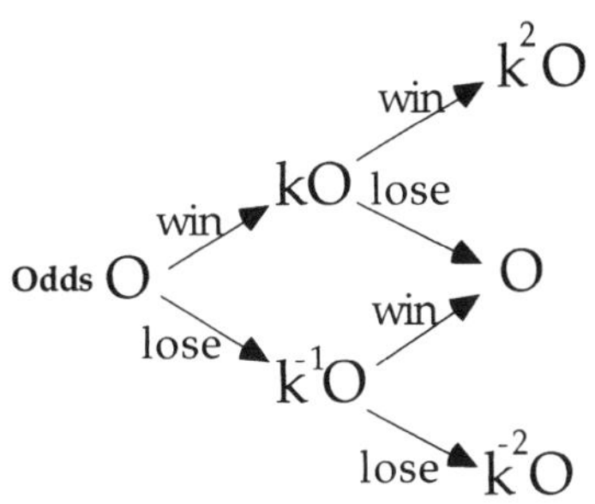

Figure 1. The odds model: Winning a set increases the odds of winning the next set by a factor *k*.

Because for this estimator only the ratio of the ranks is relevant, this implies that, for example, the highest ranked player has the same probability of success against the 4th ranked player as the 20th ranked player has against the 80th. The parameter α determines what this probability is. Small values of the parameter α imply a large random element to the outcome regardless of differences in rank, but large α implies that even small differences in rank lead to a high probability of success for the higher ranked player.

Although the relationship between ranks and odds for the elite players as given by the preceding equation has a strong theoretical basis and has been used by several authors, its usefulness depends on the accuracy of the ranking system used by the ATP. If the ranking system is poor and lesser players have been ranked above better players, then the predictive value of the estimator will suffer. It is necessary to test whether our model for odds in the first set of a match actually fits the data. Because we are dealing with individual successes or failures in each set, we need to group the Wimbledon and U.S. Open tennis data to test for goodness of fit. When this is done it can be shown that the model does indeed provide a good fit to the data. This not only validates our use of this particular estimator but also lends support to what is a widely accepted view among professional tennis players that the ranking system provides a fair and reasonably accurate guide to the relative merits of the tournament players.

The Odds Model

We now introduce a model for the odds of winning a set, the odds model, that incorporates the effect due to the "score in sets" at the commencement of the set. The odds model was one of several models that allowed for the existence of PM that were fitted to the Wimbledon and U.S. Open tennis data. We define O_{ij} to be the odds of winning the next set (the $i+j+1$st set) when the score is (i,j) in sets. Then, in simple terms, the odds model states that "Winning a set increases the odds of winning the next set by a factor k"(see Fig. 1).

For the odds model, the odds for success in the next set depend only on the difference between the number of successes (sets won) and failures up to that set and on O_{00}, the odds for success in the first set.

Because the ranks of the players in each match are known, we can exploit this information by using Equation (1a) to estimate O_{00}, the odds of winning the first set. We can then rewrite Equation (3) which defines the odds model, as follows:

$$\log(O_{ij}) = \alpha \cdot \log(\text{ratio}) + \log(k) * (i-j) \qquad (3a)$$

where (i,j) is the score in sets and ratio is the ratio of the ranks of the players in that match.

Taking $k = 1$ in the model statement eliminates the dependence on the score and therefore makes the independence model a special case of the odds model. Both the odds model and the independence model, in which the odds in the first set of each match are estimated from the ranks of the players, are fitted to the original 1987 Wimbledon and U.S. Open tennis data of Jackson (1993). The independence model provides a very poor fit, whereas the odds model, which includes the effect caused by the score, explains that data extremely well. For the odds model, the estimate for the parameter k, which is a measure of psychological momentum, is $k = 1.6$ with standard error .12. It is particularly noticeable for the independence model that it badly underestimates the number of heavy defeats in the data. Later in the article we fit both of these models and some additional models to a new Wimbledon and U.S. Open tennis dataset, which contains the results of matches from 1987 and 1988, and the improvement in the fit by introducing the effect due to the score into the model is just as marked (see Table 1) for the larger dataset as it was previously. Moreover, as before, the number of heavy defeats in the larger dataset (see Table 2) is very poorly accounted for by the independence model.

So far we have summarized the evidence for PM in tennis and mentioned some of the work that has taken place in the ongoing search for evidence of PM in other sports. Although the evidence for PM in tennis is strong, perhaps it is possible to tell a different story that explains the apparent overabundance of heavy defeats that we see in tennis. A possible candidate is a random-effects model for player ability.

Random-Effects Model for Player Ability

One may accept (we do but with certain reservations) that the ATP ranking system is adequate and also accept, as we do, that the ratio of the ranks of the players is a reasonable function to use in estimating odds in the first set of a match. Yet one may still argue that a player's ability varies from day to day. In that case one may argue that a player's ranking is only an indicator of his average ability and that the function $\alpha \cdot \log(\text{ratio})$ is an estimate of the average log-odds in the first set of a match between contestants with known ranks.

For instance a frequent occurrence in tennis is that a player inflicts a heavy defeat on his opponent on one day but himself suffers a heavy defeat by the same opponent in a subsequent match.

Results of two matches between the same players:
Day 1: A beats B 3/0
Day 2: A loses to B 0/3

In this example we have only heavy defeats.

Model 1: Psychological momentum

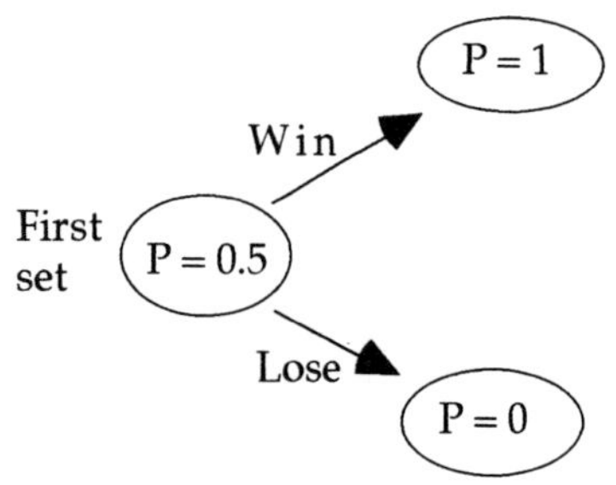

We can explain these data by PM, as before, by saying that there is a probability of .5 of winning the first set in each match but that whoever wins the first set has probability 1 of winning any subsequent set, a true dependency structure. [See Jackson (1995) for more detailed discussion of this type of dataset.]

Model 2: Player ability varies from day to day

For player A
$P = 1$ for each set on DAY 1
$P = 0$ for each set on DAY 2

Alternatively we can say that on average these players are of equal ability but that player A had probability 1 of winning every set on the first day and probability 0 on the second. In this case player ability varies substantially from day to day but remains fixed on any given day.

The second model is an example of a random-effects model for player ability, and it explains the apparent overabundance of heavy defeats that we see in the data just as well as PM; that is, the likelihood of the observation is the same for both models. It is also an independence model because the probability of winning a set does not vary within a match. If we were to predict what would happen in a subsequent match, we would say that one of the players will have a probability 1 of winning every set but that player is equally likely to be player A or B.

Random-Effects Models and Heavy Defeats

We can generalize this relationship between random effects and heavy defeats. If the true probability (P) of winning a set for Player A against Player B is a random variable from match to match with mean p, this implies that heavy defeats are more likely (for both players) than if Player A's probability of winning a set remained a constant (p) from match to match. This addition of a random effect because of an apparent overabundance of heavy defeats in the data is very similar to the introduction, in other circumstances, of a random effect into a model in an attempt to compensate for overdispersion.

Linear Logistic Model With a Random Effect

We want to adopt a linear logistic model for the relationship between the true probability P of winning a set, which is assumed to be a random variable independently chosen from day-to-day and any explanatory variables. In that case an independence model is that the log-odds for success in a set is a fixed effect, which is particular to that match and based on the relative abilities of the players, plus some random effect that is chosen independently for each match.

$$\log(\text{odds for success in a set}) = \text{fixed effect} + r \quad (4)$$

$$\log(\text{odds for success in a set}) = \alpha \cdot \log(\text{ratio}) + r \quad (4a)$$

- Independence model
- r is a random-match effect

As before, for players of known ranks, we choose to use $\alpha \cdot \log(\text{ratio})$ for the fixed effect leading to the model specified by Equation (4a). We also assume that the random component r has zero mean. In that case we associate the fixed term in the model with the average log-odds for that match. This is an independence model because the probability of success in a set is constant for a given match, although it will vary from match to match for the same two players due to the random effect. Similarly for matches between different players in which the ratio of the ranks is the same in both matches, the probability of winning a set will vary between matches because the random effects have been independently chosen.

To fit such a model we need to specify the distribution of the random variable r. Here we assume that the random

Model Formulas

Models A and B are straightforward logistic regressions; models AR and BR are logistic regressions with random effects.

Model	Formula
A	$\log(O_{ij,m}) = \alpha \cdot \log(\text{ratio}_m)$
B	$\log(O_{ij,m}) = \alpha \cdot \log(\text{ratio}_m) + \log(k)*(i-j)$
AR	$\log(O_{ij,m}) = \alpha \cdot \log(\text{ratio}_m) + r_m$
BR	$\log(O_{ij,m}) = \alpha \cdot \log(\text{ratio}_m) + \log(k)*(i-j) + r_m$

Where $\log(O_{ij,m})$ are the log-odds of the higher ranked player winning the next set when the score in sets is (i,j) in match m. Ratio_m is the ratio of the ranks of the players in match m as previously defined. The random effect $r_m \sim N(0, \delta^2)$ and the population parameters α, k, and δ^2 are to be estimated.

effects are chosen independently from a Normal$(0, \delta^2)$ distribution, although that is only one of many distributions that could reasonably have been chosen. Of course, whatever distribution is chosen one needs specialist software to fit any of these random effects models including the Normal. We have used the Multi Level modeling package (MLn) for all models, whether or not they include a random effect.

Parameter Estimation and Model Comparison: The New Wimbledon and U.S. Open Data

The results of the matches in the 1987 men's singles tournaments at Wimbledon and the U.S. Open made up the original dataset to which the odds model and some other dependent trials models (but not any random-effects models) were fitted in earlier works. Here we have added the matches from the 1988 tournaments at both venues. The dataset now consists of 1847 sets from 501 matches. In the accompanying sidebar we summarize the results of fitting the following four models to this dataset. The models that we wish to compare and which we specify in full in the sidebar are

A. The simple independence model
B. The odds model. This allows for the existence of PM, but does not preclude simple independence; that is, $k = 1$.
AR. The independence model with a $N(0,\delta^2)$ random effect. If $\delta^2 = 0$ this again reduces to simple independence.
BR. The odds model with a Normal random effect

The year and the venue were also considered as explanatory variables, however, because these did not have any significant effect we do not report the results here.

Model comparisons and interpretation of the analysis follow:

(1) AR and Â. By comparing deviances, we see from Table 1 that model AR, the independence model with a Normal random effect, is a big improvement when compared to the simple independence model. The estimate for δ^2 of .625 (s.e. = .12) is significantly different from 0, and there is a reduction in the deviance of 38.

(2) B and A. The improvement in fit for the odds model (B) over simple independence is significantly greater than that achieved by the random-effects model (AR). For the odds model there is a reduction of 65 in the deviance and the estimate for k, and the index of PM is 1.48.

(3) BR and B. The addition of a Normal random effect to the odds model does not significantly improve the fit.

The improvement in fit for the random-effects model (AR) over simple independence is not unexpected because we know that one of the main flaws in the simple independence model is that it badly underestimates the number of heavy defeats, and we suspected that the introduction of a random effect was likely to go some way toward correcting that defect. It doesn't go far enough, however. The analysis confirms that the impact of PM cannot be ignored. For the odds model, the estimate for k of approximately 1.5 is not only statistically significant but also has a large practical impact. For example, for two evenly matched players it implies that the winner of the first set will have a probability of .6 of winning the second set. If he wins that set then his probability of winning the third set is .69.

The estimate for the variance of the random effect in models AR and BR enables us to calculate the reasonable range of probabilities of winning the first set in each of these models. Even for model BR (it is much larger for model AR) this estimate of the variance, $\delta^2 = .142$, implies a considerable level of variation in player ability from day to day. In this case, for evenly matched players there is at least a 5% likelihood that the probability of winning the first set on any given day is outside the range .315 to .685.

The primary question we seek to answer is "Is it possible to rescue the concept of independent sets within matches solely by the addition of a random effect?" In other words, can we produce

Table 2—Results for Higher Ranked Player of 501 Matches From the 1987 and 1988 Wimbledon and U.S. Open Tournaments

Expected values for the following models.
(Parameters as given in Table 1.)

		Simple independence	Odds model	Independence with a Normal random effect
Result	Observed	Model A	Model B	Model AR
3/0	191	158.3	193.7	172.5
3/1	104	132.8	112.3	115.3
3/2	57	87.2	58.8	70.2
2/3	36	52.7	36.7	49.7
1/3	54	44.6	49.1	51.9
0/3	59	25.5	50.5	41.3

an independence model that is comparable to PM as an explanation of these data, or must we necessarily abandon the idea of independence, which is a much stronger statement. Well, we haven't rescued it yet, which is not to say that it cannot be rescued, perhaps by some radically different model for random variation in player ability than the one we have been considering. What is clear, however, is:

- The independence model with a normal random effect is not comparable to the odds model as an explanation of these data.
- The proposed model for variation in a player's ability contributes little to the overall fit, whereas the effect due to the score is substantial.

Heavy Defeats at Wimbledon and the U.S. Open

The number of heavy defeats that occurred at these two tournaments is an aspect of the data that is of considerable interest. Table 2 gives the results of matches for the higher ranked player in terms of sets won and lost and the expected numbers of these results for the various models. It includes both the number of heavy defeats suffered by the higher ranked player (the 0/3 results) and the number of heavy defeats suffered by the lower ranked player (the 3/0 results).

The order in which the sets were won and lost has been suppressed in the view of the data contained in Table 2, although knowledge of the order was used in estimating some of the parameters associated with the models. The expected values were obtained by using the known ranks of the players in each match, together with the fitted parameters, to calculate the probability of a 3/0, 3/1, 3/2 result for both players in each match and summing these probabilities over all 501 matches. For the simple independence model and for the odds model, this is a straightforward calculation. For the random-effects model, however, the likelihood of a 3/0, 3/1, 3/2 result in a given match is dependent on the particular value of the random effect in that match and it is necessary to evaluate (numerically) some rather inelegant-looking integrals to calculate the unconditional likelihood for each result.

The simple independence model underestimates the number of heavy defeats in these data, considerably so for the lower ranked player (the 3/0 results) and dramatically so for the number of heavy defeats suffered by the higher ranked player (the 0/3 results). The independence model with a Normal random effect, model AR, fits this aspect of the data much better but is still not comparable to the odds model, so even if we were to judge solely on the criteria of how well the models fit to the heavy-defeats aspect of the data, the proposed model for random variation in player ability from day to day is not going to rescue the concept of an "independent-sets-within-matches" model as an explanation of these data.

In Table 2, we have refrained from including the expected values for model BR—that is, the odds model with the addition of a Normal random effect. There are two main reasons for this. First, we are primarily concerned with the comparison between the odds model and an independent-sets-within-matches model. Second as we saw in Table 1, the fit to the data for this full model, which includes both the PM effect and the effect due to random variation in player ability, is not significantly better than the model for PM on its own—namely, the odds model. Indeed the expected numbers of 3/0 and 0/3 results are very similar for both models. As the expected values are extremely burdensome to compute, we did not proceed with the computations for the other possible results for this model. If one accepts the existence of PM, however, then the full model is a reasonable starting point in any investigation of the relative contributions of random variation in player ability and PM to the observed outcomes.

[Tversky and Gilovich's] analysis of consecutive shots in basketball shows that contrary to popular belief the chances of a player hitting a shot are as good after a miss as after a hit.

Fundamental Dependency in the Tennis Data

A summary of the results of many matches may provide evidence that the sets within matches were not independent. For example, when a player wins by a score of 3/1, there are three different sequences that may occur—namely, LWWW, WLWW, and WWLW. For a 3/2 result, there are six sequences. If sets within matches are independent, then each of the three sequences for a 3/1 result will have equal likelihood and similarly for the 3/2 results, because for any independence model each sequence is equally likely (irrespective of the constant probability of winning a set in a given match). Hence, we would expect approximately equal numbers for each of these sequences in our data. If this is not the case, then this is evidence of fundamental dependency in the data.

In the Wimbledon and U.S. Open dataset, 158 matches finished 3/1 and 93 matches finished 3/2 for one or the other of the players (see Table 2); the other matches were straight sets wins. A preliminary chi-squared investigation as to whether the numbers for each of the sequences leading to a 3/1 result are significantly different proved inconclusive. Similarly, for the six sequences leading to a 3/2 result. Because the categories in which the winner of the match loses a set—that is, 1st, 2nd, 3rd, or 4th set—are clearly ordinal, however, a model that includes this ordinality was fitted to the data. For both the 3/1 results and the 3/2 results, there is evidence that the set or sets lost by the winner in these matches occurred *earlier rather than later* in the match, which would not be so for independent sets. For instance for the 3/1 results there are 60 results in which the loss occurred in the first set—that is, LWWW—and 41 in which the loss occurred in the third set—that is, WWLW. Of the 93 matches that lasted five sets, in 23 the winner lost the first two sets, more than for any other sequence, and there is an overall trend for losses in earlier rather than later sets. This is a weak test for dependency in this type of data because we cannot make use of the 3/0 or 0/3 results; however, in this case it does produce evidence of fundamental dependency that implies that any independent-sets-within-matches model will provide an inadequate description of the data. The test provides evidence of dependency, although it is not immediately obvious that it is evidence of PM. If PM exists, however, then in general it is easier for the eventual loser of a match to win a set early in the match rather than later when the effect of PM is more pronounced.

Some Conclusions for the Wimbledon and U.S. Open Data

We have seen that the independence model with a Normal random effect does not rival the odds model as an explanation of the data. Indeed the evidence of fundamental dependency implies that, if we did produce an independent-sets-within-matches model that fitted as well as the odds model, we would be forced to conclude that both models were inadequate descriptions of the data. It appears then that we must abandon the idea of independence. To abandon independence, however, is not to say that one must reject the common-sense idea that player ability varies from day to day, only that on its own such a model is unlikely to be successful. Whatever the contribution of random variation in a player's ability from day to day may be, our analysis suggests that psychological momentum is certainly a major factor in the outcome of matches at the Wimbledon and U.S. Open tennis tournaments.

Table 4—Model Comparisons for the Head-to-Head Datasets: Parameter Estimates and Goodness-of-Fit Statistics

Lendl vs. Connors	McEnroe vs. Borg
(1) Simple independence, i.e., odds constant	
Estimate	Estimate
$\hat{o} = 1.87$	$\hat{o} = 1.10$
Deviance = 58.3	Deviance = 60.9
(2) Odds model, i.e., $O_{ij} = k_{i-j}\ O_{00}$	
Estimates	Estimates
$\hat{o} = 1.75$ $k = 2.01$	$\hat{o} = 1.13$ $k = .75$
Deviance = 54.1	Deviance = 60.4
(3) Random effect, i.e., log(odds) = const + r, $r \sim N(0, \delta^2)$	
Estimates	Estimates
Const. = .74 $\delta^2 = 1.02$	Const. = .09 $\delta^2 = 0$
Deviance = 56.0	Deviance = 60.9

Head-to-Head Records: Lendl/Connors and Borg/McEnroe

When a number of matches take place between the same two players over a period of time, it is reasonable, under certain circumstances, to make the assumption that the expected probability of winning the first set (at least) remains the same in each of the matches. For instance, in the middle stages of a player's career one might assume that his average ability (allowing for possible random day-to-day variation) remains constant. Because our interest is in a possibly changing probability of winning a set within a match, this simplifies matters somewhat. It is no longer necessary, by means of the ranks or other explanatory variables, to estimate a changing underlying probability from match to match. The expected probability of winning the first set is assumed to remain constant for all matches between those two players. Unfortunately, such datasets tend to be small. That is certainly true for the head-to-head records we look at here. The data themselves are interesting because they relate to some of the greatest players of all time. They are presented here mainly for that reason, but it is doubtful if the head-to-head records, on their own, of any two professional tennis players could provide sufficient data to make possible anything other than a crude assessment of the relative abilities of the players.

Table 3 contains (a) the head-to-head record of Ivan Lendl and Jimmy Connors from 1982–1985, a period when both players can be considered to be near the peak of their abilities, Lendl having just

Table 3—Matches Won by Winning Score for Ivan Lendl versus Jimmy Connors 1982–1985 (16 matches) and for John McEnroe versus Björn Borg 1978–1981 (14 matches)

Head-to-head records				
Lendl vs. Connors: 1982–1985: Matches won			McEnroe vs. Borg: 1978–1981: Matches won	
Lendl	Connors	Winner's score	McEnroe	Borg
7	1	2/0	2	3
2	1	2/1	1	3
2	0	3/0	0	0
0	3	3/1	3	0
0	0	3/2	1	1

reached his and Connors not much past his best, and (b) the lifetime head-to-head record for John McEnroe and Björn Borg, a classic series of 14 matches over a three-year period from 1978–1981, between two players of similar age, competing for the number 1 spot in their sport. As previously, the order in which the sets were won and lost has been suppressed in this view of the data.

The matches in these head-to-head records were played using either a best-of-three- or best-of-five-sets format. For these small datasets it is assumed that the parameter k in the odds model is the same for either format. Of course any evidence of PM obtained by fitting the odds model to these datasets is applicable only to matches between the two named players. This differs from the approach taken earlier where the parameter k that is estimated is a population effect, applicable to all matches between players from the population being considered.

- For Lendl/Connors there appear to be many heavy defeats. Nine of Lendl's eleven wins were in straight sets, and none of the best-of-five matches went to the fifth set.
- For Borg/McEnroe, there were no 3/0 results for either player, and approximately half of their short matches went to a final set. There is no apparent evidence of an overabundance of heavy defeats—if anything, the reverse.

By comparing deviances we see that for Lendl/Connors, the odds model is superior to the independence model with a Normal random effect (see Table 4), although both pick up the relatively high number of heavy defeats between these two players and both provide a better fit than simple independence. The estimate of the index of PM, the parameter k in the odds model, is 2.01 and for the random-effects model the estimate for the variance of the random effect is 1.02. Of the two estimates, it is only the parameter estimate for k in the odds model that is marginally significant.

For McEnroe/Borg, the estimate for the random effect is identically 0, indicating that any variation in player ability will result in a lesser fit than the simple independence model. The odds model, however, does provide a nonredundant estimate for the index of PM between these players, although the fit to the data is practically the same as for simple independence and the estimate for k is less than 1. It is worth pointing out that values of $k < 1$ in the odds model indicate that success breeds failure or, if you prefer, failure breeds success, and it is a feature of the odds model that it is equally capable of picking up that type of dependency where it exists, as well as what is generally believed to be the more common form of PM—namely, success breeds success.

courtesy of Walter Looss, Jr. / Sports Illustrated

PM may be one explanation for the stunning victory of Björn Borg (above) over John McEnroe in their Wimbledon finals match.

McEnroe and Momentum: "You Cannot Be Serious"

So for the Borg/McEnroe series of matches there is no evidence that the probability of winning a set in any of their matches was influenced by the score or that the probability of winning a set varied from match to match. For Lendl/Connors the data do lend a little weight to the conclusion that either PM ($k > 1$) or variation in player ability from day to day (or perhaps both) was a factor in that series of matches. If indeed that was the case, and it is far from proven from this analysis, it is for others to speculate as to why that might have been so.

Thanks to Bill Benter of Hong Kong who first suggested to us that it might be worth investigating whether "if players' abilities did fluctuate from their overall rankings…this might salvage the independence model." and to the editor and referees of *Chance* for some helpful comments. And finally sincere and heartfelt thanks to the odds compilers at Ladbrokes, Hills and Corals who, by their absolute reliance on independence in a series of trials, have unknowingly and unwittingly supported this research over many years. It is literally true to say that without their generous and regular contributions to the advancement of science this work would not have been possible.

References and Further Reading

Albert, J. (1993), Comment on "A Statistical Analysis of Hitting Streaks in Baseball," *Journal of the American Statistical Association,* 88, 1184–1188.

Albright, S. C. (1993), "A Statistical Analysis of Hitting Streaks in Baseball," *Journal of the American Statistical Association,* 88, 1175–1183.

Jackson, D. A. (1993), "Independent Trials Are a Model for Disaster," *Applied Statistics,* 42, 211–220.

——— (1995), "Tennis in Lilliput: A Fable on Sports and Psychology," *Chance,* 8 (3), 7–40.

Kruskal, W. (1988), "Miracles and Statistics: The Casual Assumption of Independence," *Journal of the American Statistical Association,* 83, 929–940.

Stern, H. (1995), "Who's Hot and Who's Not," *Proceedings of the Section on Statistics in Sports,* American Statistical Association, pp. 26–35.

Stern, H., and Morris, C. (1993), Comment on "A Statistical Analysis of Hitting Streaks in Baseball," *Journal of the American Statistical Association,* 88, 1189–1194.

Tversky, A., and Gilovich, T. (1989), "The Cold Facts About the 'Hot Hand' in Basketball," *Chance* 2 (1), 16–21.

Chapter 43

Who is the Fastest Man in the World?

Robert TIBSHIRANI

I compare the world record sprint races of Donovan Bailey and Michael Johnson in the 1996 Olympic Games, and try to answer the questions: 1. Who is faster?, and 2. Which performance was more remarkable? The statistical methods used include cubic spline curve fitting, the parametric bootstrap, and Keller's model of running.

KEY WORDS: Sprinting; World record; Curve fitting.

1. INTRODUCTION

At the 1996 Olympic Summer Games in Atlanta both Donovan Bailey (Canada) and Michael Johnson (United States) won gold medals in track and field. Bailey won the 100 meter race in 9.84 seconds, while Johnson won the 200 meter race in 19.32 seconds. Both marks were world records. After the 200 m race, an excited United States television commentator "put Johnson's accomplishment into perspective" by pointing out that his record time was less than twice that of Bailey's, implying that Johnson had run faster. Of course, this is not a fair comparison because the start is the slowest part of a sprint, and Johnson only had to start once, not twice.

Ato Bolton, the sprinter who finished third in both races, was also overwhelmed by Johnson's performance. He said that, although normally the winner of the 100 meter race is considered the fastest man in the world, he thought that Johnson was the now the fastest.

In this paper I carry out some analyses of these two world record performances. I do not produce a definitive answer to the provocative question in the title, as that depends on what one means by "fastest." Hopefully, some light is shed on this interesting and fun debate. Some empirical data might soon become available on this issue: a 150 meter match race between the two runners is tentatively scheduled for June 1997.

2. SPEED CURVES

The results of the races are shown in Tables 1 and 2.

A straightforward measure of a running performance is the speed achieved by the runner as a function of time. The first line of Table 3 gives the interval times for Bailey, obtained from Swiss Timing and reported in the *Toronto Sun* newspaper. These times were not recorded for Johnson. The value 7.7 at 70 m is almost surely wrong, as it would imply an interval time of only 0.5 seconds for 10 m. I contacted Swiss Timing about their possible error, and they rechecked their calculations. As it turned out, the split times were computed using a laser light placed 20 m behind the starting blocks, and they had neglected to correct for this 20 m gap in both the 70 and 80 m split times. The corrected times are shown in Table 3.

The estimated times at each distance shown in Table 4 were obtained manually from a videotape of the races. Here is how I estimated these times. I had recorded the 100 m hurdles race on the same track. Using the known positioning of the hurdles, I established landmarks on the infield whose distance from the start I could determine. Then by watching a video of the sprint races in slow motion, with the race clock on the screen, I estimated the time it took to reach each of these markings.

Table 5 compares the estimated and official split times. After the 40 m mark, the agreement is fairly good. The disagreement at 10, 20, and 30 m is due to the paucity of data and the severe camera angle for that part of the race. Fortunately, these points do not have a large influence on the results, as our error analysis later shows. Overall, this agreement gives us some confidence about the estimated times

Table 1. Results for 1996 Olympic 100 m Final; The Reaction Time is the Time it Takes for the Sprinter to Push Off the Blocks after the Firing of the Starter's Pistol; DQ Means Disqualified

Name	Time	Reaction time
1. Bailey, Donovan (Canada)	9.84	+.174
2. Fredericks, Frank (Namibia)	9.89	+.143
3. Bolton, Ato (Tobago)	9.90	+.164
4. Mitchell, Dennis (United States)	9.99	+.145
5. Marsh, Michael (United States)	10.00	+.147
6. Ezinwa, Davidson (Nigeria)	10.14	+.157
7. Green, Michael (Jamaica)	10.16	+.169
8. Christie, Linford (Great Britain)	DQ	

Wind speed: +.7 m/s

Table 2. Results for 1996 Olympic 200 m Final; "?" Means the Information was Not Available

Name	Time	Time at 100 m	Reaction time
1. Johnson, Michael (United States)	19.32	10.12	+.161
2. Fredericks, Frank (Namibia)	19.68	10.14	+.200
3. Bolton, Ato (Trinidad and Tobago)	19.80	10.18	+.208
4. Thompson, Obadele (Barbados)	20.14	?	+.202
5. Williams, Jeff (United States)	20.17	?	+.182
6. Garcia, Ivan (Cuba)	20.21	?	+.229
7. Stevens, Patrick (Belgium)	20.27	?	+.151
8. Marsh, Michael (United States)	20.48	?	+.167

Wind speed: +.4 m/s

Robert Tibshirani is Professor, Department of Preventive Medicine and Biostatistics and the Department of Statistics, University of Toronto, Toronto, Ont., Canada M5S 1A8. The author thanks Trevor Hastie, Geoff Hinton, Joseph Keller, Bruce Kidd, Keith Knight, David MacKay, Carl Morris, Don Redelmeier, James Stafford, three referees, and two editors for helpful comments, Cecil Smith for providing Bailey's official split times from Swiss Timing, and Guy Gibbons of Seagull Inc. for providing the corrected version of the Swiss Timing results. This work was supported by the Natural Sciences and Engineering Research Council of Canada.

Table 3. Official Times at Given Distances for Bailey; The "?" Indicates a Suspicious Time, Later Found to be in Error

Distance (m)	0	10	20	30	40	50	60	70	80	90	100
Original time (s)	.174	1.9	3.1	4.1	4.9	5.6	6.5	7.7?	8.2	9.0	9.84
Corrected time (s)	.174	1.9	3.1	4.1	4.9	5.6	6.5	7.2	8.1	9.0	9.84

Table 4. Estimated Times at Given Distances; Bailey Starts at the 100 m Mark; "+" Denotes Distance Past 100 m: For Example, "+12.9" Means 112.9 m

Distance (m)	0	50	100	+12.9	+40.3	+49.4	+67.7	+76.9	+86.0	+100
Bailey:	.174			2.8	5.0	5.7	7.0	7.8	8.5	9.84
Johnson:	.161	6.3	10.12	11.4	14.0	14.8	16.2	17.0	17.8	19.32

Table 5. Comparison of Official and Estimated Interval Times for Bailey

Distance (m)	0	10	20	30	40	50	60	70	80	90	100
Official	.174	1.9	3.1	4.1	4.9	5.6	6.5	7.2	8.1	9.0	9.84
Estimated	.174	2.1	3.4	4.3	5.1	5.7	6.4	7.2	8.0	8.9	9.84

Table 6. Estimated Times (seconds) for Johnson for Distances over 100 m

	100	110	120	130	140	150	160	170	180	190	200
Johnson	10.12	11.10	12.09	13.06	13.97	14.83	15.61	16.40	17.26	18.23	19.32

for Johnson, and some idea of the magnitude of their error. The speed curves were estimated by fitting a cubic smoothing spline to the first differences of the times, constraining the curves to be 0 at the start of the race. The curves for each runner are shown in the top panel of Figure 1. Because Bailey's 100 m was much faster than Johnson's first 100 m but slower than his second 100 m, it seems most interesting to make the latter comparison. Hence I have shifted Bailey's curve to start at time 10.12 s, and Johnson's time at 100 m.

If Johnson's speed curve always lay above Bailey's, then this analysis would have provided convincing evidence in favor of Johnson because he achieved his speed despite having already run 100 meters. However, Bailey's curve does rise above Johnson's, and achieves a higher maximum (13.2 m/s for Bailey, 11.8 m/s for Johnson). A 95% confidence interval for the difference between the maxima, computing using the parametric bootstrap, is $(-.062, 1.15)$. Hence there is no definitive conclusion from this comparison. The Appendix gives details of the computation of this confidence interval.

We note that the estimate of 13.2 m/s for Bailey's maximum speed differs from the figure of 12.1 m/s reported by Swiss timing. Bailey's estimated final speed is 12.4 m/s versus 11.5 m/s reported by Swiss timing. This size of discrepancy is not unexpected because the interval times are only given to within .1 of a second. When a sprinter is running at top speed, he covers 10 m in approximately .8 s, giving a speed of $10/.8 = 12.5$ m/s. Now if each of the interval times are off by .05 s, then the estimated speed ranges from $10/.9 = 11.1$ m/s to $10/.7 = 14.3$ m/s.

Who would win a race of say 150 meters? Here is a simple-minded approach to the question. Bailey's speed at the 100 m mark was 12.4 m/s, and his speed was decreasing by only .036 m/s every 10 m. Johnson's estimated time at 150 m was 14.83 s, as given in Table 6. In order to beat that time Bailey would need "only" to maintain an average speed of more than 10.02 m/s for another 50 m. Of course, it is not clear whether he could do this. In the next section we appeal to a parametric model to perform the necessary extrapolation.

For interest, in the bottom panel of Figure 1 we compare Bailey's curve to that from Ben Johnson's 1987 9.83 s world record race (he was later disqualified for drug usage). They achieved roughly the same time in quite different ways: Ben Johnson got a fast start, and then maintained his velocity; Bailey accelerated much more slowly, but achieved a higher maximum speed.

3. PREDICTIONS FROM KELLER'S MODEL

Keller (1973) developed a model of competitive running that predicts the form of the velocity curve for a sprinter using his resources in an optimal way. Here we use his model to predict the winner of a 150 m race.

According to Keller's theory, the force $f(t)$ per unit mass at time t, applied by a sprinter in the direction of motion, may be written as

$$f(t) = \frac{dv(t)}{dt} + \frac{v(t)}{\tau} \tag{1}$$

where $v(t)$ is the velocity and τ is a damping coefficient. This is just Newton's second law, where it is assumed that the resistance force per unit mass is $v(t)/\tau$.

Keller estimated τ to be .892 s from various races. Excellent overviews of Keller's work are given by Pritchard (1993) and Pritchard and Pritchard (1994).

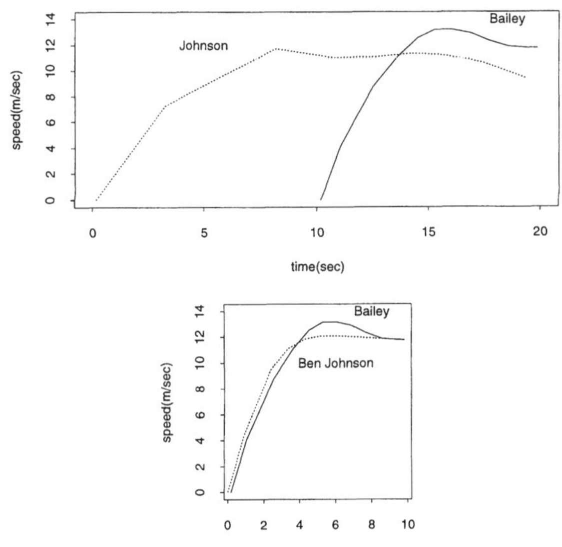

Figure 1. Top Panel: Estimated Speed Curves for Bailey and Johnson. Bailey's curve has been shifted to start at time 10.12 s, Johnson's time at 100 m. Bottom panel: estimated speed curves for Bailey and Johnson from the latter's 1987 world record race.

Starting with assumption (1) and a model for energy storage and usage, Keller shows that the optimal strategy for a runner is to apply his maximum force F during the entire race, leading to a velocity curve

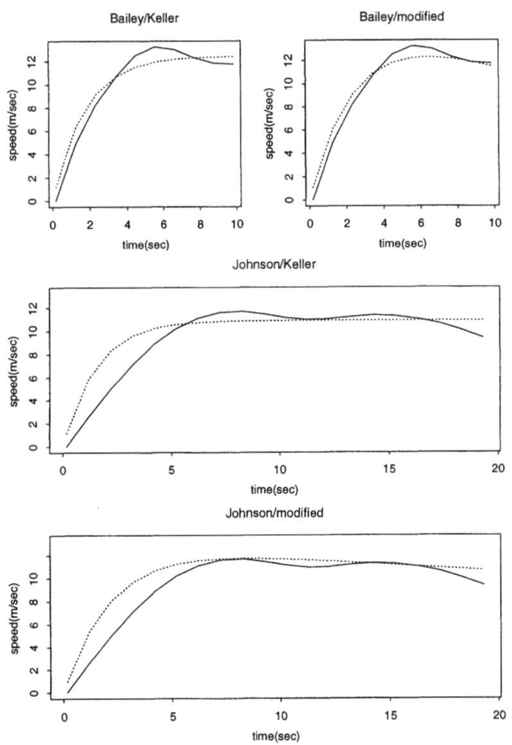

Figure 2. Top Row: Optimal Velocity Curves (Broken) for Bailey's 100 m. The top left panel uses Keller's model (2); the top right panel uses the modified model (4). The middle and bottom rows show the fit of the Keller and modified models for Johnson's 200 m. In all panels the solid curve is the corresponding actual (estimated) velocity curve from the top panel of Figure 1.

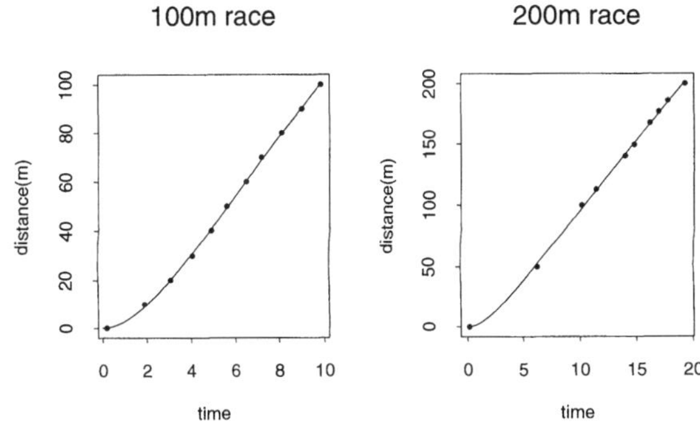

Figure 3. Estimated Distance Curves and Actual Distances (Points) from Least Squares Fit of Model (4).

$$v(t) = F\tau(1 - e^{-t/\tau}). \tag{2}$$

This applies to races of less than 291 m. For greater distances there is a different optimal strategy. By integrating (2) we obtain the distance traveled in time t:

$$D(t) = F\tau^2(t/\tau + e^{-t/\tau} - 1). \tag{3}$$

Figure 2 (top left and middle panels) shows the optimal speed curves for the 100 and 200 m races, with Bailey's and Johnson's superimposed. We used least squares on the (time, distance) measurements to find the best values of τ and F for each runner in equation (3): these were (1.74, 7.16) for Bailey and (1.69, 6.80) for Johnson.

We can use (3) to predict the times for a 150 m race; note that the reaction times must be included as well. The predictions are 13.97 s (Bailey) and 15.00 s (Johnson).

The same model also predicts a completely implausible 200 m time of 17.72 s for Bailey. One shortcoming of the model is the fact that the velocity curve (2) never decreases, but observed velocity curves usually do. To rectify this it seems reasonable to assume that a sprinter is unable to

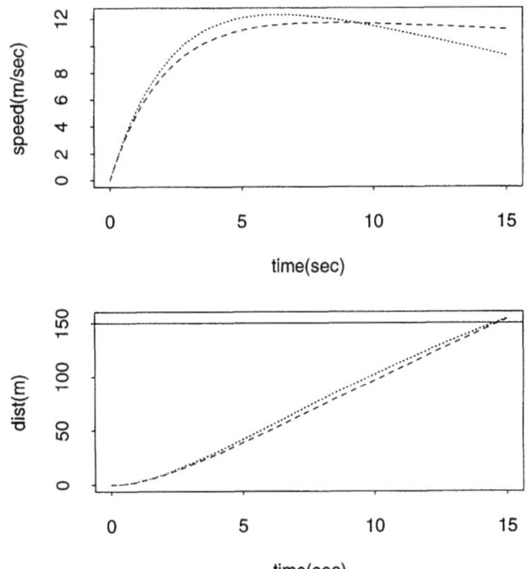

Figure 4. Estimated Optimal Velocity and Distance Curves over 150 m for Bailey (Dotted) and Johnson (Dashed).

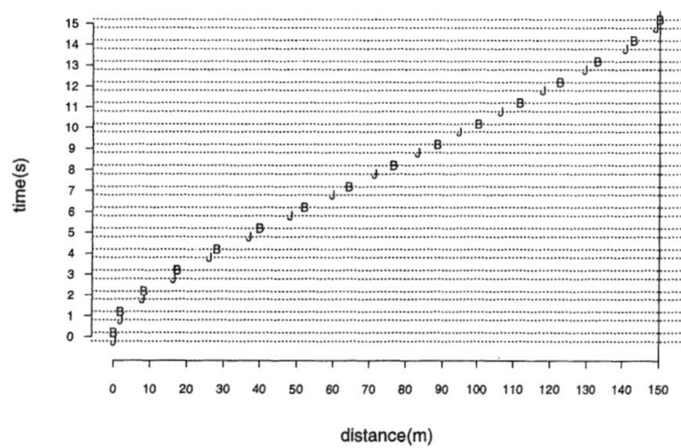

Figure 5. The Predicted Race in 1 s Snapshots. Shown is the Estimated Distance Traveled by Bailey (B) and Johnson (J) at Time = 0 s, 1 s, . . . , 14 s, and at the End of the Race (Time 14.73 s).

maintain his maximum force F over the entire race, but instead applies a force $F - c \cdot t$ for some $c \geq 0$. Using this in (1) leads to velocity and distance curves

$$v(t) = k - ct\tau - ke^{-t/\tau}$$

$$D(t) = kt - c\tau t^2/2 + \tau k(e^{-t/\tau} - 1) \qquad (4)$$

where $k = F\tau + \tau^2 c$. We fit this model to the observed distances by least squares, giving parameter estimates for (τ, F, c) of (2.39, 6.41, .20) and (2.06, 6.10, .05) for Bailey and Johnson, respectively. The fitted distance values are plotted with the actual ones in Figure 3. Note that the estimated maximum force is greater for Bailey than Johnson, but decreases more quickly. Bailey also has a higher estimated resistance.

The estimated curves are shown in the top right and bottom panels of Figure 2. The estimated 150 m times from this model are 14.73 s for Bailey and 14.82 s for Johnson. The latter is very close to the estimated time of 14.83 s at 150 m in the Olympic 200 m race from Table 6.

Figure 4 shows the estimated optimal velocity and distance curves over 150 m from the model, and Figure 5 depicts the predicted race in 1 s snapshots. Bailey is well ahead at the early part of the race, but starts to slow down earlier. Johnson gains on Bailey in the latter part of the race, but does not quite catch him at the end. The estimated winning margin for Bailey is .09 s. The bootstrap percentile 95% confidence interval for the difference is (.03 s, .26 s), and the bias-corrected 95% bootstrap confidence interval is (.02 s, .19 s). One thousand bootstrap replications were used—see the Appendix for details. Figure 6 shows a boxplot of difference in the predicted 150 m times from the bootstrap replications.

Note that this model does not capture a possible change of strategy by either runner in a 150 m race. This might result in different values for the parameters.

From Keller's theory one can also predict world record times at various distances as a function of F and τ. Keller fit his predicted world record times to the actual ones, for distance from 50 yards to 10,000 m, in 1973. From this he obtained the estimates $F = 12.2$ m/s^2, $\tau = .892$ s. The fit was quite good: for 100 m—9.9 s (actual), 10.07 s (predicted); for 200 m—19.5 s (actual), 19.25 s (predicted). (The world records that Keller reports in 1973 of 9.9 and 19.5 s are questionable. The 100 m record was 9.95 s, although 9.9 s was the best hand-timed performance. The 200 m record was 19.83 s.) It is interesting that at the time, the 100 m record was faster than expected, but the 200 m record was slower. Johnson's performance brings the 200 m world record close to the predicted value. It may be that the 200 m record has been a little "soft," with runners focusing on the more glamourous 100 m race. Note that the predictions do not include a component for reaction time: with Johnson's reaction time of .161 s, the predicted record would be 19.41 s.

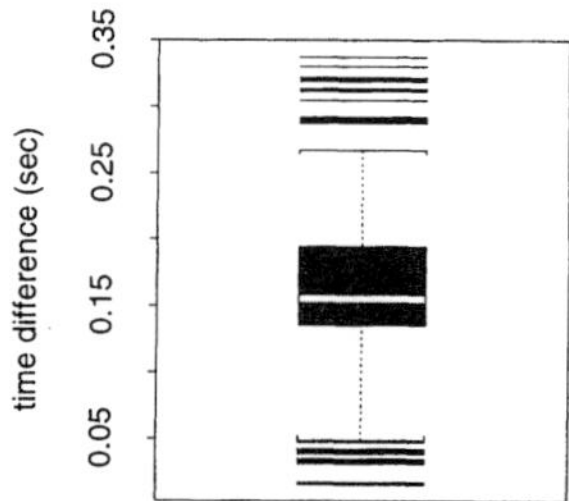

Figure 6. Boxplot of Johnson's Minus Bailey's Predicted 150 m Times from 1,000 Bootstrap Replications.

4. ADJUSTMENT FOR THE CURVE

Johnson's first 100 m (10.12 s) was run on a curve, and Bailey's was run on a straight track. Figure 7 shows the sprint track.

In the previous analysis we ignored this difference. Assuming we want to predict the performance of the runners over a straight 150 m course (the course type for the May 1997 race has not been announced at the time of this writing), we should adjust Johnson's 200 m performance accordingly. Intuitively, he should be given credit for having achieved his time on the more difficult curved course.

What is the appropriate adjustment? The centripetal acceleration running of an object moving at a velocity v around a circle of radius r is $a = v^2/r$. The radius of the circular part of the track is $100/\pi = 31.83$ m. With Johnson's velocity ranging from 0 to 11.8 m/s, his centrepital acceleration ranges from 0 to 4.37 m/s^2. We cannot simply add this acceleration to the acceleration in the direction of motion because the centrepital acceleration is at right angles to the direction of motion. However, he does biological work in achieving this acceleration, and hence spends energy. Unfortunately, just how much energy is expended is

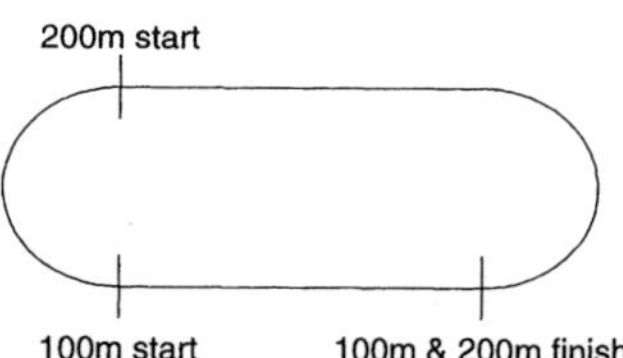

Figure 7. The Sprint Track, Showing Start and Finish Lines for the 100 and 200 meter Races.

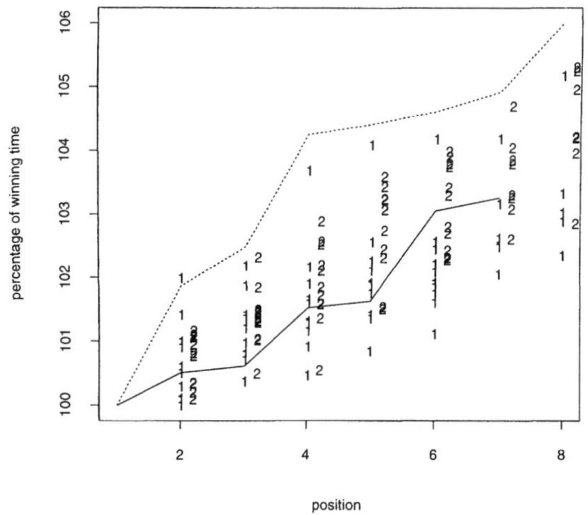

Figure 8. Percentage that Each Runner Achieved as a Function of the Winning Time in the Race for the 100 meter (Solid Curve) and 200 meter (Broken Curve) Races. The "1s" and "2s" correspond to the previous nine Olympic 100 and 200 m finals, respectively.

difficult to measure, in the opinion of the physicists that I consulted. Hence I have not been able to quantity this effect.

5. COMPARISON TO OTHER RACE COMPETITORS

In the rest of this paper I focus on the question of which of the two performances was more remarkable. These two races were particularly unique because the same two runners (Fredericks and Bolton) finished second and third in both. This suggests an interesting comparison. Figure 8 shows the percentage that each runner achieved as a function of the winning time in the race for the 100 (solid curve) and 200 meter race (broken curve). Johnson's winning margin was particularly impressive. Also plotted in the figures are the corresponding percentages achieved in the previous nine Olympic games, going back to 1952. (Throughout this analysis I restrict attention to post-1950 races because before that time races were run on both straight and curved tracks, and it was not always recorded which type of course had been used.) There has never been a winning margin as large as Johnson's in a 200 meter Olympic race, and only once before in a 100 meter race. This was Robert Hayes' 10.05 s performance in 1964 versus 10.25 s for the second place finisher. Johnson's margin over the second place Fredericks is also larger than the margin between the winner and third place in all but two of the races.

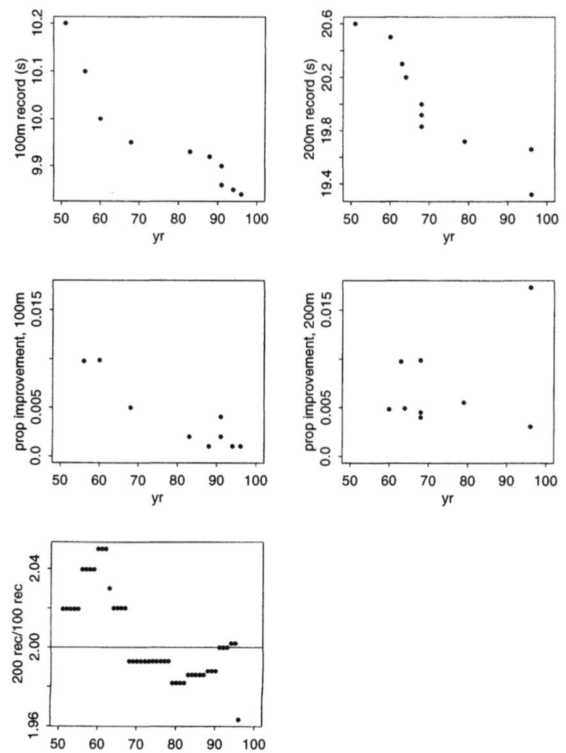

Figure 9. The Top Panels Show the Evolution of the 100 m (Left) and 200 m (Right) World Records. The middle panel shows the proportion improvement of the existing record that was achieved each time in the 100 m (left) and 200 m (right). The bottom left panel shows the evolution of the ratio of the 200 to 100 m world record times.

6. EVOLUTION OF THE RECORDS

The top panels of Figure 9 show the evolution of the 100 and 200 meter records since 1950. The proportion improvements, relative to the existing record, are shown in the middle panels of Figure 9. Johnson's 19.32 performance represented a 1.7% improvement in the existing record, the largest ever. (Tommie Smith lowered the 200 m world record to 20.00 s in 1968, a 1.0% improvement from the existing world record of 20.2 s. However, the 20.2 s value was a hand-timed record: the existing automatic-timed record was 20.36 s, which Smith improved by 1.8%.) If we include Johnson's 19.66 world record in the 1996 U.S. Olympic Trials, then overall he lowered Pietro Mennea 19.72 world record by 2.02% in 1966.

The bottom left panel of Figure 9 shows the ratio of the 200 meter world record versus the 100 meter world record from 1950 onward. The ratio has hovered both above and below 2.0, with Johnson's world record moving it to an all-time low of 1.963. The average speed in the 100 m record race was 10.16 m/s, and that for the 200 m race was 10.35 m/s, the fastest average speed of any of the sprint or distance races. A ratio of below 2.0 is predicted by the mathematical model of Keller (1973).

7. CONCLUSIONS

Who is faster, Bailey or Johnson? The answer depends on the definition of "faster," and there is no unique way of comparing two performances at different distances. Our results are inconclusive on this issue:

- It is not fair to compare the average speeds (higher for Johnson) because the start is the slowest part of the race, and Johnson had to start only once.
- Bailey appeared to achieve a higher maximum speed, although the difference in maxima was not statistically significant at the .05 level; Johnson maintained a very high speed over a long time interval.
- Predictions from an extended version of Keller's optimal running model suggest that Bailey would win a

(straight) 150 m race by .09 s. However, they do not account for the fact that Johnson's times are based on a curved initial 100 m.

It would clearly be a close race, and there are a number of factors I have not accounted for. This entire comparison is based on just one race for each runner: consistency and competitiveness come into play any race. Perhaps most important is the question of strategy. Each runner would train for and run a 150 meter race differently than a 100 or 200 meter race. The effect of strategy is impossible to quantify from statistical considerations alone.

Whose performance was more remarkable? Here, Johnson has the clear edge:

- Johnson's winning margin over the second and third place finishers (the same runners in both races!) was much larger than Bailey's, and was the second largest in any Olympic 100 or 200 m final race.
- Johnson's percentage improvement of the existing world record was the largest ever for a 100 or 200 m race. However, the 200 m record might have been a little "soft" because it was well above the record as predicted by Keller's theory.

SOURCES

- David Wallechinsky, *The Complete Book of the Summer Olympics*, Boston, New York, Toronto, London: Little, Brown, 1996.
- Mika Perkimki, csmipe@uta.fi, *World-Wide Track & Field Statistics On-Line*, [http://www.uta.fi/~csmipe/sport/index.html].
- IBM, *Official Results of the 1996 Centennial Olympic Games*, [http://results.atlanta.olympic.org].
- *Track and Field News* (Oct. 1996).

APPENDIX: ERROR ANALYSIS

The data in Table 4 were obtained manually from a videotape of the race, and hence are subject to measurement error. To assess the effect of this error, we applied the parametric bootstrap (Efron and Tibshirani 1993). The maximum amount of error in the times was thought to be around $\pm.05$ s for Bailey's times, and $\pm.20$ s for Johnson's early times and $\pm.15$ s for Johnson's last 100 m times. Therefore, I added uniform noise on these ranges to each time measurement. For each resulting dataset I estimated the speed curves for Bailey and Johnson. This process was repeated 1,000 times.

The observed difference in the maximum speeds was $13.2 - 11.8 = 1.4$ m/s. The upper and lower 2.5% points of the 1,000 observed differences was $(-.062, 1.15)$. The same parametric bootstrap procedure was used for the error analysis of the fit of Keller model, in Section 3, and was used to produce Figure 6.

[Received October 1996. Revised January 1997.]

REFERENCES

Efron, B., and Tibshirani, R. (1993), *An Introduction to the Bootstrap*, London: Chapman & Hall.

Keller, J. (1973), "A Theory of Competitive Running," *Physics Today*, 42–45.

Pritchard, W. (1993), "Mathematical Models of Running," *SIAM Review*, 359–379.

Pritchard, W., and Pritchard, J. (1994), "Mathematical Models of Running," *American Scientist*, 546–553.

Chapter 44

Do runners and cyclists enjoy an advantage in today's triathlons?

Resizing Triathlons for Fairness

Howard Wainer and Richard D. De Veaux

In the country of Brogdinnian, a free college education is offered to top scorers on a test of general academic ability. The test is made up of two parts: a verbal part and a mathematical part. Women do better on the verbal part, whereas men do better on the mathematical part. There are 100 questions worth 1 point each on the test; 80 are mathematical questions, and 20 are verbal questions. Almost all of the scholarships go to men. In a recent class action suit, women claimed that the contest was unfair. The defense attorney countered that because the variance of math tests is smaller than that of verbal tests, more math questions are needed to spread out the competition. The judge, however, saw through this argument and ruled in favor of the women, pointing out that although variance is, indeed, important, the test proportions must take into account the mean performance for each group.

Although most people would agree that such an academic competition is unfair, the most common triathlon races are grossly unfair to a large potential pool of participants. The typical triathlon is composed of three parts: swimming, bicycling, and running. The winner is determined by the total amount of time needed to complete all three parts. It is clear that to be fair to athletes with special expertise in any one of the three components, the distances for each event should be chosen carefully. A race of 100 yards of cycling, 25 yards of swimming, and 25 miles of running would be considered unfair to everyone except experienced marathon runners. Certainly, no one anticipating to compete in such a triathlon would waste much time training for anything but the run.

The triathlon distances described here illustrate our point in the extreme. Although existing distances for the triathlon components are less extreme than these, they are still a long way from being fair to all athletes who might consider participating. Of course, to make progress toward a fair triathlon, we must define what we mean by fair. The ideal distances should be symmetric in some sense, but how to measure the symmetry is unclear. We would like a relatively strong or weak performance in any one segment to be equally rewarded or penalized. If we use total time to determine the winner, some contend that we need to make the variances of the times of each segment equal. This argument has been used to justify the relatively long cycling segments of the current Iron Man and Olympic Triathlons. We maintain that the argument is specious in this context and fails to take into account the variance that would obtain if a different group of athletes, specifically those less predisposed toward cycling, were to participate.

Instead, we will take "fairness" to mean that a cyclist, runner, and swimmer, all equally proficient, can each traverse the associated segment of the triathlon in approximately equal times; that is, the best swimmer in the world can complete the swimming segment in about the same amount of time as the best runner in the world can

Dede Barry, cyclist. Photo courtesy of USA Cycling, © 2004.

complete the running segment, and the best cyclist the cycling segment.

In this article we use this definition to derive fair triathlon proportions for various total elapsed times. We discuss the equal-variance argument further in the discussion section. We conclude with a plan for the Ultimate Paris-to-London Triathlon.

How Long Is a Triathlon?

Triathlons come in all sizes and shapes; a sampling of them is shown in Table 1. Here we show two well-known triathlons, the Iron Man and the Standard International, or Olympic Triathlon, along with the Garden State Tin Man.

How were the proportions selected? Legend has it that the first triathlon came about as the result of a bar bet by some sailors stationed in Hawaii. Previously, some of them had participated in the annual events known as the Waikiki Rough Water Swim, the Around the Island Bicycle Race, and the Honolulu Marathon. One proposed the challenge to complete the equivalent of all three events in one day. Thus, the Iron Man Triathlon was born. It has now been contested every October since 1981 in the village of Kailua-Kona on the island of Hawaii. It consists of a 2.4-mile swim, a 26.2-mile run, and a 112-mile bicycle ride, precisely the distances of the preexisting events. [The record for this race of 8 hours, 9 minutes, 15 seconds (8:09:15) was set by Mark Allen in 1989.] The order of the three events, for logistical and physiological reasons, is swimming, cycling, and running. Although it would be interesting to consider the effects of other orderings, we will assume the standard order throughout this article, thus also ignoring the changing effect fatigue would play in a different sequence.

Table 1—Some Typical Triathlons

Race	Swimming	Running	Cycling	Total
Iron Man				
Distance (km)	3.9	42.1	180.2	226.2
Proportions	1.7%	19%	80%	100%
International (Olympic)				
Distance (km)	1.5	10	40	51.5
Proportions	2.9%	19%	78%	100%
Garden State Tin Man				
Distance (km)	.8	10	37	47.8
Proportions	1.6%	21%	77%	100%

Note that the ratio of distances is 1 to 11 to 48—the length of the run is 11 times the length of the swim, and the length of the cycling course is 48 times the length of the swim. Although organizers of triathlons are free to choose any distances they want, many are roughly scaled versions of the Iron Man, with a few exceptions. Are these proportions fair to all potential participants? To judge this, let us hold the running segment of the contest constant and calculate "fair" lengths of the other two segments.

The great Ethiopian runner Belayneh Densimo ran the 1988 Rotterdam marathon in just under 2 hours and 7 minutes (2:06:50), the fastest marathon ever. Marathon courses vary considerably, but there is reasonable consistency among marathon times, and 2:07 seems to be a plausible figure to represent the current best possible time.

How far can the best bicyclists go in 2 hours and 7 minutes? This year, Spain's Miguel Indurain won the 21-stage, 2,490-mile Tour de France in just under 101 hours, thus averaging almost 25 mph. This provides us with a lower bound of 53 miles for 2 hours and 7 minutes. Courses differ, and surely cyclists would be able to go faster if the race were to be only 2 hours. As we shall derive later, a good estimate for what would be a world record cycling distance for 2:07 is about 60 miles. Thus, we see that a fair Iron Man would shrink the cycling leg almost in half.

What about swimmers? How far can the best swimmers go in 2 hours and 7 minutes? This is hard to estimate because the longest pool race is 1,500 m, and the world record for that, held by Australia's Kirin Perkins, is 14 minutes, 43.48 seconds (14:43.48). This record means that for 15 minutes he can maintain a pace of just under 59 seconds per 100 m. There are open-water marathon races that take many hours, but currents, low water temperatures, and other nonstandard conditions make them a poor source of data for estimating optimal human performance. Another count against using open-water races is that marathon swimming is not a particularly popular sport, and hence most of the greatest swimmers do not participate. Instead we have opted to use a less formal source to estimate swimming ability: performance in practice. Records kept by Rob Orr, coach of the Princeton men's swimming team, revealed that sometimes in a 2-hour practice the swimmers can complete 10,000 m. Kirin Perkins, whose practices are legendary, has maintained a near 1 minute (1:02) per 100 m pace for 2 hours. Using this pace as a standard would yield a total of approximately 12,300 m in 2 hours and 7 minutes. Making a more conservative estimate (1:03:05) still yields 12,000 m (about 7.5 miles). Thus, a fair Iron Man would need to more than triple the swimming leg while simultaneously halving the cycling portion.

Most triathlons are open-water swims, rather than pool swims; therefore, our estimate of 12,000 m is quite possibly a bit optimistic (depending on the direction of the current!). However, because running is the last event, the estimate of 26.2 miles for 2:07 will also be optimistic in the context of a triathlon. We use 12,000 m as a starting point for a discussion of a fair triathlon, with a view toward reevaluating the proportions using actual split times for participants once they become available.

Consequently, a triathlon like the Iron Man, in which each leg was scaled to be a shade over 2 hours long for the best in the world (competing in peak circumstances and without the distraction of the other two events), would not be proportioned 1 to 11 to 48 as is currently the case, but rather 1 to 3.5 to 8. The current Iron Man has a cycling leg that is *six times longer* than it ought to be relative to swimming! It is no wonder that very few triathletes describe themselves as primarily swimmers: The contest is so tilted against them that it is hardly worth a swimmer's effort to compete.

So far, we have focused primarily on a single race, the Iron Man, and derived fair proportions for it. As we indicated in Table 1, not all triathlons are the same length, nor do they have the same proportions as the Iron Man. For example, the Triathlon World Championship race, which is held annually, consists of a 1.5-km swim, a 40-km

Table 2—World Records in Three Sports

Cycling records		Running records		Swimming records	
Distance (m)	Time (h:m:s)	Distance (m)	Time (h:m:s)	Distance (m)	Time (h:m:s)
1,000	0:01:05	800	0:01:42	50	0:00:21.8
4,000	0:04:30	1,000	0:02:12	100	0:00:48.4
5,000	0:05:51	2,000	0:04:51	200	0:01:46.7
10,000	0:11:53	3,000	0:07:29	400	0:03:45.0
20,000	0:24:06	10,000	0:27:08	800	0:07:47.9
40,200	0:49:24	20,000	0:56:57	1,500	0:14:43.5
80,450	1:43:46	25,000	1:13:56	12,000	2:07:00.0[a]
100,000	2:14:02	30,000	1:29:19		
160,900	3:45:17	42,195	2:06:50		

[a]World record imputed for 12,000 meters swimming based on practice performance.

Table 3—Equilateral Triathlons

		Distances (km)			Distances (miles)		
World record times for each leg (min)		Swim	Run	Bike	Swim	Run	Bike
	10	1.0	3.9	8.5	.6	2.4	5.3
	15	1.5	5.7	12.5	.9	3.5	7.8
"Olympic"	28	2.7	10.0	22.4	1.7	6.2	13.9
	30	2.9	10.8	24.2	1.8	6.7	15.0
	45	4.3	15.8	35.7	2.7	9.8	22.2
	60	5.7	20.7	47.0	3.6	12.8	29.2
	75	7.1	25.5	58.1	4.4	15.9	36.1
	90	8.6	30.4	69.2	5.3	18.9	43.0
	105	10.0	35.2	80.2	6.2	21.9	49.8
	120	11.4	40.0	91.1	7.1	24.8	56.6
"Iron Man"	127	12.0	42.2	96.2	7.5	26.2	59.7

bike ride, and a 10-km run. (The record of 1:48:20 was set by Miles Stewart of Australia in 1991.) These distances have become the standard distances for international competitions and will be used in the Olympics. The ratio of these distances is roughly 1 to 7 to 27. Although this is fairer to swimmers than the Iron Man, it is still not entirely fair, the running section being twice as long as it should be relative to the swim, and the cycling leg over three times as long.

Data and Analyses

Table 2 shows the world records for the three sports at various distances. (The cycling data are from Van Dorn 1991; running and swimming data from Meserole 1993.) We have included an imputed world record for 12,000 m swimming based on practice performance.

Fitting a mathematical function to these record times allows us to interpolate accurately between them and produce estimated distances traversed for all three sports for any intermediate time. Regression, fitting the logarithm of distance (in meters) to polynomials in the logarithm of time (in seconds), is used to estimate the relationships. The functions that were fit are shown in Fig. 1.

Table 3 provides guidelines for what might properly be called "Equilateral Triathlons" of various duration, based on world-record times for each segment. The distance proportions for equilateral triathlons are roughly 1 to 3.5 to 8 (depending on the duration of the

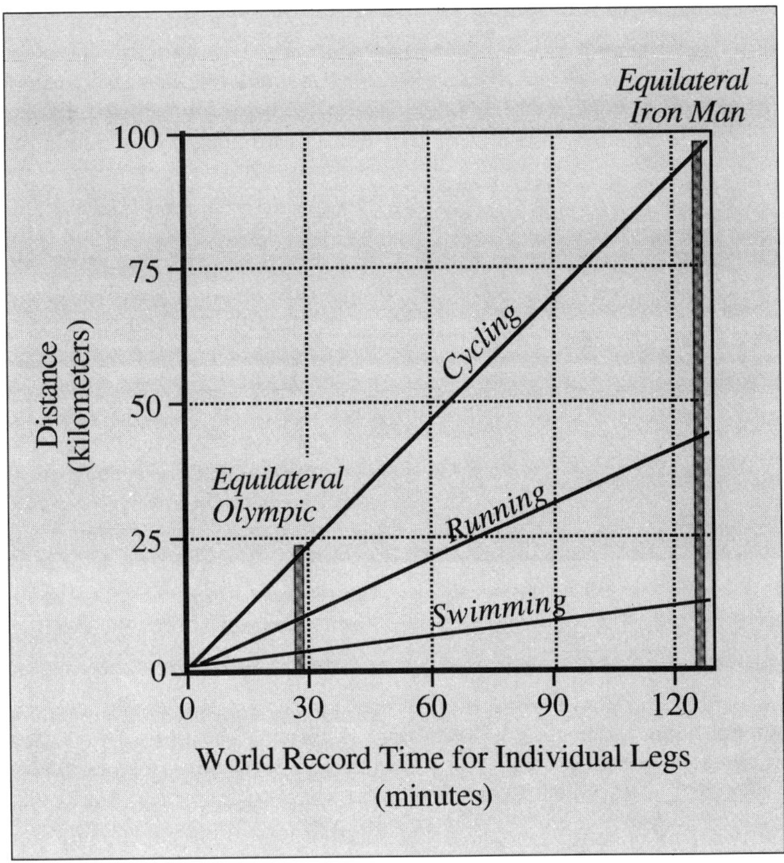

Figure 1. A graph that allows one to construct equilateral triathlons of many different durations. Equilateral versions of the two canonical durations are indicated by the vertical bars.

segments), or 8% to 28% to 64%. The range extends from a "sprint" triathlon, which most competitors should be able to complete in under 1 hour, to an Equilateral Iron Man that might take a professional triathlete 8–10 hours to finish. For example, the entry for the equilateral equivalent of the Olympic Triathlon (keeping the running leg constant) is found on the third line of the table and consists of a 2.7-km swim, a 10-km run, and a 22.4-km cycling leg. Compare these to the current Olympic distances of 1.5-km swim, 10-km run, and 40-km cycle.

Discussion

In this account we tried to expose the unfairness of triathlons as they are currently configured. We offered alternative proportions that make the three segments of the triathlon, in some real sense, equal. By instituting these proportions we feel there will be greater participation from athletes whose best sport is swimming, rather than the current domination by cyclists and runners. Our recommendations have not met with universal approbation; some feel that our definition of fairness is incorrect. One common complaint was first offered by Sean Ramsay, a well-known Canadian triathlete, who suggested that the current distances were chosen to equally spread out the competitors. Hence, the amount of discrimination among competitors that can be achieved in a relatively short swimming race required a much longer cycling race. Ramsay's observation is true for the athletes currently competing in triathlons. Would it be true if the races were proportioned differently?

Let us imagine a different pool of competitors for the equilateral triathlons. Perhaps more swimmers will compete. The variation of cycling times will become much larger because, in addition to the current variation we see in cycling times, there is the additional variation due to participants for whom cycling is not their best sport. Therefore, the decrease in variation due to the relatively shorter cycling leg of an equilateral triathlon may be compensated for by the increased variation expected for the new pool of participants. To determine the extent to which this is true requires gathering data from equilateral triathlons. It is likely these data would include some athletes who currently do not participate. We expect the pool of participants to change as the triathlons become fairer (see sidebar). The analysis of such data will then allow the iterative refinement of the triathlon proportions based on actual split times.

To conclude our exposition, we extrapolate our definition of a fair triathlon a little further (further perhaps than ought to be done). We would like to propose a new race from Paris to London.

The Ultimate Paris-to-London Triathlon

Imagine the race beginning at dawn on a sunny day in mid-August from beneath the Arc de Triomphe in Paris. Crowds cheer as competitors set out on bicycles for Calais, 250 km away. The best of them begin to arrive in Calais at about noon. They strip away their cycling clothes and grease their bodies for the 46-km swim across the English Channel to Dover. Although the weather is perfect and they have caught the

The Effects of Selection on Observed Data

The notion of a bias in inferences drawn from the results of existing triathlons due to the self-selection of athletes into the event is subtle, although the size of the bias may be profound. Even Harald Johnson, publisher of *Swim Bike Run* magazine and a perceptive observer of triathlons, missed it. In a 1982 open letter he wrote to his subscribers, Johnson discussed the difficulties in balancing triathlons. He analyzed the split results of three 1981 triathlons and arrived at proportions of 1:4:11. These are closer to our recommendations than current practice but still not quite right. The difference is accountable because he used split times from existing triathlons. "Analysis of actual triathlons is the only way to do this kind of research," he said. But, of course, the split times would be quite different if the proportions were different. The subtle nature of this effect of self-selection is best exposed if we express the problem mathematically.

Let S_i equal the time for person i to complete the swimming leg, R_i equal the time for person i to complete the running leg, and C_i equal the time for person i to complete the cycling leg. Moreover, let K_i be an indicator variable that takes the value 1 if person i decides to participate in the triathlon, and 0 if not. So far, we have chosen distances so that World Record(**S**) = World Record(**R**) =World Record(**C**) over all individuals i. Ramsay's criticism was that the distances should be chosen so that

$$\text{var}(\mathbf{S}) = \text{var}(\mathbf{R}) = \text{var}(\mathbf{C}) \quad (1)$$

Perhaps Ramsay's objection is reasonable, but all we can observe is that

$$\text{var}(S \mid K=1) = \text{var}(R \mid K=1) = \text{var}(C \mid K=1) \quad (2)$$

The variance of **S** over all individuals is composed of two sets of terms. One set is observable: $\text{var}(S \mid K=1)$ and $\mathbf{E}(S \mid K=1)$. But the second component is unobservable because it contains both the mean and the variance of performance for those potential triathletes who did not participate: $\text{var}(S \mid K=0)$ and $\mathbf{E}(S \mid K=0)$. Moreover, because the pool of potential participants is defined only loosely, we also cannot observe $\mathbf{P}(K=1)$. There is a similar term for each of the other two segments as well. If (2) is observed to be true, then (1) is certainly true if the unobserved variances are all equal [i.e., $\text{var}(S \mid K=0) = \text{var}(R \mid K=0) = \text{var}(C \mid K=0)$], and the unobserved means equal the observed means [$\mathbf{E}(S \mid K=1) = \mathbf{E}(S \mid K=0)$, $\mathbf{E}(R \mid K=1) = \mathbf{E}(R \mid K=0)$, and $\mathbf{E}(C \mid K=1) = \mathbf{E}(C \mid K=0)$]. These assumptions are too far fetched to be credible.

It is reasonable to believe that if the swimming segment is tripled, participation among swimmers would increase. It is also likely that they would be worse cyclists on average than those who are already participating [i.e., $\mathbf{E}(C \mid K=0) > \mathbf{E}(C \mid K=1)$], and so $\text{var}(C \mid K=1)$ would increase. How much the variance would increase is unknown and cannot be known until we see how changing the race proportions changes the participation rates from various subpopulations of athletes. Perhaps the change in participants would increase the variance enough to counteract the variance shrinkage that will occur from shortening the cycling distance.

tide, few will come close to Richard Davey's 1988 record of 8 hours and 5 minutes. The leaders start emerging 9–10 hours later. The sun has already set. These hardy souls grab a quick bite to eat, change into running shoes, and set off on shaky legs for London, a mere 115 km away. A good ultramarathoner, starting fresh, would cross the finish line in Trafalgar Square after about 6 hours; indeed a world-class race walker could do it in about $7\frac{1}{2}$ hours. But no one is fresh, and no one who has emerged from the Channel is a running specialist. In fact, the only ones who have finished the swimming segment so far are chubby channel swimmers. The waiting crowd looks anxiously through the darkness for the sleek cyclists and runners who traditionally win triathlons. As the sun begins to rise, none emerge. At the finish line, the winner arrives, half jogging, half walking, more than 24 hours after the beginning of the race. She gracefully accepts the trophy and prize money and then heads for a bath, breakfast, and a bed.

The careful reader will notice that the Paris-to-London swimming segment is a bit longer than our equilateral analysis suggests. (It also places the swimming segment second for logistical reasons.) This could be corrected by beginning the race a bit east of Paris and finishing the run somewhat north of London. But the disparity here is small in comparison to the disadvantage to which swimmers are ordinarily put. In addition, the appellation "The Ultimate La Queue en Brie to Chigwell Triathlon" has neither the cachet nor the euphony of our proposal.

Additional Reading

Keller, J. B. (1977), *A Theory of Competitive Running in Optimal Strategies in Sports*, eds. S. P. Ladany and R. E. Machol, The Hague: North Holland, pp. 172–178.

Meserole, M. (ed.) (1993), *The 1993 Information Please Sports Almanac*, Boston: Houghton Mifflin.

Smith, R. L. (1988), "Forecasting Records By Maximum Likelihood," *Journal of the American Statistical Association*, 83, 331–338.

Tryfos, P. and Blackmore, R. (1985), "Forecasting Records," *Journal of the American Statistical Association*, 80, 46–50.

Van Dorn, W. G. (1991), "Equations for Predicting Record Human Performance in Cycling, Running and Swimming," *Cycling Science*, September and December 1991, 13–16.